Modern Trends in Manufacturing Technologies and Equipment

ICMTMTE 2021

Modern Trends in Manufacturing Technologies and Equipment 2021, ICMTMTE 2021, September 06-10, 2021, Sevastopol, Russian Federation

Editors
Sergey Bratan and Stanislav Roshchupkin

[1]Sevastopol State University, Department of Mechanical Engineering, Russian Federation

Peer review statement

All papers published in this volume of "Materials Research Proceedings" have been peer reviewed. The process of peer review was initiated and overseen by the above proceedings editors. All reviews were conducted by expert referees in accordance to Materials Research Forum LLC high standards.

Published under License by **Materials Research Forum LLC**
Millersville, PA 17551, USA

Published as part of the proceedings series
Materials Research Proceedings
Volume 21 (2022)

ISSN 2474-3941 (Print)
ISSN 2474-395X (Online)

ISBN 978-1-64490-174-8 (Print)
ISBN 978-1-64490-175-5 (eBook)

Distributed worldwide by

Materials Research Forum LLC
105 Springdale Lane
Millersville, PA 17551
USA
http://www.mrforum.com

Manufactured in the United States of America
10 9 8 7 6 5 4 3 2 1

Table of Contents

Preface

The International Conference on Modern Trends in Manufacturing Technologies and Equipment 2021 (ICMTME 2021) was held in Sevastopol, Russia from 06 to 10 September 2021 and was organized by the Association of Machine Builders, Sevastopol State University with support from National University of Science and Technology «MISIS», Bauman Moscow State Technical University and Inlink Ltd.

ICMTMTE 2021 was organized to provide scientists and industry experts with the opportunity to summarize the latest developments in manufacturing technologies, materials science, manufacturing equipment and tools, to discuss future developments in the new technique and technologies, to build up the network and partnerships for the future collaboration.

The conferees represented collaborative scientific teams from different countries – Russia, Ukraine, Belarus, Kazakhstan, South Africa, Germany, USA, Bulgaria, Poland, China, Algeria, Mongolia, Uzbekistan, Armenia, Vietnam.

The proceedings of the ICMTMTE 2021 published in Materials Research Proceedings consists of 76 selected papers, which were presented orally or via online platform – "Zoom".

We express our deepest gratitude to all members of the scientific committee, invited speakers, reviewers, sponsors, participants and members of the ICMTMTE 2021 for their valuable contribution to the successful organization of ICMTMTE 2021. We also would like to express our special thanks to Thomas Wohlbier from Materials Research Forum LLC for his great support to ICMTMTE 2021.

With best wishes,
Prof. Sergey Bratan, General Chair of ICMTMTE 2021

Dr. Stanislav Roshchupkin, Secretary of ICMTMTE 2021

Committee

Conference Chair
Prof. Sergey Bratan (Sevastopol State University, Sevastopol, RUS)

Conference Co-Chair
Prof. Vyacheslav Prikhodko (Moscow Automob. and Road Construct. St. Technical University (MADI), Moscow, RUS)
Prof. Sergey Reznik (Bauman Moscow State Technical University, Moscow, RUS)
Prof. Sergey Gorbatyuk (National University of Science and Technology «MISIS», Moscow, RUS)
Prof. Alexander Glezer (National University of Science and Technology «MISIS», Moscow, RUS)

Co-chair of International Program Committee
Prof. Anatoliy Suslov, Honored Worker of Science and Technology of the Russian Federation, Honorary Chairman of the Presidium of the Association of Machine Builders (BSTU, Bryansk, RUS)
Prof. Yu. Novoselov (Sevastopol State University, Sevastopol, RUS)

International Program Committee
Prof. V. Bezyazychnyi (Rybinsk St. Aviation Tech. Un., RUS)
Prof. B. Bazrov (IMASH RAN, RUS)
Prof. O. Fedonin (rector of Bryansk St. Tech. University, RUS)
Prof. Yu. Tchigirinskiy (Volgograd State Tech. Un. RUS)
Prof. A. Vasilyev (Bauman Moscow St. Tech. Un., RUS)
Prof. V. Makarov (Perm St. Un, RUS)
Prof. Yu. Zubarev (SMTU, RUS)
Prof. U. Abdulgazis (Crimean. Eng. and Ped. Un., RUS)
Prof. T. Duyun (Belgorod St. Technol. Un., RUS)
Prof. G. Bahadirov (IMSS named after Urazbaeva, UZB)
Prof. A. Buchacz (Silesian Technical University, POL)
Prof. N. Danijarov (Karaganda St. Tech. Un., KAZ)
Prof. A. Kharchenko (Sevastopol State Un., RUS)
Prof. V. Smolentsev (Voronezh State Tech. Un.,RUS)
Assoc. Prof. A. Giutuni (Tunis Tech. University, TUN)
Prof. M. Heyfetc (Nat. Academy of Sc. of Belarus, BLR)
Prof. A. Handozhko (Bryansk St. Tech. University, RUS)
Assoc. Prof. G-J. Jasper (Plymouth University, GBR)
Prof. R. Kasper (OVG University Magdeburg, DEU)
Prof. S. Khristoforian (St. Eng.Un. of Armenia, ARM)
Prof. G. Barsukov (Orel State University, RUS)
Prof. O. Pruteanu (Tech. Un. "Gh. Asachi", ROU)
Prof. A. Toca (Technical University Moldova, MDA)
Assos. prof. A. Srivastava (Shamb. In. of Eng. and Technol. IND)
Assoc. Prof. K. Monkova (Tech. University Kosice, SVK)
Prof. Z. Sharifov (Azerbaijan Technical. University, AZE)

Investigation of the Relationship of the Chemical, Phase Composition with the Mechanical Properties of the Transition Zone of the Welded Joint A182 + 321

V.B. Dementyev[1], A.A. Shushkov[1], S.L. Kim[1, a *], V.V. Romanko[1]

[1]Udmurt Federal Research Center of the Ural Branch of the Russian Academy of Sciences, Izhevsk, Russia

[a]lgkim@yandex.ru

Keywords: Welded Connections, Transition Zone, Hardness, Elastic Modulus, Chemical and Phase Composition, Nanoindentation, NanoTest 600

Abstract. The physical and mechanical characteristics of the transition zone of the welded joint of steels A182 + 321, obtained by explosion welding, have been investigated. The width of the transition zone of a bimetallic joint based on A182 and 321 steels has been measured in various ways: according to the calculated and experimental data, which are in good agreement, it is about 8 μm. The dependences of the average values of hardness and the reduced modulus of elasticity of the metal of the investigated transition zone and the steels that make up the bimetal are obtained. The hardness of steels prior to explosion welding is 3.11 GPa and 5.30 GPa, respectively. The average value of the transitional hardness takes a value close to the hardness of the base metal (0.17 units lower) and is 2.94 GPa. A similar relationship was found when measuring the elastic modulus. The phase composition of the explosion welded joint made of steels A182 and 321 was studied, the width of the transition zone was determined. The dependence of the amount of retained austenite on the content of Ni and Cr in the transition zone and the relationship with the reduced modulus of elasticity and hardness are established.

Introduction

Welded joints (hereinafter referred to as bimetallic joints or bimetals) have attracted the interest of many researchers and are widely used in practice. The performance of bimetallic structures depends on the quality of joining dissimilar materials and is determined by the physical and mechanical properties of the joining zone. Currently, an urgent task is to improve the quality of bimetallic compounds at all stages of production [1].

In this regard, the aim of the work was to study the physical and mechanical characteristics (namely, hardness and elastic modulus) of bimetallic joints based on A182 + 321 steels by nanoindentation on the NanoTest 600 integrated measurement system.

The practical usefulness of the study is to determine the physical and mechanical properties of the transition zone, which will ensure the quality of bimetals.

One of the main methods for studying the physical and mechanical characteristics of the surface layers of steels and thin films is the nanoindentation method. The indentation experiment is carried out in accordance with GOST 9450-76. The measuring dynamic indentation method is the basis of the international standards ISO 14577 and ASTM E 2546-07. The Russian Federation is currently developing a modernized domestic standard for this measurement method.

The problem of determining the physicomechanical parameters of the transition zone of bimetals is little studied, since there are a number of difficulties in determining the hardness, elastic modulus and other characteristics due to the small size of the latter. The width of the transition zone of bimetals obtained by explosion welding does not exceed several microns [2]. In the works of Russian authors [2-4], the dependences of the hardness of bimetallic joints on the distance to the transition zone with an indentation step of 100 - 400 μm, an indentation load of 100 - 500 mN were constructed on the PMT-3M device. However, for a more accurate determination of the mechanical characteristics of the transition zone and the areas adjacent to it, such loads are not always acceptable, since the indentation of the indenter completely overlaps the transition zone with its dimensions.

The bimetal obtained by explosion welding has a set of properties that allows it to be used in critical structures in the energy, chemical, petrochemical and other branches of modern mechanical engineering [5]. Many studies are devoted to the study of the processes of obtaining, determining the properties of bimetals [6-10]. The main parameter that determines the characteristics of the bimetal is the transition zone, the structure and properties of which determines the quality of the bimetal as a whole. There is a lot of information on the structure, chemical, and phase composition of the transition zone of such compounds in the literature [11-13]. However, it became necessary to study the phase composition of the transition zone in a bimetallic sheet made of A182 steel with a surface clad with 321 stainless steel, and its relationship with its chemical composition and mechanical characteristics (hardness, reduced elastic modulus).

Taking into account the above, new methods are needed to study the physical and mechanical properties of both steels and the transition zone of bimetals, with an indentation step as small as possible, without disrupting the interaction of two neighboring indenter pricks, studying the phase composition of the transition zone, finding the relationship between the phase composition, chemical and physical and mechanical characteristics.

Experimental technique

A sample of the bimetallic joint A182 + 321 was obtained by explosion welding. For experiments with nanoindentation, the sample was a square plate 30 mm long and 5 mm thick. The sample surface was mechanically polished using polishing pastes.

A sample for research with a size of 30x30x5 mm was cut from a bimetallic sheet obtained on the basis of steels A182 and 321 by explosion welding. The chemical composition of the steels that make up the bimetal under study was determined on a Belec Compact Port device by the arithmetic mean of 5 measurements: steel A182-C = 0.146%, Cr = 1.016%, Mo = 0.473%; steel 321-S = 0.047%, Cr = 16.97%, Ni = 9.11%, Ti = 0.547%. Welding was carried out according to the mode: explosive [ammonium nitrate + diesel fuel in a ratio of 96: 4] with a throwing speed of 2400 m / s at a ratio of the mass of the explosive to the mass of the projectile metal of 1: 3. Steel 321 was used as the projectile material.

The width of the transition zone of the bimetal was determined by two methods: from the thermo-EMF diagrams on a device created on the basis of a PMT-3 microhardness meter [11], and X-ray spectral on a Solver-P47 scanning probe microscope with an electron beam 1 μm in diameter in points. The trajectory of scanning the surface of the material from the cladding layer towards the main one was chosen linear and perpendicular to the boundary of the joint zone. Thermo-EMF diagrams were built with a step of 5 μm and a number of measurements of 100, increased by 10 times, a dimensional grid was imposed on them to determine the width of the transition zone. The criterion was the beginning of a sharp increase in thermo-EMF in the positive direction, after the end of the growth of values. In this case, the error was 0.5 μm. In the X-ray microprobe method, chemical analysis was carried out point by point; the size of the transition zone was judged by the change in the content of nickel Ni, chromium Cr and iron Fe in microvolumes.

The phase composition of the transition zone was determined qualitatively on a D2 PHASER X-ray diffractometer with a Bragg-Brentano geometry and a LYNXEYE linear counter. The sample was taken in copper K_α radiation, the diffraction patterns were analyzed using the DIFFRAC.EVA software module, and the phases were identified using the PDF-2 / Release 2010 RDB database of the International Center for Diffraction Data ICDD (The International Center for Diffraction Data).

Microstructure analysis was performed on a digital optical microscope based on an MBI-1 and a modified color CCD matrix. The survey was carried out under illumination of the sample at an angle of 45° and 90° to their surface, the frame width corresponded to 400 μm (Fig. 1).

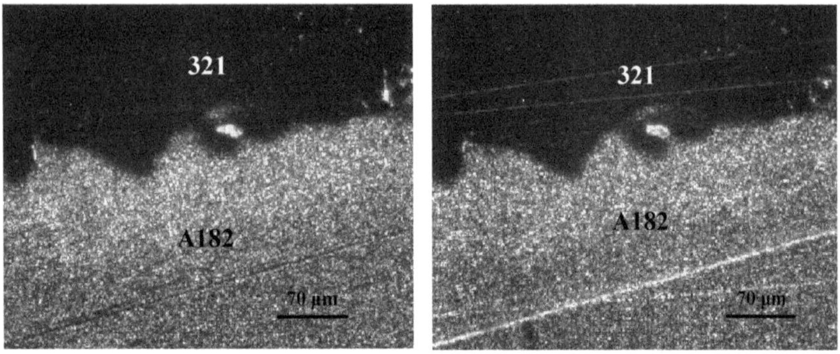

a) ***b)***

Fig. 1. *Surface structure of a bimetal sample: a) - at an angle of 45°, b) - at an angle of 90°*

Physicomechanical characteristics were determined by the indentation method on a complex measuring system NanoTest 600 according to the Oliver-Pharr method, using a Berkovich indenter (a trihedral diamond pyramid with an apex angle of 65.30 and a radius of about 200 nm) [12].

Measurement of the physical and mechanical characteristics of the surface layers of the bimetal sample of steels A182, 321 was carried out with a load of 50, 100, 250, 500 mN. Each loading force corresponds to a given penetration depth of the indenter. The loading rate of the test point of the sample was equal to the ratio of the maximum force at which the required penetration depth of the indenter was achieved, divided by 20 s. The distance between the indentation points was set to 40 μm. The boundary conditions were set in depth, i.e. upon reaching the maximum depth of penetration of the indenter into the sample (for example, 500 nm for a load of 50 mN or less), further loading of the sample was stopped. After each penetration of the indenter into the sample, when passing to the next indentation point, the indenter was retracted from the surface at a distance of 20 μm to avoid contact with the surface. In order to increase the reliability of the obtained result, the measurement procedure was performed 10 - 20 times for each load force.

The physical and mechanical properties of the transition zone of the bimetal sample were determined (for a width of 5-9 μm) with a load force of up to 50 mN (maximum penetration depth of the indenter 500 nm). At this depth of penetration of the Berkovich indenter, the calculated value of the edge of the indent of the indenter is 4.1 μm and the contact area of the indent of the indenter does not go beyond the width of the transition zone. Indentation was carried out in the center of the transition zone with a positioning error of 0.02 μm (Fig. 2).

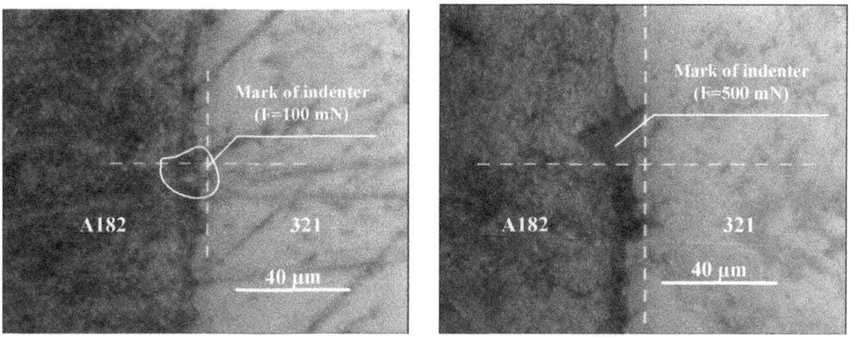

a) b)

Fig. 2. *Photographs of the marks of the Berkovich indenter in the transition zone of the A182 + 321 bimetal specimen after tests for indentation with load forces: a) - 100 mN, b) - 500 mN.*

To determine the mechanical characteristics, a thin section was prepared with a surface with a roughness parameter Ra of 40 nm. Since the calculated value of the face of the indent of the indenter under optimal boundary conditions (the depth of penetration of the indenter into the sample of 500 nm) was 1.9 μm, the indentation was carried out in the center of the transition zone with an indenter positioning error of 0.2 μm. Thus, the imprint of the Berkovich indenter at a given penetration depth did not go beyond the width of the transition zone. The roughness of the test sample was less than 10% of the penetration depth of the indenter and does not affect the accuracy of the experiment. In order to increase the reliability of the obtained result, the measurement procedure was carried out 20 times in different parts of the transition zone, its average value was taken as the result.

Results and discussion

On the investigated section of the joint of a bimetal sample of steels A182, 321, using an optical microscope with a magnification of 40X and a sensor for determining the displacement of the NanoTest 600 system, it was found that the width of the transition zone varies from 1 to 9 μm. In explosion welding, the joint boundary is formed in the form of an uneven wavy line (Fig. 1,3), which is in good agreement with the known research data on bimetallic joints [2] obtained by explosion welding.

The transition from steel base to clad metal, i.e. The transition zone is a narrow strip with uneven edges, which is most likely determined by the short-term temperature-force conditions of the welding process. According to the data of X-ray spectral studies, the width of the transition zone in the investigated area is about 7.7 μm (Fig. 3, Table 1). According to the results of measuring the thermo-EMF (Fig. 4), where the transition from the clad metal to the base metal is clearly shown in the form of a sharp increase in the values of thermo-EMF, the size of the transition zone is within (8 ±0.5) microns.

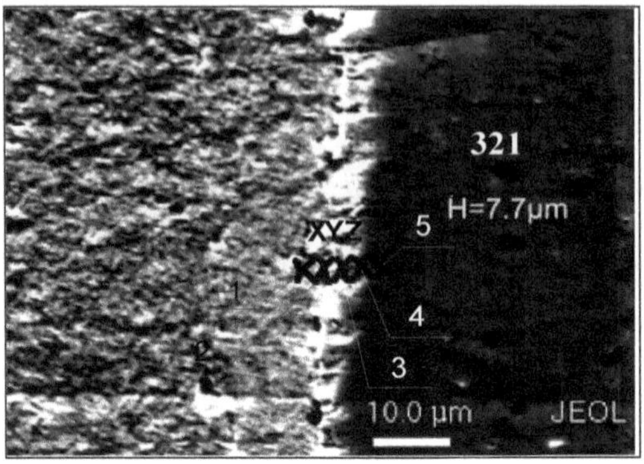

Fig. 3. Microstructure of the transition zone of the bimetallic sample from steels A182 and 321: × - measurement points (1÷5) - local microvolumes from which X-ray spectra and diffractograms X (2), Y (3), Z (4) were taken

Table 1. Results of determining the width of the transition zone in a bimetallic joint from steels A182 and 321 by various methods

Method of determination	Width, μm
Molecular dynamics calculation [13]	8,2
Thermo-EMF	8±0,5
Micro X-ray spectral analysis	7,7

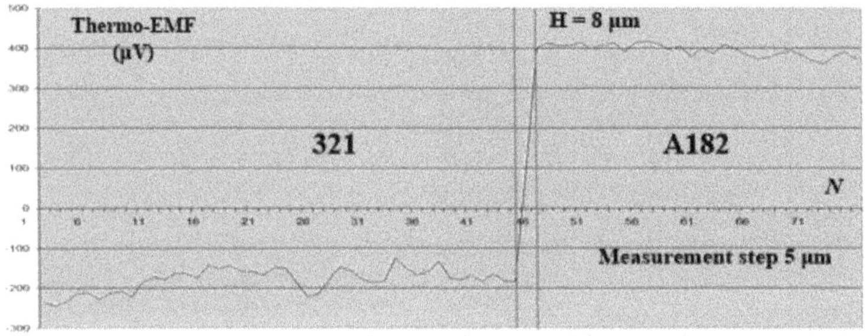

Fig. 4. Diagram of the change in thermo-EMF in the connection steel A182 + 321 from the number of measurements N

Calculation [13] using computer simulation by the molecular dynamics method for the section of the transition zone, on which the experimental studies were carried out, also showed the size of the transition zone close to the experimental one (Table 1).

The nature of the distribution of chemical elements (chromium, nickel and iron) in the transition zone, shown in Fig. 5, indicates the diffusion of chromium and nickel from the cladding layer into the base metal, the content of which decreases from 17% to 2% and from 11% to 1% with distance from the layer, respectively.

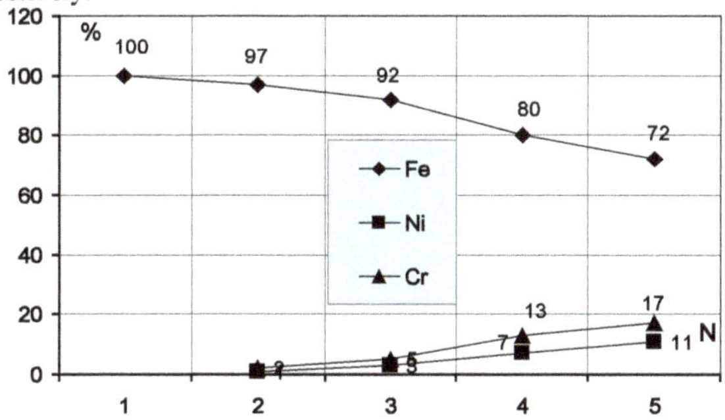

Fig. 5. Change in the content of chemical elements in the transition zone bimetallic connection in the transition from the base metal to the clad

*Table 2. Phase composition and characteristics of the crystallographic structure of the studied steels in the bimetallic compound A182 + 321. * - main lines*

	Formula, symbol	Structural type	Periods		
			a	b	c
A182	Fe_3C	*Pbnm (62), orthorhombic*	4.5144	5.0787	6.7297
	$(Cr_{0.2}Fe_{08})$*	*Im-3m (229), cubic*	2.8664	-	-
	$(Fe_{24}Mo)*_{0.08}$	*Im-3m (229), cubic*	2.8679	-	-
321	(Cr,Ni)*	*Fm3m (225), cubic*	3.591	-	-
	Cr_7C_3	*P63mc (186)*, hexogonal	14.01	-	4.532
	$Cr_{23}O_6$	*Fm-3m (225), cubic*	10.6599	-	2.98060
	Cr_3C_2	*Cmcm (63), orthorhombic*	2. 85	9.25	6.96

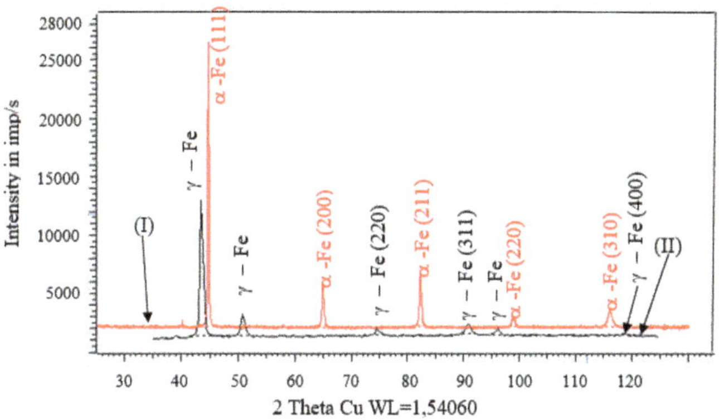

Fig. 6. Diffraction patterns of steels being welded: A182 (I) and 321 (II). Austenite - γ-Fe; ferrite-α-Fe.

The transition zone is represented by diffraction patterns (X), (Y) (Z) in Fig. 7, the analysis results of which are shown in table. 3. The numbers indicate the main lines of austenite (1, 3) and chromium iron (2, 4).

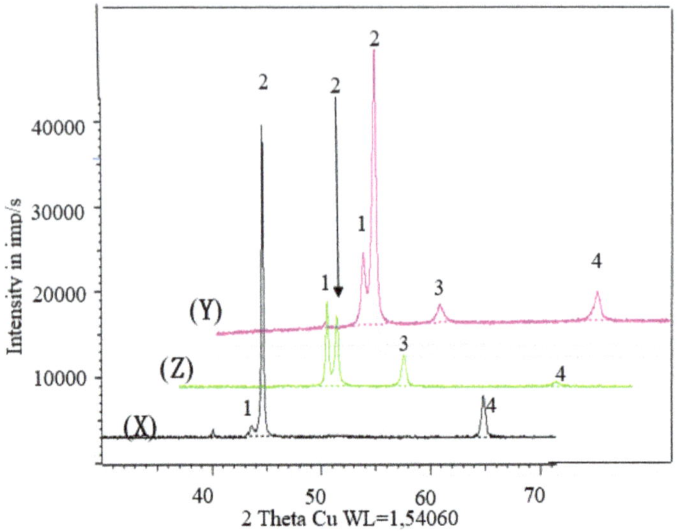

***Fig.** 7. Diffraction patterns (X), (Y) (Z), obtained from transition zone after explosion welding of steels (A182, 321): 1, 3 - γ-Fe; 2, 4 - α-Fe.*

***Table 3.** Phase composition and characteristics of the crystallographic structure of the structural components of the transition zone in the bimetallic compound A182 + 321*

Formula, symbol	Structural type
(Fe, Cr) – chromium iron*	*Im-3m (229)cubic*
(Cr, Ni) –austenite*	*Fm-3m (225)cubic*
Fe_3C- cementite*	*Pbnm (62), orthorhombic*
$(Fe,Cr)_7C_3$- carbide	*P31c (159), orthorhombic*
Cr_7C_3- tricarbide	*Pmcm (51)*
$Cr_{23}C_6$- carbide	*Fm-3m (225)cubic*
Mo_2C- carbide (at the background level)	*P-63/m2 (187_3) orthorhombic*
Fe_2C- martensite	*I4/mmm (139) tetragonal*

* - main lines;

As a result of the analysis of the diffraction patterns, it was found that with distance from the base metal zone (see points 2, 3 and 4 in Fig. 3, 5), austenite appears in the structure, the amount of which increases when passing from section (X) to section (Y) and maximum in the region (Z) -

peaks 1 and 3 from γ-Fe in Fig. 7, which indicates the diffusion of chromium and nickel from the cladding layer into the base metal. As was found by X-ray microanalysis, the concentration of nickel and chromium also changes with an increase (Fig. 5) and correlates with an increase in the amount of austenite in the joint zone. It should be noted that with an increase in the austenite phase in the structure, the amount of iron carbide Fe3C decreases (see in Fig. 6 an unmarked peak in the region of angles 2θ~40 deg.)

In addition, along with the phases of the base metal, the structure (Table 2) contains chromium carbides $Cr_{23}C_6$, Cr_7C_3 and solid solutions of Cr and Ni based on α–Fe in small amounts (very weak reflections) and even tetragonal martensite, which indicates that at the explosion welding moment, the metal of the surfaces to be joined is melted and for a short time is in a liquid state, when mixing and diffusion of alloying elements over short distances is possible. Upon cooling, these volumes crystallize and quench with the fixation of a high-temperature solid solution and the phases formed in them (Table 3). Consequently, the transition zone has a complex phase composition, represented by non-equilibrium compounds of alloying elements that make up the original joined steels.

In fig. 8 shows photographs of the imprints of the Berkovich indenter of the bimetal sample under study, the points of measurements at different load forces. The results obtained for the sample loading forces of 50 and 100 mN were significant for further analysis. According to GOST 9450-76, after measurements with loads of 250 and 500 mN (Fig. 8), it was found that the distance between the centers of the prints did not correspond to the required experimental conditions.

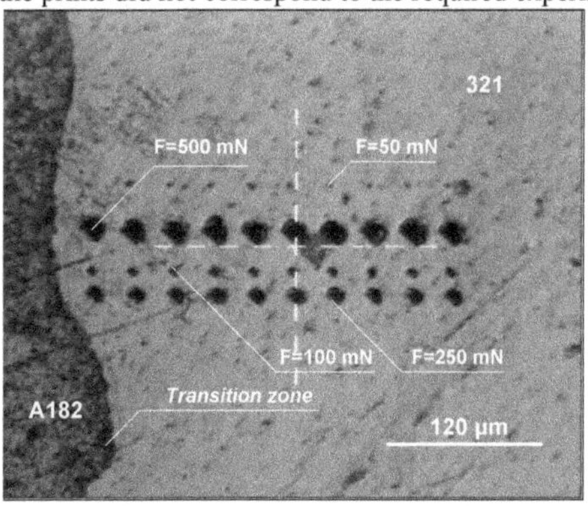

Fig. 8. *Photographs of a bimetal sample, steels A182, 321 and the transition zone after 40 tests for indentation with four loading forces.*

A detailed series of tests of 40 - 80 indenter penetrations was carried out on a A182 steel sample. Based on the results of the diagrams of the indenter penetration into the bimetal sample (Fig. 9), the physicomechanical characteristics of steels A182, 321 were obtained, namely, the reduced Young's modulus of elasticity, hardness.

Reference values of the modulus of elasticity at a temperature of 20 °C: A182 - 216 GPa, 321 - 198 GPa. The average obtained values of the reduced elastic modulus of the studied steels (for a penetration depth of 500 nm): A182 - 221.9 ± 13.2 GPa, 321 - 203.7 ± 4.4 GPa. Thus, at a given depth of penetration of the indenter into the sample (500 nm), the obtained average values of the modulus of elasticity of steels A182, 321 coincide with the reference values for the macromaterial.

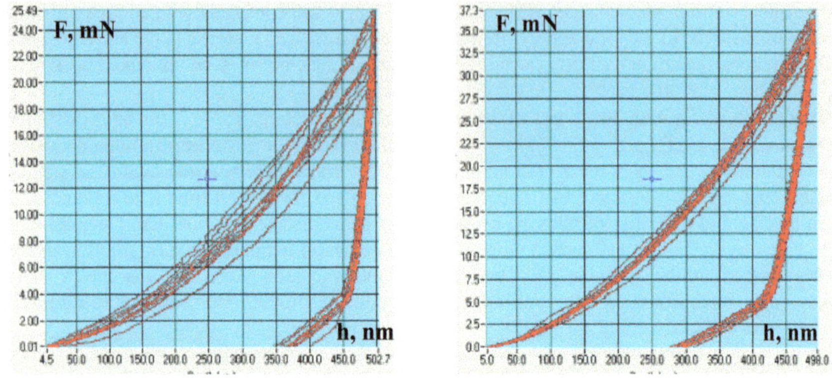

a) b)

Fig. 9. *Dependence of the applied force F, (mN), on the penetration depth of the indenter h, (nm) into the sample of the bimetal under study (load - unloading diagrams), for a maximum penetration depth of the indenter of 500 nm, 10 indentation points: a) - steel A182, b) - steel 321.*

The physical and mechanical characteristics of steels A182, 321 were investigated in a wide range of indenter penetration depths from 1 to 2 μm, with a step of increasing the penetration depth of 50 nm. A complete indentation procedure (loading - unloading) was carried out at one and the same point of the test specimen 20 times with increasing force up to the next, greater by 50 nm indenter penetration depth. The following averaged measurement results were obtained for the reduced modulus of elasticity: A182: - 204.9 ± 8.6 GPa, 321: - 185.5 ± 8.4 GPa. Thus, the values of the reduced elastic modulus obtained as a result of measurements decrease in comparison with the results obtained for an indenter penetration depth of 500 nm. With an increase in the penetration depth of the indenter, the reduced modulus of elasticity and hardness most often decreases. Different values of the reduced modulus of elasticity are a consequence of the size effect (Indentation Size Effect). The main reasons for this effect are given in [14], namely: the influence of external vibrations (the table of the NanoTest 600 system is installed on an air cushion, so they are completely excluded for these indenter penetration depths); by riveting the surface of the sample during polishing; an increase in the relative error in measuring the size of the print; the influence of grain boundaries and inclusions; an increase in the relative influence of the indenter imperfection with a decrease in the indentation, surface work hardening (hardening) during indentation, and others.

The study of the mechanical properties of the transition zone is represented by the characteristics of hardness and the reduced modulus of elasticity, obtained on the basis of the diagrams of penetration of the indenter (Fig. 10). The required depth of penetration of the indenter into the transition zone of the bimetal is achieved at different values of the load force, which leads to different values of the reduced modulus of elasticity and hardness.

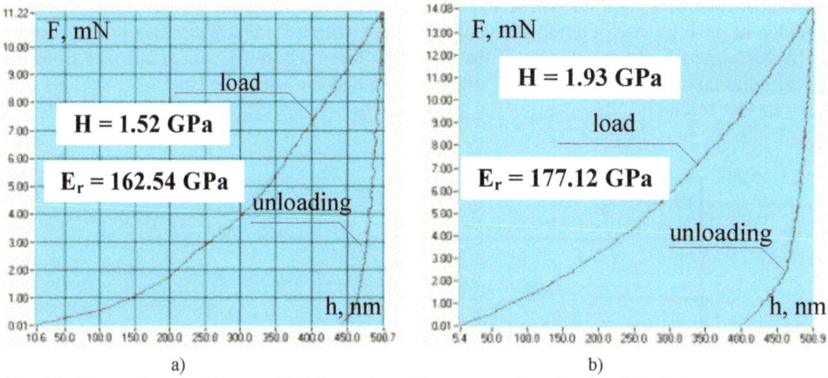

Fig. 10. *Dependence of the applied force from the penetration depth of the indenter at two points of the transition zone and the main characteristics. (a) the width of the transition zone is 6 μm, (b) - 8 μm.*

Investigations from 20 measurements on the physical and mechanical properties of the transition zone of the bimetal of steels A182, 321 with a width of 5-9 microns, a maximum indentation depth of 500 nm (maximum load of 50 mN) were carried out. Each indentation was carried out only in the center of the transition zone, the next value of the mechanical characteristics was taken in other parts of the transition zone. Figures 11, 12 show the dependences of the hardness and the reduced modulus of elasticity of the test sample.

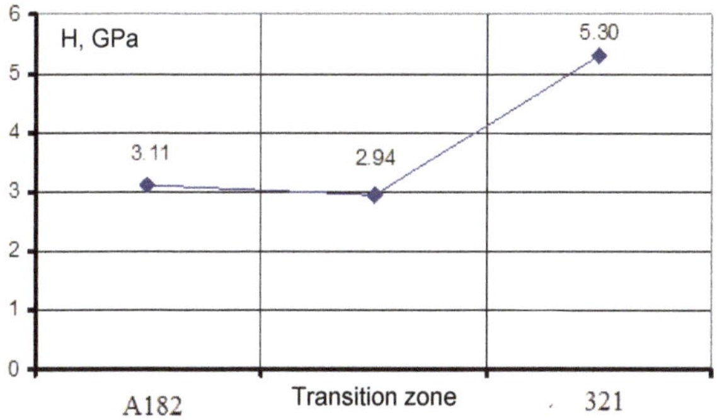

Fig. 11. *hardness average values of steels 12XM, 12X18H10T and transition zone*

The dependence of the reduced modulus of elasticity and hardness (Fig. 11, 12) is not constant in nature. The values of the hardness of the transition zone at any measurement are either greater or less than the values of the hardness of steel A182 and do not reach the values of hardness of steel 321. The values of the reduced elastic modulus of the transition zone in most cases are less than the values of the reduced elastic modulus of steels, the bimetal under study. This is due to the possible influence of a high temperature, higher than the melting point of the metal, which is formed due to the high rate of deformation upon impact, with different thickness of the transition zone. In this case, the structural state of the transition layer is formed, which changes from the chemical interaction of the molten metal after cooling and the substrate base (A182). This task requires further research.

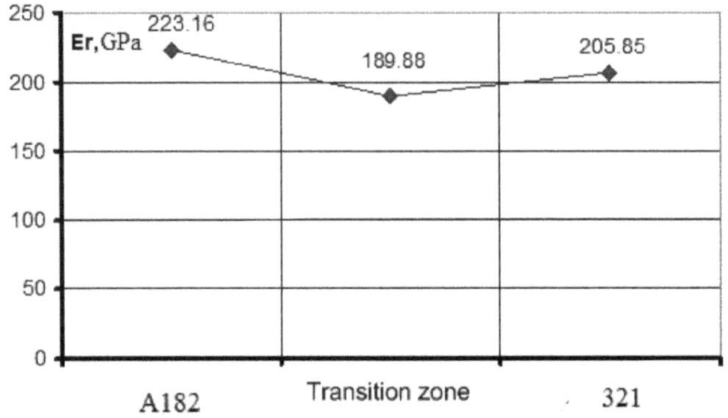

Fig. 12. *Average values of the reduced modulus of steels elasticity A182, 321 and the transition zone*

Conclusion

Investigations of the physical and mechanical characteristics of the bimetal (steels A182, 321) were carried out at different values of the load and the depth of indenter penetration. Experimental dependences of the reduced modulus of elasticity and hardness of the transition zone of a bimetallic joint were obtained in a small range of the width of the transition layer up to 8 μm. The research results have shown that the values of hardness and reduced modulus of elasticity in the transition zone have a non-constant character of change associated with different thickness of the transition zone, phase composition.

The width of the transition zone of a bimetallic joint based on steels A182 and 321 was measured in various ways: according to the calculated and experimental data, which are in good agreement, it is about 8 μm.

Determined the phase composition of the steels that make up the bimetallic compound and the transition zone. The main phases of the transition zone are chromium iron, austenite, martensite. Iron and chromium carbides are present in small amounts.

A relationship has been established between the phase composition and the presence of alloying elements in the transition zone. The dependence of the amount of retained austenite on the content of nickel and chromium in the joint zone is revealed.

The relationship between the phase composition and mechanical characteristics was not found (the hardness of steel A182 and transition zone is practically the same). A small change in the content of chromium and nickel 5%, 3% also does not cause a noticeable change in hardness and reduced modulus of elasticity.

References

[1] Moskvitin S.P., Pudovkin A.P. Method and system for quality control of bimetal // Bulletin of TSTU. 2009. T. 15, No. 2. pp. 315-320.

[2] Gurevich L.M., Trykov Yu.P., Arisova V.N. et al. Structure and micromechanical properties in BT -0 + AD1 bimetal obtained by explosion welding according to the angular scheme. Izvestiya Volgograd State Technical University. 2010. T. 4, No. 4. pp. 38-42.

[3] Trykov Yu.P., Pronichev D.V., Gurevich L.M. et al. Investigation of heat and electrical conductivity of SIK titanium-steel, Bulletin of the Volgograd State Technical University. 2010. T. 4, No. 4. pp. 17-21.

[4] Trudov AF, Trykov Yu.P., Klochkov SV, Dontsov D.Yu. Influence of heating on the structure and properties of the explosion-welded bimetal St.3 + 12X18H10T, Bulletin of the Volgograd State Technical University. 2008. T. 10, No. 2. pp. 18-22.

[5] Los I.S., Rosen A.E., Perelygin Yu.P., Pervukhin L.B., Usatii S.G., Khorin A.V. Investigation of the structure and properties of a multilayer corrosion-resistant material obtained by welding of explosions, Bulletin of the Volgograd State Technical University. 2010. T. 5, No. 4. pp. 93-96.

[6] Soloviev S. D., Dementyev V. B. Kinetics of the formation of interatomic bonds in bimetals. Izhevsk: IPM Ural Branch of the Russian Academy of Sciences, 2006. p. 152.

[7] Slyusarev M.V. Investigation of the parameters of the qualities of bimetallic sheets, Bulletin of the Volgograd State University. Series 9: Research by Young Scientists. 2007. No. 6. pp. 176-182.

[8] Denisov I.V., Pervukhin LB, Pervukhina O.L., Rosen A.E. Deformation processes in explosion welding, Bulletin of the Volgograd State Technical University. 2008. No. 6. pp. 39-45.

[9] Lysak V.I., Kuzmin S.V. Microinhomogeneity of welded explosions of joints, Bulletin of the Volgograd State Technical University. 2004. No. 6. pp. 4-22.

[10] Kim S.L., Dementyev V.B., Soloviev S.D. Mathematical modeling of the formation of interatomic bonds in bimetallic compounds, Chemical Physics and Mesoscopy. 2008. Vol.10, No. 4. pp. 419-427.

[11] Steklova E.O., Soloviev S.D., Kim S.L. Application of the thermopower method in the study of the chemical composition and structural state of welded joints, Welding and diagnostics. 2011. No. 2. pp. 10-13.

[12] Oliver W., Pharr G. An Improved Technique for Detemining Hardness and Elastic Modulus Using Load and Displacement Sensing Indentation Experiments, J. Mater. Res. 1992.№ 7(6). pp. 1564-1583. https://doi.org/10.1557/JMR.1992.1564

[13] Kim S.L. Investigation of the formation of a chemical bond in the production of bimetals based on the kinetics of interaction of atoms, Avtoref. diss. for the degree of Cand. phys-mat. sciences. Izhevsk, 2010.20 p.

[14] Fedosov SA, Peshek L. Determination of mechanical properties of materials. Modern foreign techniques. M.: Publishing house. Faculty of Physics, Moscow State University, 2004.98 p.

Estimating the Cost of Improving Quality in Production Dry Building Mixtures

Valentina Loganina[1,a*], Tatiana Uchaeva[1,b] and Maria Zaytseva[1,c]

[1]Penza State University of Architecture and Construction, Penza, st. Titova,28, Russia

[a]loganin@mail.ru, [b]uchaevatv@mail.ru, [c]zajc@yandex.ru

Keywords: Dry Building Mixtures, Quality Costs, Dispersion of Quality Indicators of Products and Raw Materials

Abstract. Provides information about the cost model associated with improving product quality. On the example of dry building mixtures, the issues of statistical quality management are considered. The cost model takes into account the cost of reducing the bias and variance of raw materials.

Introduction

Analysis of the building materials market shows an increase in the share of dry building mixtures (DBM) for various functional purposes [1,2]. In connection with the increased competition of dry mortar producers, the issues of their quality are of particular relevance.

Measuring control is an important element of product quality management. In the process of control, a comparison of the actually achieved results of quality indicators with the data of regulatory documents is carried out. Any measurement process is associated with variations in its parameters caused by a large number of factors affecting it. This undoubtedly affects the quality indicators of the final product [3-5].

It is known that the variability of the properties of building materials obeys the normal distribution law. With a normal distribution law, the probability that the deviation in absolute value will be less than three times the standard deviation is 0.9973, i.e. inconsistencies are possible only in 0.27% of cases.

Let's say the quality of a product is characterized by m properties. With the number of indicators m = 3, the probability of obtaining quality products is 0.9919. However, given that over time, even a well-regulated process can give shifts on average up to 1.5 sigma, it is relevant to take into account the variations in the components of the DBM formulation [6,7].

Materials and research methods

We analyzed dry construction mixes of the New MIX company (Penza region) (CemPlast, Bundes Koroed B2, Bundes Koroed C3). Cements Sengileevsky CEM I 42.5B (Eurocementgroup), Haldenberg CEM I 42.5N from HeidelbergCement Volga were used for the production of dry mixes. Quartz sand of the Makhalinsky deposit was used as a filler. The values of the standard deviation were calculated for the indicators of compressive strength, moisture content of the dry mixture and water retention capacity, as well as for indicators of cement activity, sand size modulus. The average value of the quality indicator and standard deviation were calculated based on the results of 50 tests.

The probability of compliance with regulatory requirements was calculated using the Laplace function [8,9]

$$P\left(\frac{UT-\bar{x}}{\sigma}\right) = 2\Phi(t) = \gamma$$

or (1)

$$P\left(\frac{\bar{x}-LT}{\sigma}\right) = 2\Phi(t) = \gamma$$

where σ is the standard deviation;
UT, LT - upper and lower tolerances.

Research results
The results are shown in Table 1.

Table 1. *The probability of defective products*

DBM type	The probability of defective products
Cem Plast	0.00007/0.00069
Bundes Koroed B2	0.09011/0.13351
Bundes Koroed C3	0.1615/0.17881

Note. Above the line are the values of the probability of the appearance of defective DBM production when using Sengileevsky cement, below the line - Haldenberg cement.

The research results (Table 1) show that the values of the probability of obtaining high-quality dry mixes depend on the type of cement. So, when using Sengileevsky cement, the value of the standard deviation (RMS) of the compressive strength of the DBM CemPlast samples was 1.8 MPa, and when using Haldenberg cement - 2.1 MPa. This difference, in our opinion, is due to the different variability of the cement. So, the values of the standard deviation of the activity of Sengileevsky cement are σ = 4.1-5.2 MPa, and Haldenberg cement - 4.5-5.7 MPa. Analysis of the data (Table 1) shows that when using Haldenberg CEM I 42.5N cement from HeidelbergCement Volga, the likelihood of defective products increases. So, the probability of the appearance of defective DBM Bundes Koroed B2 is 0.09011 and 0.13351 when using Sengileevsky cement and Haldenberg cement, respectively. Similar patterns are typical for DBM Bundes Koroed C3 and CemPlast

Thus, the choice of the type of cement, as well as other components of the DBM formulation should be carried out taking into account the variability of indicators characterizing their quality. Considering the above, it becomes necessary to analyze the functional relationship between the variability of the quality indicators of the DBM and the variability of the quality indicators of the input variables (raw materials). This will allow, knowing the required variance values of the product quality indicator, to determine the optimal variances of the input variables.

Dry mortar was considered as a system consisting of multiple input variables and one or more output variables. Let σY denote the variance of the output variable Y, and $\sigma_1, \sigma_2 \ldots .\sigma_n$ denote the variance of the input variables X_1, X_2 ,. ..., X_n. We will assume that the controlled parameters are normally distributed random variables. The results of statistical data processing are shown in Table 2.

For DBM CemPlast, the dependence of the standard deviation of the indicator (RMS) of the compressive strength on the RMS of the indicator of cement activity, the modulus of the size of the sand has the form

$$Y = a + bx + cx^2 \tag{2}$$

Fig. 1 shows a graphical model of the dependence of the standard deviation of the indicator (RMS) of the compressive strength on the RMS of the indicator of cement activity. The values of the coefficients in the model (2) for the RMSD of the cement activity are equal to a = 5.031; b = -2.3; c = 0.32. The correlation coefficient is r = 0.996

Table 2. *Variability of quality indicators for dry mixes and raw materials*

Standard deviation		
Haldenberg (activity)	Sengeleevsky (activity)	Sand (fineness module)
4.5/1.18	4.1/1.05	0.16/1.1
4.7/1.24	4.2/1.1	0.18/1.3
4.9/1.4	4.6/1.3	0.23/1.5
5.23/1.8	4.9/1.6	0.267/1.92
5.7/2.3	5.2/1.8	0.8/2.3

Note. Above the line are the values of the root-mean-square deviation of the activity of cement and the modulus of sand size, below the line - the compressive strength of DBM-CemPlast.

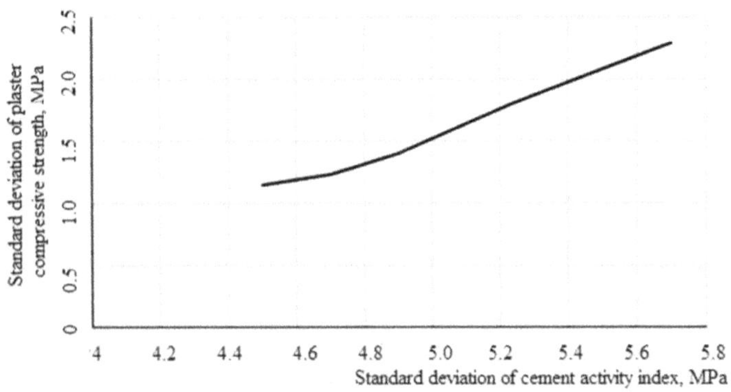

Fig. 1. *Dependence of the standard deviation of the compressive strength of CemPlast plaster on the value of the standard deviation of the cement activity indicator*

Using the proposed models, it is possible to clarify the requirements for the RMS of raw materials quality indicators. So, for example, with a compressive strength equal to 12.43 MPa, the standard deviation is σ = 1.91 MPa, and the value of the coefficient of variation is ν = 15.36%. With the tightening of the requirements for the scatter of data on the strength index of DBM to ν

= 10.0% (σ = 1.243 MPa), the value of the RMSD of the cement activity should be in accordance with formula (2) σ = 4.7 MPa.

The ultimate goal of a statistical management strategy is to minimize the overall cost to both producer and consumer, or the cost of the entire system. The cost to the consumer is related to the expected loss of quality of the output variable and is caused by the deviation from the target value. The costs to the producer are associated with the changing probability distribution of the input variables.

Let us consider a model of costs for improving product quality, which consists in taking into account the average values and variance values of quality indicators. The expected loss of quality for the customer consists of two parts: process bias and process deviation. The bias reduction strategy is to tune the workflow until the expected quality metric matches the target value. To reduce the variance of the Y exponent, the designer must reduce the variance of the input variables, which will also increase the cost. The challenge is to balance the reduced expected loss of quality with the increased cost of reducing the bias and variance of the input variables [10-12]. We denote the cost of variance control for the i-th input variable Xi as $\sum_{i=1}^{n} C_i(\sigma_i^2)$ and the control cost for the i-th input variable Xi as $D_i(\mu_i)$. The general model is:

$$TC = \sum_{i=1}^{n} C_i\,(\sigma_i)^2 + \sum_{i=1}^{n} D_i\,(\mu_i) + k[\sigma_y^2 + (\mu_i - y_o)^2] \quad (3)$$

where $k[\sigma_y^2 + (\mu_i - y_o)^2]$ is the expected loss of quality for the consumer;
k is a constant in the quality loss function;
Y_0 is the target Y value.

In equation (3), the first two terms $\sum_{i=1}^{n} C_i\,(\sigma_i^2)$ and $\sum_{i=1}^{n} D_i(\mu_i)$ are the costs of variances and the mean values of the input variables or cost to the manufacturer.

Consider the costs of the manufacturer of the DBM. If the average value of cement activity does not correspond to the target value Y_0 and the standard deviation σ does not provide a low variability of the quality indicator of the DBM, then the DBM manufacturer is not able to change anything in regulating the quality of cement, therefore, the cement supplier should be changed to another, supplied cement with a lower variability. In this case, the manufacturer of the DBM does not incur additional costs. Sand for dry mixes must comply with GOST 8736-93 "Sand for construction work. Technical conditions". As a rule, graded sand is used for DBM. If the variability of the sand does not suit the manufacturer of dry mixes from the point of view of the variability of the quality of dry mixes, then the costs of the manufacturer will be determined either by the purchase of a classifier that allows obtaining two or three fractions of sand, or by the purchase of fractionated sand.

Conclusion

A correlation has been established between the variability of the quality indicators of dry building mixtures and the variability of the quality of the components of the DBM formulation.

A cost model is proposed that takes into account the costs of reducing the bias and variance of raw materials. The proposed approach will allow regulating the quality of products already at the design stage and will minimize costs for both the manufacturer and the consumer.

References

[1] Market of building (concrete) mixtures in Russia: research and forecast until 2025 Release: October, 2021 https://roif-expert.ru/stroitelstvo/stroitel-nye-smesi/rynok-stroitel-nyh-betonnyh-smesej/rynok-stroitel-nyh-betonnyh-smesej-v-rossii-obzor-i-prognoz.html

[2] Analysis of dry mortar in Russia in 2015-2019, assessment of the impact of coronavirus and forecast for 2020-2024 https://businesstat.ru/images/demo/dry_building_mix_russia_demo_businesstat.pdf

[3] V.I. Loganina, Evaluation reliability of the control building materials and products for stability technological processes of production, Contemporary Engineering Sciences.7(33-36) (2014) 1927–1933. https://doi.org/10.12988/ces.2014.411207

[4] V.I. Loganina, J.P. Skachkov, T.V. Uchaeva, Assessment of the stability of the staining process of building products and constructions, International Journal Of Applied Engineering Research. 11(19) (2016) 9727–9729

[5] V.I. Loganina, Y.P.Skachkov., V.S. Lesovik, Quality control of building materials according to uncertainty of measurement and stability of the technological process of production.Solid State Phenomena. 299 (2020) 1161–1165. https://doi.org/10.4028/www.scientific.net/SSP.299.1161

[6] N.A. Rubichev, V.D. Frumkin, Reliability of tolerance quality control, Moscow, Publishing house of standards, 1990 .

[7] V.E. Gmurman, Probability theory and mathematical statistics: textbook for universities, Moscow, Yurayt Publishing House. 2020

[8] Yu.P Adler, B.L. Rozovsky, Operational statistical quality management. Moscow: Knowledge. 1984; 102

[9] E. Shindlovsky, O. Shchurts, Statistical methods of quality management, Moscow, Mir. 1976.

[10] L.V. Condra, Reliability Improvements with Design of Experiments, Marcel Dekker, Inc., New York, 2001

[11] M.Rausand, and A. Hoyland, System Reliability Theory: Models, Statistical Methods and Applications, 2nd ed., Wiley-Interscience, Hoboken, NJ, 2003

[12] Z.G. Stoumbos, M.R., Reynolds, T.P. Ryan and W.H Woodall, The State of Statistical Process Control as We Proceed into the 21st Century, Journal of the American Statistical Association. 95 (2000) 992–998. https://doi.org/10.1080/01621459.2000.10474292

Tribological Properties of Sapphire under Dry Friction against Chromium and Zirconium Ceramics

Valery Alisin[1,a*] and Dmitry Gutsev[2,b]

[1]Mechanical Engineering Research Institute of the Russian Academy of Sciences, 119334, 4 Bardina st, Moscow, Russia

[2]Institut metalpolymer systems of the Belarus Academy of Sciences, 246144, 32a Kirov st., Gomel, Belarus

[a]vva-imash@yandex.ru, [b]gutsev@mpri.org.by

Keywords: Tribological Properties of Sapphire, Dry Friction of Crystals, Tribological Tests, Kinetic Microindentation

Abstract. The issues of sapphire friction and wear in contact with chrome steel and zirconium ceramics are discussed in this article. The kinetic microindentation method is used to study the mechanical properties of sapphire in frictional contact in modelling experiments on the indentation of a Vickers indent. Taking into account the anisotropy of the mechanical properties of sapphire, the influence of the load and orientation of the indent on the modulus of elasticity and hardness is analyzed. It has been established that, under dry friction conditions it is promising the use of zirconium ceramics in the supports of devices operating under conditions in which the use of lubricants is impossible. The possibility of using a friction couple of sapphire-zirconium ceramics in friction units operating in open space is noted.

Introduction

Sapphire refers to superhard wear-resistant materials and is widely used in friction units of precision instruments; high-strength steels serve as a counterface. Trends in the improvement of heat engines are based on an increase in the temperatures of the operating environment, therefore, there is growing interest in the use of sensors and precision mechanics instruments operating at temperatures at which it is impossible to use lubricants. The antifriction properties of aluminum oxides are not high; therefore, many works are devoted to reducing the friction coefficient and improving their tribological properties. The tribological characteristics of sapphire are strongly influenced by the surface roughness. Due to the difficulty of machining, much attention is paid to the technological issues of processing the friction surfaces of sapphire. In [1], sapphire was tested for friction with a pin-on-disk test. The disc was coated with agglomerated diamond grit. It was found that sapphire is mainly removed by an abrasive by means of plastic deformation. In work [2], the processes of grinding single-crystal sapphire with a diamond grinding wheel with a metal bond were investigated. In the process of sapphire processing, the morphology of the processed surface was observed in the sense of the workpiece surface roughness, the mechanism of the influence of the wear of the grinding wheel on the surface quality of the workpiece was investigated. It was found that when the grinding depth reaches 240 microns, the surface quality of the workpiece deteriorates. In [3], a new abrasive tool is proposed that can provide a higher finishing efficiency, as well as a better surface quality of a sapphire base in the process of chemical-mechanical grinding. The novelty of the development lies in the sintering technology of the machining tool. In work [4], the issues of ultrasonic grinding of single-crystal sapphire are considered. A new grinding method has been proposed, in which an elliptical ultrasonic vibration is applied to the workpiece using an ultrasonic vibrator. The grinding forces and the roughness of the working surface are measured. It has been shown experimentally that improved characteristics of diamond wheel grinding can be realized using the proposed technology. In [5], the characteristics and mechanism of scratch formation on the sapphire surface were investigated, namely, the force during scratch formation, the depth and specific energy for each direction of the scratch on the sapphire plane, taking into account the anisotropy of the mechanical properties. As a result, the direction of application of the force was established, in which the specific energy of scratching is minimal. Was investigated [6] the mechanism of changing the characteristics of wear and cutting throughout the life cycle of a diamond wire saw when cutting sapphire. A new scheme for cutting sapphire with a diamond wire saw is proposed. Work [7] is devoted to the study of dry friction of amorphous carbon nitride on sapphire and steel. A strong influence of atmospheric humidity on friction and wear has been established. Various lubricants are used to improve the tribological properties of friction couples containing sapphire. In works [8, 9], tests for friction and wear of a sapphire and stainless steel contact were carried out. For contact lubrication, a water-based lubricant is developed by adding liquid crystals of ammonium carboxylate. Tests were carried out with linear reciprocating sliding at 75 ° C in the liquid crystal area and at 110 ° C above its melting point. Friction reduction up to 80% was found. At 110 ° C, for the sapphire and AISI 316L contact, there is a sharp decrease in the friction coefficient to an ultra-low steady-state value of 0.007-0.009. The influence of periodic surface structures formed by laser treatment on the coefficient of friction of silicon under dry friction was investigated [10]. Tribological tests were carried out in air using a "ball on a flat surface" nanotribometer with sapphire balls 3 mm in diameter as a counterface. It was found that the coefficient of friction of sapphire on a polished silicon surface remains approximately constant during the tests (in the range of 0.12–0.14) and does not depend on the applied load. There are few studies of the tribological properties of sapphire under dry friction. The aim of this work is to study the tribological properties of sapphire under dry sliding friction over a hard chromium coating of stainless steel and zirconium ceramics.

Materials

The samples were made from a single crystal of leucosapphire (synthetic sapphire), chemical formula Al_2O_3, density 4 g / cm3. The counterface was a chrome-plated stainless steel surface and a zirconium ceramic sintered from partially stabilized zirconia nanopowders. The chemical formula of the ceramics is ZrO_2 + 3 mol% Y_2O_3.

Equipment and methods

The mechanical properties of the materials were determined by the kinetic indentation method using a MNT_Z_AE_000 microhardness tester from CSM Instruments (Switzerland) in accordance with the ISO / DIS 14577 standard with a Vickers tetrahedral indent. Friction and wear tests were carried out on the stand of reciprocating motion with the recording of the friction force by a strain gauge device. The wear of the samples was measured by the gravimetric method on a lab balance.

Results

The stability of the mechanical properties of sapphire is an important condition for ensuring the stability of the friction units of precision instruments. With regard to friction units of machines, the following are of greatest interest: modulus of elasticity (E); hardness, in this work the hardness is determined by the Vickers method by indenting a diamond pyramid with a square base (HV); characteristic of the plastic properties of surface layers (WP), determined by the size of the area bounded by the unloading curve when the indent is pressed in and by the axes of rectangular coordinates; force is the depth of penetration of the indent. Indentation experiments were carried out at loads from 1 to 4 N.

Almost all crystals possess anisotropy of mechanical properties along the directions of the crystallographic axes, the determination of the directions of which is laborious and requires special equipment, which creates problems in the process of solving practical problems. In the experiments, non-oriented samples were studied, therefore, to obtain a more reliable result, the mechanical characteristics were measured at each load at three points with a rotation of 45 degrees, according to which the average values were determined. Table 1 shows the results of processing the indentation diagrams.

Table 1. *Characteristics of mechanical properties of sapphire at loads from 1 to 3 N*

N, [H]	1			2			3		
α, [degree]	0	45	90	0	45	90	0	45	90
E*, [GPa]	439	381	388	377	343	346	340	327	306
HV, Vickers	2702	2475	2581	2492	2320	2340	2258	2185	2122
W, [μJ]	0.58	0.60	0.59	1.69	1.73	1.75	3.24	3.31	3.46
W_P, [μJ]	0.32	0.31	0.31	0.88	0.91	0.93	1.71	1.75	1.84

There is a marked tendency towards a decrease in the values of the modulus of elasticity (Fig. 1) and hardness (Fig. 2) with an increase in the load on the indent.

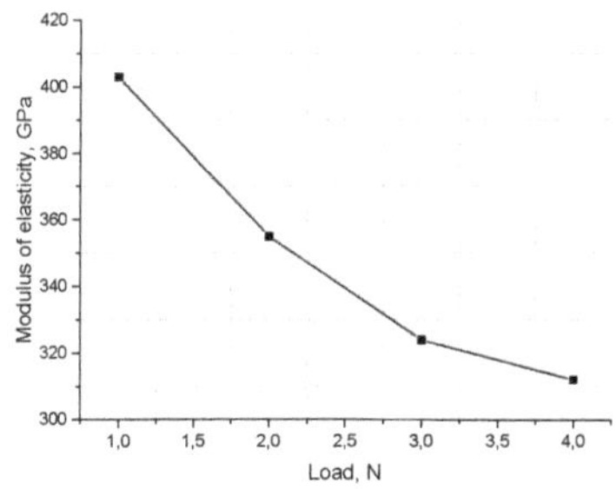

Fig. 1. *Effect of load on elastic modulus*

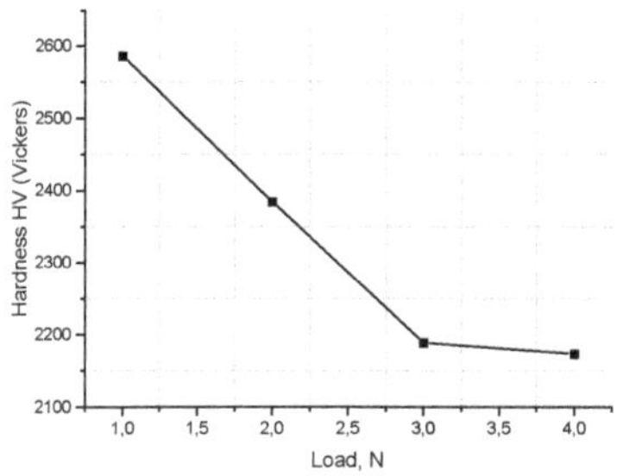

Fig. 2. *Effect of load on hardness*

The coefficient of irreversible deformations equal to the ratio of irreversible energy losses (W_P) to the total required work on the indentation of energy (W) changes little. The test equipment is facing increasing demands, taking into account that this study of the tribological properties of sapphire is aimed at ensuring the operability of instrumental mechanisms in the absence of lubrication. One of the main sources of measurement errors is the presence of a hinge in the holder, which serves for self-alignment of samples. To improve the fit of the samples, it is necessary that the center of rotation of the hinge is as close as possible to the plane of contact of the samples. For these tests, a holder was manufactured (Fig. 3) with a needle support for loading the sample [7].

Fig. 3. *General view of the holder.*

The holder of the test specimen is made in the form of two interconnected disks with a clearance between their end surfaces, whereas on the inner side of one disk located on the side of the loading element, a cylindrical protrusion with a conical tip is made.

The tribological test methodology is based on measuring the wear of the samples before and after testing. Table 2 shows the results of experiments under dry friction with reciprocating motion (p = 3 MPa, V = 4 mm/s).

Table 2. *Results of comparative tests of sapphire with friction against chromium and ceramics*

Friction couple	Density [g/cm^3]	Path friction, [cm]	Linear wear, cm	f_0	f_k	wear rate
1. Chromium 2. Sapphire	7,2 4	720	0,00299 0,00599	0,228	0,243	4,2x10^{-6} 8,3x10^{-6}
1. Ceramics 2. Sapphire	5,9 4	720	0,000031 0,0000125	0,05	0,122	4,3x10^{-8} 1,7x10^{-8}
1. Ceramics 2. Sapphire	5,9 4	1440	0,0000339 0,0000125	0,05	0,119	2,4x10^{-8} 9,0x10^{-9}
Note: f_0 - initial coefficient of friction, f_k - final coefficient of friction						

It was found that in the absence of lubrication, the use of chrome-plated steel axles is unacceptable, due to very high values of the coefficient of friction and very high rates of wear of surfaces. With friction against ceramics, the friction coefficient is 2 times less, and the wear resistance of the samples is more than 100 times higher.

Conclusion

Based on the determination of the tribological properties of the friction couple sapphire and zirconium ceramics, it is advisable in the supports of devices operating under conditions of inadmissibility of the use of lubricants, for example, in controls for high-temperature heat engines, it is advisable to recommend to sapphire thrust bearings as a counterface ceramic composites based on partially stabilized zirconia. This friction couple has an acceptable coefficient of friction (not higher than 0.2) and has shown high wear resistance. Sapphire and zirconium ceramics have a very high resistance to high vacuum and high temperatures, therefore, it is of interest to use them in friction units designed for long-run operation in open space.

References

[1] Jiapeng Chen, Nannan Zhu, Fengli Niu, Yanan Peng, Jianxiu Su, Yongwei Zhu, Influence of agglomerated diamond abrasive wear on sapphire material removal behavior, Diamond and Related Materials. 108 (2020) 107965. https://doi.org/10.1016/j.diamond.2020.107965

[2] Bin Luo, Qiusheng Yan, Jisheng Pan, Jiabin Lu, Zhanliang Huang, Influences of processing parameters on metal-bonded diamond wheel wear when grinding a sapphire wafer, Diamond and Related Materials. 113 (2021) 108275. https://doi.org/10.1016/j.diamond.2021.108275

[3] Ke Wu, Daichi Touse, Libo Zhou, Wangpiao Lin, Jun Shimizu, Teppei Onuki, Julong Yuan, Chemo-mechanical grinding by applying grain boundary cohesion fixed abrasive for monocrystal sapphire, Precision Engineering. 70 (2021) 110-116. https://doi.org/10.1016/j.precisioneng.2021.01.015

[4] Zhiqiang Liang, Xibin Wang, Yongbo Wu, Lijing Xie, Zhibing Liu, Wenxiang Zhao, An investigation on wear mechanism of resin-bonded diamond wheel in Elliptical Ultrasonic Assisted Grinding (EUAG) of monocrystal sapphire, J. of Materials Processing Technology. 212/4 (2012) 868-876. https://doi.org/10.1016/j.jmatprotec.2011.11.009

[5] Wangpiao Lin, Jun Shimizu, Libo Zhou, Teppei Onuki, Hirotaka Ojima, Investigation of nanoscratch anisotropy of C-plane sapphire wafer using friction force microscope, Precision Engineering. 73 (2022) 51-62. https://doi.org/10.1016/j.precisioneng.2021.08.011

[6] Youkang Yin, Yufei Gao, Chunfeng Yang, Sawing characteristics of diamond wire cutting sapphire crystal based on tool life cycle, Ceramics International. 47/19 (2021) 26627-26634. https://doi.org/10.1016/j.ceramint.2021.06.070

[7] Xiaoxu Liu, Noritsugu Umehara, Takayuki Tokoroyama, Motoyuki Murashima, Tribological properties of ta-CNx coating sliding against steel and sapphire in unlubricated condition, Tribology International. 131 (2019) 102-111. https://doi.org/10.1016/j.triboint.2018.10.022

[8] M.D. Avilés, F.J. Carrión, J. Sanes, M.D. Bermúdez, Effects of protic ionic liquid crystal additives on the water-lubricated sliding wear and friction of sapphire against stainless steel, Wear. 408–409 (2018) 56-64. https://doi.org/10.1016/j.wear.2018.04.015

[9] M.D. Avilés, F.J. Carrión, J. Sanes, M.D. Bermúdez, Bio-based ionic liquid crystal for stainless steel-sapphire high temperature ultralow friction, Wear. 484–485 (2021) 204020. https://doi.org/10.1016/j.wear.2021.204020

[10] Alves-Lopes, A. Almeida, V. Oliveira, R. Vilar, Influence of laser surface nanotexturing on the friction behaviour of the silicon/sapphire system, Optics & Laser Technology. 121 (2020) 105767. https://doi.org/10.1016/j.optlastec.2019.105767

Investigation of the Influence of Direct Metal Deposition Modes on Microstructure and Formation of Defects in Samples of Heat Resistant Alloy

A.V. Balyakin[1, a], E.P. Zlobin[1, b], M.A. Oleynik[1, c]* and E.S. Goncharov[1, d]

[1]Samara National Research University named after Academician S. P. Korolev, Samara, Russia

[a]a_balik@mail.ru, [b]ep.smr@mail.ru, [c]oleynik1997@mail.ru, [d]zekangon@gmail.com

Keywords: Additive Technologies, Direct Metal Deposition, Porosity of The Material, Nickel-Based Alloy, Mechanical Properties

Abstract. The article discusses the influence of technological modes of the DMD method on the macro- and microstructure of a heat-resistant nickel-based alloy to use this technology for heat-resistant materials in the manufacture of parts for combustion chambers in gas turbine plants.

Introduction

In the aerospace and energy industry, products are widely used, the parts of which are made of high-temperature alloys. Such details include flame tubes, discs, blades and turbine housings [1]. Usually, blanks for these parts are obtained by such methods as casting, metal forming and powder metallurgy methods [2]. This requires the manufacture of expensive equipment (foundry, die), the design and manufacture of which takes a long time [3, 4]. In this case, the mass of the initial billet made of the heat-resistant alloy can be 10-15 times greater than the mass of the finished part, and the waste obtained in the foundry cannot always be reused. The chips obtained in the process of machining require special technologies for recycling.

Additive technologies (AT) make it possible to obtain products of complex geometric shapes, which are impossible to obtain by traditional methods. This reduces the nomenclature of parts and the time to receive the finished product. Additive manufacturing allows you to immediately produce blanks directly from 3D models obtained from a CAD system by layer-by-layer addition of materials, without the need for design and manufacture of additional equipment.

The use of additive technologies in the production of gas turbine engines makes it possible to produce blanks by melting a metal powder or wire and obtaining a continuous solid-phase structure of functional components of engines [5, 6]. The use of AT makes it possible to obtain products with high mechanical properties, the quality of which depends on many technological parameters of the process. A distinctive feature of the additive production of metal products in comparison with foundry is the possibility of obtaining a homogeneous fine-grained structure by acting on the initial material of a point source of supplied energy and high crystallization rates, which makes it possible to avoid defects in the structure characteristic of high-chromium and intermetallic nickel alloys. However, rapid crystallization in the local volume contributes to the formation of other characteristic defects, such as lack of fusion and pores with insufficient energy supply and cracks from the action of thermal stresses in the case of its excess.

The mechanical properties of parts obtained by additive technologies strongly depend on the size and shape of the metal powder grains, as well as the geometry and parameters of the formation of the surfacing bead [7]. Depending on the temperature and power load of the parts to which they are subjected during operation, the requirements for the microstructure for parts made of heat-resistant alloys will be different. For example, during the takeoff regime, the turbine disk is exposed to high loads and experiences low-cycle fatigue; therefore, a fine-grained equiaxed microstructure is required to increase the fatigue strength [8, 9]. To ensure improved mechanical properties in castings, directional solidification technology is used, in particular, for turbine blades of gas turbine engines, methods of producing monocrystalline blades are used. Monocrystalline castings have enhanced mechanical properties due to a reduction in the number of transverse grain boundaries or their complete elimination [10].

The most promising technology for the manufacture of billets of large-sized parts from heat-resistant hard-to-machine steels and alloys is the direct energy deposition, when the product is formed from a metal powder supplied by a gas-powder jet into the zone of exposure to a laser beam (L-DED) [11, 12, 13]. Among the disadvantages of this method are high roughness, as well as defects such as discontinuities in the form of looseness, pores and adhesions.

The aim of this work is to study the influence of technological parameters of the L-DED process on the microstructure of chromium-nickel alloy samples and minimizing the appearance of characteristic defects.

Materials and methods

Material. In the course of the research work, the heat-resistant alloy Inconel 718 was used. Samples made of metal powder (the particle size of the main fraction is 40-150 μm) using the L-DED technology have the shape of a parallelepiped with dimensions of 80x30x12 mm. The chemical composition of the heat-resistant powder Inconel 718 was determined using a TESCAN VEGA 3 scanning electron microscope, the results of microspectral analysis are shown in Table 1.

Table 1. *Results of microspectral analysis of heat-resistant powder of alloy Inconel 718*

Element	Ni	Cr	Si	Mn	Nb	Al	B	Ti	Mo	Fe
Percentage	52.26	19.22	0.0	0.17	5.3	0.75	0.00	0.99	2.92	0.11
Standard	50.0-55.0	17.0-21.0	<0.35	<0.35	4.75-5.5	0.2-0.8	0.006	0.65-1.15	2.8-3.3	balance

Sample Making. The samples under study were made on a robotic L-DED machine developed at Samara University, consisting of a six-axis industrial robot-manipulator Eidos A12 with an additive module and a pedestal on which the growing process was carried out.

Argon was used as a protective and transport gas, which was supplied to the deposition zone at a flow rate of 4 L/min and 10 L/min, respectively. For a better heat exchange processes in the workpiece, the deposition process was carried out on a low-carbon steel substrate, and a sacrificial layer 10 mm high was added to the sample to equalize the deposition conditions. After growing the samples, they were separated from the build platform and processed to a size of 80x20x12 mm. Table 2 shows the parameters of the mode of laser fusion of powder material.

Table 2. *Modes of L-DED of powder material*

L-DED mode parameter	L-DED parameter value			
	Sample 1	Sample 2	Sample 3	Sample 4
Laser power, W	1400	1600	1700	1800
Powder consumption, g/min	32	29	20	46
Deposition speed, mm/s	20	20	23	40
Layer step, mm	0.4	0.4	0.6	0.4
Track width, mm	1.6	1.6	1.6	1.6

Metallographic Investigation of Sample Material. During external examination of the samples, fusion beads were clearly visible, the width of which was from 1.5 to 1.7 mm. The metallographic analysis was carried out on metallographic specimen prepared along the growth direction of the samples. Additionally, thin sections were made in the transverse direction with respect to the direction of growth of the samples in the middle part with respect to the height of the samples. Etching of specimen was carried out in Vasiliev's reagent (CuSO4 - 5 g, H2SO4 - 1.4 ml, HC1 - 50 ml, H2O - 40 ml). The macrostructure of the samples obtained by the L-DED method in modes 1...4 is shown in Fig. 1.

Sample 1 Sample 2 Sample 3 Sample 4

Fig. 1. *Macrostructure of samples obtained by the L-DED method in modes 1...4*

Microanalysis of specimen without etching revealed that the material of all samples contains small discontinuities in the form of looseness, pores and seals (Fig. 2). The total porosity of the material of the samples is: in samples No. 1, 2 - up to 0.015 mm, in samples No. 3, 4 - up to 0.01 mm.

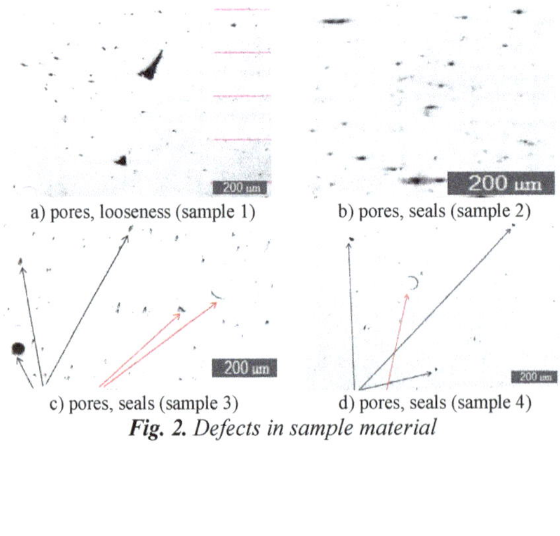

a) pores, looseness (sample 1) b) pores, seals (sample 2)

c) pores, seals (sample 3) d) pores, seals (sample 4)

Fig. 2. *Defects in sample material*

Materials Research Proceedings **21** (2022) 23-27 | https://doi.org/10.21741/9781644901755-4

After etching, cracks with a length of 0.05 to 1.1 mm were found in all samples, located along the boundaries of individual grains at a distance of 0.6 to 2.0 mm from the side surfaces. Samples 1, 2 have single cracks in the central part. The track structure is noticeable in sample No. 1, and in sample No. 2 the tracks are practically indistinguishable.

Results and discussion

Microstructure. The analysis of the microstructure of the samples, obtained with an optical microscope, was carried out. It was revealed that the microstructure of the Inconel 718 alloy is mainly represented by columnar dendrites, which grew continuously during the deposition of subsequent layers. An increase in the length of dendrites occurs in the direction of surfacing, which is explained by the co-directional temperature gradient, which creates thermodynamic conditions for solidification. It is obvious that the size of the melt pool and the characteristics of compaction of the material as a result of a decrease in its porosity and the number of non-melts (solders) differ significantly depending on the laser power. With an increase in the laser power, the depth and width of the melt pool increase, and the bonding properties of the adjacent layer increase with a constant lateral step, as a result of which the relative density of the sample increases. At the same time, the heat-stressed state increases, which increases the likelihood of cracking. At low laser power, the energy absorbed by the powder is insufficient. This leads to a mismatch in the required temperature in the melt bath and complicates the melting of the powder and the penetration of adjacent layers, which is a necessary condition for the formation of a stable bond between the layers being fused. In addition, due to the high dynamic viscosity of liquid metal at low temperatures in the molten bath, convection deteriorates and the flowability of the melt decreases, which makes it difficult to remove mixed gases and contributes to the formation of pores. At optimal laser power, the dynamic viscosity of the melt is sufficient to provide convection and gas removal from the melt bath, as a result of which the number of pores becomes minimal.

Laves phases are harmful phases, since they are very fragile compounds, and their morphology and size strongly affect the properties of the alloy [14]. The formation of Laves phases is greatly influenced by the parameters of the laser, especially the power and speed of the beads deposition [15].

Conclusion

Having studied the results of the research carried out, the following conclusions can be drawn:

1. Large discontinuities such as looseness, pores and seals were not found in the material of the samples. The maximum size of single discontinuities is: looseness - 0.03x0.06 mm, pores - from 0.03 to 0.05 mm, seals - 0.13...0.17 mm.
2. The maximum pore size is: in samples No. 1, 2, - up to 0.015 mm, in samples No. 3, 4 - up to 0.01 mm.
3. In the material of samples 1, 2, 3, cracks with a length of 0.05 ... 1.1 mm were found, located along the boundaries of individual grains at a distance of 0.6 ... 2.0 mm from the side surfaces.

Analysis of the L-DED technological modes shows that the most preferred fusion mode is mode 4. The main defects for mode 4 are single pores up to 0.03 mm. In this regard, we select the values of the 4th technological mode as the basic mode.

References

[1] Vdovin, R.A. 2019, "Improving the quality of the manufacturing process of turbine blades of the gas turbine engine", Journal of Physics: Conference Series. https://doi.org/10.1088/1742-6596/1399/4/044035

[2] Saleil, J., Mantel, M. & Le Coze, J. 2020, "Stainless steels making: History of production processes

developments. Part III. Casting methods, hot and cold forming processes", Materiaux et Techniques, vol. 108, no. 1. https://doi.org/10.1051/mattech/2020015

[3] Balaykin, A.V., Bezsonov, K.A., Nekhoroshev, M.V. & Shulepov, A.P. 2018, "Developing Parametric Models for the Assembly of Machine Fixtures for Virtual Multiaxial CNC Machining Centers", IOP Conference Series: Materials Science and Engineering. https://doi.org/10.1088/1757-899X/302/1/012009

[4] Balyakin, A.V., Dobryshkina, E.M., Vdovin, R.A. & Alekseev, V.P. 2018, "Rapid Prototyping Technology for Manufacturing GTE Turbine Blades", IOP Conference Series: Materials Science and Engineering. https://doi.org/10.1088/1757-899X/327/2/022025

[5] Sotov, A.V., Agapovichev, A.V., Smelov, V.G., Kokareva, V.V., Dmitrieva, M.O., Melnikov, A.A., Golanov, S.P. & Anurov, Y.M. 2020, "Investigation of the IN-738 superalloy microstructure and mechanical properties for the manufacturing of gas turbine engine nozzle guide vane by selective laser melting", International Journal of Advanced Manufacturing Technology, vol. 107, no. 5-6, pp. 2525-2535. https://doi.org/10.1007/s00170-020-05197-x

[6] Committee on Propulsion and Energy Systems to Reduce Commercial Aviation Carbon Emissions, Aeronautics and Space Engineering Board, Division on Engineering and Physical Sciences & National Academies of Sciences, Engineering,and Medicine 2016, "Commercial aircraft propulsion and energy systems research: Reducing global carbon emissions" in Commercial Aircraft Propulsion and Energy Systems Research: Reducing Global Carbon Emissions, pp. 1-122.

[7] Tresa M., Pollock and Sammy T. 2006, "Nickel-Based Superalloys for Advanced Turbine Engines: Chemistry, Microstructure and Properties", Journal of Propulsion and Power, vol.22:2, pp. 361-374. https://doi.org/10.2514/1.18239

[8] Winstone, M. and J. Brooks, Advanced high temperature materials: Aeroengine fatigue. Ciência & Tecnologia dos Materiais, 2008. 20(1-2): p. 15-24.

[9] Helmink, R.C., et al., Advanced superalloys and tailored microstructures for integrally cast turbine wheels. 2000: Superalloys 2000 (Ninth international symposium). p. 171-179. https://doi.org/10.7449/2000/Superalloys_2000_171_179

[10]Das, N., Advances in nickel-based cast superalloys. Transactions of the Indian Institute of Metals, 2010. 63(2-3): p. 265-274. https://doi.org/10.1007/s12666-010-0036-7

[11]Turichin, G.A., Zemlyakov, E.V., Pozdeeva, E.Y., Tuominen, J. & Vuoristo, P. 2012, "Technological possibilities of laser cladding with the help of powerful fiber lasers", Metal Science and Heat Treatment, vol. 54, no. 3-4, pp. 139-144. https://doi.org/10.1007/s11041-012-9470-y

[12]Leyens, C. & Beyer, E. 2015, "Innovations in laser cladding and direct laser metal deposition" in Laser Surface Engineering: Processes and Applications, pp. 181-192. https://doi.org/10.1016/B978-1-78242-074-3.00008-8

[13]Ocylok, S., Alexeev, E., Mann, S., Weisheit, A., Wissenbach, K. & Kelbassa, I. 2014, "Correlations of melt pool geometry and process parameters during laser metal deposition by coaxial process monitoring", Physics Procedia, pp. 228. https://doi.org/10.1016/j.phpro.2014.08.167

[14]Sui S, Chen J, Ming X, Zhang SP, Lin X, Huang WD (2017) The failure mechanism of 50% laser additive manufactured Inconel 718 and the deformation behavior of Laves phases during a tensile process. Int J Adv Manuf Technol 91:2733–2740. https://doi.org/10.1007/s00170-016-9901-9

[15] Ma MM, Wang Z, Zeng XY (2015) Effect of energy input on microstructural evolution of direct laser fabricated IN718 alloy. Mater Charact 106:420–427. https://doi.org/10.1016/j.matchar.2015.06.027

Materials Research Proceedings **21** (2022) 28-31 — https://doi.org/10.21741/9781644901755-5

Carbonized Steel-Smelting Slag Is a Promising Raw Material for the Production of Artificial Concrete

Igor Romanenko[1,a*], Alexey Fadin[1,b] and Maria Romanenko[1,c]

[1]Penza State University of Architecture and Construction, 28 German Titov, Penza, 440028, Russia

[a] rom1959@yandex.ru, [b] fadin@tulpm.ru, [c] romanenko.masha@yandex.ru

Keywords: Steelmaking Slag, Carbon Dioxide, Carbonization, Matrix, Structure, Hydration, Clinker Fund, Strength

Abstract. Obtaining an artificial stone based on steel-smelting slag is possible as a result of carbonization of the feedstock in carbon dioxide. The feedstock - slag and carbon dioxide - are by-products from steel smelting in electric furnaces, which must be disposed of in order to improve the environmental situation in the region. The condition for obtaining the cementing ability of steelmaking slag is the preparation of a charge with certain properties and maturation technology: humidity, dispersion of the fine fraction and the maximum size of the coarse fraction, the ratio between the coarse and dispersed fractions, the concentration of carbon dioxide in the gas-air environment, temperature, pressure and flow time. carbonization reactions in the reactor, the magnitude of the pressure during the production of pressed articles, the process of stone maturation in the post-carbonization period.

Introduction

In the global balance of industrial waste emissions, and in particular, steel production slag is 240-300 ml. tons per year, which negatively affects the environment [1]. It is required to develop a number of measures and on their technologies for the inclusion of waste in the production cycle of obtaining materials for construction purposes with high performance characteristics. In general, blast furnace slags [2] are widely used in the production of a hydraulic binder, since they correspond to a stable material and mineralogical composition. Steel-making slags are practically not used due to the content of calcium and magnesium oxide (> 10 %), their volumetric stability, which leads to cracking in an alkaline medium, and the pH of the medium is > 10 [2, 3].

We proposed a technological approach to obtain a durable artificial stone due to a molding composition on an optimized compositional matrix that underwent volumetric pressure and carbonization in a reactor, the starting material did not contain such substances as $Ca(OH)_2$, $Mg(OH)_2$, SiO_2.

This approach is not new. Many researchers tried to obtain a stone as a result of the action of carbon dioxide on molded products from slag in a reactor [3, 4], but the carbonization processes affected the surface zone and did not extend to the entire depth of the sample. The core remained not carbonized, and as a result of holding both in the normal hardening chamber and in the reactor, cracks formed on the surface of the samples, leading to self-destruction due to the transition of the $\beta_{\text{-c2s}}$ phase to $\gamma_{\text{-c2s}}$, which does not possess binding properties at high pH values ≥ 12, and free CaO and MgO increase in volume [5, 6].

Therefore, an important role is played by the creation of an optimal internal structure of the stone, ensuring the penetration of CO_2 into the sample to its entire thickness, and the aqueous medium played the role of an active conductor.

Materials, research methodology

The purpose of the research is to identify the technological features of the production of small-sized elements based on steel production slags, activated by carbon dioxide.

For the study, steelmaking slags of the Novolipetsk Metallurgical Plant were used, the chemical composition is presented in Table 1.

Table 1. *Chemical composition of steelmaking slag*

MgO	Al_2O_3	SiO_2	S	CaO	MnO	FeO	Fe_2O_3	TiO_2	R_2O	Zn	C
7,58	3,48	13,79	0,3	49,88	1,98	9,75	10,2	0,25	0,17	0,011	1,15

Slag particle size (SPS) 0-2,5 mm, moreover, 30 % of this slag has a particle size of 30 to 230 microns. The reaction medium was a reactor with the possibility of supplying carbon dioxide, raising the temperature (up to 200 ^{0}C) and regulating the pressure inside the reactor. Samples from the prepared mixture were molded on a press with double-sided hyper-pressing with a force of 385 kgf/cm^2.

To achieve the optimum density and carbonization of SPS, the molding mixture is moistened to a moisture content of 0,1. The reactor is preheated to 105-110 °C. After loading the samples into the reactor, evacuation takes place - 0.7 atm., 5 min, then the CO_2 pressure rises to 12 atm. at a speed of 3-4 atm./min. The temperature of the carbonization in the reactor is 105 -110 ^{0}C. Formed specimen cylinders with a diameter and height of 60 mm. placed on a pallet in the reactor chamber (Fig. 1)

Fig. 1. *Specimens molded from steelmaking slag of fraction 0-2,5 mm, molding mixture – 100 % wetted slag, mixture moisture 7,1 %*

Discussion

After carbonization of the slag in the reactor for 14 hours, the samples had no cracks on their surface (Fig. 2).

After inspecting the samples on the surface, there are 1 or 2 shots of the surface layer with a diameter of 2-5 mm and a depth of 1-2 mm. The overall surface is smooth, uniformly gray, without cracks or flaking. Exposure in a normal hardening chamber at a temperature of 20 °C and a humidity of 90 % for 28 days did not affect the quality and condition of the surface of the samples.

Fig. 2. *Samples from steelmaking slag fraction after carbonization in the reactor after 14 hours*

The compressive strength of the samples is 15,1-25,0 MPa, the density is 2340-2398 kg/m^3. The studies carried out on the split of the sample in order to determine the depth of carbonization showed by the intensity of the color of the indicator applied to the cut, that in the central part the intensity is weak and is approximately 25-30 % relative to the intensity of the outer layer.

Investigations [7-9] carried out to determine the phase composition of steelmaking slag after carbonization reactions in the reactor and in the CO_2 environment after 14 hours established the formation of the following minerals: $CaCO_3$ calcite and calcium silicate hydrate CSH, $MgCO_3$ magnesite, $Ca_3Mg(SiO_4)_2$ merwinite and FeO wustite.

Conclusions

The studies carried out have shown the possibility of obtaining low-grade concrete by carbonization of steel-making slag in a reactor in a carbon dioxide environment.

The use of steelmaking slag as a binder and filler according to traditional technology in the manufacture of building products is not acceptable due to the destruction of the structure of concrete from stresses caused by significant volumetric changes by CaO hydration products.

On the basis of steelmaking slag without adding Portland cement to the molding mixture due to carbonization, it is possible to produce wall stones of grades 100, 150 and 200 by optimizing the granular composition of the molding mixture, creating an optimal structure of artificial stone for the penetration of carbon dioxide through the molded samples.

The production of small-sized materials for road construction works based on slags from steelmaking production must strictly comply with the established technological regulations: the moisture content of the molding mixture, the temperature of the medium inside the reactor, the holding of a given pressure in the reactor, the concentration of CO_2 in the air of the reactor, the time of preliminary holding of the samples before raising the pressure in the reactor.

References

[1] S.Y. Pan, CO_2 mineralization and utilization by alkaline solid wastes for potential carbon reduction. Nature Sustainability, 3(5), (2020) 399- 405. https://doi.org/10.1038/s41893-020-0486-9

[2] Y.N. Sheen, D.H. Le, and T.H. Sun, Innovative usages of stainless steel slags in developing self-compacting concrete. Construction and Building Materials, 101 (2015) 268-276. https://doi.org/10.1016/j.conbuildmat.2015.10.079

[3] A.S. Brand and J.R. Roesler, Steel furnace slag aggregate expansion and hardened concrete properties. Cement & Concrete Composites, 60 (2015) 1-9. https://doi.org/10.1016/j.cemconcomp.2015.04.006

[4] P.S. Humbert and J. Castro-Gomes, CO_2 activated steel slag-based materials: A review. Journal of Cleaner Production, 208 (2019) 448-457. https://doi.org/10.1016/j.jclepro.2018.10.058

[5] I. Romanenko, A. Fadin, Optimization of Compositions and Hardening Conditions of Vibropress Concretes Based on Steel-Making Slag. Materials Science Forum, 1037 (2021) 715-720. https://doi.org/10.4028/www.scientific.net/MSF.1037.715

[6] I. Romanenko, A. Fadin, Steekmaking Slag – A Complex Material for the Production of Small-Size materials Using Hyper-Press Technology. Materials Science Forum, 1037 (2021) 737-742. https://doi.org/10.4028/www.scientific.net/MSF.1037.737

[7] A.A. Ruzavin, Disposal of Steelmaking Slags by Accelerated Carbonation. Bulletin of the South Ural State University. Ser. Construction Engineering and Architecture, 3(18) (2018) 68–72. https://doi.org/10.14529/build180310

[8] I.I. Romanenko, M.I. Romanenko, I.N. Petrovnina, Modification of the cement stone active silm of urban treatment facilities. IOP Conference Series: Materials Science and Engineering. International Conference on Modern Trends in Manufacturing Technologies and Equipment 2019, ICMTME 2019 (2020) 022096. https://doi.org/10.1088/1757-899X/709/2/022096

[9] I.I. Romanenko, I.N. Petrovnina, M.I. Romanenko, Modification of the binder with finely ground components. Science diary, 2(14) (2018) 10.

Influence of Pressing Temperature on Wood Properties

Maria Romanenko

Penza State University of Architecture and Construction, 28 German Titov, Penza, 440028, Russia

romanenko.masha@yandex.ru

Keywords: Hardwood, Cellulose, Lignin, Amorphous Polymer, Compression, Thermostatted, Density, Wood Processing, Finishing Material

Abstract. Obtaining wood with high performance properties on the basis of chemical and mechanical action as a result of optimization of technological processes and the use of temperature exposure. The initial raw material is hardwood (aspen, alder), which are little used in construction and in the production of finishing materials. The condition for obtaining wood with high operating properties (increasing density, strength, reducing water saturation, ensuring the dimensional stability of samples for a long time) is the ability of wood as a natural polymer to change properties under the combined effect of temperature and pressure.

Introduction

Aspen and alder are practically not used in the production of building materials such as supporting structures and finishing. Although a feature of aspen wood is the absence of knots, resistance to moisture in the sawn state does not warp, does not crack during drying and storage in the air, but at the same time the wood is prone to drying out and does not have dimensional stability. Alder warps during drying, has moderate hygroscopicity and softness, is resistant to decay, which is why it is often used for equipping wells, as well as storerooms. These types of wood have significant advantages over other hardwoods - they are fast-growing, low purchase prices, low demand in the construction market. To improve the operational properties in 1973 and 1983. researchers have proposed technologies for the heat treatment of wood [1]. Using compression and high temperatures for wood processing, the researchers significantly improved the physical and mechanical properties of samples from beech, larch and pine.

The method of hot pressing of wood treated with formaldehyde resin was widely used [2,3]. It was based on the modification of wood in the form of shavings and sawdust in the production of particle boards. Treated wood with formaldehyde resins or compositions based on them is environmentally hazardous to human life, and since the end of the 80s of the 20th century, this technology has been banned from using in Europe and Russia.

Heat treatment of wood at temperatures of 200-280 ° C for 7-12 hours allows you to increase the density, hardness of the material, but at the same time increases the tendency to brittle fracture under force. Such wood is suitable only for finishing works [4, 5].

Researchers have devoted to the study of the issue of the influence of 130 0C and pressing pressure of 7 MPa on coniferous wood [6, 7]. It has been revealed that the technology of obtaining thermally modified wood makes it possible to obtain thermally stabilized wood with increased density, hardness and high water resistance.

Materials, research methodology

The purpose of the research is to identify the technological features of the production of modified hardwood (aspen, alder) in the Penza region by hot pressing with a force of 6.0 MPa. Press plate temperature 140-160 ° C.

For the study, we used a hot press with a working surface of 600 × 600 mm at a maximum pressure of 12 MPa.

The wood underwent sawing and finishing to plates with dimensions of 250 × 250 × 16 mm. The holding time on the table of a hot press under pressure is 60 min. Table 1 shows the properties of the raw materials and technological regulations.

Table 1. *Raw material properties*

№	Wood	Density, [k/m^3]	Hardness, [MPa]	Humidity, [%]	Pressing temperature. [^{0}C]	Exposure time, [min.]	Pressing pressure, [MPa]
1	Aspen	510	23,6	3,56	140	60	12
2	Alder	486	27,3	4,00	140		12
3	Aspen	509	23,65	3,61	160		12
4	Alder	487	27,8	4,13	160		12

Determination of the hardness of wood samples was determined on samples with a size of 50 × 50 × 50 mm on a universal testing machine. The hardness of the wood was redistributed in accordance with GOST-16483.17-81, in accordance with the Brinell method. Tests for changes in the characteristics of the environment in the laboratory at a temperature of +26 ^{0}C and ambient air W-56 %.

The swelling of the wood was assessed by complete immersion in water for 24 hours, so that the top of the sample was 20-25 mm high. The temperature of the water in the chamber was automatically maintained at 20 °C.

The original dimensions of the samples before and after the tests were measured with a caliper. The results are in Table 2.

Table 2. *Properties of modified materials*

№	Wood	Reducing sample thickness, [%]	Density		Hardness		Pressing temperature. [^{0}C]	Swelling value, [%]
			[kg/m^3]	growth, [%]	[MPa]	growth, [%]		
1	Aspen	27,3	701	37,45	28,8	22,0	140	6,12
2	Alder	28,4	661	36,01	33,9	24,2	140	7,04
3	Aspen	30,2	737	44,79	28,3	19,6	160	2,98
4	Alder	31,4	684	40,45	32,9	18,4	160	3,56
control	Aspen	–	–	–	–	–	–	2,49
	Alder	–	–	–	–	–	–	2,86

Discussion

From the analysis of the data obtained (table 2), it can be seen that after the process of hot pressing of hardwood samples, the thickness changed significantly. In aspen and alder, the parameter changes significantly at a pressing temperature of 160 ° C than at 140 ° C.

Swelling also shows that the pressing temperature has a significant effect. So at a pressing temperature of 160 °C, the swelling value is 5,56 % for alder, and 2,98 % for aspen. At a pressing temperature of 140 °C, the swelling value for alder is 7.04%, for aspen is 6,12 %. All hardwood samples that passed swelling tests showed that the control samples have 2,5 times lower values.

The hardness of aspen at a pressing temperature of 140 °C is greater than at a temperature of 160 °C by 12,2 %, in alder – by 31,5 %. At the same time, the increase in wood hardness is higher at lower temperatures. It is possible to explain the processes taking place if we consider the structure and structure of hardwood.

Wood is a composite material consisting of numerous microscopic, layered tubes oriented along the trunk. Cellulose incorporated into the lignin-hemicellulose matrix acts as a reinforcing agent [8, 9]. The spatial structure of the matrix is formed by interpenetrating networks: 1 – formed by lignin and carbohydrates; 2 – lignin and hemicelluloses; 3 – three-dimensional networks of lignin itself.

It was revealed by a number of researchers that at a relative humidity of wood equal to 27-30 %, cellulose already at 22-26 °C exhibits the properties of an amorphous-crystalline polymer in a highly elastic state [10-12].

During the experiment, the moisture content of the wood was less than 5 %, the temperature of hot pressing was 140 and 160 °C, according to the researchers [10], the glass transition temperature of dry wood ranges from 130 to 190 °C. Thus, the tested wood samples acquired the thermomechanical properties of the intercellular substance. This is a sign of an amorphous polymer with characteristic physical states - glassy, highly elastic and viscous fluid. In this case, the state of the samples is closer to glassy, since the samples are highly fragile. There is no reliable information in the open press about the glass transition temperature of the ligno-carbohydrate complex, and it is not necessary to carry out indirect studies, but rather aimed at understanding the pseudo reticular structure of wood, taking into account the possible volumetric targeted change in the anisotropic structure.

Conclusions

The studies carried out have shown the possibility of obtaining modified wood with high performance properties. By optimizing the technological process, it is possible to obtain finishing materials with high aesthetic and, most importantly, competitive with respect to rare varieties of wood.

The unique structure of a natural polymer (wood) is endowed with such properties as hydrophilicity, biological compatibility, thermal stability and polyfunctionality, which makes it possible to modify wood not only through mechanical and thermal effects, but also through the use of chemicals and various complexes.

The considered technological method for changing the natural structure of wood makes it possible to obtain wood similar to the texture of plane tree, elm from alder and aspen, provided that the original wood is additionally colored.

The growing need for the development of new and obtaining new cheap materials for construction and finishing purposes is determined by the fact that cellulose, as the main component of wood, is a renewable and rapidly growing raw material.

References

[1] A. Burmester, Effect of heat-pressure-treatment of semi-dry wood on its dimensional stability. Holz Roh-Werkst, 31 (1973) 237-243.

[2] E. Giebeler, Dimensional stabilization of wood by moisture-heat-pressure-treatment. Holz RohWerkst, 41 (1983) 87-94. https://doi.org/10.1007/BF02608498

[3] A.J. Stamm, Wood and cellulose science, Ronald Press, New York, USA, 1964.

[4] B. Esteves, A.V. Marques, I. Domingos, H. Pereira, Influence of steam heating on the properties of pine (Pinus pinaster) and eucalypt (Eucalyptus globulus) wood. Wood Science and Technology, 42(3) (2007) 193-207. https://doi.org/10.1007/s00226-006-0099-0

[5] O. Unsal, S.N. Kartal, Z. Candan, R.A. Arango, C.A. Clausen, F. Green, Decay and termite resistance, water absorption and swelling of thermally compressed wood panels. International Biodeterioration & Biodegradation, 63(5) (2009) 548-552. https://doi.org/10.1016/j.ibiod.2009.02.001

[6] O. Unsal, S. Korkut, C. Atik, The effect of heat treatment on some properties and colour in eucalyptus (Eucalyptus camaldulensis Dehn.). Wood. 11th International IUFRO Wood Drying Conference – 2010, 398 MADERAS: Ciencia Y Tecnologia Journal, 5(2) (2010) 145-152. https://doi.org/10.4067/S0718-221X2003000200006

[7] V.A. Shamaev, P.A. Smirnov, N.A. Trubnikov, O.A. Kalinin, V.G. Rakhmanov, I.N. Medvedev, E.A. Dolgikh, The main directions of wood. Technology and equipment of woodworking in the XXI century: interuniversity. sat. scientific. tr., VGLTA, Voronezh, (2005) 161-167.

[8] P.P. Erinsh, Study of the structure and destruction of the ligno-carbohydrate matrix of wood. Dis. ... Dr. chem. sciences. Riga, 1976.

[9] P.P. Erinsh, The structure and properties of wood as a multicomponent polymer system. Wood Chemistry, 1 (1977) 8-25.

[10] S.S. Pesetskii, B. Jurkowski, Y.A. Olkhov, O.M. Olkhova, I.P. Storozhuk, U.M. Mozheiko. European Polymer Journal, 37 (2001) 2187–2199. https://doi.org/10.1016/S0014-3057(01)00120-3

[11] I.I. Romanenko, M.I. Romanenko, Directional modification of the properties of hardwood. Science diary, 11(11) (2017) 19.

[12] I.I. Romanenko, M.I. Romanenko, Modern technologies for the creation of modified wood from hardwood. Ural Scientific Bulletin, 10(2) (2016) 90-95.

Method for Calculating the Shape of Rolls for Helical Rolling

Sergey Gorbatyuk[1,a], Boris Belelyubskii[2,b*], Aleksei Karfidov[1,c] and Sergey Snitko[3,d]

[1]National University of Science & Technology MISIS, Leninskiy Prospekt 4, Moscow, Russia

[2]Moscow Polytechnic University, Bolshaya Semyonovskaya str. 38, Moscow, Russia

[3]Donetsk National Technical University, Shybankova Square 2, Pokrovsk, Ukraine

[a]sgor02@mail.ru, [b]nis_mgvmi@mail.ru, [c]a.korf@mail.ru, [d]snitko_sa@mail.ru

Keywords: Helical Rolling, Shape of Rolls, Calculation Method, Deformation Zone

Abstract. The paper is devoted to solving the problem of determining the shape of the rolls of helical rolling mills, depending on the specified profile of the deformation zone. A universal calculation method has been proposed, thanks to which it is possible to determine the shape of the working surface of a roll for all types of helical rolling mills (with mushroom-shaped, cup-shaped, barrel-shaped and disc rolls), any relative arrangement of the rolling axis and rolls axes, and various locations of the deformation zone on the rolling axis. The proposed method is implemented as a standalone exe-application with a simple intuitive interface. The application allows you to output the calculation results into txt-files, which can then be imported into CAD systems to create 3D roll models.

Introduction

Recently, the development of different industry sectors [1-11] requires the solution of various technological production problems [12-22] and also increases the requirements for the efficiency and quality of production technologies [23-30]. One of the most promising and profitable methods to produce high-quality seamless pipes and rods is helical rolling.

In helical rolling, metal deformation occurs between obliquely located rolls, the axes of which are inclined relative to the rolling axis by a certain angle. Points on the surface of the billet move along a helical line, performing both rotational and translational movements. This method of metal forming is widely used for the manufacture of products in the form of bodies of revolution (balls, axes, pipes, etc.)

When considering many issues related to the helical rolling process, geometric parameters are the basis. This is explained by the fact that the shape and size of the deformation zone significantly affect the conditions of rolling, the required shape of the rolls, etc.

Helical rolling mills are generally classified according to the shape of the work rolls into four main types (figure 1): *a* - with barrel-shaped rolls, *b* - with mushroom-shaped rolls, *c* - with cup-shaped rolls, *d* - with disc rolls.

Calculation model of the geometric parameters of the helical rolling process

The calculation method is based on the description of the relative position of the rolling and roll axes by the values of the angle α and the distance ρ between them (figure 2).

This was done because during helical rolling the rolling axis and roll axis are skew lines, the relative position of which is uniquely determined by the values of the angle and the distance between them. Also, this approach is interesting in that α and ρ do not depend on the method of adjusting the work rolls and, therefore, are convenient for analysing processes in stands of various designs.

Materials Research Proceedings 21 (2022) 36-42 https://doi.org/10.21741/9781644901755-7

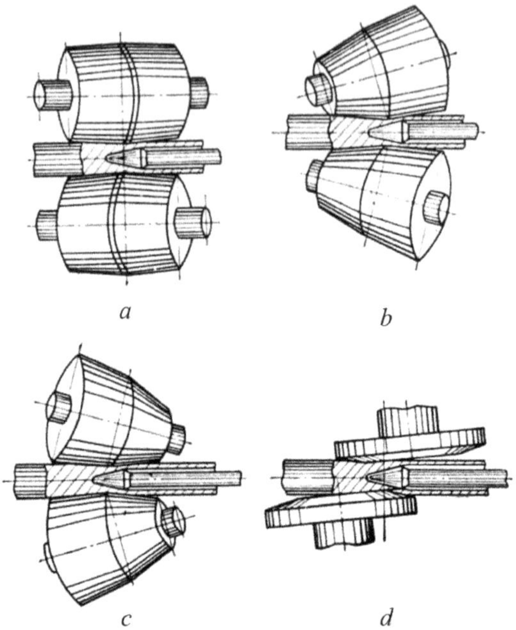

Fig. 1. *Types of rolls for helical rolling:*
a - *barrel-shaped,* b - *mushroom-shaped,* c - *cup-shaped,* d - *disc rolls.*

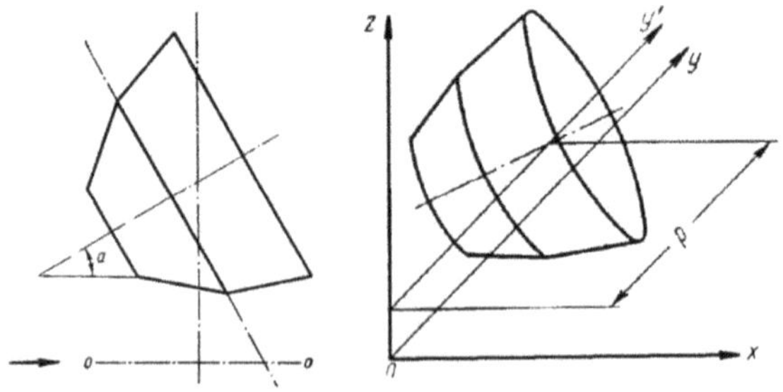

Fig. 2. *Parameters of the relative position of the rolling and roll axes.*

Thus, if we introduce the assumption that each *i*-th section of the deformation zone is processed by the *i*-th section of the roll, then to calculate the exact values of the roll radius we need to know:

α – angle between rolling and roll axes;

ρ – distance between rolling and roll axes;
x_i and y_i – coordinates of the points of the deformation zone.

For a given section of the roll, the roll radius is gradually increased by the iterative method, and the condition is checked whether the points of the circle formed by this radius are outside the billet

$$\begin{cases} \Delta Z \le |Z_{roll}| \\ \Delta Y \le |Y_{roll}| \end{cases} \tag{1}$$

As soon as the condition is met, the calculation will be interrupted.

Figure 3 represents a diagram that shows that for all points lying outside the billet, either $\Delta Y > Y_{roll}$, or $\Delta Z > Z_{roll}$. Only for one point of the circle ΔY will be equal to Y_{roll}, and ΔZ will be less than Z_{roll}, thus, at this point the condition will be fulfilled and the calculation for this section will be interrupted. The smallest possible radius at which the condition is satisfied on the entire circle will be the necessary one.

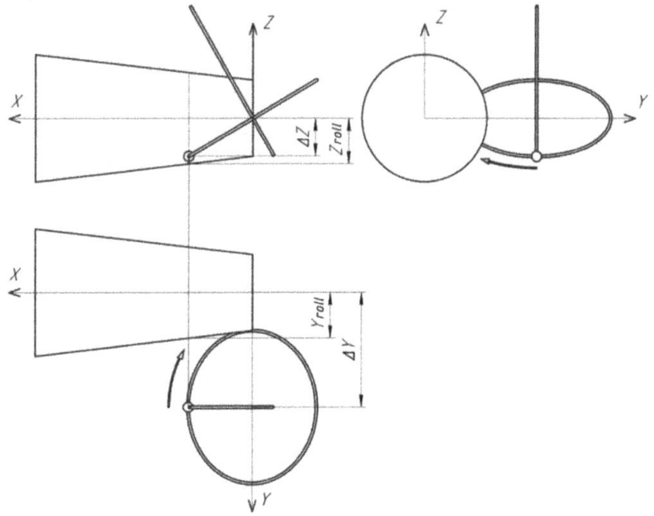

Fig. 3. *2D-scheme of calculation method.*

Figure 4 shows the current point on the circle with coordinates X_0, Y_0, Z_0, and the arrow shows the direction of reference. The roll radius R increases until one of the points of the circumference touches the surface of the billet.

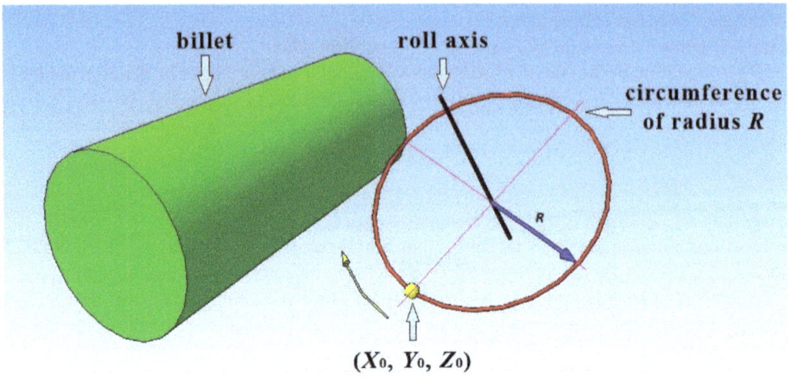

Fig. 4. *3D-scheme of calculation method.*

It is important to note that, *firstly*, the calculation is performed for each section of the roll, and not for the deformation zone, which makes it possible to get rid of additional transformations in the transition from the coordinates associated with the rolling axis to the coordinates associated with a roll axis. *Secondly*, a different mathematical model is used, based on simple formulas, which makes it possible to exclude errors of the roll caliber at values of $\alpha > 60°$ and the correction of the deformation zone associated with these errors, thus, the roll bends around the billet without penetrating anywhere ... And, *thirdly*, this method is suitable for constructing rolls of any type of mill, including even the cases of $\alpha = 0°$ and $\alpha = 90°$.

Application for calculating the shape of rolls

To implement the computational method on a computer, it was decided to use Microsoft Visual Studio 2010 with .NET Framework 4.0 as a programming environment, and C# as a language. All this allowed us to create an application with an intuitive interface

The basic procedure for the program:

1. The values of α and ρ and the geometry of the deformation zone profile are specified.
2. The necessary calculations are made: defining the boundaries of the roll and checking the data for erroneous values.
3. The *i*-th section of the roll is taken.
4. The radius of the roll in this section is calculated by the iterative method.
5. A calculation is made for each next section of the roll.
6. Data processing and output is in progress.

After entering the required data, pressing the "Roll calculation" button starts the calculation process. Progress will be reflected to the left of the button - in the progress bar. At the end of the calculation, a folder with calculation txt-files opens. The resulting txt-files can be used by CAD-systems as input data for modeling (figure 5).

Fig. 5. *Visualization of the rolling process of a 6 mm bar obtained using the application and CAD-system KOMPAS-3D V13.*

References

[1] S. Bratan, S. Roshchupkin, P. Novikov, Modeling the Grinding Wheel Working Surface State, Procedia Engineering 206 (2017) 1419–1425. https://doi.org/10.1016/j.proeng.2017.10.655

[2] Y. Gutsalenko, S. Bratan, S. Roshchupkin, V. Dyadichev, S. Menyuk, Investigation of the Structure and Properties of Copper-Tin Bonding M2-01 in Diamond Grinding Wheel Introducing Additional Energy in the Form of Electric Discharges into the Processing Zone, Materials Today: Proceedings 11 (2019) 586–590. https://doi.org/10.1016/j.matpr.2019.01.033

[3] S. Bratan, B. Bogutsky, S. Roshchupkin, Development of mathematical model of material removal calculation for combined grinding process, Lecture Notes in Mechanical Engineering 0(9783319956299) (2019) 1759–1769. https://doi.org/10.1007/978-3-319-95630-5_189

[4] S.M. Gorbatyuk, A.N. Pashkov, I.G. Morozova, O.N. Chicheneva, Technologies for applying Ni-Au coatings to heat sinks of SiC-Al metal matrix composite material, Materials Today: Proceedings 38 (2020) 1889-1893. https://doi.org/10.1016/j.matpr.2020.08.581

[5] I.I Valiullin, O.A. Kobelev, M.G. Naumova, I.G. Morozova, A.N. Pashkov, V.A. Nagovitsyn, K.V. Goloshchapov, Application of a Discrete Double-Row Scraper Cleaner as an Efficient Procedure of Cleaning of Belt Conveyors Aimed at Transporting Bulk Materials, Metallurgist 64(11-12) (2021) 1340-1346. https://doi.org/10.1007/s11015-021-01124-x

[6] A.D. Bardovsky, S.M. Gorbatyuk, S.V. Albul, N.V. Gorbatyuk, Optimization of Process Flow Diagrams for Processing of Mineral Wastes, Metallurgist 65(3-4), (2021) 465–472. https://doi.org/10.1007/s11015-021-01181-2

[7] V.E. Kondratenko, V.V. Devyatiarova, S.V. Albul, D.S. Kartyshev, Improving methodology for calculating scaffolding formwork of monolithic slabs in building constructions,

IOP Conference Series: Materials Science and Engineering 971(5) (2020) 052037. https://doi.org/10.1088/1757-899X/971/5/052037

[8] L.V. Sedykh, S.V. Albul, D.B. Efremov, M.A. Sukhorukova, Application of additive technologies for manufacturing a wear-resistant steel pipe outlet, IOP Conference Series: Materials Science and Engineering 971(2) (2020) 022002. https://doi.org/10.1088/1757-899X/971/2/022002

[9] V.E. Kondratenko, V.V. Devyatiarova, S.V. Albul, L.M. Valeeva, Method of calculating volumetric scaffold of monolithic slab formwork, IOP Conference Series: Materials Science and Engineering 971(5) (2020) 052036. https://doi.org/10.1088/1757-899X/971/5/052036

[10] A. Keropyan, S. Albul, A. Zarapin A, Problem of Increasing Tractive Effort of Railway Locomotives in Conditions of Arctic and Continental Shelf Regions, Lecture Notes in Mechanical Engineering (2020) 651-658. https://doi.org/10.1007/978-3-030-22063-1_69

[11] G. Sergey, E. Dmitriy, A. Sergei, K. Natalya, The effect of the deformation method of the billets on mechanical properties of molybdenum bars and wire, Key Engineering Materials 864 KEM (2020) 204–210. https://doi.org/10.4028/www.scientific.net/KEM.864.204

[12] V.A. Osadchiy, S.V. Albul, N.S. Kuprienko, N.L. Kirillova, Future developments of a roll forming mill design algorithm, IOP Conference Series: Materials Science and Engineering 709(3) (2020) 044079. https://doi.org/10.1088/1757-899X/709/4/044079

[13] O.A. Kobelev, S.V. Albul, N.L. Kirillova, Research and development of broaching methods on mandrel of large-sized pipe forgings, IOP Conference Series: Materials Science and Engineering 709(4) (2020) 044104. https://doi.org/10.1088/1757-899X/709/4/044104

[14] O. Kobelev, S. Albul, N. Kirillova, Particularities of Calculating Temperature Fields of Large Forging Billets, Lecture Notes in Mechanical Engineering (2021) 797–805. https://doi.org/10.1007/978-3-030-54814-8_92

[15] A.V. Aldunin, L.S. Kokhan, S.A. Guzenkov, B.F. Belelyubskii, Temperature of working rollers in hot strip-rolling mills, Steel in Translation 39(9) (2009) 751–754. https://doi.org/10.3103/S0967091209090071

[16] O.E. Kozyreva, N.I. Volgina, B.F. Belelyubskii, Characterization of flawing on feather surface of low pressure compressor blades from VT-3 titanium alloy, IOP Conference Series: Materials Science and Engineering 747(1) (2020) 012006. https://doi.org/10.1088/1757-899X/747/1/012006

[17] A. Gerasimova, B. Belelyubskii, Production Engineering of Parts by Cold Stamping, IOP Conference Series: Earth and Environmental Science 459(6) (2020) 062095. https://doi.org/10.1088/1755-1315/459/6/062095

[18] V.A. Artyukh, V.N. Borsch, V.S. Yusupov, S.Y. Zhuk, V.A. Zelensky, G.Y. Lazarenko, B.F. Belelyubsky, Peculiarities of a Solid-Phase Method for the Production of Al–Fe/SiO2 and Al–Co/SiO2 Powder Catalysts, Inorganic Materials: Applied Research 11(3) (2020) 709–712. https://doi.org/10.1134/S2075113320030041

[19] L. Jiawen, S. Timushev, D. Klimenko, A. Krivenko, Modeling pressure pulsation fields in a screw centrifugal pump, Proceedings of the 26th International Congress on Sound and Vibration, ICSV 2019 (2019)

[20] A.M. Keropyan, D.A. Kuziev, A.E. Krivenko, Process Research of Wheel-Rail Mining Machines Traction, Lecture Notes in Mechanical Engineering (2020) 703-709. https://doi.org/10.1007/978-3-030-22063-1_75

[21] D. Kouziyev, A. Krivenko, D. Chezganova, B. Valeriy, Sensing of dynamic loads in the open-cast mine combine, E3S Web of Conferences 105 (2019) 03014. https://doi.org/10.1051/e3sconf/201910503014

[22] A.E. Krivenko, Z.K. Khanh, Influence of power fluid temperature in hydraulics on operating efficiency of hydraulic mining excavators, Gornyi Zhurnal 2020(12) (2020) 78-81. https://doi.org/10.17580/gzh.2020.12.18

[23] A.Y. Albagachiev, A.M. Keropyan, A.A. Gerasimova, O.A. Kobelev, Determination of rational friction temperature in lengthwise rolling, CIS Iron and Steel Review 19 (2020) 33-36. https://doi.org/10.17580/cisisr.2020.01.07

[24] N.A. Chichenev, S.M. Gorbatyuk, M.G. Naumova, I.G. Morozova, Using the similarity theory for description of laser hardening processes, CIS Iron and Steel Review 19 (2020) 44-47. https://doi.org/10.17580/cisisr.2020.01.09

[25] L.L. Betina, A.D. Chursin, A.G. Grigoriants, M.A. Kazaryan, N.A. Lyabin, E.V. Perevedentseva, V.S. Paramonov, V.M. Zharikov, Application of Copper Vapour Lasers in the Dimensional Processing Technology, Lasers in Engineering 13(3) (2003) 167–185.

[26] M.A. Kazaryan, N.A. Lyabin, V.M. Zharikov, Technological systems based on copper vapour laser designed for measurement and material processing, Proc. of SPIE - The Int. Society for Optical Engineering 4900(2) (2002) 1094–1098. https://doi.org/10.1117/12.484509

[27] V.N. Titov, S.D. Saifullaev, M.M. Skripalenko, A.I. Ternovykh, A.A. Sidorov, Using Deform-2d Software to Study Heat-Insulation Materials as Protection of Air Tuyeres Against Burnout, Metallurgist 64(5-6) (2020) 388–395. https://doi.org/10.1007/s11015-020-01007-7

[28] M.M. Skripalenko, O.A. Kobelev, S.D. Sayfullayev, Burnout resistance assessment of air tuyere in blast furnace, IOP Conference Series: Materials Science and Engineering 971(3) (2020) 032033. https://doi.org/10.1088/1757-899X/971/3/032033

[29] S.M. Gorbatyuk, S. Sayfullaev, O.A. Kobelev, M.A. Sukhorukova, Choice of thermal insulation for burnout protection of a blast furnace blast tuyere, Materials Today: Proceedings 38 (2021) 1388–1391. https://doi.org/10.1016/j.matpr.2020.08.110

[30] A.S. Budnikov, E.A. Kharitonov, A.S. Aleshchenko, R.V. Iskhakov, Effect of plugless deformation in screw rolling three-roll mill on the change of tube wall thickness, Chernye Metally 2019(12) (2019) 41–45.

Improving the Thermal Fatigue Strength of Hot-Working Tools by Laser Treatment

Nikolai A. Chichenev[1,a*], Sergey M. Gorbatyuk[1,b], Oleg A. Kobelev[1,2,c] and Aleksei N. Pashkov[1,3,d]

[1]National University of Science & Technology (MISIS), Moscow, Russian Federation

[2] NPO TsNIITMAShI, Moscow, Russian Federation

[3] JSC "RPC "Istok" named after Shokin", Fryazino, Russian Federation

[a]chich38@mail.ru, [b]sgor02@mail.ru, [c]oakobelev@cniitmash.com, [d]a.n.pashkov@yandex.ru

Keywords: Hot-Working Die Steel, Thermal Fatigue, Laser Quenching, Laser Spot Diameter, Diameter of Laser Tracks

Abstract. A method for increasing the thermal fatigue strength of a tool used in hot forming of bearing rings by applying circular laser tracks to the working surfaces is considered. Laser treatment is carried out with a power of 2.0...2.5 kW by applying on the end face working surface of the tool in the direction from the center to the periphery of the circular tracks with a common center coinciding with the center of the circumference of the end face; the tool is rotated at a constant angular rate, the spot diameter d_s for each track is selected according to the dependence $d_{s,i}+1/d_{s,i} = 0.85...0.90$, and the laser radiation spots of adjacent tracks have a common point of contact. The results of pilot testing are presented, which confirmed the high technical and economic efficiency of the use of laser quenching for ejectors and punches. A tool made by machining, for example, an ejector of an AMP-70 automatic press, is subjected to volume quenching and tempering. The ejector material was steel 3Kh3M3F, quenching temperature in oil – 1030...1050 °C, tempering temperature – 580...610 °C. After volume quenching, additional machining is carried out, usually grinding, in order to remove the decarbonized layer of material formed during heat treatment and to give the working surface the required roughness class. The final stage in the tool manufacturing is the quenching of its working surface by laser treatment. Pilot testing showed that the use of laser treatment made it possible to increase the durability of ejectors of various types by 2 ... 3 times, of deforming punches – by 2.2 times.

Introduction

Hot forming methods using one of the important properties of metals, ductility, are widely applied in mechanical engineering due to the ability to produce parts of various sizes and diversified shapes in a highly productive and economical way [1–4]. The undoubted advantage of hot forming is that the resulting parts are characteristic of higher strength properties than similar parts made with metal-cutting machines. This is due to the peculiarity of the structure of the deformed metal in which the material fibers are located equidistantly to the part contour in hot forming, in contrast to the use of a metal-cutting tool in which the fibers are cut across. In addition, the productivity of hot forming is much higher than that of processing with metal-cutting machines; therefore, during serial and mass production, the prime cost of parts and metal losses is reduced tenfold.

The productivity of hot forming processes largely depends on the operational durability of the technological tool which fails mainly due to the heat of the working surface [5–7]. Therefore, steel used for manufacturing technological tools for metal hot working should not only have high mechanical properties at operating temperatures, but also have increased thermal fatigue strength.

In the hot forming process, the temperature of the die is characterized by repeatedly changing heating and cooling, as a result of which the processes of thermal fatigue develop, accompanied by the appearance and development of hot cracks [8–10]. Theoretical and experimental studies have shown that the greatest increase in the durability of hot-working dies is achieved with the correct choice of the die steel grade and modes of its thermal and quenching treatment [11–13].

Heat treatment methods traditionally consisting of heating, holding and cooling not only have a limitation in the level of strength of heat-treated materials, but also have largely exhausted their possibilities in obtaining new structures and improving mechanical properties. A number of papers [14–20] shows that laser heat treatment is successfully used for surface quenching of parts and tools for various purposes, including working surfaces of a technological tool for processing materials by pressure. At the same time, the issues of laser thermal quenching of die steel 3Kh3M3F (3Х3М3Ф) widely used for manufacturing deforming tools operating at elevated temperatures are insufficiently presented for the development of technological modes of laser treatment. In some papers, the issues of laser quenching of die steel 4Kh5MFS (4Х5МФС) with surface fusion are considered, which in some cases leads to an unacceptable deterioration of the surface morphology.

Objective

This article discusses a method for increasing the thermal fatigue strength of a tool used in hot forming of bearing rings by applying circular laser tracks to the working surfaces. The results of pilot testing are presented, which confirmed the high technical and economic efficiency of the use of laser quenching for ejectors and punches.

Research carried out and analysis of the results obtained

Tool steels of the martensitic class with carbide quenching of the type 3Kh3M3F, 4Kh5MFS, etc. are used for manufacturing tools for hot forming, for example, ejectors and punches [21]. The main mechanism for quenching steels of this class is quenching for martensite and subsequent tempering, during which precipitation hardening occurs due to the release of alloying elements from the carbide matrix. Therefore, the properties of these steels are determined by the composition and structure of martensite and carbide phases, as well as the nature of the distribution of carbides and their interaction with the matrix. One of the most effective ways to increase the service life of a hot forming tool by increasing its heat resistance is to create a homogeneous finely dispersed structure on the working surface. An effective solution to this problem was obtained by using continuous laser radiation.

The results of metallographic analysis showed that the most effective grinding of the carbide phase is carried out when the tool material is exposed to continuous laser radiation with a power of P=2.0...2.5 kW at power densities q=(3...5) kW/cm^2. At laser power values less than 2.0 kW, the desired structural preparation is not effective enough, which can be explained by a weak thermal effect. At powers above 2.5 kW, there may be a decrease in the steel impact strength, caused by a significant increase in austenite grain, which will lead to accelerated softening of the material during operation.

The scheme of a thermal laser effect on the flat circular surface is the most optimal, which is the application of circular tracks with a common center coinciding with the center of the circle, and each of the subsequent circular tracks adjacent to the previous one along a closed concentric line. Due to the fact that circular tracks are applied without gaps, the amount of the unhardened part of the surface is practically zero; in addition, due to the absence of overlapping tracks, there are almost no tempering zones. The alternation of pronounced tempering or unhardened zones causes surface structural heterogeneity of the tool material, which, as noted, causes accelerated tool wear.

Thus, the specified scheme of laser treatment under the condition $d_{s,i}+1/d_{s,i} = 0.85...0.90$ provides a more uniform structure of the hardened layer and the axisymmetric distribution of residual stresses in the tool material.

The ratio $d_{s,i}+1/d_{s,i} = 0.85...0.90$ was obtained experimentally; it can be justified as follows. Due to the fact that the purpose of laser treatment is to create the most homogeneous structure of the material, it is necessary that all the processed areas obtained by applying circular tracks have the same structure with a minimum difference in properties and form a generally homogeneous finely dispersed structure. Therefore, a prerequisite for laser treatment of a hot forming tool is that the energy contribution to the material must be the same for each of the circular tracks. However, at a constant angular rotation rate ω = const of the treated part with an increase in the radius r_t of the circular tracks, the linear treatment rate v increases, which is equal to the product of the angular rate ω by the radius, i.e. $v = \omega \times r_t$. As experiments have shown, the diameter of the radiation spot should be changed according to the ratio $d_{s,i}+1/d_{s,i} = 0.85...0.90$ when applying tracks in the direction from the center to the periphery in order to create the most homogeneous structure of the hardened near-surface layer due to the fact that the laser thermal effect is carried out at a constant value of the radiation power and, taking into account that when assessing the energy contribution of laser radiation, for example, to the maximum surface temperature, it is inversely related to the values of d_s and v.

The proposed method is carried out as follows. A tool made by machining, for example, an ejector of an AMP-70 automatic press, is subjected to volume quenching and tempering. The ejector material was steel 3Kh3M3F, quenching temperature in oil – 1030...1050 °C, tempering temperature – 580...610 °C. After volume quenching, additional machining is carried out, usually grinding, in order to remove the decarbonized layer of material formed during heat treatment and to give the working surface the required roughness class (Fig. 1).

Fig. 1. *Examples of ejectors used for forming of bearing rings of various sizes*

The final stage in the tool manufacturing is the quenching of its working surface by laser treatment which is carried out as follows. To increase the efficiency of laser radiation absorption by the metal, a light-absorbing coating 10–15 μm thick, for example, blue zinc gouache, is applied to the working surface of the tool. Then the tool is mounted in a manipulator capable of rotational movement, and it is possible to translocate the rotational movement in the horizontal and vertical planes.

Materials Research Proceedings **21** (2022) 43-50 https://doi.org/10.21741/9781644901755-8

After starting the laser installation, a specific value of the laser radiation power should be set, which is included in the specified interval. The ejectors were treated with laser radiation with a power of 2.1 kW using a CO_2 laser. Having established the required radiation power using a power meter, the laser beam is focused until the desired spot diameter d_s is obtained on the end surface of the tool fixed in the manipulator. The spot diameter d_s and the angular rotation rate ω of the treated part are chosen so that the laser treatment is carried out without surface fusion. The initial diameter of the laser radiation spot when treating the ejectors was $d_{s,0} = 10$ mm, while $\omega = 0.31$ s^{-1}.

By longitudinal translocation of the tool relative to the focused beam, the radiation spot is set on the surface so as to apply the first circular track of the minimum radius relative to the center of the circumference of the end face of the cylindrical part, while the radius of the circular track is considered to be the distance from the specified center to the center of the radiation spot. Experiments have shown that it is advisable to take the radius of the first circular track equal to the initial diameter of the focused spot on the treated surface, i.e. r tr,1 = $d_{s,0}$. When treating ejectors, the initial spot diameter is $d_{s,0} = 8...10$ mm.

After turning on the rotating manipulator and the laser, a thermal effect on the laser radiation material is carried out by applying a circular track, the width of which is determined by the spot diameter. Then, the spot diameter is determined using the ratio $d_{s,i} + 1/d_{s,i} = 0.85...0.90$, at which the next circular track will be applied; focusing is carried out. Focusing of each next spot is easily done either with a lens or by translocating the tool in a vertical plane. By longitudinal translocating of the tool before applying the next circular track, the focused spot is set so that it has a common point of contact with the previous circular track. The latter operation is facilitated by the fact that the boundaries of the circular heat-affected zone are clearly marked by the burnt-out absorbing coating.

Circular tracks are applied until the working surface is completely treated. Thus, for a tool with a diameter of 65 mm, laser treatment of the working surface is carried out by applying three circular tracks with a spot diameter equal to $d_{s,1} = 10$ mm, $d_{s,2} = 9$ mm and $d_{s,3} = 8$ mm respectively; the radius of the first track is equal to $r_{t,1} = 10$ mm, the second – $r_{t,2} = 19.5$ mm, the third – $r_{t,3} = 28$ mm (Fig. 2).

The proposed method has passed pilot testing in the manufacture of a working tool of an automatic press, designed for forming bearing rings 7608 (Table 1). The most worn-out element of die tooling is ejectors, the main reason for the rejection of which is hot crack formation. As a result of manufacturing and quenching of ejectors according to the proposed method, a threefold increase in durability was achieved and, therefore, the service life of the tool increased by an average of 3 times. Experiment planning methods were used in the research [22–25].

Pilot testing of the developed technology of laser quenching of working surfaces of hot deformation dies was carried out in the conditions of the forging shop of the bearing plant.

The quenching treatment was carried out using a continuous CO_2 laser. Ejectors and punches were previously subjected to volume heat treatment in the thermal section of the forging shop, including quenching from a temperature of T = 1050...1100 °C in oil and tempering at a temperature of T = 580...610 °C; as a result, the hardness of the working surface of the tool was HRC 46...48.

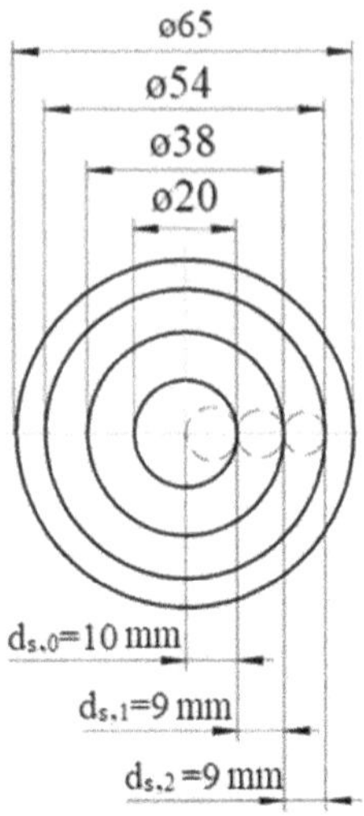

Fig. 2. *Scheme of applying circular tracks on the working surface of the ejector with a diameter of 65 mm*

Table 1. Dependence of the thermal fatigue resistance of the tool on the ratio of the diameters of the radiation spot

The ratio of the diameters of the radiation spot, $d_{sp,i+1}/d_{sp,i}$	Thermal fatigue resistance	
	Cycles until the first cracks appear	Average service life of the ejector, forgings
No laser treatment	1300	9600
1	2900	22000
0.9	4100	29500
0.85	3900	29100
0.8	3100	25300
0.7	3000	23400

Pilot testing showed that the use of laser treatment made it possible to increase the durability of ejectors of various types by 2 ... 3 times, of deforming punches – by 2.2 times. In laser treatment of punches, circular tracks were applied to the working surface by rotating the punch around its axis.

Since the number of thermal fatigue cracks in the laser-hardened tool is significantly lower compared to the tool treated according to the traditional technology, the possibility of reusing the worn-out die (ejectors and punches) has been tested. The defective layer determined by the depth of development of hot cracks was mechanically removed from the surface of laser-hardened worn-out ejectors, which had already shown a threefold increase in resistance. After grinding, the working surfaces were treated with laser radiation, while preliminary volume heat treatment was not required, and the ejectors were returned to operation. As a result of the tests, it was found that the durability of ejectors re-treated with laser radiation was twice as high as the durability of the new tool subjected to volume heat treatment according to the plant technology. It has been shown that such an operation can be performed 2 to 4 times.

In total, according to the developed technology, 580 units of working tools for hot forming of bearing rings on an AMP-70 automatic press in the conditions of the forging shop of the bearing plant were processed and tested in industrial conditions. The test results and statistical data confirmed the high technical and economic efficiency of the use of laser quenching for ejectors and punches, while the annual savings are achieved both by increasing the operational reliability of the tool and by extending the service life of the tool due to repeated sets and a decrease in the production of defective forgings.

Summary

A method is proposed for increasing the thermal fatigue strength of a tool used in hot forming of bearing rings by applying circular laser tracks to the working surfaces. Laser treatment is carried out with a power of 2.0...2.5 kW by applying on the end face working surface of the tool in the direction from the center to the periphery of the circular tracks with a common center coinciding with the center of the circumference of the end face; the tool is rotated at a constant angular rate, the spot diameter d_s for each track is selected according to the dependence $d_{s,\,i+1} / d_{s,\,i} = 0.85...0.90$, and the laser radiation spots of adjacent tracks have a common point of contact. Pilot testing showed that the use of laser treatment made it possible to increase the durability of ejectors of various types by 2 ... 3 times, of deforming punches – by 2.2 times. The possibility of reusing a worn-out die (ejectors and punches) by removing the defective layer with subsequent laser treatment of the working surface is shown.

Taking into account the high technical and economic efficiency of the use of laser quenching for ejectors and punches, it is planned to apply the developed approach for another die for hot deformation [26–30].

References

[1] Kovka i shtampovka: spravochnik. V 4 t. T. 2 Goryachaya ob"emnaya shtampovka / Pod obshch. red. E. I. Semenova. - M.: Mashinostroenie, 2010. - 720 s.

[2] Titov Yu. A., Kokorin V.N., Morozov O.I. Tekhnologiya kovki i ob"emnoi shtampovki: uchebnoe posobie. – Ul'yanovsk: UlGTU, 2016. – 111 s.

[3] B. K. Konstantinov. Tekhnologiya kovki i goryachei ob"emnoi shtampovki: uchebnoe posobie. – M. : NITs INFRA-M, 2019. – 104 s.

[4] E. I. Semenov. Kovka i goryachaya shtampovka: uchebnik. – M. : MGIU, 2011. – 414 s.

[5] A. Steniko, V. Tami. Sovershenstvovanie protsessa pryamoi zakalki na zavode kompanii Nucor Tuscaloosa. Chernye metally. 2018 № 12 s. 41-43.

[6] D. I. Vasil'ev, M. A. Tylkin, G. P. Teterin. Osnovy proektirovaniya deformiruyushchego instrumenta. - M.: Vysshaya shkola, 1980. - 223 s.

[7] Ya. M. Okhrimenko, L. I. Antonenko, L. I. Mironov. Shtampy dlya goryachei obrabotki metallov i ikh ekspluatatsiya. - M.: Mashinostroenie, I971. - 50 s.

[8] V. A. Korotkov. Issledovanie stoikosti shtampovogo instrumenta iz stali 5KhNM // Byul. «Chernaya metallurgiya». – 2010. – № 3 – s.72–76.

[9] Shtampy dlya goryachego deformirovaniya metallov / pod obshch. red. M. A. Tylkina. - M.: Vysshaya shkola, 1977. - 496 s.

[10] S. N. Aqida, F. Calosso, D. Brabazon, S. Naher, M. Rosso. Thermal fatigue properties of laser treated steels. Int. J. Mater. Form. 2010. № 3. pp. 797–800. https://doi.org/10.1007/s12289-010-0890-1

[11] A. E. Smirnov. Upravlenie fazovym sostavom kompleksno-legirovannykh teplostoikikh stalei pri vakuumnoi tsementatsii i zakalke // Metallovedenie i termicheskaya obrabotka metallov. 2020 .- № 9(783) .- s. 45 – 52.

[12] G. V. Levchenko, S. B. Bobyr', E. G. Demina, S. A. Zdorovets, Ya. Yu. Yakimenko. Vliyanie mikrolegirovaniya na strukturu i termotsiklicheskuyu stoikost' nizkolegirovannykh shtampovykh stalei // Metaloznavstvo ta termichna obrobka metaliv. 2008. № 1. s. 51–57

[13] A. B. Akinin, S. V. Boev. Povyshenie stoikosti chekanochnogo instrumenta iz stali Kh12MF provedeniem zakalki v azote vysokogo davleniya i termotsiklirovaniya // Metallovedenie i termicheskaya obrabotka metallov. 2020. № 2 (776). s. 31 - 36

[14] N. Maharjan, W. Zhou, Y. Zhou, N .Wu. Underwater laser hardening of bearing steels. Journal of Manufacturing Processes. 2019. Vol. 47, pp. 52-61. https://doi.org/10.1016/j.jmapro.2019.08.020

[15] A. Khorram, Jamaloei A.Davoodi, A. Jafari, M .Moradi. Nd:YAG laser surface hardening of AISI 431 stainless steel; mechanical and metallurgical investigation. Optics and Laser Technology. 2019/ Vol. 119, Article № 105617. https://doi.org/10.1016/j.optlastec.2019.105617

[16] Balajaddeh M. Bahrami, Moosavy H. Naffakh. Pulsed Nd:YAG laser welding of 17-4 PH stainless steel: Microstructure, mechanical properties, and weldability investigation. 2019. Optics and Laser Technology. Vol. 119, article № 105651. https://doi.org/10.1016/j.optlastec.2019.105651

[17] S. G. Gornyi, Yu. B. Grechko, M. I. Patrov, K. V. Yudin, V. I. Yurevich. Lazernaya markirovka materialov // Fotonika, 2007. № 3. s. 16-22.

[18] T. Dikova. Surface modification of 3Ch2W8F and 4Ch5MFS steels by CO2 laser // J. of the Technical University at Plovdiv "Fundamental Sciences and Applications". 2006. Vol. 13 (7). pp. 81–90.

[19] Ho Jun Shin, Young Tae Yoo. Microstructural and hardness investigation of hot-work tool steels by laser surface treatment // J. of Materials Processing Technology. 2008. Vol. 201 pp. 342–347. https://doi.org/10.1016/j.jmatprotec.2007.11.232

[20] V. V. Morozov, A. N. Shlegel', A. V. Aborkin. Empiricheskie modeli lazernogo uprochneniya rabochikh kromok detalei formovykh komplektov stekloformuyushchikh mashin // Fundamental'nye i prikladnye problemy tekhniki i tekhnologii. 2011. № 2 (286). s. 79–84.

[21] Marochnik stali i splavov. http://splav-kharkov.com/mat_start.php?name_id=287 (data obrashcheniya 24.02.2021).

[22] V. A. Lunev. Matematicheskoe modelirovanie i planirovanie eksperimenta: uchebnoe posobie. - SPb.: Izd-vo SPbGPI, 2012. – 153 s.

[23] Yu. A. Makarichev, Yu. N. Ivannikov. Metody planirovanie eksperimenta i obrabotki dannykh: uchebnoe posobie. – Samara: Samar. gos. tekhn. un-t, 2016. – 131 s.

[24] N. I. Sidnyaev. Teoriya planirovaniya eksperimenta i analiz statisticheskikh dannykh : uchebnik i praktikum dlya vuzov. — M.: Izdatel'stvo Yurait, 2019. — 495 s.

[25] N. A. Chichenev, S. M. Gorbatyuk, M. G. Naumova; I. G. Morozova. Using the similarity theory to describe laser hardening processes. CIS Iron and Steel Review. 2020. Vol. 19. pp. 44–47. https://doi.org/10.17580/cisisr.2020.01.09

[26] A. G. Grigor'yants, I. N. Shiganov, A. I. Misyurov. Tekhnologicheskie protsessy lazernoi obrabotki: Uch. posobie. - M.: Izd-vo MGTU im. N. E. Baumana, 2006. - 663 s.

[27] Lazernye tekhnologii obrabotki materialov: sovremennye problemy fundamental'nykh issledovanii i prikladnykh razrabotok / Pod red. V. Ya. Panchenko. – M.: FIZMATLIT, 2009. – 664 s.

[28] I. A. Pinakhin, V. A. Chernigovskii. Osnovy ob"emnogo impul'snogo lazernogo uprochneniya instrumental'nykh i konstruktsionnykh materialov [Elektronnyi resurs]: monografiya. — Stavropol': izd-vo SKFU, 2014. — 160 s. Rezhim dostupa: https://rucont.ru/efd/304170.

[29] Eron'ko, S.P., Oshovskaya, E.V., Tkachev, M.Y. Fast replacement of submersible tundish nozzles in a continuous slab-casting machine (2016) Steel in Translation, 46 (1), pp. 33-38. https://doi.org/10.3103/S0967091216010034

[30] A. V. Aborkin, V. E. Vaganov, A. N.Shlegel', I. M. Bukarev. Effect of laser hardening on die steel micro hardness and surface quality // Metallurgist. 2015. T. 59. № 7-8. s. 619-625. https://doi.org/10.1007/s11015-015-0148-8

Predicting the Thickness of the Hardening Coating during Diffusion Metallization of Cast Iron

Alexander Veselovsky[1,a], Irina Troyanovskaya[2,b*] and Valery Erofeev[2,c]

[1] Production Company "Travel Systems", 52 Karl Marx Street, Chelyabinsk, Russia

[2] South Ural State Agrarian University, 75 Lenin Avenue, Chelyabinsk, Russia

[a]a_a_ves@mail.ru, btripav63@mail.ru, [c]ervv52@mail.ru

Keywords: Diffusion, Coating, Saturation, Cast Iron, Carbides

Abstract. The article presents the calculations of diffusion indices of saturation of high-strength cast iron VCh 60 from powder filling. Carbide-forming elements were used as diffusers: vanadium, chromium and manganese. As a result of the research, empirical equations have been established for predicting the thickness of strengthening diffusion coatings depending on the temperature and saturation time.

Introduction

Methods of chemical-thermal treatment are the most promising methods for hardening the surface of products made of high-strength cast iron. Methods of diffusion metallization are widely used among them [1, 2, 3, 4]. Diffusion metallization restores parts with low wear, high surface hardness and working under conditions of friction with a variety of mating materials [5]. The diffusion metallization process is most widely used for parts of diesel fuel equipment. It is used for parts with fixed joint seating surfaces and which are characterized by low wear. For example, shafts, bearing cups of gearboxes, rear axles, gearboxes and others. Diffusion metallization is also used in movable joints, such as rocker shafts, idler shafts, water and oil pump shafts, and others.

Diffusion metallization repairs cast iron parts with wear within the coating thickness using pastes and coatings [6, 7]. Their composition is identical to the powder reaction mixture containing diffusing components.

Materials and methods

For the thermal diffusion hardening of specimens of high-strength cast iron VCh 60, the following components were taken from the powder medium [8, 9]:

1. Ferrovanadium with a vanadium content of at least 50% by weight;
2. Ferromanganese with an average manganese content of 88% by weight;
3. High-carbon ferrochrome with a chromium content of more than 65%, carbon, respectively 8-8.5% and silicon not more than 2%;
4. Aluminum oxide;
5. Ammonium chloride.

These ferroalloys are widely used in the metallurgical industry for alloying cast iron [10, 11, 12]. Ammonium chloride is an activator. It has a low temperature of thermal dissociation (350^{0}C), as a result of which ammonia and hydrogen chloride are formed. Hydrogen chloride reacts with the main diffusers in the charge to form chlorides and subchlorides, which are directly involved in diffusion saturation. Ammonia thermodissociates with the formation of hydrogen and nitrogen at temperatures of 450-500 ^{0}C. Aluminum oxide is highly refractory and inert to most reagents in direct contact with it at high temperatures. Therefore, it is used as an inert filler.

Materials Research Proceedings **21** (2022) 51-55
https://doi.org/10.21741/9781644901755-9

The preparation of charge materials was carried out in the following sequence. Crushed into powder ferroalloys were sieved through a sieve with a mesh size of 1 mm. In the beginning, all the components used were heated. Aluminum oxide was heated for one hour at a temperature of 700 ^{0}C. Ammonium chloride was heated for 2 hours at a temperature of 200 ^{0}C.

The initial components of the mixture were weighed with an accuracy of 0.5 %, mixed until homogeneous and poured into an airtight container before use. Thermal diffusion saturation was carried out in a muffle furnace using a container with a fusible seal. Samples of high-strength cast iron VCh 60 were placed in a furnace and covered with a powder reaction mixture. Then the container was packed. The saturation temperature was 1020-1050 ^{0}C. The saturation time was 5-6 hours for vanadium coatings and 10-12 hours for chrome and manganese coatings.

Changes in the overall dimensions of the samples were carried out using a scanning electron microscope after chemical thermal treatment [13].

Results and discussion

Based on the first law of diffusion, the thickness of the diffusion layer δ at a constant saturation concentration is

$$\delta = K\sqrt{D\tau}, \quad (1)$$

where K is coefficient depending on the type and conditions of saturation; D is diffusion coefficient; τ is saturation time.

Taking into account the temperature dependence of the diffusion coefficient D, we have

$$\delta^2 = K^2\tau D_0 e^{-E/RT}, \quad (2)$$

where E is diffusion activation energy; R is gas constant; T is absolute temperature; D_0 is pre-exponential factor.

At a constant temperature $T = const$, the thickness of the diffusion coating δ depends only on the time of the saturation process τ. We introduced the designation $2p = K^2D_0e^{-E/RT}$ and received $\delta^2 = 2p\tau$.

We have built graphical dependences $\delta^2 = f(\tau)$ of the thickness of the diffusion layer on the saturation time (Figure 1). Then the quantity $2p$ is the tangent of the slope of the straight lines. The calculation results are presented in Table 1.

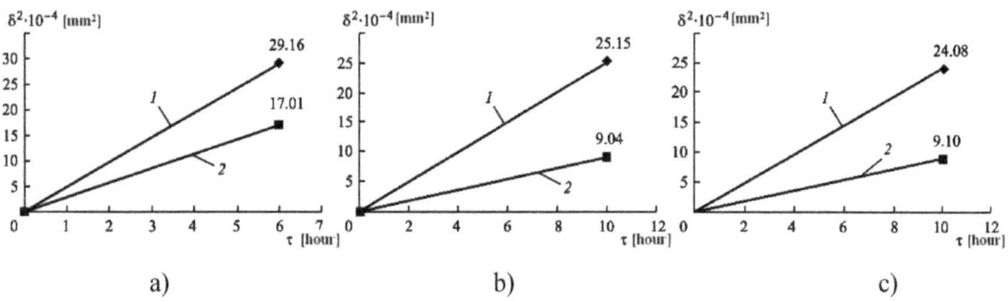

Fig. 1. *Dependence δ^2 on the saturation time τ of VCh 60 cast iron (a) with vanadium, (b) chromium (c), manganese at a temperature of (1) 1050 ^{0}C and (2) 960 ^{0}C*

Table 1. *Calculation results* 2p

Diffusing element	Temperature [^{0}C]/ [K]	$1/T \cdot 10^{-3}$	2p·10^{-4}	$ln\ (2p)$
Vanadium	960/1233 1050/1323	0.811 0.756	2.90 4.86	-8.14 -7.63
Chromium	960/1233 1050/1323	0.811 0.756	1.02 2.51	-9.19 -8,29
Manganese	960/1233 1050/1323	0.811 0.756	0.94 2.46	-9.32 -8.33

After taking the logarithm, we got

$$ln\ (2p) = ln(K^2 D_0) - E/RT \tag{3}$$

At $T = \infty$ or $1/T = 0$ equation (3) takes the form

$$ln\ (2p) = ln(K^2 D_0) \tag{4}$$

We have built a graph $ln\ (2p) = f(1/T)$ (Figure 2).

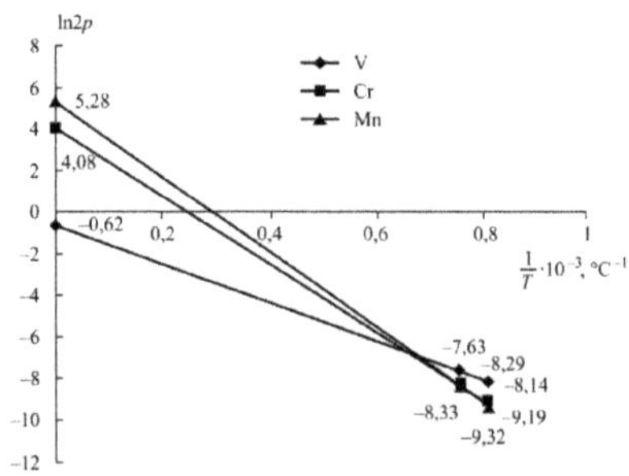

Fig. 2. *Dependence* $ln\ (2p) = f(1/T)$

The coordinates of the points of intersection of straight lines with the $ln\ (2p)$ axis are obtained on the basis of the equation of a straight line passing through two points with known coordinates

$$\frac{(x-x_1)}{(x_2-x_1)} = \frac{(y-y_1)}{(y_2-y_1)}. \tag{5}$$

Equation (3) implied

$$ln(K^2 D_0) - ln(2p) = E/RT \tag{6}$$

From Figure 2, taking into account $ln\ (2p) = f(1/T)$ was follows $tg\alpha = \frac{E}{RT}/\frac{1}{T} = \frac{E}{R}$. Then we got

$$E = Rtg\alpha \tag{7}$$

Substitute equation (7) into equation (8) and obtain

$$tg\alpha = (lnK^2D_0 - ln2p)T. \quad (8)$$

The found values of $ln(K^2D_0)$ and $ln(2p)$ were substituted into equation (9) and found the values of $tg\alpha$ and E for the corresponding temperatures (Table 2).

Table 2. *Calculation results of $tg\alpha$ and E*

Saturated element	Equation of a straight line	$ln(K^2D_0)$	$tg\alpha$	E [J/mol]
Vanadium	$y = -9.27 \cdot 10^3x - 0.62$	-0.62	9 274.23	77 106
Chromium	$y = -16.36 \cdot 10^3x + 4.08$	4.08	16 365.5	136 059
Manganese	$y = -18 \cdot 10^3x + 5.28$	5.28	18 006.1	149 702

The calculated values of the activation energy of diffusion E made it possible to obtain an empirical formula for the dependence of the thickness of the diffusion layer δ on the temperature T and the time of diffusion saturation τ.

To determine the unknown values of K^2D_0, experimental values of δ (μm) were substituted into formulas (2) at the corresponding values of temperature T [K] and time τ [sec]

$\delta^2 = 192.3\tau e^{-9274/T}$ is empirical equation for vanadium coating;

$\delta^2 = 20161.0\tau e^{-16365/T}$ is empirical equation for chrome plating;

$\delta^2 = 73529.1\tau e^{-18006/T}$ is empirical equation for manganese coatings.

Thermal diffusion saturation of iron-carbon alloys with carbide-forming elements is characterized by an increase in the linear dimensions of products due to carbide formation. The increment of linear dimensions in cast iron is close to the thickness of the formed carbide layer.

Only a slight diffusion of vanadium, chromium and manganese into cast iron is possible at the initial moment of saturation, when the concentration of diffusant condensed on the surface to be hardened has not reached the stoichiometric value for the formation of carbides. However, the concentration of the strengthening components is low. In the vanadium process of gray cast iron, the concentration of vanadium diffusing deep into the part at the boundary with the cast iron base does not exceed 10% by weight. After reaching the stoichiometry of carbide formation, the diffusion mobility of vanadium, chromium, and manganese atoms is lost and the diffusion front shifts towards an increase in the size of the part, since the settling diffusant is bound into carbides due to the diffusion of carbon from the central layers. The counter diffusion of carbon significantly exceeds the diffusion fluxes of vanadium, chromium, and manganese. Therefore, the study of the kinetics of thermal diffusion vanadium, chromium plating, and manganization must be estimated from the diffusion mobility of the counter flow of carbon through the created carbide layer to the diffusion (reaction) front.

Conclusion

The obtained empirical regularities of the thickness of diffusion vanadium, chromium and manganese coatings make it possible to predict the thickness of the created diffusion layers under given saturation conditions and using the appropriate type and concentration of reagents according to the known control parameters - saturation temperature and time.

References

[1] E.E. Bobylyov, I.D. Storozhenko, Diffusion metallization of carbide cutting tools as a way to improve the surface treatment quality, *Journal of Physics: Conference Series*. 1399(4) (2019) 044084. https://doi.org/10.1088/1742-6596/1399/4/044084

[2] P.O. Rusinov, Z.M. Blednova, Formation of the surface layers of a material with shape-memory-based tinico diffusion metallization, *Materials Science Forum*. 818 (2015) 3-6. https://doi.org/10.4028/www.scientific.net/MSF.818.3

[3] Y.V. Zaulychny, V.G. Hignjak, N.A. Harchenko, T.P. Hovorun, O.V. Hignjak, V.Y. Dolgikh, Influence of interatomic interaction processes on the mechanical properties of carbide coatings based on Ti, V and Cr, obtained by diffusion metallization, *Journal of Nano- and Electronic Physics*. 8(4) (2016) 04008. https://doi.org/10.21272/jnep.8(4(1)).04008

[4] A. Pastukhov, O. Sharaya, E. Timashov, D. Bakharev, Method of justification of machine parts hardening modes, *Engineering for Rural Development*. 20 (2021) 74-79. https://doi.org/10.22616/ERDev.2021.20.TF016

[5] X-J. Liu, H-C. Wang Microstructure and hardness of vanadium carbide coatings prepared by thermal diffusion process on different steel substrate, *Cailiao Rechuli Xuebao/Transactions of Materials and Heat Treatment*. 31(1) (2010) 150-154

[6] A.A. Veselovskii, Improving the wear resistance of high-strength cast-iron gears by applying a thermodiffusional vanadium coating, *Russian Engineering Research*. 32(1) (2012) 45-47 https://doi.org/10.3103/S1068798X12010273

[7] O.J. Akinribide, S.O. Olusunle, S.O. Akinwamide, B.J. Babalola, P.A. Olubambi, Impact of heat treatment on mechanical and tribological behaviour of unalloyed and alloyed ductile iron, *Journal of Materials Research and Technology*. 14 (2021) 1809-1819 https://doi.org/10.1016/j.jmrt.2021.07.077

[8] B. Wang, F. Qiu, G.C. Barber, Y. Pan, W. Cui, R. Wang, Microstructure, wear behavior and surface hardening of austempered ductile iron, *Journal of Materials, Research and Technology*. 9(5) (2020) 9838-9855. https://doi.org/10.1016/j.jmrt.2020.06.076

[9] A.A. Veselovsky, R.R. Dema, O.B. Kalugina, S.P. Nefedyev, Determination of parts tolerances in friction knots of grey and nodular cast iron with thermal diffusion platings vanadium and chrome, *MATEC Web of Conferences*. 129 (2017) 02043. doi:10.1051/matecconf/201712902043

[10] A.I. Degula, T.P. Govorun, N.A. Kharchenko, V.G. Khyzhnyak, M.V. Karpets, O.M. Myslyvchenko, R.S. Smetanin, Investigation of the phase and chemical compositions of complex carbide coatings. *Metallofizika i Noveishie Tekhnologii*. 37(11) (2016) 1461-1476. https://doi.org/10.15407/mfint.37.11.1461

[11] O.I. Shinskij, V.I. Litovka, V.A. Maslyuk, V.Ya. Kurovskij, N.V. Borovik, Production of high-strength cast iron with the use of briquetted modifiers, *Litejnoe Proizvodstvo*. 8 (2003) 23-26.

[12] V.G. Hignjak, G.Y. Calashnicov, N.A. Harchenko, T.P. Hovorun, O.V. Hignjak, V.Y. Dolgikh, O.O. Holyshevskiy, The structure, composition and properties of nitrided alloys after diffusion metallization, Journal of Nano- and Electronic Physics. 7(4) (2015) 04033.

[13] A.A. Veselovsky, R.R. Dema, O.B. Kalugina, S.P. Nefedyev, Determination of parts tolerances in friction knots of grey and nodular cast iron with thermal diffusion platings vanadium and chrome, *MATEC Web of Conferences* 129 (2017) 02043 https://doi.org/10.1051/matecconf/201712902043

Materials Research Proceedings 21 (2022) 56-60 https://doi.org/10.21741/9781644901755-10

Reducing the Fire Hazard of Wooden Structures

Asgat Gazizov[1,2,a], Aisylu Sagitova[1,b*] and Anton Krasnov[1]

[1]Ufa State Petroleum Technological University, 1 Kosmonavtov str., Ufa, Republic of Bashkortostan, 450064, Russia

[2]Ural State Forest Engineering University, 37 SibirskyTrakt, Yekaterinburg, 620100, Russia

[a]ashatgaz@mail.ru; [b]sagitovaais@gmail.com

Keywords: Fire Danger, Wooden Structures, Fire Protection, Fire-Retardant Composition

Abstract. Increasing the fire resistance of wooden structures is relevant because of their increased flammability. The article describes an experimental study using NEOMID flame retardant paint and silica powder as a flame retardant for wooden structures. According to the results of the experiment, a group of fire-retardant effectiveness of the tested fire-retardant composition was established with different methods of its use.

Introduction

Wood is one of the most common materials in the construction of buildings and structures for various purposes. Increased flammability of wooden structures is one of the main problems in the fire safety system of buildings. The main task is to reduce the fire hazard of wooden structures and increase their fire resistance [1].

Fireproofing of wooden building structures using coatings is the simplest method. Fire-retardant coatings protect wooden structures from fire, can release gases and water vapor during heating, which phlegmatize the flame and prevent combustion. Unlike impregnation, the use of fire-retardant coatings allows to provide the I group of fire protection of large structures, the autoclave impregnation of which is impossible to obtain the same group of fire protection [2-5].

Methods and results of research

For the study, wood samples in the form of rectangular bars with a cross section of 50x50mm were used, the deviation from the dimensions did not exceed 1 mm. Each bar before processing was weighed on electronic scales.

The fire retardant composition was NEOMID flame retardant acrylic paint. Fire retardant paint has a certificate of conformity No. C-RU.KB03.V.00216 with a validity period from 14.11.2017 to 13.11.2022.

Silicon dioxide (silica, SiO_2) [6] was also used as a fire-retardant composition together with paint.

A layered flame retardant was applied to the wood samples and dried at room temperature (Fig. 1).

Fig. 1. *A sample coated with a fire retardant*

The total fire-retardant composition flow rate applied to the wood sample was determined by summing the flow rates after each layer was applied over the surface area of the sample (Table 1).

Table 1. *Sample mass data before and after treatment*

Sample number	Drying time, [h]	Number of layers	Weight of bar before treatment, [g]	Weight of bar after treatment, [g]
1	24	1	49,5	50,0
2		2	60,5	61,0
3		3	50,0	50,5
4		1 layer + sprinkling silica 1 time	60,5	62,5
5		2 layers + sprinkling silica 1 time	61,0	62,5
6		3 layers + sprinkling silica 1 time	56,5	57,5
7		1 layer + sprinkling silica 2 times	58,0	58,5
8		2 layers + sprinkling silica 1 time	51,5	52,0
9		3 layers + sprinkling silica 1 time	66,0	67,0

After 24 hours of drying, fire tests were carried out on the Ceramic Pipe unit of samples No. 1-9. The samples were exposed to a temperature of 200 ° C ± 5 ° C for 2 minutes (Fig. 2).

***Fig. 2.** Sample burning*

After cooling the sample, the remainder of the sample was removed from the ceramic box (Fig. 3).

***Fig. 3.** Samples after incineration*

The composition flow rate, after each treatment, was determined by weight from the mass difference of the sample before and after fire-retardant composition application. Weighing of the coated samples was carried out after stopping the runoff from the surface of the uneaten solution. At the end of the sample, fire-retardant composition residues were removed with filter paper.

Total absorption R, kg/m^3, was determined by formula (1):

$$R = \frac{m_1 - m_2}{V}. \quad (1)$$

Where m_1 is the mass of the sample after coating, kg; m_2 - sample weight before coating, kg; V- sample volume, m^3.

Materials Research Proceedings **21** (2022) 56-60
https://doi.org/10.21741/9781644901755-10

Weight loss of P_i samples, %, was calculated (Table 2) by formula (2):

$$P_i = \frac{(m_{1i} - m_{2i}) * 100}{m1i} \quad . \tag{2}$$

Where m_{1i} - is the mass of the sample before the test, g; m_{2i} - is the mass of the sample after the test, g; i - is the sample number.

After the test, the arithmetic mean of the weight loss of the samples tested was determined (Table 2) by formula (3, 4).

$$|P_{avrg} - P_i| \leq 3 \text{ при } P_{avrg} \leq 9 \quad . \tag{3}$$
$$|P_{avrg} - P_i| \leq 5 \text{ при } 9 < P_{avrg} \leq 25. \tag{4}$$

Where P_{avrg} -is the arithmetic mean weight loss of the nine samples tested, %; P_i - is the weight loss value of one of the nine samples tested, %.

Table 2. *Results of the obtained calculations*

Sample number	Sample weight, [g]			Loss of sample weight		Modulus of arithmetic mean weight loss, [%]
	Before treatment	After treatment	After combustion	[g]	[%]	
1	49,5	50,0	40,5	9,5	19,0	7,14
2	60,5	61,0	54,0	7,0	11,48	0,38
3	50,0	50,5	45,0	5,5	10,89	0,97
4	60,5	62,5	54,0	8,5	13,6	1,74
5	61,0	62,5	57,0	5,5	8,8	3,06
6	56,5	57,5	51,5	6,0	10,43	1,43
7	58,0	58,5	55,0	3,5	5,98	5,88
8	51,5	52,0	44,0	8,0	15,38	3,52
9	66,0	67,0	59,5	7,5	11,19	0,67
Arithmetic mean:			51,16		11,86	

Based on the results of experiments, a fire-retardant efficiency group was established for each sample of the tested fire-retardant composition (Table 3). As such, mass loss for our samples is not more than 9%, which corresponds to the I fire-retardant efficiency group.

Table 3. *Fireproofing group of samples*

Sample number	Fire protection group
1	I
2	I
3	I
4	I
5	I
6	I
7	I
8	I
9	I

Let us present the dependence of the mass loss of the tested samples in the form of a graph (Fig. 4).

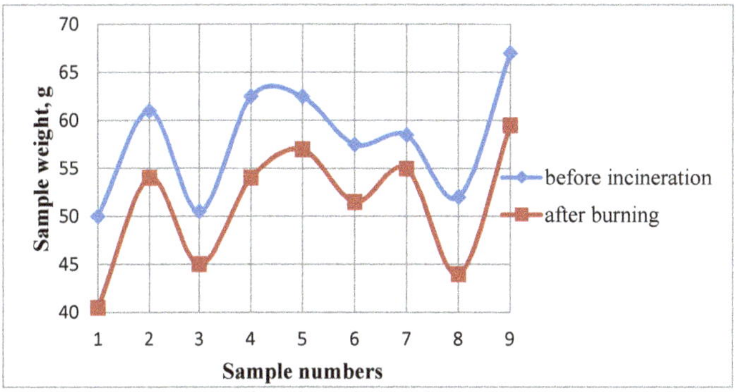

Fig. 4. *Plot of weight loss for each sample*

Conclusion

In the course of the study, it was found that all samples coated with fire-retardant paint and silicon dioxide belong to group I of fire-retardant efficiency. Thus, an fire-retardant composition consisting of a flame retardant acrylic decorative NEOMID paint and silica is an effective means for increasing the fire resistance of wooden structures.

Sample No. 7 has the highest fire retardant efficiency of all test samples, since this sample has a minimum weight loss. Thus, treatment of wood 1 with a layer of flame retardant acrylic decorative paint NEOMID + sprinkling with silica twice provides the highest fire retardant efficiency of the nine test pieces.

References

[1] F. Sh. Khafizov, A. M. Gazizov and D. F. Muratov, Fire Retardants Application for Wood Material, Oil and Gas Business, number 2, 2019, p. 16-31. https://doi.org/10.17122/ogbus-2019-2-16-31

[2] A.M. Gazizov, O.V. Kuznetsova, A.A. Sharafutdinov and K.M. Shaimuhametova, Improvement of fire retardant properties of wood materials, IOP Conference Series: Materials Science and Engineering, Volume 919 (2020). https://doi.org/10.1088/1757-899X/919/6/062014

[3] A. Y. Korolchenko and O. N. Korolchenko, Fire protection equipment, Moscow, 2006, p 258.

[4] A.M. Gazizov, E.V. Popova, O.V. Kuznetsova and A.A. Akhmadullin, Method of increasing the fire resistance of wood IOP Conference Series: Materials Science and Engineering, Volume 971(3) (2020). https://doi.org/10.1088/1757-899X/971/3/032029

[5] A. A. Leonovich and G. B. Shalun, Fire Protection of Wood Boards and Laminated Plastics, Moscow: Lesnaya promyshlennost Publ., 1974.

[6] A. M. Gazizov, I. I. Dudnichenko and F. Sh. Khafizov, Application of silica filler to increase fire resistance of wood, Oil and Gas Business, number 1, 2020, p. 42-52. https://doi.org/10.17122/ogbus-2020-1-42-52

Electromechanical Surfacing Technique

Alla Gerasimova[1,a*], Alexey Karfidov[1,b], Boris Belelybskii[2,c] and Kirill Goloshapov[1,d]

[1]National University of Science and Technology "MISIS", street the Lenin prospect,4, Moscow,119049, Russia

[2]Department of Mechanical Engineering, Moscow Polytechnical University, Bolshaya Semyonovskaya str., 38, Russia

[a]gerasimova.aa@misis.ru, [b]a.korf@mail.ru, [c]nis_mgvmi@mail.ru, [d]goloshapov_nooby@mail.ru

Keywords: Structure, Surface Layer, Part, Roller-Type Tools, Surface Treatment

Abstract. The paper considers an electromechanical surfacing technique of parts' work surfaces. It was found out that depending on strain temperature, rate and degree there were special type structures formed on a surface layer with a phase change (a bright layer), i.e. a hot deformed structure with and without recrystallization and a cold deformed structure. Depending on the original structure and hardening conditions, during surfacing of parts a thickness of this area may achieve 0.3 mm max. Impact of steel electromechanical treatment on the structure and the surface layer was found out. Depending on properties of surfaces being treated it is possible to use plate-like and roller-type tools for electromechanical hardening. It was shown that roller- type tools possessed enhanced durability and enlarged areas of surfaces being treated, though ensured poorer finish of surfaces being treated.

Introduction

Electromechanical surface hardening technique is based on a combination of thermal actions and forces applied to the surface layer of the part being treated. The essence of the technique is that in the course of part treatment, large current is passing under low voltage through the contact place of tools with the article; due to this fact protruding surface crests are subject to vigorous heating, are deformed under a pressure of tools and polished while the metal surface layer hardens [1-8].

Electromechanical surface hardening technique is used for hardening and reconditioning worn surfaces of parts. For example, this technique is utilized for reconditioning surfaces of antifriction bearings.

Electromechanical surface hardening technique process may be considered as a special type of metal surface thermomechanical treatment. High temperature thermomechanical treatment is one of thermomechanical treatment techniques. In this case depending on strain temperature, rate and degree there are special type structures formed on a surface layer with a phase change (a bright layer), i.e. a hot deformed structure with and without recrystallization and a cold deformed structure. Depending on the original structure and hardening conditions, during surfacing of rotary parts a thickness of this area may achieve 0.3 mm max. [9-16].

A bright strip was identified for the first time by V.P. Kravz-Tarnosky during impact tests of steel specimens. N.N. Davidenkov and I. N. Mirolubov explain Kravz-Tarnosky effect so that there is substance destruction and crushing along one shear area due to local deformation. With rapid sliding due to high friction, primarily large amounts of heat are generated which are transferred at high speed to the specimen basic mass and then to the steel anvil and hammer head. That's why at places of strain localization where the temperature is likely to exceed its limit level, there is primarily austenitic transformation and intense quenching further on. The layer substance has martensite texture with no specific needlelike structure since it has been formed under special conditions of overpressure and velocity of transformation that have not been explored yet [17-20].

Similar structures were found in surface layers of various machine parts. However, there is no consensus yet to the nature of this structure. Some persons attribute the formation of non-etching bright area to the saturation of friction surface with air nitrogen. Other persons consider the bright area to be a consequence of oxygen diffusion into the metal surface layer that results in formation of solid solutions and chemical compounds. The third ones conclude that the bright area structure consists of martensite, retained austenite and alloyed cementite. Differences of opinion about the nature of bright area formation are apparently explained by friction dissimilar conditions [21-24].

Methods and materials

Depending on impact on the surface layer structure and properties, the following electromechanical treatments are distinguished:

- thermal treatment when there is the bright area formed at a significant depth while there is no ductile deformation in the transition layer. This treatment features phase changes and complete rest;
- incomplete thermal treatment with homogeneous area viewed in the surface layer; the transition sub-layer is severely deformed and features partial rest viewed in the homogeneous surface layer.

Depending on properties of surfaces being treated it is possible to use plate-like and roller-type tools for electromechanical surface hardening technique.

A holder with plate-like carbide tip (refer with: Fig. 1) works well for electromechanical surface hardening technique of relatively small areas. Working surfaces of polishing plates should be adequately clean. That's why after pattern filling with a carborundum wheel they should be semi-finished with boro-carbon on cast iron disk. A lapping compound should be made of boron carbide powder and wax at a ratio of 7: 3. Due to lack of boron carbide some small enterprises provide smoothening plates with semi-finishing on common emery wheels.

Fig. 1. *Holder with plate-like tools*

When processing large areas with the necessity for profound hardening under high current of 900-1,200 A the use of roller-type tools is more rational (refer with: Fig. 2). The roller-type holder is similar to the common holder of polishing tools. The roller made of carbide material should be installed on the axis and secured in the holder body with the help of retention screws.

Fig. 2. *Holder with roller-type tools*

Test conducted on these tools demonstrated their high performance during continuous and long term operation under high current. However the surface finish obtained in this case somewhat poorer in comparison with its surfacing with the plate. It should be stated that holders (refer with: Fig.1, Fig.2) may be adapted for fixed mounting of carbide rollers if structurally changed a little bit. In this case the roller operates as the circular plate; while the best possibility for changing contact surfaces by means of periodic rotation of the roller, dozen times increases wear-resisting properties of these tools in comparison with the plate. Comparative analysis of roller-type and plate-like tools (refer with: Table 1).

Table 1. *Comparison of properties of roller-type and plate-like tools [25]*

Parameters	Plate-like tools	Roller-type tools
Surfaces being treated	Relatively small	Very large
Current mode, [A]	300-800	900-1,200
Reduction in roughness	2-3 class	1-2 class
Wear-resisting properties (treatment till the repeated semi-finishing, [mm])	2,500-3,000	> 30,000

Conclusions

Summing up what has been said, roller-type tools possess enhanced durability and enlarged areas of surfaces being treated, however, they ensure poorer finish of surfaces being treated.

References

[1] V.V. Eltsov, Reconditioning and hardening machine parts, Togliatti, Togliatti State University (TGU), 2015.

[2] B.M. Askinazi, Hardening and reconditioning machine parts by electromechanical surfacing, Moscow, Mashinostroyenie, 1989.

[3] A. Keropyan, S. Gorbatyuk, Impact of Roughness of Interacting Surfaces of the Wheel-Rail Pair on the Coefficient of Friction in their Contact Area, Procedia Engineering. 150 (2016) 406–410. https://doi.org/10.1016/j.proeng.2016.06.753

[4] A.G. Radyuk, S.M. Gorbatyuk, Y.S. Tarasov, A.E. Titlyanov, A.V. Aleksakhin, Improvements to mixing of natural gas and hot-air blast in the air tuyeres of blast furnaces with thermal isulation of the blast duct, Metallurgist. 63(5-6) (2019) 433–440. https://doi.org/10.1007/s11015-019-00843-6

[5] L.V. Sedykh, S.V. Albul, D.B. Efremov, M.A. Sukhorukova Application of additive technologies for manufacturing a wear-resistant steel pipe outlet, IOP Conf. Series: Materials Science and Engineering. 971(2) (2020), 022002. https://doi.org/10.1088/1757-899X/971/2/022002

[6] V.E. Kondratenko, V.V. Devyatiarova, S.V. Albul, D.S. Kartyshev, Improving methodology for calculating scaffolding formwork of monolithic slabs in building constructions, IOP Conf. Series: Materials Science and Engineering. 971(5) (2020), 052037. https://doi.org/10.1088/1757-899X/971/5/052037

[7] V.V. Devyatiarova, E.E. Balakhnina, L.M. Valeeva, Development of a mathematical model for a workpiece heating and cooling in a protective medium while treatment under pressure, Defect and Diffusion Forum. 410 (2021) 115–122. https://doi.org/10.4028/www.scientific.net/DDF.410.115

[8] A.D. Bardovsky, L.M. Valeeva, I.I Basyrov, A plant with a rotary jet grinder to produce small fractions of mineral raw material, IOP Conf. Series: Materials Science and Engineering. 971 (2020), 052004. https://doi.org/10.1088/1757-899X/971/5/052004

[9] O.E. Markov, V.V. Kukhar, V.N. Zlygoriev, A.A. Shapoval, A.S. Khvashchynskyi, R.U. Zhytnikov, Improvement of upsetting process of four-beam workpieces based on computerized and physical modeling, FME Transactions. 48(4) (2020) 946–953. https://doi.org/10.5937/fme2004946M

[10] A.A. Fomin, V.G. Gusev, Spindle rigidity in milling blanks with nonuniform properties, Russian Engineering Research. 33(11) (2013) 646–648. https://doi.org/10.3103/S1068798X13110087

[11] S.A. Snitko, A.V. Yakovchenko, V.V. Pilipenko, N.I. Ivleva, Modeling of ring billets rolling on radial-axial ring-rolling mill, Izvestiya Ferrous Metallurgy. 63(8) (2020) 665–673. https://doi.org/10.17073/0368-0797-2020-8-665-673

[12] A.A. Fomin, Determining undeformed chip thickness models in milling and its verification during wood processing, Solid State Phenomena. 265 (2017) 598–605. https://doi.org/10.4028/www.scientific.net/SSP.265.598

[13] V. Artiukh, V. Mazur, L. Shilova, Device for making horizontal wedgethrust of rolling stand, MATEC Web of Conferences. 106 (2017), 03002. https://doi.org/10.1051/matecconf/201710603002

[14] V. Artiukh, V. Kukhar, E. Balalayeva, Refinement issue of displaced volume at upsetting of cylindrical workpiece by radial dies, MATEC Web of Conferences. 224 (2018), 01036. https://doi.org/10.1051/matecconf/201822401036

[15] A.O. Karfidova, M.V. Vasilyev, I.G. Morozova, Modernization of an industrial passive exoskeleton prototype for lower extremities using rapid prototyping technologies, IOP Conf. Ser.: Mater. Sci. Eng. 971 (5) (2020), 052049. https://doi.org/10.1088/1757-899X/971/5/052049

[16] S.M. Gorbatyuk, A.A. Gerasimova, A.G. Radyuk, Using the coating for the diffusion layer OBTAINING on the walls of the mold (CCM), Metallurgical and Mining Industry. 7 (9) (2015) 1085-1088.

[17] A.Y. Albagachiev, A.M. Keropyan, A.A. Gerasimova, O.A. Kobelev, Determination of rational friction temperature in lengthwise rolling, CIS Iron and Steel Review. 19 (2020) 33-36. https://doi.org/10.17580/cisisr.2020.01.07

[18] A.M. Busygin, Self-adjusting differential rotary feeding system of the drilling rig, IOP Conf. Ser.: Mater. Sci. Eng. 971(4) (2020), 042099. https://doi.org/10.1088/1757-899X/971/4/042099

[19] A. Bardovsky, I. Basyrov, L. Valeeva, Systematization of the mineral raw material waste recycling process, MATEC Web of Conferences. 346 (2021), 03040. https://doi.org/10.1051/matecconf/202134603040

[20] V.N. Titov, A.I. Ternovikh, P.V. Baranov, T.Y. Sidorova, Development of coatings for protection of blast furnace air tuyeres, Metallurgist. 65 (3-4) (2021) 439–445. https://doi.org/10.1007/s11015-021-01174-1

[21] O.A. Kobelev, V.A. Tyurin, Production of large plates, Steel in Translation. 37(9) (2007) 727–729. https://doi.org/10.3103/S096709120709001X

[22] K.N. Solomonov, L.I. Tishchuk, Simulation of deformation processes in upsetting, Journal of Physics: Conf. Series. 1348 (1) (2019), 012020. https://doi.org/10.1088/1742-6596/1348/1/012020

[23] S.V. Bashkin, A.O. Karfidov, V.N. Kornienko, et al., An imaging Fourier transform spectroradiometer with a multi-element photodetector for the spectral range of 7–14 μm, Optics and Spectroscopy. 121(3) (2016) 449 – 454. https://doi.org/10.1134/S0030400X16090058

[24] L.R. Lamberson, K.S. Kapur, Reliability in engineering design, Weliy India Pvt. Limited, 2009.

[25] Elektromekhanicheskaya-obrabotka. URL: http://www. stroy-technics.ru (date: 01.08.21)

Materials Research Proceedings **21** (2022) 66-70 | https://doi.org/10.21741/9781644901755-12

Obtaining Luminous Phosphate Coatings on Steel by Cold Method

Viktoriya S. Konovalova[1,a*] and Varvara E. Rumyantseva[1,b]

[1]Ivanovo State Polytechnic University, department of natural sciences and technosphere safety, 153000, Sheremetevsky av., 21, Ivanovo, Russian Federation

[a]kotprotiv@yandex.ru, [b]varrym@gmail.com

Keywords: Protective Coatings, Phosphate Coatings, Protection of Steel, Cold Phosphating, Color Phosphating, Luminous Coatings

Abstract. The possibility of obtaining luminous phosphate coatings on steel by cold method has been studied. Modified cold phosphating solutions containing organic additives (glycerin, trilon B, OS-20 emulsifier) were selected as the basis to maintain the pH, stabilize the solution and improve the quality and structure of the deposited coatings. To obtain the glow effect, a green phosphor based on Zn_2SiO_4 containing manganese as a sensitizer was added to the phosphating solution. During deposition, phosphate coatings are obtained that glow with spots, but constant mixing of the solution during deposition contributes to the uniform distribution of phosphor in the phosphate film. Luminous phosphate coatings have good protective properties, they can be used as an independent protection of steel surfaces from corrosion.

Introduction

All conventional phosphating solutions are based on dilute phosphoric acid solutions based on alkaline or heavy metal ions, which mainly contain free phosphoric acid and primary phosphates of metal ions [1-4]. When a steel product is introduced into a phosphating solution, a topochemical reaction occurs in which the dissolution of iron is initiated by free phosphoric acid on microanodes present on the surface of the steel [5, 6]. Hydrogen is released on microcathode areas of the surface:

$$Fe + 2H_3PO_4 \rightarrow Fe(H_2PO_4)_2 + H_2$$

The formation of soluble primary ferrous phosphate leads to simultaneous local depletion of the concentration of free acid in the solution, leading to an increase in pH at the «metal – solution» interface. This change in pH shifts the hydrolytic equilibrium that exists between soluble primary phosphates and insoluble tertiary phosphates of metal ions present in the phosphating solution, which leads to rapid conversion and deposition of insoluble tertiary metal phosphates [7-9]. In a zinc phosphate bath , these equilibria can be represented as [10]:

$$Zn(H_2PO_4)_2 \leftrightarrow ZnHPO_4 + H_3PO_4$$

$$3Zn(H_2PO_4)_2 \leftrightarrow Zn_3(PO_4)_2 + 2H_3PO_4$$

A certain amount of free phosphoric acid must be present in order to suppress hydrolysis and maintain the stability of the bath for effective precipitation of phosphates on microcathode areas of the metal surface [11-13].

The composition of the phosphating solution and the concentration of its components determine the nature of the coatings formed [14]. The color of the traditional phosphate coating can be from light gray to gray, depending on the material and pre-treatment of the surface of the phosphated product [15, 16]. The development of phosphate films of colors other than gray is of particular interest. Luminous phosphate coatings with good anti-corrosion and performance characteristics can be used to protect metal structures, fences, pillars.

Materials and Methods

In order to obtain luminous phosphate coatings, a green luminophor based on Zn_2SiO_4 containing manganese as a sensitizer was introduced into modified cold phosphating solutions (Table 1) [15] at the rate of 20 g/l. Luminophors are synthetic substances capable of converting various types of energy into light, that is, luminescent [17, 18].

Table 1. *Composition of modified solution for cold phosphating of iron-carbon alloys*

Components	Ammount [g/l]
«Majef»	35-45
$Zn(NO_3)_2$	50-65
$NaNO_2$	3-4
Glycerin	1-2
Trilon Б	6-8
Emulsifier OS–20	5-10

Coatings from solutions containing luminophor were deposited on samples of steel grade St3 (the chemical composition of steel is shown in Table 2). The surface of the samples is pre-sanded, degreased and etched.

Table 2. *Chemical composition of steel samples*

Components	C	Si	Mn	Ni	S	P	Cr	N	Cu	As	Fe
Ammount [%]	0,14-0,22	0,15-0,3	0,4-0,65	up to 0,3	up to 0,05	up to 0,04	up to 0,3	up to 0,008	up to 0,3	up to 0,08	~97

The samples were hung in a container with a solution placed in a thermostat (Fig. 1). The deposition of the coating was carried out at a temperature of 30 °C.

The precipitated phosphate films were developed in the rays of an ultraviolet lamp.

The protective ability of oxide-phosphate films was determined by the drop method. As an indicator, a solution containing $CuSO_4 \cdot 5H_2O$ 80 g/l, NaCl 33 g/l, HCl (0,1 N) 13 ml/l was used. A drop of the solution was applied to a surface covered with a phosphate film, and after turning on the stopwatch, its color was monitored, which changed from blue to red. If at 15-20 °C on a steel sample with a phosphate film free of organic substances, the color change occurred earlier than 2 minutes, then the protective ability of the film was considered low, at 2-5 minutes – average, longer than 5 minutes – high. When tested under a drop of reagent, the release of the smallest particles of metallic copper was sometimes observed. This phenomenon, if it was not accompanied by a drop change, was not taken into account when assessing the corrosion resistance of the phosphate coating.

To calculate the speed and characteristics of the metal corrosion process, polarization diagrams are constructed graphically: $\varphi_a - i_a$ – the curve of anodic polarization of the anode areas of the metal surface and $\varphi_k - i_k$ – the curve of cathodic polarization of the cathodic areas of the metal surface. The analysis of the constructed corrosion diagrams consists in determining the controlling process, i.e. the stage of the process of electrochemical corrosion of the metal, which has the greatest resistance compared to other stages. For the highest value of the achieved current on corrosion diagrams, corrosion rate indicators are calculated.

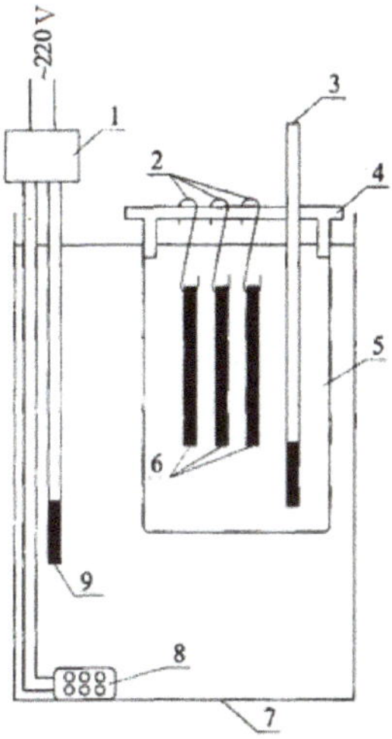

Fig. 1. *Scheme of the deposition device: 1 - 1 – electronic relay; 2 – glass hooks for hanging samples; 3 – thermometer; 4 – rod; 5 – glass vessel with electrolyte (0.1 N H_2SO_4 solution); 6 – test samples; 7 – water thermostat; 8 – electric heater; 9 – contact thermometer.*

Results and Discussion

As can be seen in Fig. 2-a, phosphate coating with luminophor glows unevenly and is not bright enough, so it was proposed to carry out precipitation with constant stirring of the solution. The luminophor in the phosphate coating, when the solution is mixed, is distributed more evenly over the surface of the sample and glows brighter when developed under ultraviolet light (Fig. 2-b).

a) b)

Fig. 2. *Luminous phosphate coatings: a – coating obtained without mixing the solution; b – coating obtained by mixing the solution*

Tests of coatings for corrosion protection by the drip method have shown that the resulting luminous coating has similar protective properties with modified phosphate coatings without luminophor [19, 20] (Table 3).

Table 3. *Results of testing of steel samples with phosphate coating by drip method*

Test time, sec	Type of coating	
	Modified	Luminous
Control time immediately after phosphating	135	137
Control time after 24 hours	375	360

By the method of contact corrosion, data were obtained to determine the corrosion rate of steel samples with phosphated coatings (Table 4). The coating obtained from a solution with the addition of luminophor is not inferior in protective properties to the modified phosphate coating.

Table 4. *Corrosion rate of steel samples protected by phosphate coatings in 10 % NaCl solution, $g/(m^2h)$*

Coating without the additive	Coating with luminophor additive
0,614	0,623

The results obtained indicate that luminous phosphate coatings can be used as an independent protection of steel products from corrosion.

Summary

Obtaining luminous phosphate coatings at room temperature is possible with constant stirring of the solution. In this case, the luminophor introduced into the phosphating solution is evenly distributed in the deposited film, resulting in a uniform glow of the phosphate coating. Phosphate coatings obtained at room temperature have good quality and high protective properties.

References

[1] D.B. Freeman, Phosphating and metal pretreatment: A guide to mod. processes a. practice, Cambridge: Woodhead – Faulkner in assoc. with Pyrene chem. services, 1986, IX, 229 p.

[2] W. Rausch, The Phosphating of Metals, London: Finishing Publications Ltd., 1990, 416 p.

[3] G. Lorin, *Phosphating of Metals*, London: Finishing Publications, 1974, 229 p.

[4] V.N. Selivanov, Yu.P. Perelygin, The universal phosphating solution, University proceedings. Volga region. Natural sciences. 1 (1) (2013) 135-143.

[5] C.-C. Jiang, G.-Y. Xiao, X. Zhang, R.-F. Zhu, Y.-P. Lu, Formation and corrosion resistance of a phosphate chemical conversion coating on medium carbon low alloy steel, New Journal of Chemistry. 40 (2) (2016) 1347-1353. https://doi.org/10.1039/C5NJ02245B

[6] F. Zucchi, G. Tranbelli, Anodic behaviour of Fe in phosphate solutions, Corrosion Science. 11 (3) (1971) 141-151. https://doi.org/10.1016/S0010-938X(71)80090-2

[7] C. Jiang, Z. Gao, H. Pan, X. Cheng, The initiation and formation of a double-layer phosphate conversion coating on steel, Electrochemistry Communications. 114 (2020) 106676. https://doi.org/10.1016/j.elecom.2020.106676

[8] T.S.N. Sankara Narayanan, Surface pretreatment by phosphate conversion coatings – a review, Reviews on advanced materials science. 9 (2) (2005) 130-177.

[9] I.Ya. Kiselev, Formation of a phosphate coating on carbon steel, Protection of Metals. 26 (2) (1990) 215-218.

[10] D. Ivanova, Phosphatazing of mild steel in zinc-manganese-nickel phosphates in different correlations, Journal of Chemical Technology and Metallurgy. 54 (5) (2019) 1072-1078.

[11] T.K. Akaeva, O.P. Akaev, O.N. Rodin, E.A. Naiman, Steel products phosphating with compositions based on phosphoric acid, Vestnik of Kostroma State University. 4 (2013) 8-12.

[12] A.G. Orlov, I.G. Antonova, N.N. Grechishkin, Rapid, low-temperature phosphating of steel items for consumer goods, Protection of metals. 34 (1) (1998) 62-65.

[13] Gh. Barati Draband, M. Liofkhazraei, Electrochemical phosphate conversion coating: a review, Surface review and letters. 24 (3) (2017) 1730003. https://doi.org/10.1142/S0218625X17300039

[14] G. Górecki, Iron phosphate coatings – composition and corrosion resistance, Corrosion. 48 (7) (1992) 613-616. https://doi.org/10.5006/1.3315980

[15] V.S. Konovalova, V.E. Rumyantseva, Production of colored phosphate coatings on steel, Materials Science Forum. 1037 (2021) 457-463. https://doi.org/10.4028/www.scientific.net/MSF.1037.457

[16] F. Gao, Z. Xiong, H. Xue, Y. Liu, Improved performance of strontium aluminate luminous coating on the ceramic surface, Journal of Physics Conference Series. 152 (1) (2009) 012082. https://doi.org/10.1088/1742-6596/152/1/012082

[17] L. Liu, X.F. Yan, Z.M. Zhang, L.M. Yu, Novel Functional Coating: Luminescent Coating, Advanced Materials Research. 337 (2011) 37-40. https://doi.org/10.4028/www.scientific.net/AMR.337.37

[18] M.N. Getz, P.-A. Hansen, H. Fjellvåg, O. Nilsen, Luminescent properties of europium titanium phosphate thin films deposited by atomic layer deposition, RCS Advances. 7 (13) (2017) 8051-8059. https://doi.org/10.1039/C6RA27644J

[19] L. Fedrizzi, F. Deflorian, S. Rossi, L. Fambri, P.L. Bonora, Study of the corrosion behaviour of phosphatized and painted industrial water heaters, Progress in Organic Coatings. 42 (1-2) (2001) 65-74. https://doi.org/10.1016/S0300-9440(01)00153-9

[20] V.S. Konovalova, V.E. Rumyantseva, Corrosion Protection of Reinforcement with Phosphate Coatings, IOP Conference Series: Materials Science and Engineering. 890 (2020) 012091. https://doi.org/10.1088/1757-899X/890/1/012091

Effect of Heat Treatment on Structural and Mechanical Properties of Alloyed Fused Alumina

Yu. Bagaiskov[1,a*] and N. Ushakov[1,b]

[1]Volzhsky Polytechnic Institute (branch) of the Volgograd State Technical University, 43a Engelsast., Volzhsky, Volgograd Region, 404130, Russia

[a] bagaiskov@post.volpi.ru, [b] ushakov@post.volpi.ru

Keywords: Fused Alumina, Alloying, Heat Treatment, Structure, Strength

Abstract. Heat treatment of ceramic-bond grinding tools during manufacture takes place at temperature of abrasive grains up to 1250 °C, which results in variation of their phase composition, structure, and strength. Similar phenomena may take place during grinding caused by the occurring high-temperature processes. It primarily pertains to alloyed fused alumina. X-ray and topography testing, as well as studying of micromechanical properties made it clear that structural performance, material strength, and fracture pattern of single grains of Ti-containing fused alumina affected by temperature depend on the type and ratio of alloy additives in alumina and on the decomposition level of the Ti_2O_3 solid solution. In terms of aggregate properties, purple varieties of Cr-Ti fused alumina featuring the minimum TiO_2/Cr_2O_3 ratio of 1.5 in the solid solution after heat treatment are preferable.

Introduction

In the course of manufacturing ceramic-bond grinding tools, abrasive-material grains are heated up to 1250 °C [1], which unavoidably results in variation of their phase composition, structure, and, subsequently, structural-mechanical properties. In operation, similar phenomena may take place during grinding caused by the occurring high-temperature processes [2].

It is understood that high-temperature heating of Ti-bearing fused alumina results in breakdowns of the Ti_2O_3 solid solution [3]. It is assumed to be a relationship between breakdown of the solid Ti solution in alumina and the structural-mechanical properties of alloyed fused alumina [4].

Studies were conducted, using 400-μm isometric-shape clear transparent alumina grains sampled from industrial batches of Cr-Ti, Ti, Cr, and white fused alumina. A binocular stereoscopic microscope was applied to select grains homogeneous in color: brown (Ti and Cr-Ti fused alumina), purple (Cr-Ti fused alumina), crimson (Cr fused alumina), colorless (white fused alumina) [1, 5, 6]. Such sampling minimized the influence of the grain size and shape and inclusions of associated phases on the grain physical and mechanical properties.

Breakdown of the solid Ti_2O_3 solution was achieved by baking the sampled grains in a silicon-carbide kiln within the range of 1150–1350 °C for 2 hours. The breakdown level of the solid solution depended on the heat-treatment temperature and duration.

The chemical composition of the initial (before heat treatment) fused alumina in terms of alloying oxides is given in Table 1, their structural condition — in Table 2. Purple and brown grains selected from the same batch of Cr-Ti fused alumina (samples 5–1 and 5–2) feature different ratios of Ti and Cr oxides in the Ti_2O_3/Cr_2O_3 solid solution $\leq$ 1.5 (this figure has been confirmed by analyzing more than 50 samples obtained in the course of studying Cr-Ti fused alumina at different times). The spectral method was applied to determine the content of Cr_2O_3 and TiO_2. The

Materials Research Proceedings **21** (2022) 71-75 https://doi.org/10.21741/9781644901755-13

solid solution of alumina contains Ti^{3+} ions, so the spectral analysis data were recalculated in terms of TiO_2 [3].

Table 1. *Chemical composition of initial alumina*

Sample No.	Type of fused alumina	Color	Mass content (%)			
			TiO_2		Cr_2O_3	
			Initial	Solid solution	Initial	Solid solution
1	White fused alumina	Colorless	-	-	-	-
2	Ti fused alumina	Brown	0.34	0.73	-	-
3	Cr fused alumina	Crimson	-	-	0.65	0.50
4	Cr-Ti fused alumina	Brown	0.75	0.64	0.26	0.23
5-1	Cr-Ti fused alumina	Brown	1.06	0.98	0.40	0.28
5-2	Cr-Ti fused alumina	Purple	0.83	0.51	0.38	0.33
6	Cr-Ti fused alumina	Purple	0.83	0.49	0.66	0.62

Study results

The Laue X-ray and topography methods were applied to study the structural condition of single-crystal grains of alloyed alumina prior to and after high-temperature heating; besides, the powder method was applied to determine the structural parameters of the lattice (interplanar spacing $d_{41.6}$ and the value of micro strain ε) of alloyed alumina (Table 2). The table comprise the structural analysis data of alloyed fused alumina after heat treatment at 1250 °C and 1350 °C as compared to the initial condition [4]. The single-crystal grains of white and Cr fused alumina are close in terms of the substructure and feature a Laue diffraction pattern with single spots inherent in perfect mosaic crystals. The substructure of Ti-bearing fused alumina (Cr-Ti and Ti fused alumina) has no significant difference. According to the Laue diffraction pattern, the single-crystal grains of such fused alumina feature a mosaic block structure; the topograms feature inhomogeneous intensity contrast typical for non-uniform distribution of the defects that form strain centers within the grain volume. The contrast pattern obtained in the topograms is associated with piling-up of dislocations caused by non-uniform distribution of the Ti_2O_3 and Cr_2O_3 solid solutions in a crystal and segregation of the mixture of the alloying agents that failed to completely penetrate into the solid solution. Besides, dislocation density ρ is significantly higher than that of white and Cr fused alumina. Thus, Ti affects straining distortion of the alumina lattice to a higher extent as compared to Cr.

Table 2. *Alloyed alumina structure prior to and after heat treatment*

Initial				1250°C			1350°C	
$d_{41,6}\cdot(\overset{0}{A})$		$\varepsilon\cdot(10^4$	$\rho,(см^{-2}$	$d_{41,6}\cdot(\overset{0}{A})$	$\varepsilon\cdot(10^4)$	$\rho,(см^{-2})$	$d_{41,6}\cdot(\overset{0}{A})$	$\varepsilon\cdot(10^4)$
1. White fused alumina	0.83057	3	$5\cdot10^6$	0.83057	3	-	-	-
2. Ti fused alumina	0.83086	8	$7\cdot10^7$	0.83072	6	$9\cdot10^7$	0.83056	3
3. Cr fused alumina	0.83069	8	$5\cdot10^6$	0.83060	5	-	0.83072	5
4. Cr-Ti fused alumina	0.83111	9	$7\cdot10^7$	0.83081	6	$15\cdot10^7$	0.83093	7
5-1. Cr-Ti fused alumina	0.83086	8	$5\cdot10^7$	0.83080	6	$15\cdot10^7$	0.83069	5
5-2. Cr-Ti fused alumina	0.83085	5	$5\cdot10^7$	0.83075	6	$15\cdot10^7$	0.83080	5
6. Cr-Ti fused alumina	0.83103	6	$6\cdot10^7$	0.83081	6	$15\cdot10^7$	0.83074	6

Alumina changes its color in the course of heat treatment in air within the temperature range of 1150–1350 °C. The Ti alumina color (sample 2) changes most abruptly: initially brown grains become milky-blue/milky-white after heating up to 1350 °C. Brown grains of Cr-Ti fused alumina (samples 4 and 5-1) become milky-pink/milky-blue; purple grains of Cr-Ti fused alumina (samples 5-2 and 6) lose transparency. Cr fused alumina (sample 3) retains its color after heat treatment and remains crimson, yet the color becomes more intensive. In Ti-bearing fused alumina, color changing is accompanied by inclusions of a new phase — TiO_2 rutile — and caused by breakdown of the Ti^{3+} solid solution in Al_2O_3 and Ti^{3+} oxidation to Ti^{4+}. The early breakdown stage of the Ti_2O_3 solid solution features occurrence of thin-plate formations that are structurally linked to the alumina matrix and cause diffused scattering in Laue diffraction patterns. Analysis of the diffused scattering pattern enabled calculating the thickness of bidimensional formations equal ~ 50–100 A. As the heating temperature rises, the thin-plate formations grow and turn into rutile pins consistently located in alumina crystals.

Breakdown of the Ti_2O_3 solid solution with the release of rutile results in variation of the alumina chemical composition: the content of Ti_2O_3 in the solid solution gradually decreases as the heating temperature rises (by 20–35 % after heating up to 1350 °C); the content of Cr_2O_3 in the solid solution remains nearly the same.

Breakdown of the Ti_2O_3 solid solution in purple grains of Cr-Ti fused alumina takes place within the range of higher temperature values, approximately by 100 °C. This is due to the fact that alumina with higher content of Cr_2O_3 in the structure is nearly always saturated with beads of the Cr-bearing alloy resulted from reduction of chromium oxide during melting. Metallic Cr oxidizes during alumina heat treatment (besides, the Cr content in the solid solution increases); oxidizing Cr creates oxygen deficit and thus prevents transfer of trivalent Ti into tetravalent.

Heating of alloyed alumina is accompanied by monotonous decrease of the parameters and a reduced micro strain level of the lattice. As the heating temperature increases up to 1350 °C, the parameters of interplanar spacing $d_{41.6}$ and micro strain value ε of Ti alumina approximate such parameters of white alumina; in Cr-Ti alumina, the parameters decrease to a lesser degree; as for Cr alumina, parameter $d_{41,6}$ even increases in a measure.

Superfine rutile inclusions regularly oriented in related to the matrix and resulted from breakdown of the Ti_2O_3 solid solution do not cause strain distortions of alumina, which is confirmed by the obtained Laue diffraction patterns and topograms.

Elastic stresses release at the inclusion-matrix/rutile-alumina interface due to formation of a dislocation structure eliminating lattice misfit between the inclusions and matrix and thus reducing elastic distortions. According to the contrast pattern in the topograms, heating of Ti fused alumina results in a more balanced lattice condition and stress relief [3, 4].

Relation between the TiO_2 content in the solid solution of alumina and the heating temperature is shown in Fig. 1.

After heating at 1150-1350 °C, the substructure of Cr alumina grains (sample 3) is not so different from that of initial grains: the topograms feature inhomogeneous intensity contrast typical for non-uniform distribution of defects (strain centers) within the grain volume.

Cr-Ti fused alumina (samples 4, 5, and 6) after high-temperature heating behaves variously, depending on the Ti^{3+} content in the solid solution: the structure of brown grains (the ratio in the TiO_2/Cr_2O_3 solid solution ≥ 1.5) becomes more balanced; as for purple grains ($TiO_2/Cr_2O_3 \leq 1.5$), defects are distributed non-uniformly, so the structure remains unbalanced similarly to Cr alumina.

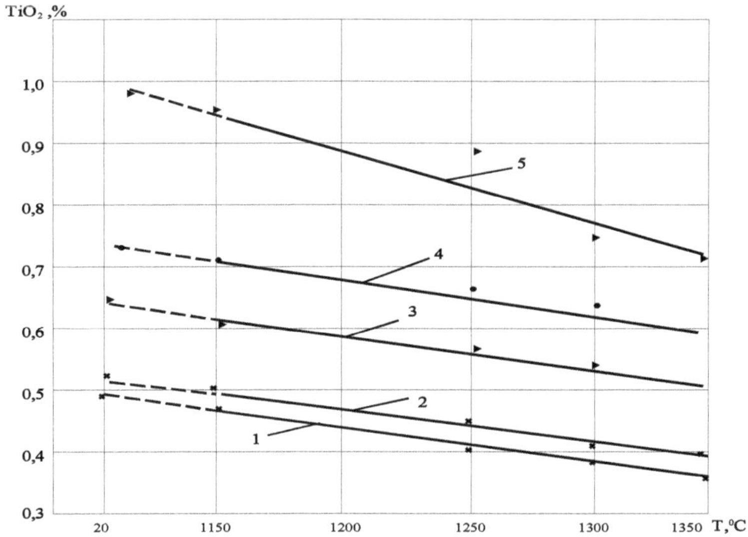

Fig. 1. Relation between Ti $^{3+}$ content in the solid solution of alumina (in terms of TiO_2) and heating temperature T: 1 — sample 6; 2 — sample 5-2; 3 — sample 4; 4 — sample 2; 5 — sample 5-1

The study revealed higher values of Cr-Ti fused alumina micromechanical parameters [4, 7, 8, 9], besides, purple grains have the maximum values; such Cr-Ti fused alumina reaches the maximum values of micro hardness and crack resistance after heat treatment at 1250 °C.

The wear resistance study [10] under micro cutting conditions made it clear that the wear resistance of all initial fused alumina involving macro shears (cutting depth, depth of the last notch before shearing) is nearly the same and fluctuates within 15–17 μm; after heating at 1250 °C, it remains at the previous level for white fused alumina and increases up to 20 μm for Ti and Cr-Ti fused alumina. This parameter describes structural performance of materials (the stronger the material the deeper the notch).

Conclusion

Thus, the conducted study made it clear that structural performance, material strength, and fracture pattern of single grains of Ti-containing fused alumina depend on the type and ratio of alloy additives in alumina and on the decomposition level of the Ti_2O_3 solid solution resulting from grain heat-treatment in open air. Purple varieties of Cr-Ti fused alumina featuring the minimum TiO^2/Cr_2O_3 ratio of 1.5 in the solid solution after heat treatment at 1250 °C are preferable.

References

[1] A. Garshin, S. Fedotova, Abrasive Materials and Tools. Manufacturing Procedure: Instructional, Aid, SPb.: Polytechnic University Press, 2008.

[2] V. Ostrovsky, Theoretical Basis of Grinding, Leningrad: Leningrad State University Press,1981.

[3] I. Lavrov, I. Rivlin and T. Nikitina, Physical and Chemical Studies of Abrasive Materials and Tools, Collected Papers, L.: VNIIASh, 1981, pp 69 — 75.

[4] O. Pushkarev, V. Schumyacher, Methods and Equipment for Monitoring of Abrasive Materials' Physical and Mechanical Characteristics: Monograph, Volgograd: VolgGASU, 2004.

[5] E. Gevorkyan, Super Hard Materials. 5 (2004) 54-57.

[6] A. Ovchinnikov, Materials for Abrasive Tools. Review Science and Education, 7, 2013. https://doi.org/10.7463/0713.0577449

[7] V. Berdikov, A. Diulin, V. Efimchuk, O. Pushkarev and G. Finogenov, Device for Studying Durability of Materials, Plant Laboratory. 7 (1986) 78 - 81

[8] O. Pushkarev, Yu. Bagaiskov, Investigation of micromechanical properties of fused alumina and silicone carbide grains, Journal of Physics: Conference Series. 2020. - Vol. 1515 : International Scientific Conf. «Metrological Support of Innovative Technologies – ICMSIT-2020» (Krasnoyarsk, Russia, 4 March, 2020). https://doi.org/10.1088/1742-6596/1515/5/052016

[9] O. Pushkarev, O. Kulik and L. Nikuiko, Plant Laboratory. Material Diagnostics. 83 (2017) 49-52.

[10] O. Pushkarev, Yu. Bagaiskov, Refinement of the procedure to determine durability of abrasive materials, IOP Conference Series: Materials Science and Engineering. Vol. 862 : II International Conference «MIP: Engineering-2020: Modernization, Innovations, Progress: Advanced Technologies in Material Science, Mechanical and Automation Engineering» (Krasnoyarsk, Russia, 16-18 April, 2020). https://doi.org/10.1088/1757-899X/862/3/032019

Materials Research Proceedings 21 (2022) 76-80 https://doi.org/10.21741/9781644901755-14

Effect of UV-Irradiation and Ozone Exposure on Thermal and Mechanical Properties of PLA/LDPE Films

Maria Podzorova[1,2,3,a*], Yulia Tertyshnaya[1,2,3,b], Anastasia Khramkova[4,c] and Alina Ivanitskikh[3,d]

[1] Emanuel Institute of Biochemical Physics, Russian Academy of Sciences, 119991, 4 Kosygina str., Moscow, Russia

[2] Plekhanov Russian University of Economics, 117997, 36 Stremyanny per., Moscow, Russia

[3] Federal Scientific Agroengineering Center VIM, 109428, 1st Institutskiy proezd, 5, Moscow, Russia

[4] Politecnico di Milano, Piazza Leonardo da Vinci, 32 20133, Milan, Italy

[a] mariapdz@mail.ru, [b] terj@rambler.ru, [c] nastiel@mail.ru, [d] alinen@yandex.ru

Keywords: Polyethylene, Polylactide, Films, Biodegradation, UV-Irradiationon

Abstract. The mechanical properties of polymer composites based on polylactide vary significantly over a wide range of values. It has been established that photodegradation of low-density polyethylene – polylactide blends occurs both in the amorphous and in the crystalline phase of the PLA matrix, which leads to deterioration of the mechanical properties of the studied mixtures. Ozonolysis affects the strength parameters of polylactide-polyethylene samples as well as photodegradation. By the differential scanning calorimetry it is determined that the melting point of polylactide decreases by 2-4 °C, the glass transition temperature - by 1-3 °C, while the degree of crystallinity increases by 3-6%. In the process of ozonolysis, the thermophysical characteristics of PLA/LDPE have changed.

Introduction

Biodegradable composite materials are obtained from a combination of synthetic polymers with the addition of natural additives that are susceptible to biodegradation. Among biodegradable polymers, a special place is occupied by polylactide (polylactic acid, PLA). One of the disadvantages of polylactide is its relatively high fragility. Unfortunately, not all materials currently being developed are able to meet all the necessary technological requirements for full use (mechanical characteristics, biodegradability).

A semi-crystalline PLA is preferable to an amorphous polymer when higher mechanical properties are required. In turn, it is worth noting that the main feature of the PLA is strength and toughness, and the development of mixed materials based on polymers that differ in their mechanical characteristics is relevant, since it is possible to create a material with specified properties [1-4].

It is also important to study changes in the structure and physico-mechanical characteristics of the resulting blends when exposed to environmental factors, for example, ultraviolet radiation. Earlier in [5, 6] it was shown that UV- radiation affects the structure and crystallinity of PLA. Some researchers suggest compositions based on boards with components that increase its resistance to UV-radiation [7-9].

The urgency of the problem of giving biodegradability properties to well-mastered multi-tonnage industrial polymers (polyethylene, polypropylene, polyethylene terephthalate, etc.) is explained by the fact that currently the share of biodegradable plastics in the global consumption of polymer materials does not exceed 2-3%, and in the foreseeable future it is not expected to overcome the threshold of 5-7%.

The effect of ultraviolet on the mechanical properties of PLA/LDPE film samples has been studied. It is shown that changes in thermophysical characteristics are observed both in the polylactide phase and in the polyethylene phase [7, 10-12].

Several degradation studies have shown that ultraviolet irradiation of polyethylene demonstrates a decrease in elongation due to embrittlement. It is known that embrittlement in semi-crystalline polymers is caused by a decrease in molecular weight as a result of the chain splitting process [13, 14].

Experimental

Blends of LDPE (TM 15803-020 from Neftekhimsevilen, OJSC, Russia) with PLA (4032D from Nature works, USA) in a wide range of compositions were studied. The PLA content in the polymer matrix was 30, 50, 70 wt % and pure polymers PLA and LDPE. The blends were prepared by mixing at Plasticorder PLD-651 (Brabender, Germany) for 5 minutes at a temperature of 180°C. After cooling, the samples of the obtained material were milled using a knife mill RM-120 (Vibrotechnik, Russia) and then compressed in a manual hydraulic press PRG-10 (VNIR, Russia) at 180°C. As a result, film round samples with a diameter of 7 cm and thickness of 100±10 μm were obtained [15].

The degree of crystallinity of polymers in the blends was studied using differential scanning calorimeter Netzsch DSC 214 Polyma at the heating rate of 8K/min and sample weight of 10±0.1 mg. The temperature scale was calibrated against an indium standard (Tm = 156.6°C, ΔH = 28.44 J/g). The value of the degree of crystallinity χ_c was calculated by the formula (1):

$$\chi_c\,(\%) = 100 \times \Delta H_0 / \Delta H_i, \qquad (1)$$

χ_c - degree of crystallinity, the heat of melting of an ideal crystal PLA ΔH_0 = 93,1 J/g [16], PE ΔH_i = 293 J/g [17].

Mechanical properties were examined by tensile compression testing machine Devotrans DVT GP UG 5 (Turkey) according to ISO 527-1:2012. The tensile strength and elongation at break were determined. The data were averaged by six samples.

The influence of ultraviolet on the obtained samples was studied by the lamp Vilber Lourmat VL-6.LC at λ=254 nm for 5 -100 hours.

Ozonolysis of film samples for 8 and 16 hours was carried out at atmospheric pressure and T = (20 ± 2) ° C in an atmosphere of an ozone-oxygen mixture with a partial concentration of ozone equal to 5.0×10^{-5} mol/L.

Results and discussion

Under ultraviolet irradiation, degradation processes occur differently in the thickness of the material, demonstrating changes in the deep profiles of the carbonyl index of polymers, which can lead to various changes in mechanical properties.

In this paper, changes in the mechanical characteristics of mixed compositions based on polylactide during exposure to UV radiation λ = 254 nm are investigated. Due to the interest in research of this kind, this wavelength is often used as an accelerated test method [6]. In composite materials, there is a decrease in thermal and mechanical characteristics after exposure to UV radiation. The melting point *T*m of PLA decreased by 12°C after exposure to UV radiation for 25 and 50 h and by 30°C after exposure for 100 h. Mechanical characteristics are represented in Figure 1.

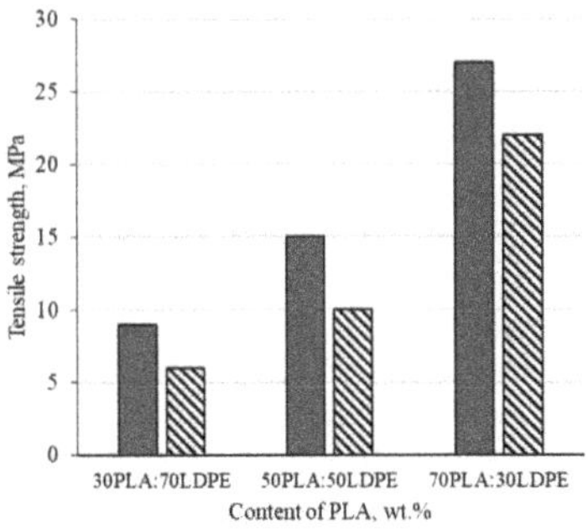

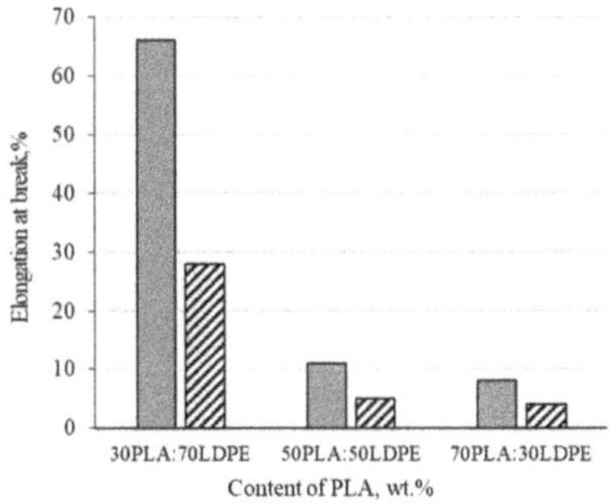

Fig. 1. *Mechanical characteristics of PLA after 5 hours of exposure to UV radiation: a – Tensile strength, b – Elongation at break (gray – before UV radiation; dashed line – after UV radiation).*

Tests have shown that ultraviolet radiation affects the polylactide, after 5 hours of exposure λ=254 nm, the tensile strength and elongation at break of the PLA decrease. These results are also confirmed by literature data, in [10] it was shown that after exposure to this wave for 12 hours, the strength of the polylactide decreases by 5 times.

It is known that photodegradation begins with an attack of the radical of the macromolecular chain of the PLA. By the same mechanism, ozone oxidation of polylactide occurs, only the activator of the process is an O$^{\bullet}$-radical formed as a result of the decomposition of an ozone molecule.

In the case of polymers, ozone treatment of the surface layer improves its adhesive properties, and also allows to insert some specific functional components into the polymer [18-20], due to which the latter acquires new properties. The degradation process initiated by ozone was considered in [18, 21] in relation to various types of polymers.

Table 1. *Thermophysical characteristics of samples of PLA /LDPE blends – initial and after their ozonation for 8 and 16 hours*

Composition [wt. %] PLA/LDPE	T_m [°C] ±0,5 °C			χ_c [%] ±0,5%		
	Initial	After 8 h ozon.	After 16 h ozon.	Initial	After 8 h ozon.	After 16 h ozon.
100/0	165	165	164	45	47	50
70/30	163/103	161/102	160/101	39/17	43/14	45/13
50/50	164/103	162/102	160/100	51/19	53/17	56/16
30/70	164/104	162/103	160/99	50/16	52/14	56/12
0/100	104	102	100	21	17	14

When ozone acts on a polymer material, many characteristics of the polymer matrix can change, ranging from thermophysical to visible surface defects (submicro- and microcracks) with deep degrees of destructive processes. In amorphous-crystalline polymers, as is known, the degradation process begins in the amorphous phase. The disintegration of the amorphous and crystalline phases leads to a decrease in the strength of the material, a change in mechanical properties.

To control the thermophysical characteristics of PLA/LDPE mixture samples: was used the DSC method. After an 8-hour action of ozone, a decrease of 1-3 °C in the melting point of PLA for mixtures is observed, while for pure PLA, the T_m does not change (Table 1). After ozonolysis for 16 hours, the tendency to increase the degree of crystallinity (χ_{cr}) of polylactide persists. Melting temperatures in blends of PLA/LDPE decrease.

Along with the degradation of the amorphous phase, exposure to ozone may contribute to the continuation of crystallization due to the rupture of strained flow chains, therefore, the degree of crystallinity of polylactide increases. In turn, the χ_{cr} of pure LDPE is reduced by 5%, and mixtures are slightly less, which may be caused by the course of an oxidative destructive process, which also leads to deterioration of mechanical properties.

Conclusion

This study examined the processes of photodegradation and ozonolysis of PLA/LDPE films. Thus, the effect of both UV radiation and ozone affects the structural characteristics of the components of the mixture and reduces the mechanical properties of PLA/LDPE samples.

Acknowledgments

The work was carried out using the scientific equipment of the Center of Shared Usage "New Materials and Technologies" of Emanuel Institute of Biochemical Physics RAS and Joint Research Center of Plekhanov Russian University of Economics.

References

[1] S. Farah, D.G. Anderson, R. Langer, Physical and mechanical properties of PLA, and their functions in widespread applications — A comprehensive review, Adv. Drug Deliv. Rev. 107 (2016) 367-392. https://doi.org/10.1016/j.addr.2016.06.012

[2] Y. Tertyshnaya, S. Karpova, M. Moskovskiy, A. Dorokhov, Electrospun polylactide/natural rubber fibers: effect natural rubber content on fiber morphology and properties, Polymers. 13 (2021) 2232. https://doi.org/10.3390/polym13142232

[3] M.P. Arrieta, J. Lopez, D. Lopez, J.M. Kenny, L. Peponi, Effect of chitosan and catechin addition on the structural, thermal, mechanical and disintegration properties of plasticized electrospun PLA-PHB biocomposites, Polym. Degrad. Stab. 132 (2016) 145-156. https://doi.org/10.1016/j.polymdegradstab.2016.02.027

[4] Y.V. Tertyshnaya, M.V. Podzorova, Composite materials based on polylactide and poly-3-hydroxybutyrate "Green" polymers, Russ. J. Appl. Chem. 91 (3) (2018) 417–423. https://doi.org/10.1134/S1070427218030126

[5] M.V. Podzorova, Yu.V. Tertyshnaya, S.G. Karpova, A.A. Popov, Impact of UV treatment on polylactide–polyethylene film properties, IOP Conf. Series: Materials Science and Engineering 525 (2019) 012043. https://doi.org/10.1088/1757-899X/525/1/012043

[6] E. Olewnik-Kruszkowska, I. Koter, J. Skopinska-Wisniewskab, J. Richert, Degradation of polylactide composites under UV irradiation at 254 nm, J. Photochem. Photobiol. A: Chemistry. 311 (2015) 144-153. https://doi.org/10.1016/j.jphotochem.2015.06.029

[7] A. Ashrafi, M. Jokar, N.A. Mohammadi, Preparation and characterization of biocomposite film based on chitosan and kombucha tea as active food packaging, Int. J. Biol. Macromol. 108 (2017) 444. https://doi.org/10.1016/j.ijbiomac.2017.12.028

[8] H. Wang, L. Wang, Developing a bio-based packaging film from soya by-products incorporated with valonea tannin, J. Clean. Prod. 143 (2017). https://doi.org/10.1016/j.jclepro.2016.12.064

[9] Y. Han, M. Yu, L. Wang, Bio-based films prepared with soybean by-products and pine (Pinus densiflora) bark extract, J. Clean. Prod. 187 (2018) 1–8. https://doi.org/10.1016/j.jclepro.2018.03.115

[10] H. J. Jeon, M.N. Kim, Biodegradation of poly(L-lactide) (PLA) exposed to UV irradiation by a mesophilic bacterium, Int. Biodeterior. Biodegrad. 85 (2013) 289-293. https://doi.org/10.1016/j.ibiod.2013.08.013

[11] M. V. Podzorova, Yu. V. Tertyshnaya, Degradation of polylactide—polyethylene binary blends in soil, Rus. J. Appl. Chem. 92 (6) (2019) 767-774. https://doi.org/10.1134/S1070427219060065

[12] L. Jiang, J. Zhang, M.P. Wolcott, Comparison of polylactide/nano-sized calcium carbonate and polylactide/montmorillonite composites: reinforcing effects and toughening mechanisms, Polymer. 48 (26) (2007) 7632-7644. https://doi.org/10.1016/j.polymer.2007.11.001

[13] B. Fayolle, E. Richaud, X. Colin, J. Verdu, Degradation-induced embrittlement in semi-crystalline polymers having their amorphous phase in rubbery state, J. Mater. Sci. 43 (22) (2008) 6999-7012. https://doi.org/10.1007/s10853-008-3005-3

[14] A. Fairbrother, H.-C. Hsueh, J. H. Kim, De. Jacobs, L. Perry, D. Goodwin, C. White, S. Watson, L.-P. Sung, Temperature and light intensity effects on photodegradation of high-density polyethylene, Polym. Degrad. Stab. 165 (2019) 153-160. https://doi.org/10.1016/j.polymdegradstab.2019.05.002

[15] Yu. V. Tertyshnaya, M. V. Podzorova, Effect of UV irradiation on the structural and dynamic characteristics of polylactide and its blends with polyethylene, Russ. J. Phys. Chem. B. 14 (1) (2020), 167–175. https://doi.org/10.1134/S1990793120010170

[16] L-T. Lim, R. Auras, M. Rubino, Processing technology for poly(lactic acid), Prog. Polym. Sci. 33 (2008) 820-852. https://doi.org/10.1016/j.progpolymsci.2008.05.004

[17] Q. Zhou, M. Xanthos, Nanoclay and crystallinity effects on the hydrolytic degradation of polylactides, Polym. Degrad. Stab. 93 (8) (2008) 1450-1459. https://doi.org/10.1016/j.polymdegradstab.2008.05.014

[18] R.O.F. Verkuijlen, M.H.A. van Dongen, A.A.E. Stevens, J. van Geldrop, J.P.C. Bernards, Surface modification of polycarbonate and polyethylene naphtalate foils by UV-ozone treatment and mu Plasma printing, Appl. Surf. Sci. 290 (2014). 381-387. https://doi.org/10.1016/j.apsusc.2013.11.089

[19] A. Mahfoudh, F. Poncin-Epaillard, M. Moisan, J. Barbeau, Effect of. dry-ozone exposure on different polymer surfaces and their resulting bio-. cidal action on sporulated bacteria, Surf. Sci. 604 (2010) 1487-1493. https://doi.org/10.1016/j.susc.2010.05.013

[20] A. Delplanque, E. Henry, J. Lautru, H. Leh, M. Buckle, C. Nogues, UV/ozone surface treatment increases hydrophilicity and enhances functionality of SU-8 photoresist polymer, Appl. Surf. Sci. 314 (2014) 280-285. https://doi.org/10.1016/j.apsusc.2014.06.053

[21] W. Yue, R. He, P. Yao, Y. Wei, Ultraviolet radiation-induced accelerated degradation of chitosan by ozone treatment, Carbohydr. Polym. 77 (2009) 639-642. https://doi.org/10.1016/j.carbpol.2009.02.015

Surface Morphology Investigations of Nanocrystalline $R_2Fe_{14}B$ (R = Y, Nd, Gd, Er) by Atomic Force Microscopy

Ivan Pelevin[1,a*], Dmitriy Ozherelkov[1,b], Tatiana Kaminskaya[2,c] and Irina Tereshina[2,d]

[1] Catalysis Lab, National University of Science and Technology MISIS, Moscow, Russia

[2] Faculty of Physics, Lomonosov Moscow State University, Moscow 119991, Russia

[a]i.pelevin@misis.ru, [b] d.ozherelkov@gmail.com, [c]ktp53@mail.ru, [d]irina_tereshina@mail.ru

Keywords: Atomic Force Microscopy, Hard Magnetic Materials, Nanostructured Materials, Rapid Quenching, Melt Spinning, Additive Manufacturing

Abstract. The study was aimed at microstructure investigations of melt-spun rare-earth intermetallic compounds using atomic force microscopy. Surface morphology of $R_2Fe_{14}B$ (R = Y, Nd, Gd, Er) was studied with nanometric resolution. Grain structure features were discovered depending on the rare-earth element composition and quenching regime. Grain size dependence on rare earth elements' composition decreased with the metal's serial number and atomic weight. Wherein structural size dependence on quenching wheel speed had non-linear character: increase the speed from 20 to 30 m/s led to 3 times decrease of the grain size and significant surface roughness reduction.

Introduction

Usually, modern hard magnetic materials are multi-component systems with hysteresis properties provided by precise composition selection and formation of their proper microstructure [1–3]. That is why identifying the relationship between magnetic properties and microstructure continues to be relevant and of great scientific importance applying to both known $R_2Fe_{14}B$ materials (where R – rare earth metal (REM)) and multicomponent alloys $(R,R')_2(Fe,T)_{14}B$ (T = Co, Mn and other) with partial substitution in REM and iron sublattices. The rapid development of synthesis technologies for new nanocrystalline materials requires new analysis methods. There are various synthesis techniques such as powder metallurgy, mechanical synthesis, rapid quenching, severe plastic deformation, and their combinations [4]. Synthesis of permanent magnets by 3D-printing arised recently involving different methods of additive manufacturing. This filed may be divided into two main branches: creating bonded magnets [5–8] and selective laser sintering/melting (SLS/SLM technologies) of powders layer by layer [9–12]. Both approaches could be combined with other technologies such as grain boundary infiltration [10,11,13]. The application of modern additive technologies opens new prospects for permanent magnet manufacturing without any shape limitations. In this regard, SLS/SLM technology is one of the most promising; however, there are still many challenges. In particular, the SLM process assumes specific morphology of the sample surface, which may significantly differ from the internal one [14–17]. Microstructure variation with depth arises due to different cooling conditions and temperature gradients and depends on laser scanning strategy, which usually involves special processing of the layer contour. All described features affect crystallization behavior and thus final grain structure. SLM process assumes high cooling rates of the melting pool ($\sim 10^4$-10^6 K/s) and thus rapid solidification resulting in superfine microstructure [18–20]. Another conventional rapid solidification technique providing superfine microstructure applying to Nd-Fe-B-type material is melt-spinning [21–23]. Such technique is rather more used than SLM for now.

It should be noted that all new methods of nanomaterials synthesis require the development of precise structure analysis methods and approaches. The most common metallic materials microstructure characterization method is scanning electron microscopy (SEM), distinguished by relatively easy sample preparation. Transmission electron microscopy (TEM) is usually involved for nano-sized objects observation, but this analytical method requires complicated and thorough sample preparation that limits its usage. Another method of structural (mostly surface) research is atomic force microscopy (AFM). Key advantages of AFM are the possibility of analysis on air or in a liquid medium and the absence of strict requirements for samples (their electrical conductivity, composition, size, magnetic properties).

Previously, the authors carried out a study of multi-component alloys $(Nd,Ho)_2(Fe,Co)_{14}B$ using AFM, which revealed the main features of the microstructure and morphology of their surface in different structural states [24]. The melt-spun samples were obtained, and a study of the topography of their contact and free surfaces was carried out. On the contact side, micron-scale cracks were observed oriented in one preferential direction. The width and depth of the cracks varied from hundreds of nanometers to a few micrometers. The morphology of nano-sized grains and clusters, their size distribution, and homogeneity were determined. The study of the magnetic properties of this material made it possible to establish their relationship with microstructural features.

In studies performed by Jakubowicz [25, 26] $Nd_2Fe_{14}B/\alpha$-Fe-type nanocomposites were investigated using atomic force microscopy. The grain size of samples with various compositions was estimated, and AFM identified their differences. High resolution of the AFM method allowed to discern nanocrystalline grains with an average size of ~19 nm. Besides, small grains tended to form agglomerates with a size of ~250 nm. Height and deflection imaging modes of AFM were used for structural investigations.

Present research work is devoted to investigating the effect of the melt-spinning method on the microstructure of $R_2Fe_{14}B$ alloys with various REM (R = Y, Nd, Gd, Er) using AFM. A comparative study of the variation in the composition of $R_2Fe_{14}B$ alloys (light (Nd) and heavy (Gd, Er) REMs, as well as magnetic (Nd,Gd, Er) rare-earth ions and their nonmagnetic analog (Y) on surface morphology was investigated.

These studies aim to develop a technique for studying the microstructure and surface morphology of melt-spun nanocrystalline magnetic materials based on REMs.. Subsequently, the scope of application of this technique can be extended to similar materials obtained using additive technologies, namely, by the method of selective laser sintering or melting.

Materials and methods

Bulk $Nd_2Fe_{14}B$, $Gd_2Fe_{14}B$, $Er_2Fe_{14}B$ and $Y_2Fe_{14}B$ intermetallic samples were obtained using induction furnace followed by remelting in quartz ampoule and melt supply to the quenching wheel surface (melt spinning technique). Speed of the quenching wheel surface was 20 and 30 m/s. Ribbon-like samples with 10 mm length, 2-5 mm width and ~0.3 mm thickness were obtained. Elemental and phase composition were controlled by means of X-ray fluorescence spectrometry and X-ray diffraction using REAN spectrometer and Difrey-401K (both by "Nauchnie pribory", Russia), respectively. Surface morphology was investigated by atomic force microscopy using an Solver P-47H (NT-MDT, Russia) at room temperature using standard HA_NC Etalon silicon probes (cantilevers) with 94 and 124 μm length and 140-235 kHz frequency range. Obtained data were processed in Nova RC1 software package. The magnetic hysteresis loops were measured on a standard SQUID magnetometer in fields up to 10 kOe at a temperature of 300 K.

Results and discussions

The analysis of the sample's grain structure was carried out by the AFM method on the initial rapidly quenched samples without any special preparation. This initial state may have a significant effect on the surface morphology and complicate the analysis. Investigation of $Nd_2Fe_{14}B$, $Gd_2Fe_{14}B$, $Er_2Fe_{14}B$ samples was carried out in semi-contact method and in tapping mode using resonance properties of the cantilever.

The structural elements of the surface revealed by this method were considered as grains. The grain size was defined as the diameter of a circle equal to the cross-sectional area of the grain. The average grain size was determined as the square root of the cross-sectional area of the particle at a height corresponding to a given value (0.8). The ratio of the maximum grain section length to the maximum width gives the aspect ratio.

Investigations and comparative analysis of the obtained contact surface structure in the case of samples with a quenching rate of 20 m/s using AFM revealed that the largest average grain size among the investigated intermetallic compounds was for the $Nd_2Fe_{14}B$ compound (see Figure 1a).

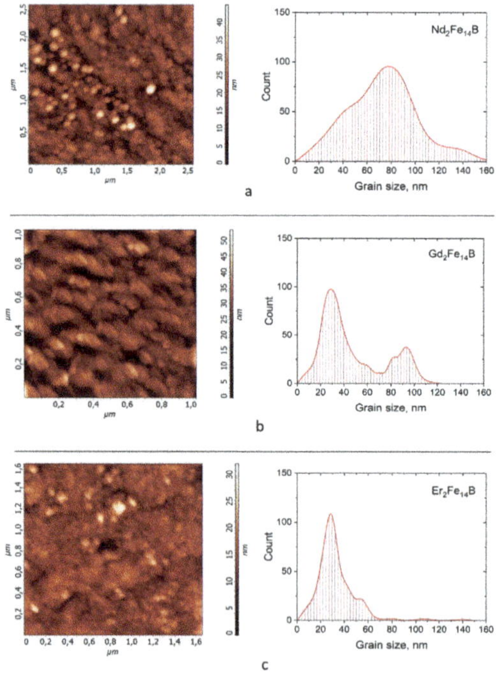

Fig. 1. *AFM images of the contact surface of melt-spun samples obtained at 20 m/s quenching wheel speed (left) and their grain size distributions (right): a – $Nd_2Fe_{14}B$, b – $Gd_2Fe_{14}B$, c – $Er_2Fe_{14}B$.*

The sample structure consisted of regular round shape grains with an average grain diameter of about 70 nm. Additionally, smaller grains with sizes of 30-40 nm were obtained. Partial agglomeration of small grains was observed, giving the total size of structural components up to 150 nm. The contact surface structure of the rapidly quenched $Gd_2Fe_{14}B$ compound was significantly different - see Figure 1b. Grains of $Gd_2Fe_{14}B$ had a highly elongated elliptical shape with dimensions of 30-35 × 90 nm, i.e. aspect ratio of 1:3. The grains' long axis is predominantly oriented along the movement direction of the quenching wheel surface. All grains were collected in agglomerates (bunches), combining around 8-12 grains each and the sizes of agglomerates vary from 120 to 300 nm. The finest grain structure was observed for the $Er_2Fe_{14}B$ melt-spun sample, where grains had a slightly elongated shape with an aspect ratio close to unity. Most grains have a round shape, and the average grain size of 25-30 nm was observed.

Also, some agglomerations up to 100 nm in size were detected. The observed difference in obtained microstructures of melt-spun samples can be explained by the different melting temperatures of the investigated compounds. Different kinetics during quenching, i.e., different temperature gradients, leads to differences in cooling rate and crystallization speed. It is known [27] that crystallization of $Nd_2Fe_{14}B$-type material during rapid quenching proceeds in two stages: at first, $TbCu_7$-type metastable phase arises along with a small amount of α-Fe phase. Then decomposition of the $TbCu_7$-type phase occurs with $Nd_2Fe_{14}B$-type and α-Fe phases formation. Such a two-stage crystallization process complicates the analysis of its kinetics. However, it can be assumed that different rare earth metals that vary from light (Nd) and heavy (Gd and Er) demonstrate different nucleation behavior due to differences in atom radius, weight, and crystal lattice parameters.

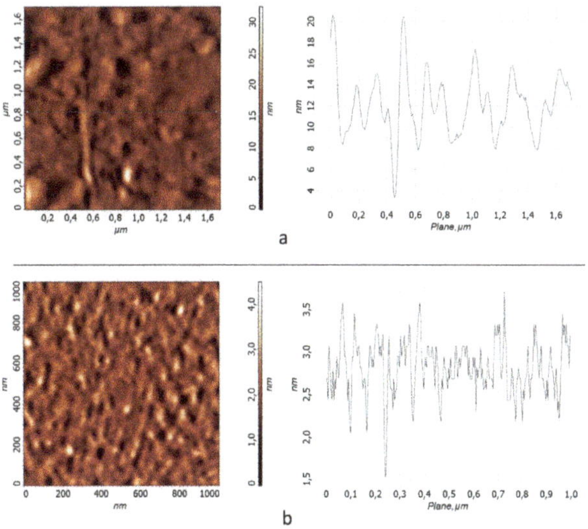

Fig. 2. *AFM images of surface topography (left) of melt-spun $Y_2Fe_{14}B$ obtained at 20 (a) and 30 (b) m/s quenching wheel speeds and profile of their surfaces (right).*

It should be noted that such behavior provides observed nano-sized structure with grain size lower than critical single-domain size. Potentially, it could create optimal conditions for high magnetic properties due to strong exchange coupling between $Nd_2Fe_{14}B$-type and α-Fe phases. Moreover, described crystallization behavior could be expected during synthesis by the novel SLM method, which is planned to attract further research.

Comparative studies of the surface topology of melt-spun $Y_2Fe_{14}B$ samples were carried out with quenching wheel speeds of 20 m/s and 30 m/s. The effect of various quenching speeds is visible in Figure 2. The studies were carried out in a contact mode using the lateral force method. In the contact mode at a distance of a few tenths of a nanometer from the tip of the probe to the sample surface, repulsive forces prevailed. The used method allowed to construct a picture of lateral forces (friction forces) by recording the torsion angle of the probe during scanning in contact with the surface. Moreover, the lateral twisting of the probe was registered simultaneously with topography. In this case, features of the surface relief, which are not resolved in other modes, are visualized due to the different frictional characteristics of the surface. The surface of the sample obtained at a higher speed of 30 m/s is much smoother, and the size of its structural inhomogeneities (i.e., grains) is much smaller. For the 20 m/s sample, the average surface roughness is around 8 nm, and the maximum deviation is 18 nm. The size of structural elements (grains) varied from 70 to 150 nm with agglomerates up to 200 nm. In the case of a 30 m/s sample, the average surface roughness was 1.5 nm, and the maximum emissions were up to 2.5 nm. The size of structural elements (grains) varied from 20 to 50 nm, about 3 times smaller than in the 20 m/s sample. Thus 1.5 times increasing of quenching wheel speed reduced structural element size by 3 times. Such obtained behavior was due to the increased heat dissipation at higher quenching wheel speeds.

Conclusions

The conducted study revealed the structural features of the melt-spun $R_2Fe_{14}B$ samples with various rare earth metals and quenching speeds. The shape and size of the structural elements (grains) were defined by the AFM method. The obtained dependence on rare earth elements in composition demonstrated a decrease of the grain size with the increasing number and atomic weight of the metal. $Nd_2Fe_{14}B$ compound with the lightest rare-earth metal possessed the largest grains among investigated samples. Wherein $Er_2Fe_{14}B$ compound with the heavy rare earth element, on the contrary, quenched with the finest microstructure. An increase of the quenching wheel speed for the $Y_2Fe_{14}B$ compound from 20 to 30 m/s resulted in a significant reduction in surface roughness. Furthermore, the grain size decreased by about 3 times compared to lower quenching speed. Thus, the atomic force microscopy application to nanostructured intermetallic $R_2Fe_{14}B$-type compounds was proposed and worked out with the prospect of using this method for additively manufactured magnetic materials.

Acknowledgments

The work was financially supported by the Russian Science Foundation (RSF), grant › 21-79-10239.

References

[1] J. Mohapatra, J.P. Liu, Rare-Earth-Free Permanent Magnets: The Past and Future, in: Handb. Magn. Mater., 2018: pp. 1–57. https://doi.org/10.1016/bs.hmm.2018.08.001

[2] D. Li, Y. Li, D. Pan, Z. Zhang, C.J. Choi, Prospect and status of iron-based rare-earth-free permanent magnetic materials, J. Magn. Magn. Mater. (2019). https://doi.org/10.1016/j.jmmm.2018.09.032

[3] O. Gutfleisch, M.A. Willard, E. Brück, C.H. Chen, S.G. Sankar, J.P. Liu, Magnetic Materials and Devices for the 21st Century: Stronger, Lighter, and More Energy Efficient, Adv. Mater. 23 (2011) 821–842. https://doi.org/10.1002/adma.201002180

[4] I.S. Tereshina, I.A. Pelevin, E.A. Tereshina, G.S. Burkhanov, K. Rogacki, M. Miller, N. V. Kudrevatykh, P.E. Markin, A.S. Volegov, R.M. Grechishkin, S. V. Dobatkin, L. Schultz, Magnetic hysteresis properties of nanocrystalline (Nd,Ho)-(Fe,Co)-B alloy after melt spinning, severe plastic deformation and subsequent heat treatment, J. Alloys Compd. (2016). https://doi.org/10.1016/j.jallcom.2016.04.228

[5] I.C. Nlebedim, H. Ucar, C.B. Hatter, R.W. McCallum, S.K. McCall, M.J. Kramer, M.P. Paranthaman, Studies on in situ magnetic alignment of bonded anisotropic Nd-Fe-B alloy powders, J. Magn. Magn. Mater. (2017). https://doi.org/10.1016/j.jmmm.2016.08.090

[6] B.M. Ma, J.W. Herchenroeder, B. Smith, M. Suda, D. Brown, Z. Chen, Recent development in bonded NdFeB magnets, J. Magn. Magn. Mater. (2002). https://doi.org/10.1016/S0304-8853(01)00609-6

[7] K. Mungale, T.N. Lamichhane, H. Wang, B.C. Sales, M.P. Paranthaman, U.K. Vaidya, Compression molding of anisotropic NdFeB bonded magnets in a polycarbonate matrix, Materialia. (2021). https://doi.org/10.1016/j.mtla.2021.101167

[8] B.G. Compton, J.W. Kemp, T. V. Novikov, R.C. Pack, C.I. Nlebedim, C.E. Duty, O. Rios, M.P. Paranthaman, Direct-write 3D printing of NdFeB bonded magnets, Mater. Manuf. Process. (2018). https://doi.org/10.1080/10426914.2016.1221097

[9] F. Bittner, J. Thielsch, W.G. Drossel, Laser powder bed fusion of Nd–Fe–B permanent magnets, Prog. Addit. Manuf. (2020). https://doi.org/10.1007/s40964-020-00117-7

[10] C. Huber, H. Sepehri-Amin, M. Goertler, M. Groenefeld, I. Teliban, K. Hono, D. Suess, Coercivity enhancement of selective laser sintered NdFeB magnets by grain boundary infiltration, Acta Mater. (2019). https://doi.org/10.1016/j.actamat.2019.04.037

[11] A.S. Volegov, S. V. Andreev, N. V. Selezneva, I.A. Ryzhikhin, N. V. Kudrevatykh, L. Mädler, I. V. Okulov, Additive manufacturing of heavy rare earth free high-coercivity permanent magnets, Acta Mater. (2020). https://doi.org/10.1016/j.actamat.2020.02.058

[12] F. Bittner, J. Thielsch, W.G. Drossel, Microstructure and magnetic properties of Nd-Fe-B permanent magnets produced by laser powder bed fusion, Scr. Mater. (2021). https://doi.org/10.1016/j.scriptamat.2021.113921

[13] L. Li, A. Tirado, B.S. Conner, M. Chi, A.M. Elliott, O. Rios, H. Zhou, M.P. Paranthaman, A novel method combining additive manufacturing and alloy infiltration for NdFeB bonded magnet fabrication, J. Magn. Magn. Mater. (2017). https://doi.org/10.1016/j.jmmm.2017.04.066

[14] I. Yadroitsev, I. Smurov, Surface morphology in selective laser melting of metal powders, in: Phys. Procedia, 2011. https://doi.org/10.1016/j.phpro.2011.03.034

[15] G. Strano, L. Hao, R.M. Everson, K.E. Evans, Surface roughness analysis, modelling and prediction in selective laser melting, J. Mater. Process. Technol. (2013). https://doi.org/10.1016/j.jmatprotec.2012.11.011

[16] S. Gao, X. Yan, C. Chang, Eric Aubry, M. Liu, H. Liao, N. Fenineche, Effect of Laser Energy Density on Surface Morphology, Microstructure, and Magnetic Properties of Selective Laser Melted Fe-3wt.% Si Alloys, J. Mater. Eng. Perform. (2021).

https://doi.org/10.1007/s11665-021-05591-w

[17] M. Zheng, L. Wei, J. Chen, Q. Zhang, J. Li, S. Sui, G. Wang, W. Huang, Surface morphology evolution during pulsed selective laser melting: Numerical and experimental investigations, Appl. Surf. Sci. (2019). https://doi.org/10.1016/j.apsusc.2019.143649

[18] P.A. Hooper, Melt pool temperature and cooling rates in laser powder bed fusion, Addit. Manuf. (2018). https://doi.org/10.1016/j.addma.2018.05.032

[19] Y. Li, D. Gu, Parametric analysis of thermal behavior during selective laser melting additive manufacturing of aluminum alloy powder, Mater. Des. 63 (2014) 856–867. https://doi.org/10.1016/j.matdes.2014.07.006

[20] M. Majeed, H.M. Khan, I. Rasheed, Finite element analysis of melt pool thermal characteristics with passing laser in SLM process, Optik (Stuttg). (2019). https://doi.org/10.1016/j.ijleo.2019.163068

[21] J.J. Croat, Manufacture of NdFeB permanent magnets by rapid solidification, J. Less-Common Met. (1989). https://doi.org/10.1016/0022-5088(89)90006-4

[22] A. Zaluzka, Y. Xu, Z. Altounian, J.O. Strom-Olsen, Effects of quench rate on the texture in melt-spun NdFeB alloys, Mater. Sci. Eng. A. (1991). https://doi.org/10.1016/0921-5093(91)90193-Q

[23] H.C. Hua, G.Y. Wang, C.H. Zheng, G.X. Huang, Q.Z. Xu, L.H. Wu, S.Y. Shi, Microstructure of melt-spun NdFeB magnet, Mater. Lett. (1988). https://doi.org/10.1016/0167-577X(88)90085-7

[24] N. V. Andreeva, A. V. Filimonov, A.I. Rudskoi, G.S. Burkhanov, I.S. Tereshina, G.A. Politova, I.A. Pelevin, A study of nanostructure magnetosolid Nd–Ho–Fe–Co–B materials via atomic force microscopy and magnetic force microscopy, Phys. Solid State. (2016). https://doi.org/10.1134/S1063783416090079

[25] J. Jakubowicz, Application of atomic force microscopy in microstructure analysis of mechanically alloyed Nd2Fe14B/α-Fe-type nanocomposites, J. Alloys Compd. (2003). https://doi.org/10.1016/S0925-8388(02)01075-7

[26] J. Jakubowicz, Application of AFM to study functional nanomaterials prepared by mechanical alloying, in: J. Mater. Sci., 2004. https://doi.org/10.1023/B:JMSC.0000039249.21167.a5

[27] K. Men, K. Li, Y. Luo, D. Yu, K. Zhang, J. Jin, Y. Mao, The crystallization behavior of as-quenched Nd9Fe85Nb0.5B5.5 alloys, J. Alloys Compd. (2015). https://doi.org/10.1016/j.jallcom.2015.02.046

The Improvement of the Operational Characteristics of the Ventilated Front Brake Discs

Denis Boldyrev[1,a], Maxim Kharchenko[2,b*], Sergey Nefedyev[2,c] and Elena Suvorova[2,d]

[1]Togliatti State University, 445020 Togliatti, Russia

[2]Nosov Magnitogorsk State Technical University, 455000 Magnitogorsk, Russia

[a]DA.Boldyrev@vaz.ru, [b*]kharchenko.mv@bk.ru, [c]ruslan246@mail.ru, [d]suvorlen@yandex.ru

Keywords: Ventilated Brake Disc, Wear Resistance, Cast Iron, Sulfur Content

Abstract. Wear and crack resistance are important operational characteristics of brake discs. The paper presents the most optimal concentration of sulfur in cast iron, which ensures its least wear, and discusses the implementation of the front brake discs manufacture from Gh190 cast iron having 0.11 ... 0.13% sulfur content at contrast to the 0.01 ... 0.03% sulfur content, and proves the change leads to a significant increase in wear and frictional properties of the discs. In the course of research, it is found that the increase in the crack resistance of brake discs is possible due to the improvement of the thermophysical properties of cast iron with the increase in the carbon content (up to 3.55 ... 3.60%) and the decrease in the silicon content (up to 1.45 ... 1.50%), while the carbon equivalent is constant.

Introduction

The working conditions of car brake discs are considered to be especially difficult. These are caused by the high contact pressures (up to 10 MPa) and intense frictional heating (up to 600 ° C), regarding dry abrasive friction [1-4]. An obstacle to further optimization of the materials in the manufacture of the "brake disc - pad" pair is the lack or insufficient information on the influence of the composition and technological factors on the operational properties of the friction pair parts, as well as the information on the mechanisms of friction and wear in the "brake disc - pad" pair [5-9].

The purpose of the work is the improvement of the operational characteristics of ventilated front brake discs by changing the composition of the material used for their manufacture (gray pearlitic cast iron Gh190).

Methods and materials

The tribotechnical testing of samples made from the materials of the friction pair under study is carried out on a universal machine testing friction and wear according to the "roller - plate" scheme. Samples in the form of plates are made from the materials of the brake pads, and samples in the form of rollers are made from cast iron. The test mode involves: 27 mm sample diameter; 72 H standard load;3.12 m/s linear velocity;10-minute time test. The wear of the cast iron samples is determined by the gravimetric method. Brinell hardness is determined using a TB 5006 device. The ultimate resistance $\sigma_B$is determined by a TIRA test 2300 machine. The full-scale tests are carried out by an inertial chassis dynamometer system. The test mode for the friction pair under study involves: J = 4.18 kg x m2moment of inertia; N = 200 number of brakes at each initial heating temperature of the brake disc. The initial temperatures of heating a brake disc before braking are: 100°C, 150°C, 200°C, 250°C, 300°Cand 350°C;the deacceleration is j_a=3 м/s^2; the

Materials Research Proceedings **21** (2022) 88-94
https://doi.org/10.21741/9781644901755-16

frictional torque isM_T=38 kg×м; the initial braking velocity is $V_{intitisl}$=100 km/h; the final braking velocity is V_{final}=60 km/h.

Results and Discussion

Gray pearlitic cast iron Gh190 is used for the manufacture of ventilated automobile brake discs. According to its mechanical properties and microstructure it corresponds to the SCh25 grade (1412-85 industry standard). The average chemical composition of Gh190 cast iron in a casting includes 3.3 ... 3.4% of C; 2.1 ... 2.2% of Si; 0.5 ... 0.7% of Mn; 0.1% of P; 0.03% of S; 0.1 ... 0.2% of Cr; 0.10 ... 0.15% of Ni; 0.45 ... 0.50% of Cu (Cu + Ni ≤ 0.60%). Let us introduce a conventional value for this chemical composition of cast iron for ventilated brake discs as composition 1.

However, in comparison with cast irons of the same class, Gh190 cast iron has a lower wear resistance. To increase the operational properties and service life of a given part, components increasing its physical, mechanical and tribotechnical characteristics [10] can be added to the cast iron in certain concentrations. In particular, the chemical composition of alternative grades of cast iron contains such a non-scarce element as sulfur.

A comparative wide-ranging analysis of the materials to manufacture ventilated brake discs for some domestic and foreign cars is carried out. Despite the fact that all the ventilated brake discs are made of pearlitic gray cast iron, their chemical composition, microstructure, mechanical and wear-friction properties have significant differences. It is found that, though the strength properties (σ_B and HB) of gray cast iron for ventilated brake discs of domestic cars are higher in comparison with the materials of the imported ventilated brake discs, the tribological characteristics are noticeably inferior to the latter.

The metallographic analysis shows that manganese sulfides MnS (Fig. 1) play a significant role on the wear-friction properties of gray cast iron. The manganese sulfides appear when sulfur is bound by manganese, having an affinity to it. The influence of sulfur on wear resistance is also studied.

In laboratory tribotechnical tests, the dependence of the wear decrease of Gh190 cast iron and its imported counterparts on the sulfur content is established. With an increase in the sulfur content up to 0.12%, there is a twice increase in its wear resistance.

It should be noted that the sulfur content in the materials of imported ventilated brake discs is in the range of 0.06 ... 0.12%. Therefore, to confirm the assumption about the positive influence of sulfur, experimental melting of brake discs are carried out, the controlled sulfur content in cast iron are: 0.03%, 0.10%, 0.12% and 0.13% respectively.

According to the results of the comparative laboratory tribotechnical tests of samples made from experimental castings of ventilated brake discs, the optimal concentration of sulfur is determined. The least wear of the material (up to 2 times) corresponds to the sulfur content of 0.11 ... 0.13%, which verifies the results obtained in the study of imported analogues. The higher is the sulfur content, the greater is the amount and size of MnS. On the one hand, manganese sulfides act as a solid lubricant. On the other hand, when the sulfur content is over 0.13%, the coarsening of sulfides occurs, which negatively affects the contact strength of the surface [11].

Materials Research Proceedings **21** (2022) 88-94
https://doi.org/10.21741/9781644901755-16

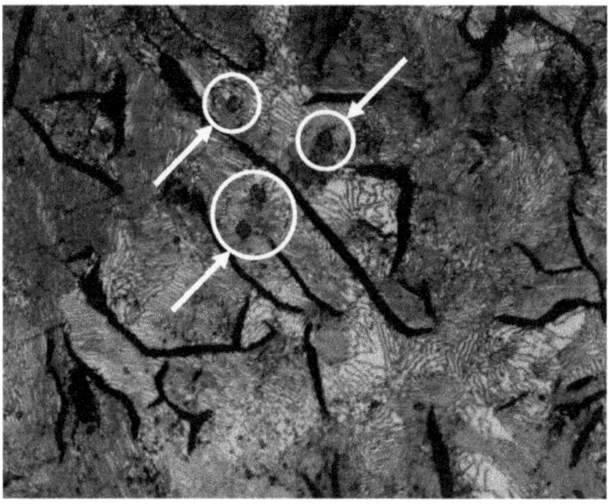

Fig. 1. *Microstructure of Gh190 cast iron, when the sulfur content is 0.12% (x500) (manganese sulfide inclusions are indicated by arrows)*

Bench true-to-life tests on the experimental ventilated brake discs show that the increase of the sulfur concentration in gray pearlitic cast iron Gh190 to 0.12%,when the manganese content is constant, makes it possible to increase the wear resistance of ventilated brake discs and to reduce the increase in the thickness variation by 30% respectively, taking into consideration that it directly influences the negative operational effect "vibration at braking"[11]. Let us designate this chemical composition of cast iron for ventilated brake discs as composition 2.

On the one hand, the decrease of the negative operational effect "vibration during braking" is associated with the decrease in the growth of the thickness difference of the working surfaces of the brake disc. On the other hand, it is caused by the improvement of the vibration-absorbing properties of cast iron. Regarding that cast iron is inherently a material with a pronounced heterogeneous structure (it has a metal matrix and graphite inclusions), it becomes clear that with the increase of sulfur content in gray cast iron (above 0.04 ... 0.05%), the appearing spherical manganese sulfide inclusions become additional structural components that enhance the damping properties of the cast iron. The mechanism for increasing the damping capacity of the cast iron with additional inclusions lies in the occurrence of local plastic deformation in the inclusion sat the soft plastic phase, that are manganese sulfides, as well as at the interphase boundary "manganese sulfide - perlite" [12].

The necessity to implement the 2nd stage of the research is due to the fact that ventilated brake discs are inferior in crack resistance to non-ventilated, solid brake discs. Due to the presence of cavities between the stiffening ribs, ventilated brake discs have improved cooling properties, but this factor, as a rule, also causes the appearance and spreading of cracks on the working surfaces of the disc when there is a lack of thermophysical characteristics of the material.

When developing cast iron for brake discs, one of the main requirements is to obtain the lowest tendency to crack. It is known that the properties of cast iron are determined by the content of carbon and silicon, as well as by a value which characterizes their combined content as a carbon equivalent [13]:

$$C_э = \%C + \frac{\%(Si+P)}{3}, \quad (1)$$

where C is carbon content, %; $Si$is Si-content, %; $P$is P-content, %.

Gray cast iron with lamellar graphite, having a constant carbon equivalent, possesses improved thermophysical characteristics (thermal conductivity, coefficient of linear expansion) with an increased content of free graphite and a decreased content of silicon [14]. At the same time, a decrease of the silicon content in the metal matrix makes it possible to weaken the process of graphitization of cast iron during heating and to increase the ferrite content.

In the range of carbon content from 3 to 4% and silicon from 1 to 3% [15], the thermal conductivity of cast iron,λ, $W(м{\cdot}K)$,can be accurately estimated using the following regression equation:

$$\lambda = 17{,}4(\%C) - 6{,}3(\%Si). \quad (2)$$

Thus, the increase in the fracture toughness of ventilated brake discs is possible due to the improvement in the thermophysical properties of cast iron with the increase in the carbon content and the decrease in the silicon content, when the carbon equivalent is constant. The optimization of the carbon equivalent of cast iron involves the correct balance of carbon and silicon content. It is verified that the higher is the carbon content and the lower is the silicon content, the greater is the strength and thermal conductivity, and due to this, the crack resistance of cast iron is better while its other characteristics and indicators are the same. The mechanism of this phenomenon is explained by the depletion of pearlite silico ferrite in silicon with a simultaneous increase in the content of free graphite, which has a beneficial effect on both the thermophysical (due to the high thermal conductivity of graphite and ferrite with a low silicon content) and the wear-frictional properties of cast iron (due to the reduced brittleness of ferrite with a low silicon content and a high content of free graphite, being a solid lubricant).

It is known that the reduction of the carbon equivalent in gray cast iron implies the use of higher grades. In this case, under the same conditions of the smelting process, the structure of gray iron undergoes qualitative changes. increase in the base metal content in gray cast iron at one and the same cooling velocity causes free cementite (chill) to appear. During the experimental work it is found that the decrease in the silicon content from 1.30 ... 1.62% to 0.95% regarding a relatively constant carbon content (3.30 ... 3.39%) dramatically changes the morphology of graphite inclusions. If uniformly distributed (type A) and rosette (type B) and interdendritic non-directional (type D) graphite inclusions initially prevail in cast iron, then the lack of silicon results in the shift of the types of graphite, they change their places, types B and D prevail, while type A is present.

Thus, during the experimental work, it is found that in practice the silicon content in the casting should not be lower than 1.3% in order to avoid the appearance of structurally free cementite (chill) in the cast iron structure. After carrying out a complex of studies of standard ∅30×220 mm samples, a work program is prepared and implemented to develop the smelting technology and optimize the carbon equivalent of gray cast iron of the pearlitic class Gh190 for ventilated brake discs.

The purpose of the research at this stage is to change the content of carbon and silicon in cast iron within the range of 3.55 ... 3.60% of C and 1.45 ... 1.50% of Si, regarding a constant carbon equivalent and the same content of other elements, in order to increase the operational characteristics of brake discs. A pilot lot of ventilated brake disc castings have the following actual chemical composition of cast iron: C – 3.59%, Si – 1.46%; Mn – 0.64%; P – 0.034%; S – 0.134%; Cr – 0.12%; Ni – 0.13%; Cu – 0.53%. Let us designate it as composition 3.1.

Comparative bench true-to-life tests of experimental ventilated brake discs and discs made from Gh190 cast iron having 0.12% sulfur content show an additional reduction in the wear of the experimental brake disc by 10%. Consequently, the increase in the wear resistance of the experienced ventilated brake discs in comparison with the serial ones made of Gh190 cast iron with a low sulfur content (up to 0.03%) accounts for 40%.

With a satisfactory value of thermal distortion and according to the results of the tests with the experimental ventilated brake discs, no cracks of a length exceeding 10 mm are found, and no through cracks are found as well. Thus, the results obtained during the tests show that the experienced ventilated brake discs have advantages in wear resistance and resistance to fatigue cracking.

However, in practice, the achievement of such carbon and silicon content in ventilated brake disc castings requires strict adherence to all technological parameters of smelting and modification of cast iron, that is why, manufacturer 1 refuses to introduce the worked out chemical composition of cast iron.

To eliminate the demonopolization of the supply of ventilated brake discs, a work with manufacturer 2 is carried out to optimize the chemical composition of brake discs in order to increase the stability of the technological process and improve the complex of operational characteristics of brake discs. Taking into account the peculiarities of the used charge materials for smelting, the following chemical composition of cast iron is proposed: C – 3.55 ... 3.60%, Si – 1.85 ... 1.90%; Mn– 0.50 ... 0.55%; P – 0.1%; S –0.11 ... 0.13%; Cr – 0.1%; Ni – 0.05%; Cu –0.65 ... 0.70%. Let us designate it as composition 3.2.

The results of the bench true-to-life tests show that the experienced ventilated brake discs in comparison with the serial brake discs made of Gh190 cast iron with a low sulfur content (composition 1) have 40% lower wear. Thus, composition 3.1 and composition 3.2 for ventilated brake discs can be considered equivalent in terms of wear resistance.

Ventilated brake discs of composition 3.2 have the same thermal distortion and crack resistance as those of brake discs of composition 3.1. Consequently, these compositions of ventilated brake discs can be considered equivalent in terms of thermophysical characteristics. A higher silicon content and a lower nickel content in composition 3.2 compared to composition 3.1 is compensated by a lower content of manganese, chromium and a higher content of copper.

The consistent development of the chemical composition of the cast iron of ventilated brake discs to improve the operational characteristics of the brake discs is shown in the diagram (Fig. 2).

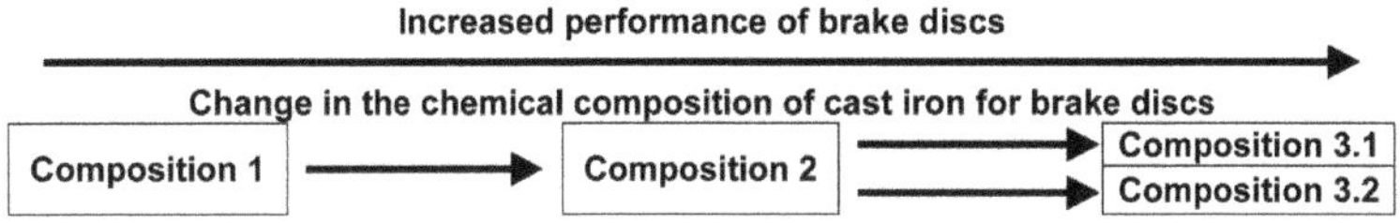

Fig. 2. *The diagram of improving operational characteristics of brake discs*

Conclusion

The optimal concentration of sulfur in cast iron is determined, which ensures its lowest wear (composition 2). The manufacture of front brake discs from Gh190 cast iron with a sulfur content of 0.11 ... 0.13% instead of 0.01 ... 0.03% is introduced, and it results in a significant increase in wear and friction properties. According to the results of bench tests showing satisfactory crack resistance, the decrease in the wear rate and the increase in the thickness difference of the brake discs is 30%.

It is established that the increase in the crack resistance of brake discs is possible due to the improvement in the thermophysical properties of cast iron with the increase in the carbon content (up to 3.55 ... 3.60%) and the decrease in the silicon content (up to 1.45 ... 1.50%) when the carbon equivalent (composition 3.1) is constant. The mechanism of this phenomenon lies in the depletion of pearlite silico ferrite with silicon while having a simultaneous increase in the content of free graphite, which has a beneficial effect both on the thermophysical (due to the high thermal conductivity of graphite and ferrite with a low silicon content) and the tribotechnical properties of cast iron (due to the reduced brittleness of ferrite with low content of silicon and a high content of free graphite, being a solid lubricant).

The chemical composition of cast iron for ventilated brake discs is optimized (it is composition 3.2), which makes it possible to obtain the same level of wear (up to 40%) and thermophysical properties (as for composition 3.1) due to the difference in the content of silicon, manganese, chromium, nickel and copper. This chemical composition is used for the manufacture of cast blanks for ventilated brake discs.

Acknowledgments

This work is carried out within a framework of the government order (No. FZRU-2020-0011) of the Ministry of Science and Higher Education of the Russian Federation.

References

[1] M.J.Haigh, H.Smales, M. Abe, Vehicle judder under dynamic braking caused by disc thickness variation, ImechE, C444/022 (1993) 247-258.

[2] G. Gigan, V. Norman, J. Ahlström, T. Vernersson, Thermomechanical fatigue of grey cast iron brake discs for heavy vehicles, Proc. of the Institution of Mechanical Engineers, Part D: J. Automobile Engineering, 233(2) (2017) 453-467. https://doi.org/10.1177/0954407017739723

[3] S.B. Istoc, H. Winner, Heat cracks in brake discs for heavy vehicles, Automotive and Engine Technology, 3 (2018) 61-68. https://doi.org/10.1007/s41104-018-0027-y

[4] F. Bagnoli, F. Dolce, M. Bernabei, Thermal fatigue cracks of fire fighting vehicles gray iron brake discs, Engineering Failure Analysis, 16(1) (2009) 152-163. https://doi.org/10.1016/j.engfailanal.2008.01.009

[5] D.A. Boldyrev, R.R. Dema, O.B. Kalugina, The microstructure and hardness of casting a solid brake disc after late graphitizing modification IOP Conf. Series: Materials Science and Engineering, 966 (2020) 012021. https://doi.org/10.1088/1757-899X/966/1/012021

[6] D.A. Boldyrev, A.N. Shapovalov, S.P. Nefedev, R.R. Dema, V.N. Kononov, M.V. Kharchenko, E.V. Suvorova and I.Yu. Markova, The electron-microscopic and x-ray spectral analysis of phase composition of CGI inoculant structure, J. Chem. Technol. Metall. 54 (2019) 348-361.

[7] D.A. Boldyrev, S.I. Platov, N.V. Urtsev, D.V. Terentyev and O.R. Latypov, Research and features of preliminary graphitizing processing of melt of iron with silicon carbide on the structure and properties of cast iron castings, IOP Conf. Series: Materials Science and Engineering, 1008 (2020) 012047. https://doi.org/10.1088/1757-899X/1008/1/012047

[8] Erokhin Yu.V., Zakharov A.V., Leonova L.V. Izuchenie veshchestvennogo sostava shlakov proizvodstva vanadievogo chuguna alapaevskogo metallurgicheskogo zavoda, Vestnik of NMSTU 18(2) (2020) 13-21. https://doi.org/10.18503/1995-2732-2020-18-2-13-21

[9] V.A. CHajkin, D.A. Boldyrev, N.V. CHajkina, Osobennosti grafitiziruyushchego modificirovaniya vysokoprochnogo chuguna smesevymi modifikatorami v usloviyah OAO "AVTOVAZ", Vestnik of NMSTU, 4 (2007)53-58.

[10] A.P. Gulyaev, Metallovedenie, Mashinostroenie, Moscow, 1977.

[11] D.A. Boldyrev, Povyshenie rabotosposobnosti i resursa pary treniya «tormoznoj disk - kolodka», PhD thesis, Togliatti, Russia, 2004, 26 p.

[12] M.S. Blanter, Yu.V. Piguzov, Metod vnutrennego treniya v metallovedcheskih issledovaniyah, Metallurgiya, Moscow, 1991.

[13] A.D. Sherman, A.A. Zhukov, CHugun, Metallurgiya, Moscow, 1991.

[14] K.Bungardt, W.Spyra, Wärmeleitfähigkeit von legierten und unlegierten Stählen und Legierunden bei Temperaturen zwischen 20 und 700°C, Arch. Eisenhüttenwesen, 26 (1971).

[15] D.Severin, U.Franke, M.Lampic, Steigerung der Lebensdauer von Bremsscheiben, ATZ,11 (2002) 1016–1023. https://doi.org/10.1007/BF03223482

Diagnostics of Polymer Composite Materials by Waveguide Method

German V. Dmitrienko[1,a*], Dmitry V. Mukhin[1] and Aleksandr A. Fedorov[1]

[1]Ulyanovsk Technical University. Institute of Aviation Technologies and Management, 13A, Sozidateley Avenue, Ulyanovsk, Russia, 432072

[a]dmitrienko.german@yandex.ru

Keywords: Polymer Composite Materials, Carbon Plastics, Dielectric Permittivity

Abstract. This article considers the use of radio wave method for testing polymer composites used as structural material for manufacture of aircrafts. Measuring process, analysis and calculation by integral equation method are described.

Introduction

Since the middle of the last century polymer composite materials are gaining widespread use and occupy an increasingly wide scope of application across the full range of technology. The most important feature of these materials is that their properties depend not only on the properties of the components they are composed of, but also on the structure of the material being created, as well as on the technology of manufacturing parts from them. This leads to the steady growth of the range of new PCMs in order to obtain the properties most desirable for use in new areas. This feature also has a downside: any violations of the production process when manufacturing parts from PCM can lead to an undesirable change in the characteristics of future parts. In order to prevent this, the process of manufacturing parts from PCM includes various methods of non-destructive testing (NDT), which allow to control the parameters of the production process of PCM parts manufacturing, to adjust the process itself to eliminate arising defects and to reject defect parts at an early stage of manufacturing, if the revealed defect does not seem possible to be eliminated. The most informative NDT methods are radio wave methods with the information parameter being the complex reflection coefficient or the complex transmission coefficient of an electromagnetic wave, depending on the structure of the polymer material.

The complex dielectric constant $\underline{\varepsilon}$ of the PCM is measured by the wave guide technique according to the method and the measuring scheme described in [1-3]. A sample of the material to be measured is placed on the open end of the waveguide as a short-circuit, and the values $\underline{\varepsilon}$ and $\tan\delta$ are calculated using the Fresnel formulas from the measured complex reflection coefficient.

The complex reflection coefficient has the form

$$\underline{R} = \frac{1-\sqrt{\underline{\varepsilon}}}{1+\sqrt{\underline{\varepsilon}}}$$

, where $\underline{R}$ is the complex reflection coefficient, $\underline{\varepsilon}=\varepsilon(1-j\tan\delta)$,

from the formula, the values of the dielectric constant - ε and the tangent of the angle of dielectric losses - $\tan\delta$ are expressed:

$$\varepsilon = \frac{(1-R^2)^2 - 4R^2\sin^2\phi}{(1+R^2+2R\cos\phi)^2}, \qquad \tan\delta = \frac{4R(1-R^2)\sin\phi}{(1-R^2)^2 - 4R^2\sin^2\phi}. \qquad (1)$$

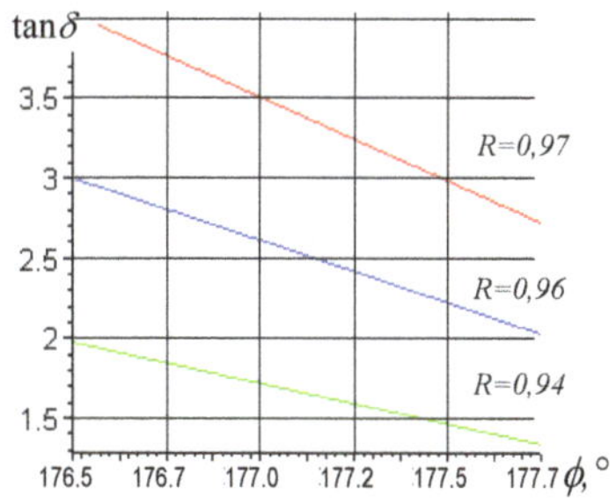

a b

Fig. 1. *Dependences of ε and $\tan\delta$ on the modulus $|R|$ and the phase ϕ of the reflection coefficient*

According to the formulas (Eq.1), the dependences of the relative permittivity ε and the tangent of the dielectric loss angle $\tan\delta$ on the modulus of the reflection coefficient and the phase of the reflection coefficient ($|R|$, ϕ) are plotted, as shown in Fig. 1. It is seen from the analysis of the dependences (Fig. 1) that the phase of the reflection coefficient tends to the phase of total reflection $\phi \to 180^\circ$; the graphs ascend very rapidly, which means that even a slight instrument error in measuring the phase ϕ will lead to a large final error in determining the relative permittivity ε.

In order to improve the accuracy of calculating and processing the obtained measurement results, a more rigorous method from an electrodynamic point of view is applied – that is the method of integral equations (IEM). The procedure for deriving integral equations, specified in [1-3], is realized by studying the surface currents induced on the surface of the PCM sample and the metal structure of the waveguide sensor by a probing electromagnetic wave.

The design of the measuring sensor is determined by the capabilities of the IEM [4-8], which allows you to numerically calculate the complex reflection coefficient in case of a sensor created on the basis of a round waveguide with a conical horn (Fig. 2). It is possible to measure the dielectric parameters of electrically thin PCM samples (conductive carbon plastics with a thickness of 1-2 mm, or fiberglass used as protective caps for the radar of an aircraft) by transmitting electromagnetic wave through the material with a receiving antenna added at the output of the measuring sensor.

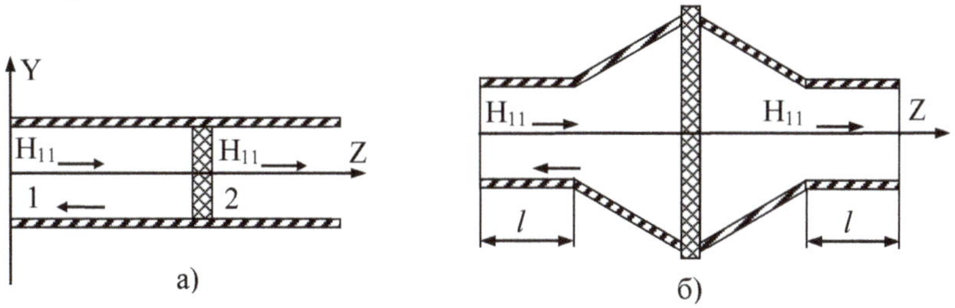

a) б)

Fig. 2. *Waveguide Transducer Design for Transmission Measurements:*
a) - on a round waveguide; b) - on a round waveguide with a conical horn.

To measure the PCM dielectric characteristics for the transmitting an electromagnetic wave through the material, an emitter with an increased power is used to facilitate such an intensity of radiation at the output from the sensor, sufficient for measuring instruments. The reason for the increase in the measurement power is caused by high attenuation of the electromagnetic wave in the PCM.

When measuring by the transmitting electromagnetic wave, the measured PCM sample is represented as an electrically thin body. The measured parameter is the complex transmission coefficient [9,10].

A system of integral equations for the outer (illuminated) and rear (shaded) boundaries of a thin body is compiled using the equations. In integral equations the described current on the illuminated surface is represented as $J^M = J^M_{пад} + J^M_{отр}$, and on the shaded surface as $J^M = J^M_{про}$. On the outer surface (*L1*) the probing wave field is created by the current, on the rear surface (*L2*) the transmitted wave will be described by the induced currents J_2^M and J_2^E. Integral equations are derived in the same manner as in [9, 10], taking into account the boundary conditions [*n*,*E*]=*Z*[*n*,[*n*,*H*]] at the boundaries *L1* and *L2*, where *n* is the outward normal directed into the PCM sample, *Z* is the diagonal matrix of the surface impedance $Z = \begin{vmatrix} Z_{11} & 0 \\ 0 & Z_{22} \end{vmatrix}$. The matrix character of the impedance shows that it can be different in different directions, in case of PCM anisotropy. The system of equations has the form:

$$-\frac{1}{2}\tilde{J}_1^M(p_1) + \int_{L1} \xi_1^M \tilde{J}_1^M(q_1)dl + \int_{L2} \xi_1^E J_1^E(q_1) - \xi_1^M \tilde{Z}\left[\vec{n}(q_1), J_1^E(q_1)\right]dl = 0 \quad (2)$$

$$\frac{1}{2}J_2^M(p_2) + \int_{L1} \xi_1^M \tilde{J}_1^M(q_2)dl + \int_{L2} \left(\xi_2^E - \xi_1^E\right)J_2^E(q_2) - \left(\xi_2^M - \xi_1^M\right)\tilde{Z}\left[\vec{n}(q_2), J_2^E(q_2)\right]dl = 0 \quad (3)$$

$$\frac{1}{2}J_2^E(p_2) + \int_{L1} h_1^M \tilde{J}_1^M(q_2)dl + \int_{L2} \left(h_2^M - h_1^M\right)J_2^E(q_2) - \left(h_2^M - h_1^M\right)\tilde{Z}\left[\vec{n}(q_2), J_2^E(q_2)\right]dl = 0 \quad (4)$$

where p_1, q_1 are points on the boundary *L1*, and a p_2 q_2 on the boundary *L2*.

Equation (Eq.2) for the outer surface L1 is similar to the equations for a homogeneous semi-infinite body. The last integral term on the left side of it considers the reflection from the rear surface. Equations (Eq.3) and (Eq.4) describe the field of the transmitted wave, with the first integral terms considering the influence of the outer surface. The boundary conditions $\tilde{J}^M = -\tilde{Z}\left[n, J^E\right]$, $\tilde{Z} = Z/\sqrt{\mu_0/\varepsilon_0}$ are satisfied on both surfaces. The resulting equations are Fredholm equations of the second kind.

Integral equations are solved by a numerical method (Galerkin's method) [4,11-16].

The PCM sample was tested according to the proposed theory. Based on the measurement results, the value of the relative dielectric constant ε=940±30% and the tangent of the dielectric loss angle tan*δ*=1,3±40%, with a reflection coefficient modulus being 0.94, were determined.

References

[1] A. A. Brandt, *Investigations of Dielectrics at Ultrahigh Frequencies* [in Russian], Moscow: GIFML (1963) p.404

[2] G.V. Belov, *Technical quality control of products made of carbon materials* [in Russian], Moscow: *Izdatel'stvo standartov* (1991) p.92.

[3] G.V. Dmitrienko *Method for measuring the complex reflection coefficient of low-impedance composite materials in the microwave range using intermediate standard*s [In Russian], Metrologia (2008) No 4. pp. 31-38.

[4] *E.N. Vasiliev Excitation of bodies of revolution* [In Russian] Moscow, Radio and communication (1987) p. 271.

[5] M. Zh. Kislyuk, *Method of Secondary Waves in Problems of Electrodynamics* [in Russian], LÉTI, Leningrad (1970) p.151.

[6] G.V. Dmitrienko, A.A. Fedorov and G.L. Rivin, *Radio-Frequency Method for Diagnostics of Aeronautical Polymer Composite For Newly Launched Products* [in Russian] FGUP „VIAM" (2018) ISBN 978-5-905217-30-2. https://conf.viam.ru/conf/284/proceedings

[7] C.A. Fletcher, *Computational galerkin methods* [in Russian] *Izdatel'stvo Mir*, (1988) p. 347

[8] V.A. Dandarov, *Numerical solution of the 2-dimensional integral Fredholm equation in the problem of radio wave propagation over an irregular earth surface* [in Russian] *Radiotekhnika i elektronika* (2002) Vol.47, No.11, pp.1329-1334

[9] G.V. Dmitrienko, V.G. Anisimov, *Measurement of the dielectric constant of low-impedance composite materials at microwave frequencies* [in Russian] *Izmeritel'naya tekhnika* (2009) No. 2. https://doi.org/10.1007/s11018-009-9236-0

[10] G.V. Dmitrienko, N.A. Trefilov, RF Patent No 2194285 IPC[7] G01 R 27/04. *A method for determining large values of the complex dielectric constant of impedance materials*, applied for 13.03.2001; published on 10.12.2002, Bul No.34.

[11] G.V. Dmitrienko, L.V. Fedotov, *Quality control process of low-impedance composite materials at microwave frequency* [in Russian] *Kachestvo, innovatsii, obrazovanie* (2009) No.10

[12] G.D Malushkov, *Methods of solving problems of the electromagnetic excitation of bodies of revolution* (January 1975) · *Izvestiya Vysshikh Uchebnykh Zavedenij*. https://doi.org/10.1007/BF01037102

[13] A. F. Verlan and V. S. Sizikov, *Integral Equations: Methods, Algorithms, Programs* [in Russian], *Naukova Dumka*, Kyiv (1986) p.544

[14] N. A. Khizhnyak, *Integral Equations of Macroscopic Electrodynamics* [in Russian], *Naukova Dumka*, Kiev (1986) p.256

[15] VV Nikol'sky, Projection methods in electrodynamics [in Russian] Prikladnaya elektrodinamika (1977)

[16] *P.V. Novitsky* and I.A. Zograf, *Evaluation* the *Errors of Measurement Results*, Moscow, Energoatomizdat Publishing (1985) p.247.

Mechanism and Kinetics of Thermal Embrittlement of Austenitic Chromium-Nickel Steel during Long-Term High-Temperature Holding

Kirill A. Okhapkin

NRC "Kurchatov Institute" - CRISM "Prometey", 49 Shpalernaya str., St. Petersburg, Russia, 191015

kirill.okhapkin@mail.ru

Keywords: Austenitic Steel, Thermal Embrittlement, Mechanism, High-Temperature Holding

Abstract. This article describes a model of structural changes in austenitic chromium-nickel steel during long-term thermal soakings. The time for reaching equilibrium content of $Cr_{23}C_6$ carbide at a temperature of 600 °C is calculated. The calculation results are compared with experimental data on the kinetics of thermal embrittlement of steel at a temperature of 600 °C after thermal holding for up to 15000 hours.

Introduction

Long-term high-temperature operation under conditions of high diffusion mobility of alloying elements in steel leads to changes in the structure and, as a consequence, in the mechanical properties of structural materials. Periodic destructive testing of equipment elements of reactor facilities makes it possible to assess the residual level of properties at the time of testing. To be able to predict the level of mechanical properties during further operation (after destructive testing), it is necessary to predict the structural and phase transformations occurring in the material.

The purpose of this work is a computational and experimental assessment of the kinetics of the process of structural transformations and thermal embrittlement in austenitic steel grade 09Cr18Ni9 during long-term high-temperature holding.

Mechanism of thermal embrittlement and a description of the calculation model

To describe the mechanism of thermal embrittlement, it is necessary to establish the relationship of structural changes with a decrease in the characteristics of impact toughness over time.

Secondary phases in steel grade 09Cr18Ni9 during long-term high-temperature holding are formed in the following order. First, due to the limited solubility of carbon, it is released from the solid solution with the formation of chromium carbide $Cr_{23}C_6$. This leads to the formation of areas with a low chromium content near the carbide precipitates. It is known [1] that chromium with its content in steel ~ 17% in combination with nickel, in contrast to its usual ferritizing action, strongly contributes to the formation of an austenitic structure as a result of complex interatomic interaction. Therefore, in zones depleted in chromium, austenite is transformed into ferrite (α). In this case, the excess nickel content by diffusion can pass into the forming G-phase.

To confirm the stated mechanism, an experiment was carried out on a metal with different carbon content: 0.07 and 0.09 wt. %. The samples were exposed to heat holding for various durations at a temperature of 600 °C, during aging, the content of the α-phase was determined (Fig. 1). In the material with a carbon content of 0.07 %, the α-phase was formed significantly less, which confirms the relationship between the precipitation of chromium carbide and the formation of the α-phase, since at a carbon content of 0.07 and 0.09 wt. %, the amount of $Cr_{23}C_6$ at a temperature of 600 °C in an equilibrium state according to the data of thermodynamic modeling is also significantly different and amounts to 0.87 and 1.57 wt. % respectively.

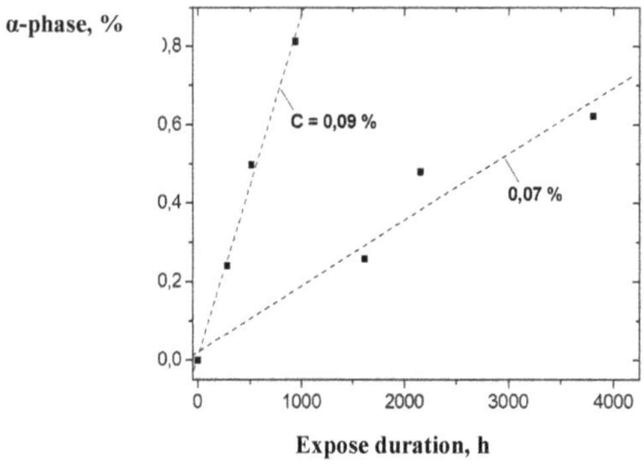

Fig. 1. *Influence of carbon content on the kinetics of α-phase precipitation at a temperature of 600 °C.*

Of the secondary precipitates found in steel, only chromium carbide has an embrittling effect. Therefore, the formation of chromium carbide will determine the kinetics of thermal embrittlement. The formation of particles of the carbide phase will proceed until the equilibrium content is reached at a given temperature. According to the results of thermodynamic modeling, the equilibrium content of $Cr_{23}C_6$ at a temperature of 600 °C is 1.57 %. Let's assume that all grains in the material structure are spherical and of the same size. Under these conditions, the process of carbide formation in each of the grains will proceed in the same way. In the range of operating temperatures, the carbide phase is formed mainly along the grain boundary. Fig. 2 shows a diagram of the formation of a layer of carbide with a thickness of δ along the grain boundary γ - austenite in the direction v from the boundary to the center of the sphere.

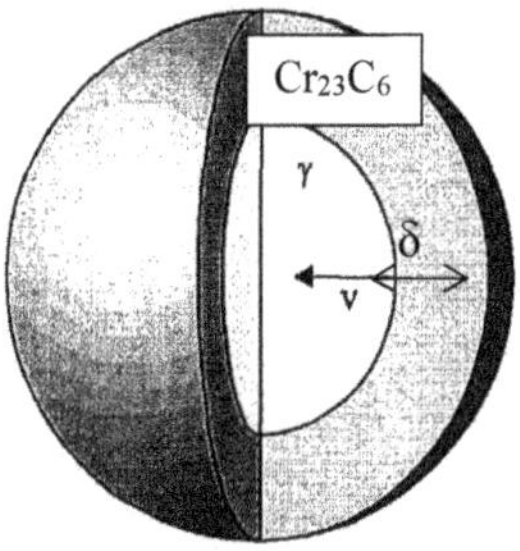

Fig. 2. *Scheme of the growth of the carbide phase along the grain boundary of austenite.*

The thickness of the carbide layer δ is related to the volume fraction by the following relationship [2]:

$$V^{\kappa} = 1 - \left(\frac{d_{\gamma} - 2\delta}{d_{\gamma}} \right)^3, \tag{1}$$

where V_{κ} is the volume fraction of the carbide phase, d_{γ} is the austenite grain diameter.

The calculation of the thickness of the layer of the carbide phase at each moment of time can be carried out on the basis of the analysis of the physical and mathematical model of chromium diffusion. A diagram of the growth process of the carbide phase, limited by chromium diffusion, is shown in Figure 3 [3]. There is a transition region between carbide and ferrite, but for simplicity, a sharp phase boundary is assumed in the scheme, at which the chromium concentration changes abruptly from the equilibrium value in ferrite (C^{α}_{Cr}) to a similar value in $Cr_{23}C_6$ carbide (C^{κ}_{Cr}). At a considerable distance from the center of the new phase, the chromium concentration retains its initial value (C^{γ}_{Cr}).

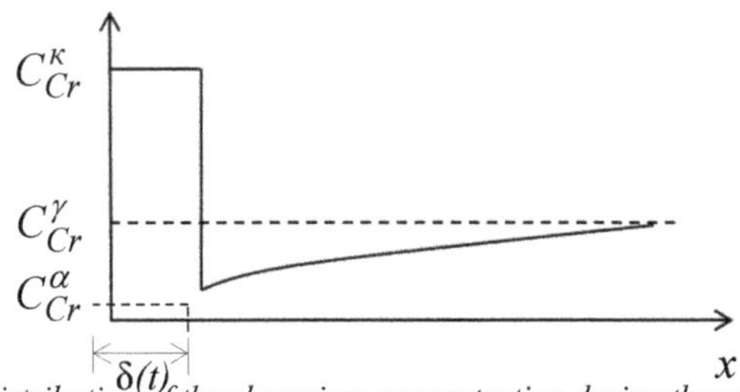

Fig. 3. *Schematic distribution of the chromium concentration during the growth of the carbide phase layer [3].*

The mathematical formulation of the growth problem for a layer of a new phase is given, for example, in [4]. The redistribution of chromium in the region $\delta(t) < x \le \infty$ is described by the diffusion equation:

$$\frac{1}{D}\frac{\partial C}{\partial t} = \frac{\partial^2 C}{\partial x^2} \tag{2}$$

with the following boundary conditions:

$$\lim_{x \to \infty} C(x,t) = C^{\gamma}_{Cr};\ C(\delta(t),t) = C^{\alpha}_{Cr};\ C(x,t) = C^{\kappa}_{Cr},\ \text{at } x < \delta(t);\ \frac{\partial C(\infty)}{\partial x} = 0.$$

A mass balance must be observed on the surface of the new phase layer:

$$\left(C^{\kappa}_{Cr} - C^{\alpha}_{Cr}\right)\frac{d\delta(t)}{dt} = D\left.\frac{\partial C}{\partial x}\right|_{x=\delta(t)} \tag{3}$$

Materials Research Proceedings **21** (2022) 99-103 https://doi.org/10.21741/9781644901755-18

Based on the solution of the indicated boundary value problem in [5], an expression was obtained for calculating the thickness of the new phase:

$$\delta(t) = 2\frac{\left(C_{Cr}^{\gamma} - C_{Cr}^{\alpha}\right)}{\left(C_{Cr}^{\kappa} - C_{Cr}^{\alpha}\right)}\sqrt{\frac{Dt}{\pi}} \quad (4)$$

$$t_p = \frac{\pi}{D}\left(\frac{d_{\gamma}\left(1-\sqrt[3]{1-V^{\kappa}}\right)}{2\frac{C_{\text{гр}} - C_{Cr}^{\gamma}}{C_{Cr}^{\kappa} - C_{\text{гр}}}}\right)^2 \quad (5)$$

Results and discussion

For the calculation, the following values of the parameters included in the expression (5): D_{Cr} = 8,1*10^{-20} m^2s^{-1} [6]; d_{γ} = 66,5 μm; $C_{Cr}^{\gamma} = 14{,}76\%$; $C_{\text{гр}}$ = 0,1* C_{Cr}^{γ}=1,476 %; $C_{Cr}^{\kappa} = 81{,}95\%$. The time to reach the equilibrium content of $Cr_{23}C_6$ carbide at a temperature of 600 °C, according to the calculation results, was 12164 hours. To verify the calculated data, an experiment was carried out, during which thermal soakings were carried out at a temperature of 600 °C for various durations, followed by determination of the impact toughness (Figure 4). The cessation of the precipitation of the $Cr_{23}C_6$ carbide leads to the stabilization of the values of the impact toughness. According to the calculation data, the time to reach the equilibrium carbide content is 12164 h, according to the experimental data, the thermal embrittlement process is stabilized at a holding time of 11000 h. Thus, the calculation results are in good agreement with the experimental data, the discrepancy is 11 %, while the calculated estimate shows a longer value of time reaching an equilibrium content of the carbide phase, therefore, it is conservative.

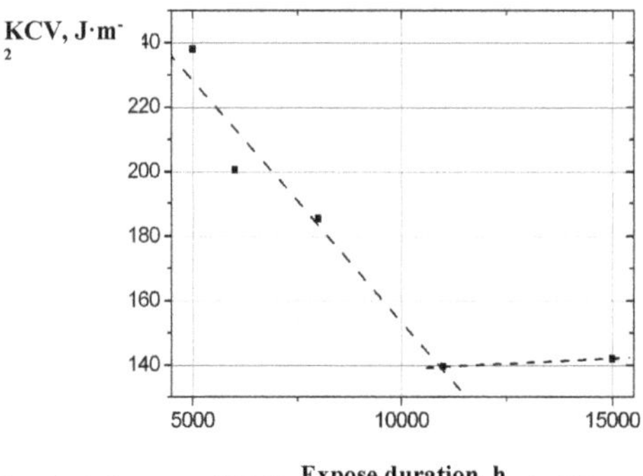

Fig. 4. *Thermal embrittlement of 09C[illegible] holding at a temperature of 600 °C.*

Summary

The time for reaching equilibrium content of $Cr_{23}C_6$ carbide at a temperature of 600 °C is calculated. According to the calculation data, the time to reach the equilibrium carbide content is 12164 h, according to the experimental data, the thermal embrittlement process is stabilized at a holding time of 11000 h. Thus, the calculation results are in good agreement with the experimental data, the discrepancy is 11 %, while the calculated estimate shows a longer value of time reaching an equilibrium content of the carbide phase, therefore, it is conservative.

Acknowledgements

Experimental studies were carried out on the equipment of the Core shared research facilities "Composition, structure and properties of structural and functional materials" of the NRC «Kurchatov Institute» - CRISM "Prometey" with the financial support of the state represented by the Ministry of Education and Science of the Russian Federation under agreement No. 13.CKP.21.0014 (075-11-2021-068). The unique identifier is RF----2296.61321X0014.

References

[1] Basics of alloying steel. Meskin V.S. Moscow, State Scientific and Technical Publishing House of Literature on Ferrous and Non-Ferrous Metallurgy, 1959. - 688 p.

[2] Y. van Leeuwen, J. Sietsma, S. Vooijs, S. van der Zwaag Effect of the Geometry in Modelling the Kinetics of Diffusional Transformations in Substitutional Iron Alloys // Proceedings from Materials Solutions '97 on Acceleraed Cooling/Direct Quenching Steels 15-18 September 1997, Indianapolis, Indiana, p. 59 – 66.

[3] Formation of chemical microheterogeneity in cast alloys. G.P. Anastasiadi. - SPb.: Polytechnic, 1991. - 148 p.

[4] Lyubov B.Ya. Diffusion processes in inhomogeneous solid media. - M.: Science. Main edition of physical and mathematical literature, 1981.

[5] A.I. Rudskoy, G.P. Anastasiadi, A.S. Oryshchenko, S.Yu. Kondratyev, M.D. Fuchs Features of structural changes in heat-resistant alloy 45Cr26Ni33Si2Nb2 at operating temperatures. Message 3: Mechanism and kinetics of phase transformations // Scientific and technical bulletin of SPbSPU. Science and education 3 - 2'2012, p. 143 - 150.

[6] Shao Bo Ping, Fei Xie, Rui Kun Wang, Zhi Jun Zheng, Yan Gao Diffusion Kinetics of Chromium in a Novel Super304H Stainless Steel // High Temperature Materials and Processes, 2017, Volume 36, Issue 2, pp. 175 – 181. https://doi.org/10.1515/htmp-2015-0227

Materials Research Proceedings 21 (2022) 104-108 https://doi.org/10.21741/9781644901755-19

Plastic Folding of Bar Stocks of Round and Barrel-Shaped Sections

O.S. Zhelezkov[1,a], A.A. Laktyushin[1,2,b], R.R. Dema[1,c*] and B.B. Makarov[1,d]

[1] Nosov Magnitogorsk State Technical University, 455000, Magnitogorsk, ave. Lenin, 38, Russia

[2] Holding ARS LLC, 127254, Moscow, Ogorodny proezd 5, building 6, Russia

[a]ferumoff@mail.ru, [b] Lotos77707@inbox.ru, [c]demarr78@mail.ru, [d]latolegraf@list.ru

Keywords: Plastic Bending, Plastic Folding, Bar Stock, Round and Barrel-Shaped Section, Bending Angle, Tool Movement, Finite Element Modeling, Deformation Force

Abstract. Research has been carried out on the process of plastic folding of bar stocks with round and barrel-shaped cross-sections. The dependence of the movement of the movable tool on the bending angle has been established. The force parameters of the deformation process and the stress-strain state in the bending workpiece are determined based on the results of finite element modeling of plastic bending.

Introduction

Rod products formed using plastic bending operations are widely used in various industries, railway transport, construction, etc. In particular, spring clips are widely used in modern railway track structures [1, 2], bent reinforcing wire in the form of clamps is used in reinforced concrete structures [3], and bent welded links are used in lifting devices for various purposes [4].

Either standard presses with special technological equipment in the form of bending dies or special bending machines and mechanisms are used to implement the bending processes.

Plastic bending is a shaping operation in which there is a change in the curvature of the bar axis of the workpiece [5,6]. In the process of bending, only a part of the workpiece in the form of a deformation zone is subject to plastic deformation. The rest of the bar of the workpiece either does not deform or deform elastically. Uneven stresses and deformations occur along the section of the deformation zone. Linear deformation along the workpiece axis is practically absent.

Materials and Methods

Folding finds wide application among the variety of plastic bending methods [7-10]. In fig. 1 shows a diagram of plastic folding of a bar stock. The original workpiece 1 with a section radius r is fixed on the supporting surface of the stationary tool 2, the working cylindrical surface of which has a radius R_1. Deformation of the workpiece is carried out by a movable working tool 3, the working surface of which is made in the form of a cylinder with a radius of R_2. The original length of the deformable section of the workpiece is l. When moving the working tool by the value h, the workpiece is bent through an angle α.

Based on the analysis of the geometric relationships of the sizes and parameters of the bending process, the dependence of the movement of the working tool h on the angle α was established

$$h = [l - (R_1 + r)\cdot \mathrm{tg}(\alpha / 2)]\mathrm{tg}\alpha - (R_2 + r)(1 / \cos\alpha - 1), \quad (1)$$

where l is the initial length of the deformable section of the workpiece;
R_1 is the radius of the bearing surface of the stationary tool;
R_2 is the radius of the working surface of the movable tool;

Materials Research Proceedings **21** (2022) 104-108 https://doi.org/10.21741/9781644901755-19

r is the radius of the cross section of the original workpiece.

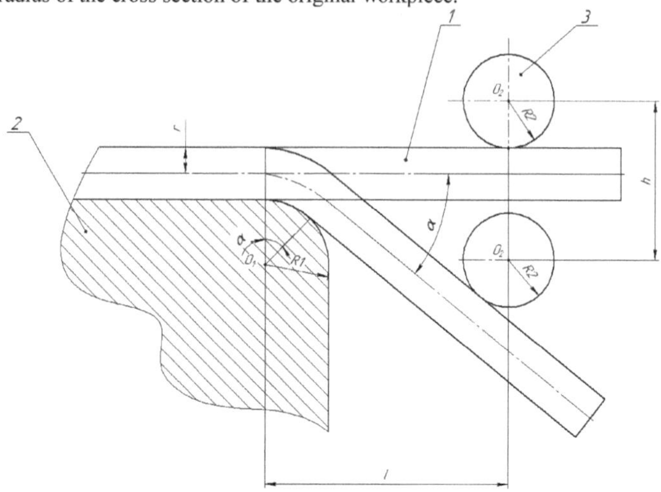

Fig. 1. *Folding scheme*

Computer modeling of the process of folding of bar stocks with round and barrel-shaped sections using the finite element method based on the widely used DEFORM-3D software package [11-16] has been performed. Fig. 2 shows the cross-sections of bent workpieces. The main dimensions of the sections are d = 17 mm; h = 13 mm; a = 12 mm; b = 21 mm. The cross-sectional areas are close in size for given dimensions. The moment of inertia of the area relative to the x-axis is minimal, and relative to the y-axis is maximum for a barrel-shaped section.

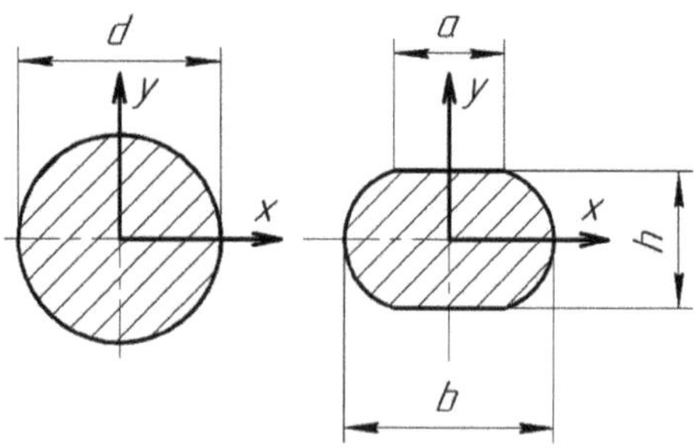

a *b*

Fig. 2. *Cross-sections of bar stocks:*
a – round section; b – barrel-shaped section.

The workpiece material is steel 40S2A (GOST 14959-2016), which is used in the manufacture of spring clips for fastening rails. The mechanical properties and the hardening curve of 40S2A steel were determined from the results of tensile tests [3].

Results and Discussion

Fig. 3 shows the position of the workpiece and the tool at the intermediate stage of the cantilever bending process and the stress intensity distribution field σ_i in the deformation zone.

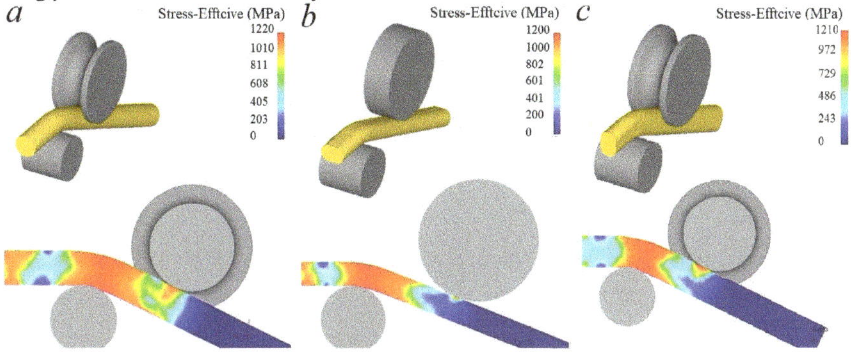

Fig. 3. *The position of the workpiece and the tool at the intermediate stage of deformation and the stress intensity distribution field* σ_i *in the plastic zones:*
a - round section; b - barrel-shaped section in bending about the x-axis;
c - barrel-shaped in bending about the y-axis

The graphs of the dependence of the bending force *P* on the movement of the movable tool *h* were built, which are shown in Fig. 4 based on the results of the calculations. Curve *a* is bending of a round section, curve *b* is bending of a barrel-shaped section about the *x*-axis, curve *c* is bending of a barrel-shaped section about the *y*-axis.

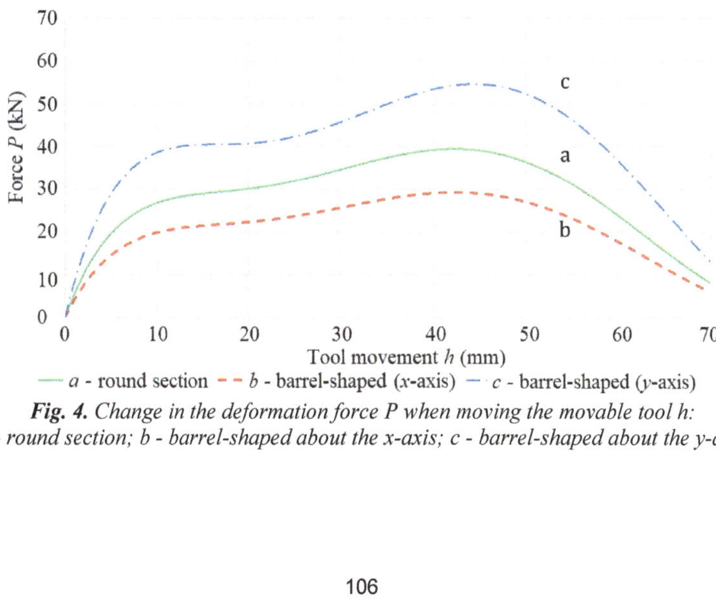

Fig. 4. *Change in the deformation force P when moving the movable tool h:*
a - round section; b - barrel-shaped about the x-axis; c - barrel-shaped about the y-axis

The polynomial of the 6th degree was used to describe the nature of the change in deformation forces depending on the movement of the movable tool.

$$P = a_1h^6 + a_2h^5 + a_3h^4 + a_4h^3 + a_5h^2 + a_6h. \quad (2)$$

Equation (2) quite accurately describes the nature of the change in the bending force P from the movement of the movable tool h, since the coefficients of determination have a high value (R^2 = 0.995…0.998). The values of the coefficients of the polynomial (2) are presented in the table 1.

Table 1. *Coefficients of equation (2)*

Section	Coefficient values					
	a_1	a_2	a_3	a_4	a_5	a_6
Round	-10^{-8}	$3 \cdot 10^{-6}$	-0,0004	0,0187	-0,4692	5,8317
Barrel-shaped (x-axis)	-10^{-8}	$4 \cdot 10^{-6}$	-0,0005	0,0277	-0,7138	8,699
Barrel-shaped (y-axis)	$-8 \cdot 10^{-9}$	$2 \cdot 10^{-6}$	-0,0003	0,0138	-0,3472	4,3154

Conclusion

Based on the research results, the following conclusions can be drawn.

1. Based on the results of the analysis of the geometric relationships of the sizes and parameters of the bending process, the dependence of the movement of the working tool h on the bending angle α was established.
2. Finite element modeling of the process of plastic folding of bar stocks with round and barrel-shaped sections has been carried out. It was found that when bending the barrel-shaped section relative to the x-axis with a minimum moment of inertia, the maximum bending force is 2.4 times less than when bending the round section and 1.88 times less than when bending a barrel-shaped section relative to the y-axis with a maximum moment of inertia.
3. The obtained research results are recommended to be used when choosing equipment for plastic bending of rod products, when designing a bending tool and assessing its service life.

Acknowledgments

This work is carried out within a framework of the government order (No. FZRU-2020-0011) of the Ministry of Science and Higher Education of the Russian Federation.

References

[1] I.N. Eisenmann, Improvement of the upper structure of the railway track, ZHeleznye dorogi mira, 12 (1997) 61-65.

[2] V.M. Fedin, A.I Borts., V.V. Kuznetsov et al., Production technology of elastic terminals for rail fasteners ZhBR-65: ways of improvement, Vesti VNIIZhT, 5 (2005) 21-27.

[3] A.N. Malakhova, Reinforcement of reinforced concrete structures, Stroitel'stvo, Moscow, 2014.

[4] B.P. Dodonov, V.A. Lifanov, Hoisting and transport devices, Mashinostroenie, Moscow, 1990.

[5] M.I. Lysov, Theory and calculation of the processes of manufacturing parts by bending methods, Mashinostroenie, Moscow, 1971.

[6] S.I. Vdovin, N.V. Petrov, Engineering calculations of bending operations, Kuznechno-shtampovochnoe proizvodstvo. Obrabotka materialov davleniem, 12 (2004) 26-28.

[7] V. Kobelev, Elastoplastic Stress Analysis and Residual Stresses in Cylindrical Bar Under Combined Bending and Torsion, J. Manuf. Sci. Eng. 133(4) (2011) 044502. https://doi.org/10.1115/1.4004496

[8] S. Mróz, A. Milenin, Numerical modelling of the metal flow and stock bending during the rolling of unequal angle bar, J. Materials Processing Technology, 177(1–3) (2006) 561-565. https://doi.org/10.1016/j.jmatprotec.2006.04.027

[9] A.V. Bobylev, A.V. Kozlov, S.P. Maksimov, E.V. Khaliulin, Study of thermal phenomena during pipe bending with rolling, Vestnik of NMSTU, 16(3) (2018) 87-97. https://doi.org/10.18503/1995-2732-2018-16-3-87-97

[10] K.A. Mironov, A.V. Kozlov, V.G. Sherkunov, A.L. Suvorov, The study of power characteristics during pipe bending with burnishing, using an automated module and a PC, Vestnik of NMSTU, 50(2) (2015) 45-48.

[11] D.H. Norrie, G. de Vries, The Finite Element Method – Fundamentals and Applications, Academic Press, New York, 1973.

[12] R.R. Dema, R.N. Amirov, O.R. Latypov, Mathematical model for assessing the management of quality parameters of hot-rolled strips according to the criterion of local thickness variation, Materials Today: Proceedings. ICMTMTE 2019, (2019) 2417-2421. https://doi.org/10.1016/j.matpr.2019.08.047

[13] S.I. Platov, R.R. Dema, M.V. Kharchenko, R.N. Amirov, Experience of application of liquid lubricating materials during wide strip hot rolling, IOP Conference Series: Materials Science and Engineering, 287 (2018) 012011. https://doi.org/10.1088/1757-899X/287/1/012011

[14] R.R. Dema, V.V. Alontsev, O.B. Kalugina, A.N. Shapovalov, Computer simulation and research of the hot rolling process in Deform-3D, Materials Today: Proceedings. ICMTMTE 2019, (2019) 2312-2315. https://doi.org/10.1016/j.matpr.2019.07.677

[15] A.G. Vinogradov, O.S. Zhelezkov, K.Yu. Morozov et al., Investigation of the process of bending of ICR spring terminals using the finite element method, Processing of solid and layered materials, NMSTU, Magnitogorsk, (36) 2010 68-70.

[16] V.A. Arzamastseva, O.S. Zhelezkov, A.A. Laktyushin, Geometric parameters of the non-contact zone during double-support plastic bending of rod blanks, Theory and technology of metallurgical production, 32(1) (2020) 31-34.

Locomotive Reliability Assessment Method

Pavel Gubarev[1,a*], Dmitry Glazunov[1,b] and Vladimir Krotov[1,c]

[1]Rostov State Transport University, Rostov-on-Don, 344038, Russia

[a]pavel.gybarev@yandex.ru, [b]glazunovdm@yandex.ru, [c]tehn_met@rgups.rul

Keywords: Rolling Stock, Diagnostic Methods, Electrical Equipment, Reliability, Repair, Technical Diagnostics

Abstract. The authors describe the analysis of the current state of the problem under consideration. A definition of "averaged failure flow parameter" is given. The periods of traction rolling stock life cycle are considered. The assumption of event distribution laws exponentiality is introduced, which makes it possible to obtain expressions of the main reliability indices in the analytical form. The work of depot service locomotives to ensure the required reliability and readiness of the rolling stock during their normal operation has been assessed. The introduction of the term "readiness" into the modern practice of traction rolling stock reliability estimation is considered. The initial data for calculating the indexes of locomotive uptime and readiness are presented. Calculated values of readiness and no-failure indices of electric locomotives in operation are obtained. The calculated values of internal and technical availability coefficients are compared with similar indicators established by technical specifications. Control procedures were performed to determine the compliance of each set of locomotives (EP1, 2ES4K) with the uptime requirements. As a result of comparing the calculated values of internal and technical availability factors (for electric locomotives EP1 and 2ES4K with analogous values set by specifications (EP1 and 2ES4K) it was determined that the surveyed locomotives comply with the established availability requirements. As a result of control procedures to determine the compliance of each set of EP1 and 2ES4K locomotives with the uptime requirements, it was determined that the set of 2ES4K electric locomotives for the run in question does not fully comply with the uptime requirement. And the set of EP1 electric locomotives meets the reliability requirements, but the error value is higher than 20%. To clarify both events, it is necessary to increase the mileage interval of the locomotives and repeat the procedure for determining compliance with the uptime requirements. The method of assessing the uptime and readiness of locomotives during their normal operation makes it possible to identify existing shortcomings in the operation of rolling stock and to form measures to improve the quality of rolling stock operation.

Introduction

In recent years, a lot of work has been done to improve reliability of locomotives and to improve their maintenance and repair [1-4]. However, ensuring the required level of reliability for complex equipment and apparatus with the existing means of control is a very difficult task. The majority of equipment failures nowadays are detected visually during direct external inspection. In this regard, the need has arisen to create methods and devices that allow detecting and eliminating defects before the failure of an individual unit or the entire locomotive. Technical diagnostics makes it possible to detect defects and predict the condition of a monitored unit or the whole system. The greatest effect can be achieved by creating and introducing computer-based diagnostic tools in each locomotive depot and improving the system of preventive maintenance [5-7].

Reliability of rolling stock is manifested as an objective result of a variety of acting random factors, the laws of change of which cannot be established only by classical methods of functional analysis or mathematical physics. These laws need special research methods. The existing methods of research and calculation of electric locomotive uptime can provide serious assistance in the operation of traction rolling stock for a comprehensive technical and economic assessment and development of measures to improve their quality [8-10]. Therefore, the use of special methods of calculating locomotives for reliability and non-failure is important. One of such methods is the assessment of reliability and readiness of locomotives during normal operation. This method consists in evaluating the impact on the reliability and readiness of the main parts and assemblies of a locomotive during its normal operation, the data on the reliability of which allows calculating the total reliability of a locomotive

Statement of the essence of the solution to the problem

After the locomotives have been built during the warranty period and subsequent operation, each of them experiences failures. Scheduled and unscheduled repairs restore their serviceability, and the locomotives continue their operational work. The alternating failures and restores of locomotives create some streams of failures and restores as a kind of streams of random events [11, 12]. As a quantitative indicator of the locomotive failure flow, the "averaged failure flow parameter" - the mathematical expectation of the number of failures N of locomotives per run - is used [13]. The averaged parameter of the flow of failures ω(L) is used as the main indicator of locomotive uptime. It characterizes the average number of locomotive failures over a sufficiently small run interval ΔL and is defined as:

$$\omega(L) = \lim_{\Delta L \to 0} \frac{\Sigma\{N(L+\Delta L)\} - \Sigma\{N(L)\}}{\Delta L} \quad (1)$$

The locomotive life cycle period is divided into two periods. The first, established between the Supplier and the Recipient, is the period of warranty operation of the locomotive. The second is the period of normal post-warranty operation. Failure flow during the period of post-warranty operation is assumed to correspond to the simplest Poisson flow of failures, in which the distribution of operating time between failures is quite close to the exponential one. The simplest Poisson flow is characterized by constancy of density of distribution of failures, mutual independence of failures, the failures come not in groups, but one by one. Introduction of an assumption about exponentiality of event distribution laws allows us to get expressions of the main reliability indices in an analytical form. The exponential law plays the same role in reliability theory as Ohm's law in electrical engineering [14,15]. Most electric circuits contain non-linear elements, which complicates the calculation of such circuits. However, the assumption of a linear character of the current-voltage relationship for all circuit elements, which reflects Ohm's law, greatly simplifies the calculations. Of course, the results of such calculations will be approximate. It will require more detailed specification of the results by other methods and circuits and reliability indicators [16].

A characteristic sign of exponentiality of locomotive failure distribution law will be:

$$T = \sigma_t$$

where T - is the mean value of the locomotive's operating time before failure, $T = 1/\lambda$;

σ_t -is the mean square deviation of failures, $\sigma_t = \sqrt{D}$;

λ - failure rate;

D_t - dispersion of operating time to failure, $T = 1/\lambda^2$.

The statistical estimation of the parameter of failure flow for changing number of locomotives during the run interval ΔL is defined as [3, 6]

$$\acute{\omega} = \Delta N / \sum_{i=1} L_i \tag{2}$$

- ΔN - the number of failures of all M locomotives in the interval ΔL;

L_i - total mileage of the set of locomotives;

The upper bound of the one-way interval for the parameter of the failure flow

$$\omega_{1-\beta} = \omega \cdot \frac{x_\gamma^2}{2 \cdot \Delta N}, \tag{3}$$

where x_γ^2 - value of the "xu-square" function of the distribution at a confidence probability γ and number of degrees of freedom $n-1$.

The error of an estimation of parameters of a stream of failures

$$\varepsilon = \frac{\omega_{1-\beta} - \omega}{\omega} \cdot 100\%, \tag{4}$$

where $\omega_{1-\beta}$ - is the value of the locomotive failure flow parameter, corresponding to the upper boundary of the one-sided confidence interval.

The conditions for making a decision on the compliance of locomotives with the established requirement of no-failure performance

$$\omega_{1-\beta} \le \omega_\beta \text{ and } \varepsilon \le \varepsilon_3,$$

where $\omega_\beta = \omega_{TY}$ is the rejection value of the locomotive failure rate parameter according to the technical specifications (TS);

ε_3 - estimation error for making a decision on compliance of locomotives with the established requirement of no-failure performance.

According to RD 50-690-89 [19]

-error of parameter estimation ; $\omega \le \varepsilon_3 \le 20\%$ (5)

-confidence probability ; $\gamma = 0,8$ (6)

The term "readiness" is widely used in the modern practice of locomotive reliability assessment. Before 2011, domestic normative documents on terminology in the field of technical reliability and reliability of traction rolling stock (TCS) did not contain the term "readiness" [17,18]. According to the "Methodological provisions..." [13], introduced in 2008 in Russia, the readiness of locomotives, taking into account the European practice, is defined as the ability of a locomotive to perform a required function within a given time interval (year, half a year etc.) on condition it is provided with required resources (materials, spare parts etc.). Quantitatively, the readiness of a locomotive, in general terms, can be defined through the readiness factor k_g

$$k_g = T_{pc} \Big/ \left(T_{pc} + T_{nc}\right) \tag{7}$$

where T_{pc} - is the dwell time of the locomotive in an operable condition;

T_{nc} - is the time of the stay of the locomotive in a non-operational state;

$\left(T_{pc} + T_{nc}\right)$ - time fund of the period under consideration.

The readiness of a locomotive is characterized by statistical estimates of values of three coefficients - internal k_{vp}, technical k_{tg} and operational (logistical) $k_{во}$ readiness. If the summand in the time $\left(T_{pc} + T_{nc}\right)$ budget T_{nc} corresponds to the downtime of the locomotive for scheduled repairs, then $k_g = k_{vg}$; if the summand T_{nc} includes the downtime for scheduled and unscheduled repairs, then $k_g = k_{tg}$; if the logistic delays of the locomotives are added to the mentioned downtime, then $k_g = k_{vO}$. Coefficients of internal k_{vg} and technical k_{tg} readiness in accordance with requirements of Russian Railways are normalized and specified in technical specifications (TS) for the locomotive. For example, for freight locomotives 2ES4K, 2ES5K the coefficient of internal availability in accordance with TS is 0,97 and the coefficient of technical availability k_{tg} for passenger electric locomotives (EP1, EP2K) must not exceed 0,96. Let's use the initial data in Table 1 to calculate readiness indicators.

***Table 1**. Input data for calculating indicators of locomotive uptime and readiness*

No	Parameter name	EP1	2ES4K
1	Annual stock of the locomotive Tf, h 1	32334	187133
2	Idle time on scheduled repairs Tpl, h	4589.2	2824,4
3	Average idle time of one locomotive on unscheduled repairs Tpr, h	247	2564
4	Number of locomotives in operation M, units	83	41
5	Number of refusals (unscheduled repairs)	20	99
6	Total mileage Li, million km	2.519	1.859
7	Internal readiness coefficient by specifications,	0.98	0.96
8	Technical readiness coefficient by specifications,	0.97	0.95

Initial data were provided by Locomotive Technologies group of LLC TMH-Service branch "North-Caucasus". The calculation is made by expressions (1)...(4) with regard to (5) and (6). Criterion x^2 (K. Pirsov) was defined according to Appendix A in [14] by confidence probability $\gamma = 80\%$. The results of the calculations are summarized in Table 2.

Materials Research Proceedings **21** (2022) 109-116 https://doi.org/10.21741/9781644901755-20

Table 2. *Calculated values of availability and uptime indicators of electric locomotives in operation*

No	Parameter name	EP1	2ES4K
1	Total operating time of the locomotive in serviceable condition , h	127497.8	181744.6
2	Internal availability coefficient of electric locomotive	0.998	0.986
3	Technical readiness coefficient	0.963	0.97
4	Failure rate of electric locomotives in operation (unscheduled failures), km-1	7.94	47.9
5	Upper limit of confidence interval of failures flow, km-1	9.82	52.67
6	Error in evaluation of failure flux parameter	23.7	10

Let's compare the calculated values of internal and technical readiness coefficients (Table 2) with similar indicators established by technical specifications (TS). The conditions for deciding whether the EP1 and 2ES4K locomotives meet the established readiness requirement [19-20] are met if

$$K_{vg} \geq K_{vg}^{tu}; \mathrm{K}_{tg} \geq K_{tg}^{tu} \tag{8}$$

The calculated values of the internal and technical readiness coefficients of the EP1 and 2ES4K locomotives with similar values set (TU) are shown in Figure 1.

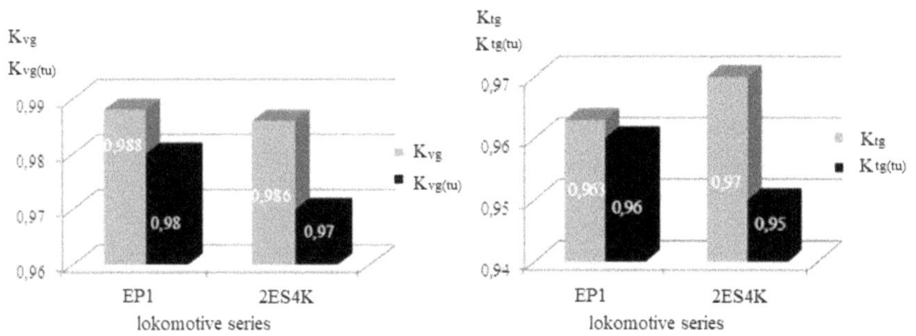

a) b)

Fig. 1. *Calculated values of internal (a) and technical availability (b) coefficients of EP1 and 2ES4K locomotives with similar indicators established (TU)*

For electric locomotive EP1 $K_{vg} = 0,988$, $K_{vg}^{tu} = 0,98$, $\mathrm{K}_{tg} = 0,963$, $K_{tg}^{tu} = 0,96$ for electric locomotive 2ES4K $K_{vg} = 0,986$, $K_{vg}^{tu} = 0,97$, $\mathrm{K}_{tg} = 0,97$, $K_{tg}^{tu} = 0,95$.

Requirements (19) are met - this is a sufficient condition for making a consistent assessment of locomotive compliance in terms of reliability. However, a necessary condition is that the locomotives meet the established requirement for reliability. We perform control procedures to determine whether each set of locomotives (EP1, 2ES4K) meets the reliability requirement according to the inequalities given in [14].

There are three compliance events:

-full compliance $\left(\omega_{1-\beta} \le \omega_{\beta}\right) \wedge \left(\omega \le \omega_{\beta}\right) \wedge \left(\varepsilon \le \varepsilon\right)$; (9)

- nonconformance $\left(\omega_{1-\beta} > \omega_{\beta}\right) \wedge \left(\omega > \omega_{\beta}\right) \wedge \left(\varepsilon > \varepsilon\right)$; (10)

- "limited" mismatch $\left(\omega_{1-\beta} > \omega_{\beta}\right) \wedge \left(\omega \le \omega_{\beta}\right) \wedge \left(\varepsilon > \varepsilon\right)$; (11)

where $\wedge$ - reads as "AND"

ω_{β} - the failure flow parameter according to the TU, $\omega_{\beta} = 10 \cdot 10^{-6}$ km (EP1),

$\omega_{\beta} = 11 \cdot 10^{-6}$ km (2ES4K).

Let us use the data of table 2 and for each set of electric locomotives in accordance with inequalities (8)...(11) determine the conformity of requirements for no-failure rate.

For the set of electric locomotives EP1: $(9{,}82 < 10)^{-6} \wedge (7{,}94 < 10) \cdot 10^{-6} \wedge (23{,}7 > 20)$

For the set of electric locomotives 2ES4K: $(52{,}67 > 11)^{-6} \wedge (47{,}9 > 11) \cdot 10^{-6} \wedge (10 < 20)$

Thus, the control procedures to determine the compliance of each set of EP1 and 2ES4K locomotives with the reliability requirements according to the inequalities (8)...(11) showed that the set of 2ES4K electric locomotives for the considered run does not fully meet the reliability requirement. In order to clarify this event, it is necessary to increase the interval of the locomotives' run and repeat the procedure of determining compliance with the reliability requirements, as well as to develop and implement measures to improve the reliability of the considered set of 2ES4K electric locomotives. The EP1 set of electric locomotives meets the reliability requirements, but the error value is higher than 20% . This means that the tabulated value of the chi-square distribution function at confidence probability is not 80%, but 76.3%. In this case, an increase in the mileage L of the locomotives is required. And then the error of the estimate of the failure flow parameter can be refined . The reliability of the 2ES4K electric locomotives was significantly affected by the number of unscheduled repairs. Unscheduled repairs increased the flow of failures, from which the value of the upper limit of the one-sided interval was higher than the rejection value of the parameter of the flow of failuresv . The most informative is the availability factor , which characterizes such reliability properties of EP1 and 2ES4K electric locomotives, as fault tolerance and maintainability. These characteristics can be increased by reducing downtime before, during and after maintenance and repairs, advance training of operating and maintenance personnel, saturation of locomotive facilities with modern equipment, devices and technologies, increasing the role of transport science, etc.

The results of research and their discussion

Taking into account the importance of the research and calculation of the electric locomotives uptime, as well as the absence of a universal method for assessing the work of service locomotives in depots to ensure the required uptime, the sequence of assessing the uptime and readiness of locomotives during their normal operation is given. As a result of comparing the calculated coefficient values of the internal and technical readiness (for electric locomotive EP1 ; for electric locomotive 2ES4K) with the analogous values, established by the technical conditions (EP1 corresponds ; ; 2ES4K corresponds), it is determined that the investigated locomotives meet the established readiness requirements.

Conclusions

As a result of the control procedures to determine the compliance of each set of locomotives EP1 and 2ES4K with the reliability requirements, it was determined that the set of electric locomotives 2ES4K for the run in question does not fully comply with the requirement for uptime. And the set of EP1 electric locomotives meets the reliability requirements, but the error value is higher than 20%. To clarify both events, it is necessary to increase the mileage interval of the locomotives and repeat the procedure for determining compliance with the uptime requirements. The method of assessing the uptime and readiness of locomotives during their normal operation makes it possible to identify existing shortcomings in the operation of rolling stock and to form measures to improve the quality of rolling stock operation.

References

[1] Teptikov N.R., Mashnin D.V. Principles of maintenance and repair processes organization for electric locomotives // Actual problems of railway transport development and young scientists' role in their solution: Collection of scientific papers. Rostov n/D.: RGUPS, 2000. C. 79-80.

[2] Lyabakh N. N., Teptikov N. R., Matvienko V. P. Problems of Technical Diagnostics of Electric Rolling Stock // Integration of Branch and University Science: Problems of Modern Machine Building: Materials of the International Scientific and Engineering Conf. C. 126-127.

[3] Nagovitsyn V.S. Method and algorithm for making decisions on the rolling stock maintenance (repair) according to the state // Train Traffic Safety: Proc. IV scientific-practical conf. / MIIT - Moscow, 2003. - P. IV- 64 - IV - 69.

[4] Stepanov, S.I., Buyanov, A.B. Technique of Calculating Thermal State of Power Semiconductor Device with Coolant //Emproving of Power Semiconductor Converters Cooling Systems of Railway Transport: Collection of Scientific Papers of LIIZhT. Л.: 1988. C. 105-116.

[5] Khasyanova D.U., Khasyanov U. Design technology of thermomechanical joints using combined couplings and their application. In the collection: Basic research and innovative technologies in mechanical engineering. Proceedings of the V-th International Scientific Conference. Editorial board: Ganiev R.F., Glazunov V.A., Sigov A.S., Albagachiev A.Y., Romanov A.N., Khruschov M.M., 2017. C. 299-300.

[6] Babkov Yu.V. Guaranteed locomotives - the lost readiness / Yu.V. Babkov, V.A., Perminov, E.E. Belova et al. // Lokomotiv, 2013, №5 . - c. 34-35.

[7] Khasyanova D.U. Increasing the reliability of fixed joints of pipelines through the use of shape memory effect in couplings of thermomechanical joints. Problems of mechanical engineering and machine reliability. 2019. № 1. C. 103-108.

[8] Kokhanovskii V.A., Glazunov D.V. Composite fluoroplastic coatings in reciprocating circular motion. Russian Engineering Research. 2020. T. 40. № 2. C. 130-132. https://doi.org/10.3103/S1068798X2002015X

[9] Mayba I.A., Glazunov D.V. Optimization of tribotechnical characteristics of wheel–rail friction modifiers. Journal of Friction and Wear. 2020. T. 41. № 6. C. 517-520. https://doi.org/10.3103/S1068366620060136

[10] Basics of theory and practice of reliability of technical devices / A.V. Gritsenko, V.V. Strekopytov.-Petersburg: Sudarynya, 2004. -272c. ISB5-88718-050-1.

[11] Kokhanovskii V.A., Glazunov D.V., Zoriev I.A. Macrocompositional polymer-powder bearings. Journal of Machinery Manufacture and Reliability. 2019. T. 48. № 2. C. 130-135. https://doi.org/10.3103/S1052618819020080

[12] Kokhanovskii V.A., Maiba I.A., Glazunov D.V., Bol'shikh I.V. Lubricator casings for locomotive wheel rims. Russian Engineering Research. 2016. T. 36. № 5. C. 364-365. https://doi.org/10.3103/S1068798X16050099

[13] Chetcherugov V.A. Reliability of locomotives / V.A. Chetcherugov, A.D. Puzankov -M.: Route 2003.-415 p.

[14] Appendix to the order of OAO Russian Railways from 12.11.2008 № 2367r/Methodical provisions for calculating the reliability of locomotives on the results of their operation. - Moscow: OAO "Russian Railways", 2010.

[15] Gorsky A.V. Reliability of Electric Rolling Stock / A.V. Gorsky, A.A. Vorobyov. - Moscow: Marshrut, 2005 - 303 p.

[16] Pugachev V.S. Probability Theory and Mathematical Statistics.

[17] Kochanowski V.A., Mayba I.A., Glazunov D.V., Zoriev I.A. Powder bearings with polymer inserts. Journal of Friction and Wear. 2019. T. 40. № 3. C. 229-233. https://doi.org/10.3103/S1068366619030048

[18] Kokhanovsky V.A., Glazunov D.V. A lubricant for rotaprint lubrication of the wheel–rail system. Journal of Friction and Wear. 2020. T. 41. № 6. C. 531-537. https://doi.org/10.3103/S1068366620060100

[19] RD50-690-89. Technical specifications for locomotives building. Guiding document for standardisation. Group T51- Reliability in machinery. Methods for evaluation of reliability indices according to experimental data. Introduced on 01.01.1999.

[20] GOST P 56046-2014. Indicators of use of locomotives. Terms and definitions.

Investigation of the Efficiency of Manufacturing Polymer Molds by FDM-Printing

Oleg Krupennikov[1,a], Mihail Nazarov[1,b] and Evgeniy Kiselev[1,c*]

[1]Ulyanovsk State Technical University, Ulyanovsk, Russia

[a] krupennikov_oleg@mail.ru, [b] nazarov.mv@inbox.ru, [c] kec.ulstu@mail.ru

Keywords: Additive Technologies, 3D-Printing, Fused Deposition Modeling (FDM), Polymeric Materials

Abstract. The results of experimental studies of the efficiency of manufacturing polymer foundry models by the FDM method are presented. Rational consumables for 3D-printing are revealed. The optimal values of the shrinkage value of the 3D-model during printing and the parameters of the 3D-printer operation mode are revealed. It has been established that polymer casting molds are rationally manufactured by FDM printing, using ABS-plastic as a consumable, while making a correction to linear dimensions at the stage of creating a 3D-model to eliminate plastic shrinkage.

Introduction

The use of additive technologies in foundry is a promising direction, since over the past few years the cost of 3D-printers has significantly decreased, the print quality has improved and the technology of printing parts of technological equipment has been simplified [1-5].

Currently, several varieties of 3D-printing are used to prototype parts of foundry equipment. At the same time, each of the methods of 3D-printing has a number of advantages and disadvantages that determine the suitability of its use for prototyping various parts of technological equipment.

For example, the method of layer-by-layer deposition of FDM, based on extrusion through a calibrated nozzle of heated and molten fiery thermoplastic, is characterized by the relative cheapness of the equipment and consumables used, as well as high printing speed while ensuring the strength characteristics of the products. At the same time, due to the shrinkage of thermoplastic during its cooling, warping and delamination of parts can occur, which, in turn, can lead to a decrease in the accuracy of their geometric parameters and surface quality.

To assess the effectiveness of the manufacture of polymer casting molds by FDM-printing, the authors conducted a number of experimental studies.

The experiments were carried out by printing on the FDM-printer Anycubic i3 Mega S STL-models of parts of casting molds, pre-processed in the Ultimaker Cura slyser.

In order to find out how the dimensions of parts change in the process of 3D-printing due to plastic shrinkage, at the *first stage* of research samples were printed in the form of a parallelepiped with dimensions of 120*15*10 mm from four types of plastic: ABS, PLA, PETG and SBS. The obtained samples measured the actual overall dimensions and compared with similar dimensions set in the 3D-model of the sample during printing.

It is established that the shrinkage of plastic, regardless of its brand, is significantly influenced by the value of the nominal value of the overall size. In this case, the lowest shrinkage is recorded along the length of L samples, and the highest shrinkage value is obtained by the height of H samples. In addition, the amount of shrinkage of the samples is influenced by the degree of plastic filling of the 3D-model, which was installed in the 3D-model of the sample during the preparation for printing and ranged from 40% to 100%.

It was revealed (fig.1) that with an increase in the degree of filling the model with plastic, there is a slight change in the shrinkage value on the one hand, and on the other hand, as the degree of filling the sample with plastic increases, there is an increase in shrinkage fluctuations in samples made of different materials. Obviously, a large amount of plastic filling the sample during the printing process leads to large temperature deformations and fluctuations in the properties of plastics.

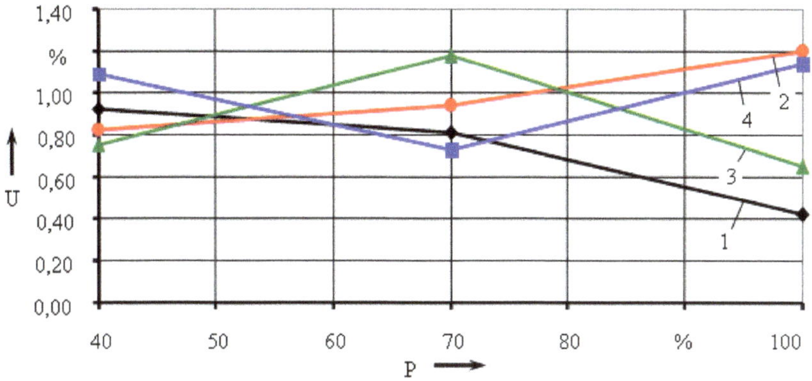

***Fig. 1**. Effect of the degree of filling of the 3D-model P on the shrinkage of U samples from the studied plastics: 1,2,3,4 - respectively ABS-, PLA-, PETG - and SBS-plastic*

In the process of research, it was established (fig. 2) that the average shrinkage value with an increase in the degree of filling the sample with plastic decreases from 0.9% to 0.7%.

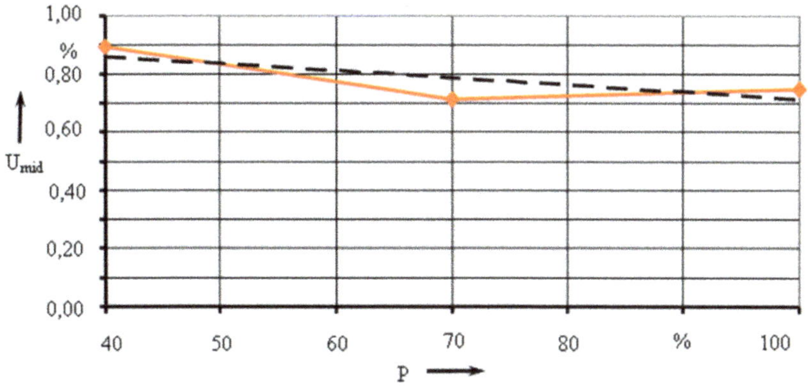

***Fig. 2.** Effect of the degree of filling of the 3D-model P on the average shrinkage U_{mid} samples from the studied plastics*

Thus, when developing 3D-models of casting mold parts for the manufacture of wax blanks, it is necessary to make a clarification equal to 1.07 - 1.09, depending on the degree of filling of the model.

In accordance with the developed methodology, at the *second stage* of research, rational materials were selected for 3D-printing of parts of foundry molds. Varied four plastic materials (ABS, PLA, PETG and SBS) under constant printing modes corresponding to the recommended values from the reference literature.

As can be seen from figure 3, the lowest values of the height parameters of the surface profile are achieved in samples made of ABS-plastic, slightly worse indicators in samples from PLA- and PETG-plastics. The worst parameters were shown by samples made of SBS-plastic.

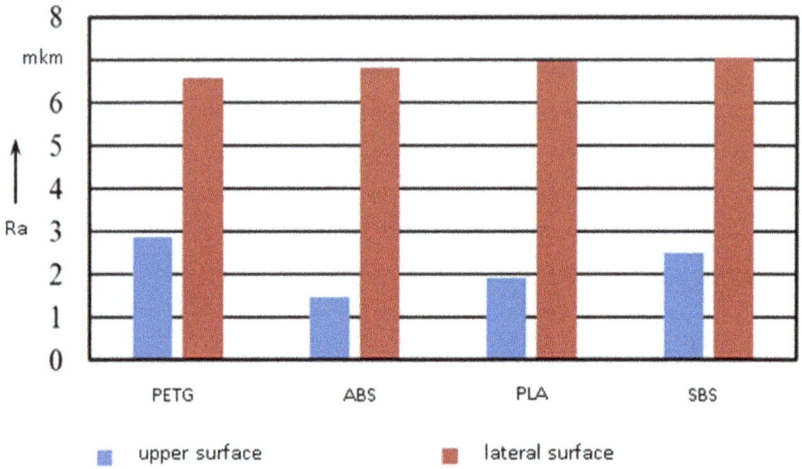

Fig. 3. *Influence of plastic material on the arithmetic mean deviation of the surface profile Ra*

In order to draw definitive conclusions in favor of choosing a rational material for 3D-printing of casting mold parts, studies were conducted to determine the toughness of the samples.

As can be seen from the results presented in figure 4, samples made of ABS-plastic have the greatest toughness. PETG-plastic samples have a toughness (10-15) % less than that of ABS-plastic samples. The worst result was shown by samples from PLA- and SBS-plastics - their toughness is (40-50) % less than that of samples made of ABS-plastic. Thus, the best quality of 3D-printing is provided by this indicator in samples printed from ABS- and PETG-plastics, which have the highest toughness.

However, the cost of PETG-plastic is 1.4-1.5 times higher than ABS-plastic with the same degree of shrinkage of the material. Therefore, based on the results of experimental studies, ABS-plastic should be recommended as a rational material for 3D printing of casting mold parts.

Materials Research Proceedings **21** (2022) 117-121
https://doi.org/10.21741/9781644901755-21

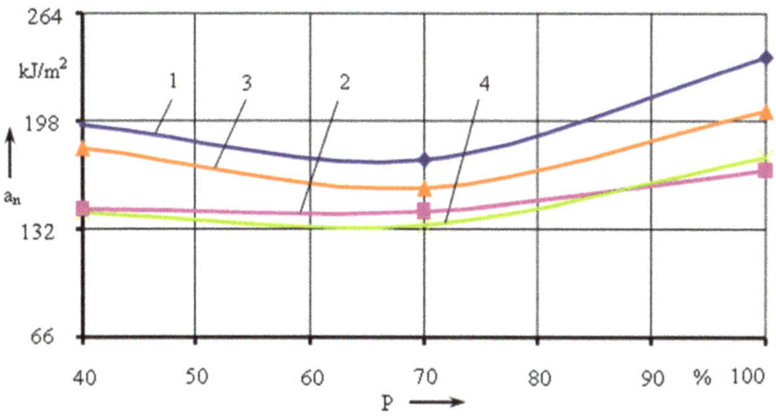

Fig. 4*. Effect of the degree of filling of the 3D-model P on the toughness a_n samples depending on the grade of material: 1, 2, 3, 4 -respectively ABS-, PLA-, PETG- and SBS-plastics*

Increasing the degree of filling P to 100 % leads to an increase in the toughness of an samples by (15-20) % (see figure 4). Therefore, for the manufacture of parts of casting molds with increased toughness, they must be printed at 100 % filling of the 3D-model.

In the *third stage* of the research, the influence of the elements of the 3D-printing mode of casting mold parts on the efficiency of the printing process was evaluated. The samples were made of ABS-plastic, chosen as a rational material in the second stage of research.

It was found that the increase in the temperature of the extruder slightly affects the altitude parameters of the irregularities of the upper and lateral planes of the samples. In this regard, Te = 240 °C was taken as the recommended temperature of the extruder nozzle, which corresponds to the recommended reference data.

In the course of research, it was revealed that small values of the temperature of the 3D-printer table in the range (100-110) °C lead to significant fluctuations in the values of the altitude parameters of the surfaces of the printed sample. This indicates unfavorable conditions for the formation and subsequent cooling of the layers of the product. With a further increase in the table temperature to 120 °C, the print quality stabilizes, which indicates an exit to the rational mode of 3D-printing.

In studies of the influence of the extruder displacement Ve on the efficiency of 3D-printing, it was established (see figure 5) that the Ve velocity has significant changes in the roughness (according to the Ra parameter) of the upper and lateral surfaces of the samples under study.

As can be seen from the results of the studies presented in figure 5, it is most rational to move the extruder in the process of 3D-printing at a speed of (35-40) mm / s. When the specified limit is exceeded, the roughness parameter Ra of the lateral plane of the samples increases significantly, which indicates that the plastic does not have time to solidifies and it is deformed by the newly applied layer of molten filament..

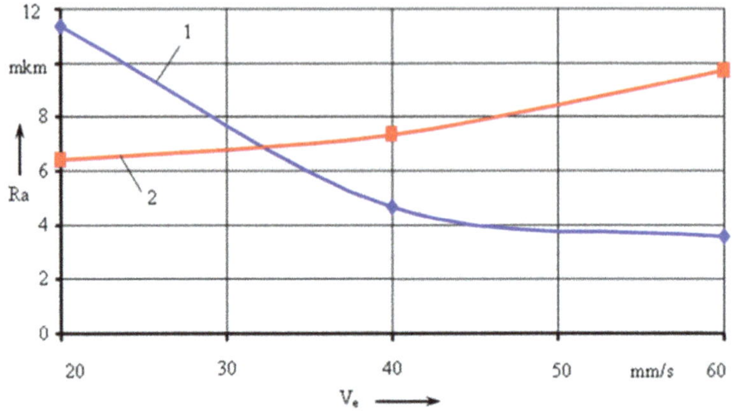

Fig. 5. *Influence of the velocity of the extruder V_e on the arithmetic mean deviation of the surface profile Ra: 1 and 2 - respectively the upper and lateral planes of the sample*

Thus, experimental studies have shown that polymer casting molds are rationally made by FDM-printing, using ABS-plastic as a consumable, making correction to linear dimensions at the stage of creating a 3D-model to eliminate shrinkage of plastic.

References

[1] Singh, S. Precision investment casting: A state of art review and future trends / S. Singh, R. Singh // Proceedings of the Institution of Mechanical Engineers, Part B: Journal of Engineering Manufacture.– 2016. – T. 230. – № 12. – C. 2143-2164. https://doi.org/10.1177/0954405415597844

[2] Bourell, D.L. Roadmap for additive manufacturing: identifying the future of freeform processing / D.L. Bourell, M.C. Leu, D.W. Rosen. - The University of Texas at Austin, 2009. – C. 11-15.

[3] Harun, W.S.W. Characteristic studies of collapsibility of ABS patterns produced from FDM for investment casting / W.S.W. Harun [et al.] // Materials Research Innovations. – 2009. – T. 13. – № 3. – C. 340-343. https://doi.org/10.1179/143307509X441513

[4] Piekło, J. Methods of Additive Manufacturing used in the Technology of Skeleton Castings / J. Piekło, M. Maj // Archives of Metallurgy and Materials. – 2014. – T. 59. – № 2. – C. 699-702. https://doi.org/10.2478/amm-2014-0114

[5] Olkhovik, E.O. Use of additive technologies for practical working with complex models for foundry technologies / E.O. Olkhovik, A.A. Butsanets, A.A. Ageeva // IOP Conference Series: Materials Science and Engineering. – IOP Publishing, 2016. – T. 140. – № 1. https://doi.org/10.1088/1757-899X/140/1/012013

Improving the Efficiency of Turning Processing of Heat-Resistant Alloys by Introducing Ultrasonic Field Energy into the Cutting Zone

Aleksandr Khramov[1,a], Maksim Gorchcov[1,b], Nguyen Khan Toan[2,3,c] and Evgeniy Kiselev[3,d]

[1]GK "HALTEC", Ulyanovsk, street Narimanova, 1,b,Russian Federation

[2]Hyngien University of Education and Technology, Hanoi, Socialist Republic of Vietnam

[3] Ulyanovsk State Technical University, Ulyanovsk, street Nord Venec,32, Russia Federation

[a]alexander_khramov@haltec.ru, [b]maksim_gorshkov@haltec.ru,
[c]Nguyen_Khan_Toan@gmail.com, [d]kec.ulstu@mail.ru

Keywords: Turning, Heat-Resistant Alloys, Cutting Zone, Cutting Tool, Ultrasonic Energy

Abstract. Studies have been carried out to assess the effectiveness of dry processing by the current preparations from the heat-resistant alloy CN45MVTUBR with mineral ceramic incisors with the introduction of the ultrasound-field energy treatment zone. It has been established that the use of ULTRASOUND in the rough treatment of mineral ceramic tools without coolant allows to reduce the depth of the defective layer in 1.5 times.

A feature of modern machining production is the need to manufacture an ever-increasing number of complex and non-rigid, thin-walled parts in various machines and mechanisms, primarily in aircraft. The latter is explained, on the one hand, by an increase in the speeds of their executive movements (and, consequently, by the desire to reduce inertial loads), and on the other, by the removal of almost complete previous restrictions on designers in creating complex parts due to the existing technological capabilities of modern 3–5 axial machines and CNC machining centers. One more thing can be added to this: a decrease in structural rigidity, as a rule, is accompanied by the simultaneous use of high-strength and, therefore, hard-to-work steels and alloys as structural materials to maintain the necessary strength while reducing the cross sections of manufactured parts. A feature of modern machining production is the need to manufacture an ever-increasing number of complex and non-rigid, thin-walled parts in various machines and mechanisms, primarily in aircraft. The latter is explained, on the one hand, by an increase in the speeds of their executive movements (and, consequently, by the desire to reduce inertial loads), and on the other, by the removal of almost complete previous restrictions on designers in creating complex parts due to the existing technological capabilities of modern 3–5 axial machines and CNC machining centers. One more thing can be added to this: a decrease in structural rigidity, as a rule, is accompanied by the simultaneous use of high-strength and, therefore, hard-to-work steels and alloys as structural materials to maintain the necessary strength while reducing the cross sections of manufactured parts.

As materials for such parts, hard-to-work corrosion-resistant and heat-resistant steels are used, as well as alloys based on titanium and nickel. Thus, the share of titanium and heat-resistant alloys in the design of fourth-generation aircraft engines for the year was about 89% [1], in the fifth-generation engines this percentage is even higher. The problem of ensuring the specified operational properties in the manufacture of non-rigid parts from difficult-to-process materials, in which residual stresses of any sign arising due to heat-stress and structural-phase transformations can cause significant changes in shape and spatial orientation, is much more acute than for other structural materials. This is due to the fact that their heat and temperature conductivity is 4-5 times less than that of structural carbon steels and all the heat that occurs in the cutting zone is localized in the surface layer (SL) of the workpiece being processed.

In practice, this translates into the need to significantly reduce the elements of the cutting condition, and in some cases, even the need for empirical selection of the manufacturing sequence. The machining of workpieces of parts with complex spatial shapes is most often carried out at expensive machining centers for which such performance losses are unacceptable. It is obvious that a high-performance removal of stocks from blanks from any structural materials is accompanied by a high heat-power intensity of the processing process. This leads to the appearance in SL of preforms from difficult to process materials of significant technological residual stresses (TRS), the negative impact of which on the operational characteristics of manufactured parts is well known.

Traditional technologies for processing workpieces made of heat-resistant alloys belonging to the category of difficult to work, do not allow to achieve high values of the cutting condition elements (cutting speed, feed). This is due to their physico-mechanical properties, such as high hardening of the material during deformation by cutting, the ability of the material to maintain its original strength and hardness at high temperatures arising in the cutting zone, low thermal conductivity, which contributes to insufficient heat removal from the contact zone of the tool and the workpiece, as well as the presence of intermetallic or carbide inclusions in their composition, leading to abrasive wear of the tool cutting edge.

The most efficient processing of workpieces made of heat-resistant nickel-based alloys can be carried out with a tool made of such a material, which is characterized by high cutting properties, such as high values of red resistance, stability of cutting properties over a wide temperature range and resistance to abrasive wear. Carbide tools currently used as tool materials do not possess the required characteristics.

The low machinability of heat-resistant alloys is determined by their physical and mechanical properties. In these conditions, it is very necessary to disclose the reasons affecting their machinability, and to find ways and means to increase the productivity of their processing on metal cutting machines.

The basic structure of most heat-resistant alloys is usually an austenitic-grade solid solution with a face-centered cubic lattice. Moreover, most of the deformable heat-resistant alloys belong to the type of dispersion hardening, i.e.in these alloys, a structural component is released from the solid solution — the second phase, which is different from its base and is dispersed throughout the alloy in finely divided form.

High dispersion of the structure prevents the occurrence and development of slip processes, while the creep resistance of the alloy increases.

Heat-resistant nickel-based alloys are characterized by a significant loss of strength and facilitation of the cutting process at temperatures above 800 degrees C. [2]. Earlier experiments at HALTEC Group found that the processing efficiency of workpieces made of heat-resistant nickel-based alloys can be increased by replacing conventional cutting tools with SSY ceramic tools (30 times more productivity compared to using carbide tools) with a simultaneous increase in speed cutting up to 600 and more m / min (fig.1) [3].

Materials Research Proceedings **21** (2022) 122-126 https://doi.org/10.21741/9781644901755-22

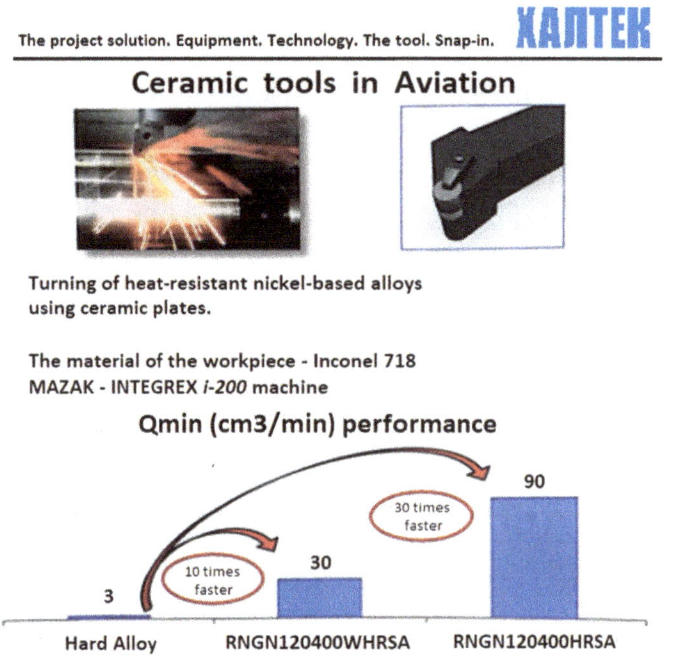

Carbide plate	Ceramic plate RNGN120400WHRSA	Ceramic plate RNGN120400HRSA
Lubricating and cooling liquid Ap=1,5 mm V=20 m/min (n=32 rpm) F=3 mm/min (f=0,1 mm/rev)	Lubricating and cooling liquid Ap=1,5 mm V=200 m/min (n=318 rpm) F=32 mm/min (f=0,1 mm/rev)	Air blowing Ap=1,5 mm V=600 m/min (n=955 rpm) F=96 mm/min (f=0,1 mm/rev)

The SSY ceramic tool showed a 30-fold increase in performance relative to the carbide!

Fig. 1. Results of pilot tests of new types of mineralokeramika at the production site of HALTEK Group of Companies

The latter is explained by high values of contact temperature (800-850 degrees Celsius), which leads to the implementation of the process of forming new surfaces under thermoplastic conditions. Such processing is accompanied by a significant decrease in cutting forces, which explains the

Materials Research Proceedings **21** (2022) 122-126 https://doi.org/10.21741/9781644901755-22

reasons for the sharp increase in the productivity of stock removal. Obviously, this method can be used only during preliminary (rough) processing, since high contact temperatures in the cutting zone will inevitably cause the appearance of a defective layer. Further research by HALTEC employees and industrial testing at a number of aircraft engine companies established the high efficiency of cermet tools for rough milling and turning of billets made of heat-resistant alloys without cutting fluid (coolant). However, tests at the Rybinsk Motor-Building Association established the depth of the defective layer at (0.2-0.3) mm. This is unacceptable for final machined parts, increases the duration of semi-finishing and final processing and is unacceptable in the manufacture of thin-walled non-rigid parts.

Further intensification of the machining processes by cutting, reducing the depth of propagation of TRS and reducing the likelihood of other defects in the SL of the machined parts can be carried out by external energy influences on the contacting objects [4]. At the same time, this will facilitate the process of forming new surfaces by reducing the cost of dislocation movement, provide a reduction in cutting force, but also reduce contact temperatures. It can reduce the depth of the defective layer. However, the latter, on the contrary, can complicate the processes of shaping. This will inevitably lead to a decrease in the productivity of stock removal, but can reduce the depth of the defective layer of manufactured parts.

From the large number of types, forms and patterns of energy effects on the cutting process [5], the most simple and economical is the introduction of the energy of an ultrasonic (ultrasonic) field into the contact zone. Considering that the introduction of ultrasonic vibrations helps to reduce frictional processes, and hence contact temperatures, the feasibility of such an effect when processing billets of heat-resistant alloys with ceramic tools requires experimental confirmation.

The result of numerous studies performed in various countries was the possibility of practical implementation of dimensional machining with ultrasonic testing by using special devices and installations for applying the energy of the ultrasonic field to the workpiece (tool, coolant, machine parts and tooling, etc.) [5-7]. At present, DMG MORI has developed and mass-produced a line of 18 models of the ULTRASONIC series machines, in which a device for superimposing ultrasonic cutting on the cutting tool is integrated in the design of the spindle unit. AXILE Machining machines are similar. At the same time, ultrasonic mandrels (Altrasonic, CRENO Industry, Pulchertool companies) are commercially available, the use of which is permissible on CNC machined centers equipped with spindles with ISO40 cone (SK40, BT40) [8]. Under these conditions, it is extremely important to identify the possibility of reducing contact temperatures without reducing the intensity of stock removal by introducing the ultrasonic field energy into the contact zone of the cutting tool and suggest new effective means of reducing the cost of manufacturing complex non-rigid parts from heat-resistant chromium-nickel alloys.

During the research, preliminary results were obtained on evaluating the efficiency of turning blanks from the heat-resistant CN45MVTUBR alloy (analog of the Inconel 718 alloy) dry (air cooling) with RNGN120400HRSA mineral ceramics cutters SSY series (South Korea) with the introduction of ultrasonic field energy into the processing zone. The experiments were performed on a Quantum Opti D420x 1000 DPA screw-cutting lathe. During the processing, the components of the cutting force Pz, Py, Px and the torque M cr. were measured using a UDM-600 dynamometer equipped with Kyowa kfg-5-120-c1-11 strain gages. In addition, using the TemPro-1200/1600/2200 thermal imager-pyrometer, the temperature on the front surface of the ceramic-mineral plate was monitored as close as possible to the contact zone with the workpiece. In the first series, workpieces were processed without introducing ultrasonic vibrations into the shaping zone, in the second, ultrasonic testing from an ultrasonic generator UZG 641 A was applied to the cutting tool, with a frequency f = 18.6 kHz and an amplitude A = 10 μm. In both series, elements of the cutting condition were used: cutting depth t = 1.5 mm, feed S = 0.1 mm / rev, cutting speed V = 600 m / min.

The introduction of the energy of ultrasonic vibrations into the shaping zone, according to the authors, can change the nature of the movement of dislocations in the shaping zone and reduce the size of the defective layer. To verify this assumption, before and after processing, the samples for research were evaluated by the non-destructive method of phase transformations in the surface layer of metals and alloys on the RIKOR-7 X-ray measuring complex and technological residual stresses of the first kind on the SITON ARM device.

As the analysis of research results shows, the assumptions made are fully confirmed. The introduction of the energy of ultrasonic vibrations into the turning zone reduces the contact temperature on the front surface of the cutter by (12-17) %, however, no reduction in processing productivity (estimated by the cutting force components) was observed. The control of the quality parameters showed that the depth of the defective layer (evaluated by the magnitude and depth of propagation of the technological residual stresses and phase changes) decreased in 1.45 times.

Thus, the performed studies have proved the feasibility of introducing ultrasound field energy into the turning zone of billets from heat-resistant alloys.

References

[1] A.Y. Kachan, V.F. Mozgovoy, S.B. Belikov: Main directions of technology development and metal cutting machines with CNC In:Visnik of dwigunobuding. 2007 №2, pp. 82-85.

[2] Kiselyov E.S.: Modern ultrasonic dimensional technologies of mechanical processing of hard-to-process and composite materials. In: Science-intensive technologies in mechanical engineering 2020. № 9, pp. 33-39. https://doi.org/10.30987/2223-4608-2020-9-33-39

[3] Kiselev E.S., Blagovskiy O.V. Management of the formation of residual stresses in the manufacture of critical / parts: monograph..- St. Petersburg: Lan, 2020, 140 p

[4] Kumabe D.: Vibrational cutting. Translation from yap. S.L. Maslennikova. Edited by I.I. Portnova and V.V. Belov. M.: Mechanical Engineering (1985).

[5] Mikhailov A.N.: Basics of synthesis of functionally oriented technologies. Donetsk: DonNTU (2009).

[6] Nguen Han Toan, Kiselyov E.S.: Heatproof Relaxation of Welding Stresses Openwork Designs of Large Size Ultrasonic Field. Solid State Phenomena. In: Vol. 303 (05). 2020, pp 124-131 (2020). https://doi.org/10.4028/www.scientific.net/SSP.303.104

[7] Nazarov M. V., Savelev K. S. and Kiselev E. S. .Automatically search of a rational cutting mode for milling non-rigid blanks with the specified quality parameters ICMTMTE IOP Conf. Series: Materials Science and Engineering 709(2020) 044069, IOP Publishing. https://doi.org/10.1088/1757-899X/709/4/044069

[8] Nguyen Khanh Toan, Mezin N. N. Study of the Efficiency of Ultrasonic Turning of Heat-Resistant Alloys with Tools from Mineral Ceramics. In: MATEC Web of Conferences 297, 01011(2019). https://doi.org/10.1051/matecconf/201929701011

Materials Research Proceedings **21** (2022) 127-132 https://doi.org/10.21741/9781644901755-23

Methodology for the Study of Residual Temperature Stresses in the Butt Contour of a Welded Joint Made of Carbon and High-Alloy Structural Steels during Multi-Pass Welding

Lubov Mironova[1,a*], Ruslan Nigay[2,b] and Elena Yakusheva[3,c]

[1]Moscow Aviation Institute (State National Research University), 4, Volokolamskoe sh., Moscow, 125993, Russia

[2]Russian University of Transport (MIIT), 9, Obraztsova st, Moscow, 127994, Russia

[3]National Research Nuclear University MEPhI, 31, Kashirskoe sh., Moscow, 115409, Russia

[a]mironova_lub@mail.ru, [b]ruslan1010@mail.ru, [c]elena.yakusheva77@gmail.com

Keywords: Temperature Field, Temperature Stresses, Welding, Welded Joint, Thermal Conductivity Equation, Residual Stresses

Abstract. The method of investigation of residual temperature stresses in the butt contour of the welded joint "shell – plate" made of carbon and high-alloy structural steels during multi-pass welding is described. The temperature problem was solved on the basis of the obtained experimental data of temperature measurement at the reference points of the structure during the application of the rollers. A solution is proposed that takes into account several stages, each of which corresponds to a specific intermediate temperature field due to the peculiarities of welding technology. The evaluation of residual stresses was carried out on the basis of the energy theory of plasticity, taking into account the dependences of the yield strength and elastic modulus of welded metals on temperature.

Introduction

In engineering practice, when designing welded structures made of low-carbon or alloy steel, it is assumed that local deformations themselves cannot exhaust the plasticity reserve of mild steel and, consequently, cracks cannot form [1]. In fact, a sharp drop in thickness in the places of the welded joint of individual parts of the bearing body leads to the formation of local zones with high values of temperature stresses and deformations. The different thickness of the shell structure leads to an uneven distribution of the temperature gradient between the front and inner surfaces, one of which experiences tensile stresses, the other, on the contrary, compressive. In addition, the mechanisms of the occurrence of shrinkage stresses have not been sufficiently studied, which, together with the acting force factors during operation, can be "dangerous" and contribute to the formation of microcracks and further crack development [2].

In many cases, when welding, studies of thermal processes with sufficient accuracy are carried out using the results obtained in the theory of thermal processes [3, 4].

However, exact quantitative dependences cannot be obtained only theoretically, since the assessment of welding stresses and deformations includes consideration of issues of a number of scientific disciplines: thermal conductivity, elasticity, plasticity and creep of metals in a wide range of rapidly changing temperatures [5].

In this paper, methods for estimating residual stresses and deformations caused by welding processes are proposed based on an experimental study of the thermal process of multi-pass welding.

Materials Research Proceedings **21** (2022) 127-132 https://doi.org/10.21741/9781644901755-23

Characteristics of the investigated welded structure Fig. 1. shows the welded joint and the scheme of applying the weld. A welded joint from the position of mechanics can be conditionally considered as a "shell – plate" joint. A pipe of considerable length is welded to the base of sufficient thickness. The material of the plate is low-carbon steel 20, the material of the shell (tube) is high-alloy austenitic steel. The technological properties of these steels in terms of weldability are good.

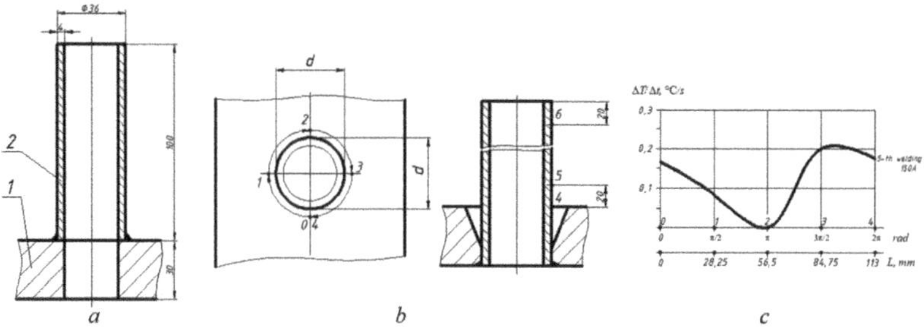

***Fig. 1**. Sample (a) and scheme of welding seam (b): 1 – footing; 2 – tube; 3 –distribution of the temperature gradient along the contour of the welded joint "shell - plate"*

A distinctive feature of the welded joint is the welding of a pipe of considerable length, which, under conditions of thermal loading by a movable heat source, can lead to significant warping of the structure. The welded joint is non-reversible. The welding method is manual electric arc melting of the electrode. The roller was applied manually with the breakdown of the welded contour into sectors "0-1", "1-2", "2-3", "3-4".

In accordance with the typical welding process, the recommended welding current values range from 20 to 70 A per welding pass. Thus, in one pass of the electrode when welding butt joints, depending on the value of the welding current, the effective thermal power can vary between 22 and 32%.

Purpose and methodology of the experiment

The purpose of the experiment was to study temperature fields and deformation processes in the zone of technological influence during multi-pass welding of the shell –plate joint under various conditions of thermal loading by welding current: with a decrease and increase in welding current by 6%. within the normal mode.

The experimental work included two stages. The first stage included multi-pass application of the welding seam in accordance with the accepted welding modes.

The second stage consisted in measuring the temperature in the vicinity of the points corresponding to the sector welding in one pass (points 1, 2, 3, 4, Fig. 1 (b)). Temperature measurement was carried out at these points three times with a time interval of 2, 5, 5 minutes after each measurement. The measuring device is a contact digital thermometer of the TK 5-0.1 brand. The measurement accuracy is 1°C. The welded joint was cooled in natural conditions in air at a temperature of 20°C.

Table 1 shows the obtained data on the temperature distribution in the welded circuit after the fifth pass.

Table 1. *Temperature change in the contour connection after the fifth pass*

Time, s		Temperature, °C	Characteristic points				
t_i	Δt		0	1	2	3	4
120	120	T_1	200	250	245	270	303
420	300	T_2	250	225	245	210	250
ΔT, °C			+50	–25	0	–60	–53
Rate of temperature change $\|\Delta T\| / \Delta t$, °C/sec			0.166	0.083	0	0.2	0.176
Average value of the rate of temperature change, °C/ sec			0.125				

The rate of temperature change was determined by the formula

$$\frac{\Delta T}{\Delta t}=\frac{|T_2-T_1|}{t_2-t_1}. \tag{1}$$

Here T_1 and T_2 are the temperature at the first and second measurements, t_2 and t_1 are the intermediate values of the temperature measurement time: $t_1 = 120$ s, $t_2 = 420$ s. The time interval during which the rate of cooling of the welded joint in air was determined was $\Delta t = 300$ s.

The graph of the temperature gradient distribution along the contour of the welded joint at the fifth pass is shown in Fig. 1 (c). The results of the conducted studies of the distribution of high-gradient temperature fields and modeling of the thermal process are presented in [6-8].

For welding processes, in approximate calculations of welding stresses, it is necessary to have dependences of the yield strength and elastic modulus of the metals being welded on temperature. Moreover, residual stresses occur after the complete cooling of the welded structure. The reason for the formation of residual stresses is uneven heating and plastic deformation of the metal during heating [5]. Approaches and methods to the determination of residual stresses in the butt contour of the welded joint will be based on the work [5, 9, 10]

To determine the level of residual stresses in the contour of the weld and the near-weld zone, we will use the methodology described in [10], where the value of the yield strength of the material is taken as the boundary above which residual stresses appear. In accordance with this , we will determine the residual stresses from the equality

$$\sigma_{res}=\sigma_{\max}-\sigma_T, \tag{2}$$

Where σ_{res} are residual stresses; $\sigma_{\max}$ are maximum stresses; σ_T is the yield strength of the material.

The determination of the maximum values of temperature stresses will be determined from the solution of the temperature problem by the graphoanalytic method, using experimental data of temperature measurement at reference points (Fig. 1), taking into account the methodology of work [5]. Thus, the solution of the problem is divided into two stages: the solution of the temperature problem and the problem of stresses in the elastic-plastic formulation.

The graphoanalytical method for determining temperature fields as a function of $T(r, \theta, t)$ in the welded contour of the shell–plate joint is based on a step-by-step solution, each of which corresponds to a specific intermediate value of temperature in time measured experimentally,

Table 1. Figure 2 (a, b, c) shows the graphical distribution of temperature fields in the radial direction according to the results of the experiment.

Fig. 2 (a) shows the distribution of the temperature field in the contour of the "shell – plate" welded joint two minutes after applying the roller along the entire contour in accordance with the scheme of Fig. 1. There is a relative uniform temperature redistribution from point " 0 "to point" 4». After 7 minutes after applying the roller over the entire contour, the distribution pattern changed, Fig. 2 (b). The greatest heat capacity is observed in the sector of the contour "2-4".

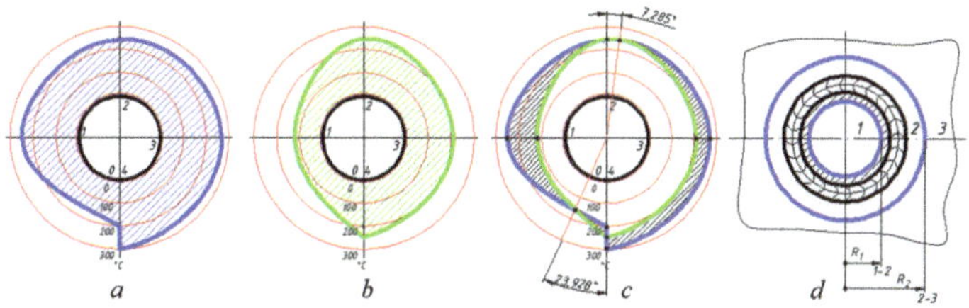

Fig. 2. *a, b, c – distribution of temperature fields in the radial direction of the contour; d – Stress distribution zones*

The distribution of the temperature field in the contour of the welded joint for the time Δt = 300 s = 5 min is shown in Fig. 2 (c). The highest rate of temperature redistribution is observed in the vicinity of point "1", and in the sector of the contour "0 – 1" at the intersection of the two curves, the temperature is redistributed to its increase in the coordinate φ = 23.928°C.

In accordance with the theory of welding deformations and stresses [5, 11, 12], it should be assumed that the vicinity of the near-weld zone 0-1-2 is most susceptible to the formation of hot cracks. The stress problem will be considered in the first approximation as axisymmetric, despite the fact that the circular seam is not welded at the same time and the conditions for the formation of metal deformations are different in the zones around the circumference. We divide the section of the welded joint into three zones (Fig. 2 (d)): inner, middle and outer, we have three rings with conditional boundaries of transition from one ring to another in the radial direction 1-2, 2-3.

Zone 2 corresponds to the tensile stress zone in a conventional straight seam and is characterized by high stresses σ_{t2}, close to the yield strength. The voltage σ_{r1-2} acts on the inner zone, the voltage σ_{r2-3} acts on the outer zone. Since ring 2 is balanced, then the inequality is valid

$$\sigma_{r2-3} > \sigma_{r1-2} \,. \tag{3}$$

In low-carbon and austenitic steels, zone 2 is in a plastic state. Let's write down the plasticity condition [5]

$$\sigma_{r2}^2 + \sigma_{t2}^2 - \sigma_{r2}\sigma_{t2} - \sigma_T^2 = 0 \,. \tag{4}$$

Taking into account (4), we write the differential equilibrium equation for our case in the for

$$r\frac{d\sigma_{r2}}{dr} = -\frac{\sigma_{r2}}{2} - \sqrt{\sigma_T^2 - \frac{3}{4}\sigma_{r2}^2} \,. \tag{5}$$

We write down the boundary conditions at the boundaries of the transition from one ring to another in the radial direction 1-2, 2-3

$r = r_1$; $\sigma_{r2} = \sigma_{r1-2}$; and $r = r_2$; $\sigma_{r2} = \sigma_{r2-3}$. (6)

Taking into account (6), we find the solution of the differential equation (5) in the form

$$\ln\frac{r_2}{r_1} = -\frac{1}{2}\ln\left(\frac{\sigma_{r2-3} - 2\sqrt{\sigma_T^2 - \frac{3}{4}\sigma_{r2-3}^2}}{\sigma_{r1-2} - 2\sqrt{\sigma_T^2 - \frac{3}{4}\sigma_{r1-2}^2}}\right) - \sqrt{\frac{3}{4}}\left[\arcsin\left(\sqrt{\frac{3}{4}}\frac{\sigma_{r2-3}}{\sigma_T}\right) - \arcsin\left(\sqrt{\frac{3}{4}}\frac{\sigma_{r1-2}}{\sigma_T}\right)\right]. \quad (7)$$

It follows that a specific value of σ_{r1-2} in the radial direction corresponds to a specific value of the voltage σ_{r2-3} and a very specific distribution of stresses in zone 2. The temperature stresses in the radial direction are determined by the formula [9]

$$\sigma_r(t) = \frac{\alpha E \Delta T(t,r)}{2(1-\nu)}, \quad (8)$$

where $\Delta T(t, r)$, is the rate of temperature change in the radial direction.

To estimate the temperature stresses in the welded circuit, we will take into account the distribution of the temperature field in the contour of the welded joint, using the proposed graphoanalytic method, Fig. 2 (a, b, c). The temperature stresses obtained by calculation, which exceed the value of the yield strength, can be considered residual stresses in the weld contour.

References

[1] Mel'nikov, Design forms and methods of calculation of nuclear reactors, Atomizdat, Moskow, 1972.

[2] L.I. Mironova and I.I.Fedik, The local thermal loading of two intersecting cylindrical rotational shells with a variable wall thickness, J. of Machinery Manufacture and Reliability 44 (8) (2015) 720-725. https://doi.org/10.3103/S1052618815080026

[3] H.S. Carslaw and J.C. Jaeger, Conduction of Heat in Solids, Oxford, 1959.

[4] N. Rykalin, Calculation of thermal processes during welding, GLTIMN, Moskow, 1951.

[5] V.A. Vinokurov, Welding strains and stresses, Mashinostroenie, Moscow, 1968.

[6] R. Nigay, E. Nigay, L. Mironova, Investigation of thermal and deformation processes in the welding of shell structures made of carbon and high-alloy structural steels. J. of Physics: Conference Series, 1, (2020) 1431. https://doi.org/10.1088/1742-6596/1431/1/012039

[7] L. Mironova, R. Nigay, E. Nigay, About one method of modeling high-gradient temperature fields in the welding of shell structures made of carbon and high-alloy structural steels, IOP Conference Series: Materials Science and Engineering, 971(3) (2020) 032064. https://doi.org/10.1088/1757-899X/971/3/032064

[8] L. Mironova, R. Nigay, E. Nigay. Experimental and Computational Method for Determining Temperature Stresses in the Welding of Structures Made of Carbon and High-Alloy Structural

Steels. J. Materials Science Forum, 1037 (2021) 343-348. https://doi.org/10.4028/www.scientific.net/MSF.1037.343

[9] S. Timoshenko, Theory jf plates and shells, McGraw-Hill, New York, 1959.

[10] G.N. CHernyshev, A.L. Popov, V.M. Kozincev, I.I. Ponomarev, Residual stresses in deformable solids, Nauka, Fizmalit, Moscow, 1996.

[11] K. Gatovskij, V. Karhin, Theory of welding strains and stresses, LKI, Leningrad, 1980.

[12] G. Nikolaev, S. Kurkin, V. Vinokurov, Welded construction. Strength of welded joints and deformation of structures, Vysshaya shkola, Moscow, 1982.

Investigation of the Physical and Mechanical Properties of ABS Plastic for the Additive FDM Process in the Manufacture of Models for Experiments in Wind Tunnels

Alexander Volkhonsky[1,a*], Kirill Dudkov[1,b] and Reza Norollahisomarin[1,c]

[1]Moscow Aviation Institute (National Research University), Moscow, Russian Federation

[a]volkhonsk@mail.ru, [b]jdg4917@gmail.com, [c]Norollahi_reza@mail.ru

Keywords: Additive Technologies, FDM Layer-By-Layer Synthesis Technologies, Layer-By-Layer Overlay of Molten Polymer Filament, 3D Printer, ABS Plastic, Physical and Mechanical Properties, Sample, Tests, Temperature, Fracture, Models for Experiments in Wind Tunnels

Abstract. Work has been carried out to determine the physical and mechanical properties of ABS plastics for the manufacture of aerodynamic models using the additive technology of layer-by-layer synthesis FDM for testing in wind tunnels. The samples were tested at normal and negative temperatures. The features of the structure of the samples and the features of their destruction are revealed. The possibility of using ABS plastic for the manufacture of various models used in experiments in wind tunnels is evaluated.

Problem statement

The analysis of aircraft and rocket and space technology products shows that modern aircraft, and in particular objects for testing in wind tunnels, are characterized by a variety of materials used, small series of manufacture, consist of parts with a wide range, complex geometric shape and use a wide range of technologies that carry a huge fleet of various equipment. The FDM printer, which performs 3D printing by layering molten polymer filament (Fused Deposition Modeling – FDM), allows for the manufacture of these models to print plastic parts of a given configuration, taking into account their design features (holes, grooves, etc.), increase the accuracy of the dimensions of the working surface, expand functionality, increase reliability, resource, and improve other performance characteristics of the product as a whole, not to mention reducing production time [1,2].

One of the most popular thermoplastic copolymers in the production of complex molded products requiring high manufacturing accuracy is ABS plastic (acrylonitrile butadiene styrene) – thermoplastic triple copolymer of styrene, acrylonitrile and butadiene, a resistant and durable polymer suitable for the manufacture of parts for various purposes using FDM technology and is the best material for working with a 3D printer [3].

Introduction

The material of the models obtained by FDM technology consists of a set of volumes with unidirectional stacked threads. Therefore, to determine the strength properties of the material, it is necessary to investigate the strength of a model material with unidirectionally laid threads and to assess the anisotropy of the properties, tests should be carried out along and across the direction of laying the threads. The physical and mechanical characteristics of the selected ABS plastics for samples are given in Table 1.

The samples were printed on a MakerBot Replicator 2x Experimental 3D printer with a print area of 24.6×15.2×15.5 cm, a layer resolution of 100 microns and an extruder nozzle diameter of

0.4 mm [6]. REC ABS plastic samples for tensile testing (at normal, elevated and lowered temperatures) they were manufactured according to GOST 11262-2017 "Plastics. Tensile test method" (Figure 1). For impact strength tests, ten samples were made in accordance with GOST 4647-2015 "Plastics. The method of determining the impact strength by Sharpie". A sample of type B (with an incision) is selected, the parameters of the samples are shown in Figure 2.

Table 1. *Physical and mechanical characteristics of REC ABS plastic in comparison with analogues*

Indicator Name	ABS-plastic REC	ABS-M30	Injection Molded ABS
Density [g/cm3]	1.03*	1.04	1.04
Tensile strength [MPa]	33±7	36	35–50
Elongation [%]	1,8	–	4
Young's modulus [MPa]	1168±515	2413	1700–2930

* This density was obtained using the hydrostatic weighing method and is not a valid value due to the presence of pores/voids in the samples.

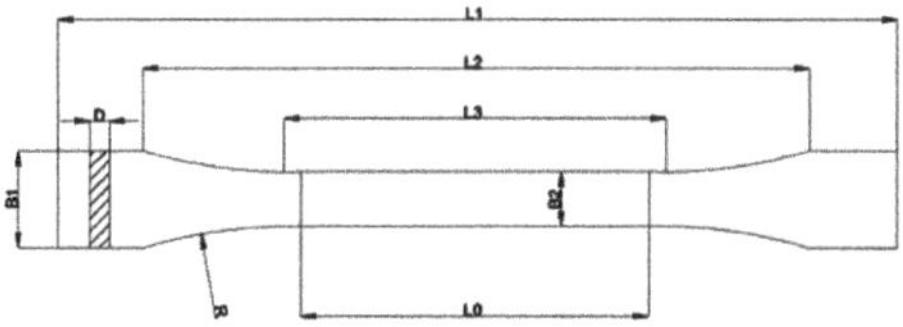

Fig. 1. *Sketch of a sample for conducting tensile tests*

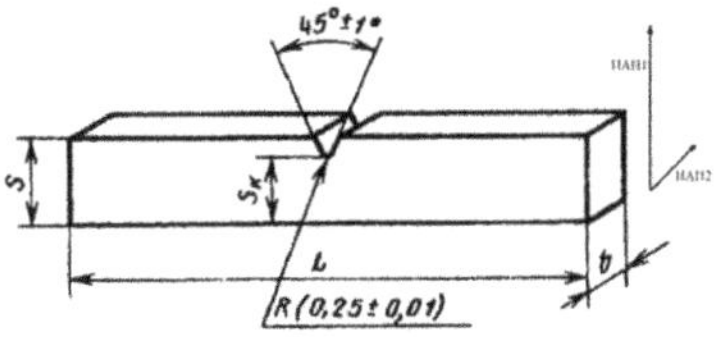

Fig. 2. *Sketch and parameters of the impact strength test sample*

Tensile testing of ABS plastic samples at normal, elevated and lowered temperatures

The input control of the samples was carried out using an electronic caliper MATRIX with a measurement accuracy of ±0.01 mm (GOST 166-1989).

The tests were carried out in accordance with the methods of GOST 11262-2017 "Plastics. Tensile test method" and GOST 9550-81 "Plastics. Methods for determining the modulus of elasticity in tension, compression and bending". Tests at normal temperature were carried out on the IR breaking machine-5047-50-11. Tests at elevated and lowered temperatures were carried out on a bursting machine IR-5047-50-11. To ensure the set temperature in the test area of the samples, a thermal cabinet STI-1 was connected to the bursting machine.

Three samples of REC ABS plastic were subjected to tensile tests at normal temperatures from +22 °C to +27 °C. The tensile strength was 33±7 MPa, the tensile strength was 32±6 MPa, the modulus of elasticity was 1168 ±515 MPa.

Tensile tests at elevated temperatures from +70 °C to +80 °C (the temperature was selected based on the literature data [4, 5]) with an exposure time of 15 minutes were subjected to samples from REC ABS plastic in the amount of 3 pieces.

The tensile strength was 21±3 MPa, the tensile strength was 17±8 MPa.

Tensile tests at a negative temperature of –60 °C in nitrogen vapor with an exposure time of 15 minutes (the mode is selected based on literature data) samples from REC ABS plastic were subjected in the amount of 4 x.

The tensile strength was 41±18 MPa, the modulus of elasticity was 1550±308 MPa.

When trying to test the sample at a temperature of –196 °C in a snap-in that ensures a constant stay in liquid nitrogen with an exposure of 15 minutes, the sample collapsed during its installation in a bursting machine without fixing the moment of destruction. An optical study of the destruction sites was carried out on the destroyed sample.

Testing of samples made of ABS plastic for impact strength

The MK-30 pendulum coper was used for testing. In accordance with the GOST 4647-2015 methodology, when the sample is destroyed, the minimum reduction in impact energy should be at least 10%. To do this, the impact mechanism of the pendulum copra was installed at different impact heights to perform the technique. Due to the design features of the MK-30 pendulum copra, the reduction in impact energy, even from the lowest possible height, was less than 10%.

It was not possible to obtain reliable values of impact strength. To obtain reliable data, it is necessary to test samples on pendulum copra with lower impact energy and higher sensitivity.

Identification of structural features of ABS plastic samples

The visual inspection of the samples made of ABS plastic from REC company that passed the tests was carried out using optical devices (a camera and an Olympus SZX10 digital microscope).

During the visual inspection, it was revealed that the surface of the samples is white with a gloss that makes visual inspection difficult, while:

- the bottom layer, which was applied first to the technological platform of the 3D printer, is smooth, repeats its surface. If dirt particles or irregularities were present on the surface of the platform, then depressions and defects are formed on the lower layer, repeating the irregularities;
- the upper layer has a ribbing, the lateral surface of the samples also has this defect (Figure 3). The number of layers on the side surface of the sample corresponds to twenty, which corresponds to the application of the material in layers 0.2 mm thick.

a) *b)*

Fig. 3. *Photos of the ribbing on the side and upper surfaces of the sample*
a) – the lateral surface of the sample with a thickness of 4 mm; b) – swellings on the upper surface

Materials Research Proceedings **21** (2022) 133-139 https://doi.org/10.21741/9781644901755-24

In order to determine the interlayer contact, some samples were destroyed by striking with a tool that creates delaminating forces in two directions: perpendicular and parallel to the plane of the layers. The material withstands a blow perpendicular to the plane of layer construction with deformation, a blow parallel to the plane of layer construction led to the destruction of the sample into two parts (Figure 4). The interlayer boundary and the ordered placement of internal structures are clearly visible, which makes it possible to declare the non-monolithicity of the samples and the presence of oriented pores in the sample structure.

a) b)

Fig. 4. *Photos of parts of samples after impact with a tool that creates delaminating forces in two directions: a – impact perpendicular to the layers; b) – impact parallel to the layers*

Figure 5 shows the place of destruction of the sample during a tensile test at a temperature of –196 °C and an exposure time of 15 minutes. At very low temperatures, there are no plastic deformations in polymer materials during destruction, and they break down brittle. Accordingly, the internal structure of the sample is visible in the places of destruction without plastic deformations. Triangular-shaped pores with a base from 60 to 90 microns are clearly visible. These pores are formed during the creation of the layer and are regular throughout the entire volume of the sample. A single applied structural element (a thread with a D-shaped profile) is clearly fixed. The lower platform of the structural element occupies approximately 320 microns, the upper platform is ≈380 microns, the lateral contact between the structural elements is up to a pore of160160 microns with a layer thickness of ≈200 microns.

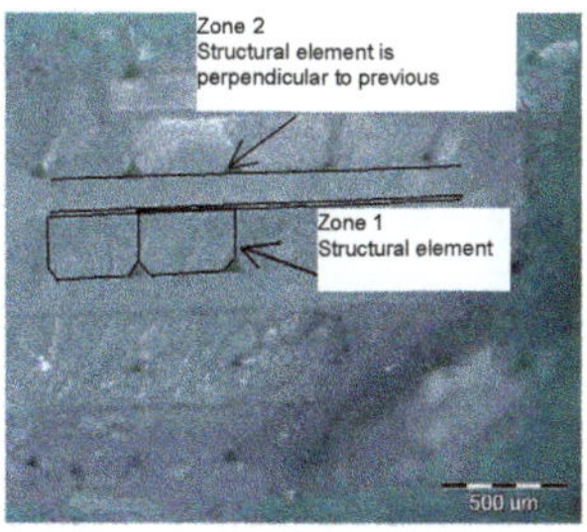

Fig. 5. *Photo of the place of destruction of the sample after mechanical action during installation in the bursting machine at a temperature of –196 °C*

In addition, the orientation of individual elements is clearly fixed. You can easily allocate 2 zones:

– zone 1 – a chain of threads (shaped like the letter D), the layer has a direction to the observer;

– zone 2 is a strip with a light (side surface of a single element) and a dark (pore area) area, the layer has a direction along the drawing.

The hydrostatic weighing method determined the density 1.03 g/cm^3, however, due to the presence of air pores in the material, it is likely that the obtained value is not correct. The pores pass through the entire material in directions that coincide with the orientation of the structural elements, so the declared technological mode of 3D printing "100% filling" actually corresponds to approximately ≈95%. It was not possible to obtain photos of the destruction site at the edge of the sample due to the glossy surface of the material and large height differences (focusing optical devices does not give a clear image).

Figure 6 shows photos of the destruction of REC ABS plastic samples at normal and elevated temperatures. The internal structure is clearly visible, which (in comparison with Figure 5) is distorted by deformation from the mechanical load of the sample during testing.

a) b)

Fig. 6. *Places of destruction of samples during tensile tests at positive temperature: a) – photo of the place of destruction near the lateral edge, the separation of structural elements after stretching is visible; b) – the place of destruction, capturing all layers, layers with different orientations are visible*

Analyzing Figure 6, it is possible to determine the distribution and orientation of structural elements along the cross-section of the sample (Figure 7).

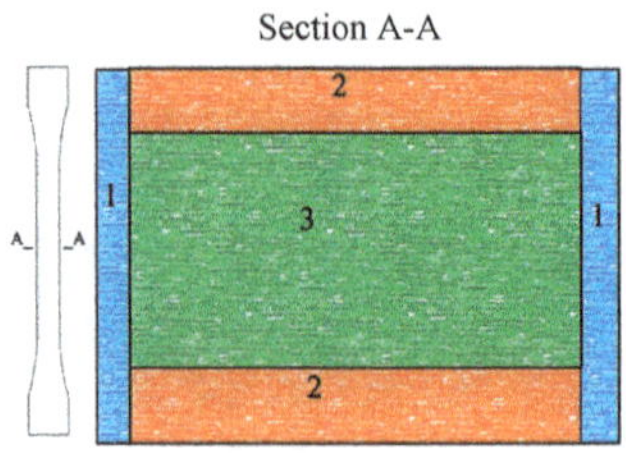

Fig. 7. *Distribution and orientation of structural elements in the sample (the sample has 20 layers with a thickness close to 0.2 mm):*

Zone 1. Structural elements forming the contour of the layer during the construction of the sample.

Zone 2. Outer surface, thickness is 4 layers with alternating orientation of layers of structural elements ± 45°.

Zone 3. The internal structure of the sample, the thickness is 12 layers with alternating orientation of layers of structural elements 0° and 90°

Results and discussion

Based on the tests of ABS plastic, the following conclusions can be drawn. Using 3D printing, it is possible to obtain plastic parts of a given configuration, taking into account structural features, in relation to the manufacture of models for testing in wind tunnels with different flow rates. The mechanical characteristics and features of the internal structure of the ABS plastic material of various brands are determined. During the tensile testing of the samples at normal temperature, an average tensile strength of 30 MPa was obtained, which is approximately 70% of the strength of the injection grades (the average tensile strength of injection grades is 42.5 MPa) and 83% of the Polish equivalent of the ABS-M30. The average value of the tensile modulus of elasticity is 1570 MPa, which is approximately 70% of the modulus of elasticity of injection molding grades (the average value of the modulus of elasticity of injection molding grades is 2300 MPa) and 65% of the Polish equivalent of the ABS-M30. The lag of ABS plastic from injection-molded ABS brands is due to a feature of 3D printing technology, namely, layer-by-layer application of the material.

In the process of tensile testing of samples at elevated temperatures, an average tensile strength of 21 MPa was obtained (it drops by 30% compared to the tensile strength at normal temperature).

When testing the samples for tensile strength at a reduced temperature, an average tensile strength of 41 MPa was obtained (it increases by 37% compared to the tensile strength at normal temperature) and the average tensile modulus of 1550 MPa (compared to the elastic modulus at normal temperature practically does not change).

The temperature range of the tests (from –60 ° C to +70 ° C) was wider than the operating interval from –40 ° C to +75 ° C and shows that when the temperature drops to –60 °C the material is operable. The peculiarity of this material is the presence of several links in the molecular chain of the material (styrene, butadiene and acrylonitrile), which complicates the construction of a complete picture of the behavior of the material. To clarify the temperature of frost resistance, a number of different tests are required at lower temperatures.

In the process of optical examination of the destruction sites , the following structural features of the samples were revealed:

The temperature range of the tests (from –60 °C to +70 °C) was wider than the operating interval from –40 °C to +75 °C and shows that when the temperature drops to –60 °C, the material is operable. The peculiarity of this material is the presence of several links in the molecular chain of the material (styrene, butadiene and acrylonitrile), which complicates the construction of a complete picture of the behavior of the material. To clarify the temperature of frost resistance, a number of different tests are required at lower temperatures.

In the process of optical examination of the destruction sites , the following structural features of the samples were revealed:

- during the application of the layer, pores are formed along the thread of the material throughout the part (the presence of pores raises the question of water entering the pores and turning it into ice at negative temperatures with the creation of tension from the inside);
- the pores have a shape close to triangular, with the side of the base of the triangle from 60 to 90 microns and a height of 40 microns;
- filling of samples with 100% 3D printing settings is actually approaching 95% due to the peculiarity of the technology – the pores formed;
- anisotropy of the strength index is observed depending on the direction. Interlayer strength is the weakest part of this material, which must be taken into account when manufacturing products using 3D printing technology.

Conclusions

The conducted studies have shown the possibility of using additive technologies and ABS plastic for the manufacture of various geometrically similar models of aircraft and rocket and space technology for testing in wind tunnels. The use of such models significantly reduces the actual cycle of their manufacture – from one year to several days, reduces manufacturing costs by more than 60%, reduces the volume of documentation for the manufactured model by an order of magnitude, and also increases the reliability of their design. However, due to the presence of pores in the structure of ABS plastic, which is a feature of FDM technology, it is impossible to manufacture drained aerodynamic models, since the material does not ensure their tightness. Samples made of ABS plastic have a rough surface and require additional processing to ensure a given purity. To create models with high requirements, it is necessary to use a different type of technology, or the use of a material that ensures the absence of internal defects and high quality geometry.

References

[1] Volkhonsky, A.E., Gudkov, K.V. Methods of manufacturing prototypes and parts of aggregates of various industrial products using additive technologies. // Educational Technologies, No.1, 2014, pp. 127–143.

[2] Panferova, E.A., Strukov, A.N., Strukova, A.V., Mishchenko, V.Yu. Methods of prototyping products in the aerospace industry and the equipment used. // Flight, 2013, No. 7. pp. 29–33.

[3] Volkhonsky, A.E., Dudkov, K.V. Manufacturing of "pad" type parts by FDM prototyping. // Polymer Materials, No.3 (178), 2014. pp. 42–45.

[4] Astapov, V.Yu., Khoroshko, L.L., Giordani, M.S., Khoroshko, A.L. CAD in modeling the modes of technological processes for the production of structural elements of aircraft // Proceedings of MAI. 2016. No. 87. URL: http://trudymai.ru/published.php?ID=69638

[5] Laboratory of three-dimensional printing. [Electronic document] – Access mode: http://www.lab3dprint.ru/support/425-abs-vs-pla.

[6] MakerBot Replicator+3D printer. [Electronic document] – Access mode: https:/www.makerbot-gr/shop-makerbot-products/makerbot–replivator-2x-experimental-3d-printer/

Materials Research Proceedings **21** (2022) 140-143 https://doi.org/10.21741/9781644901755-25

Development the Graphical Method of Calculating Structural Elements for Fixing a Gear in a Self-Centering Chuck through the Involute

Alexander S. Serkov

Omsk State Technical University, Mira 11, Omsk, 644050, Russia

Sanya_93@bk.ru

Keywords: Gear Manufacturing, Planetary Gears, Gearwheel Fixing, Grinding, Boring

Abstract. The article discusses the developed graphical method of calculating the size rolls diameter and the radius of their contact with the jaws.

Introduction

In the manufacture of gears (planetary gears, gearwheels), in the process of their production, there are difficulties related to ensuring the shape and size accuracy of the gearwheel hole involute profile, as well as their relative position relation to each other [1]. During chemical-thermal and heat treatments (cementation, hardening, etc.) of the gear, warping of its surfaces appears, an error of eccentricity arises. Therefore, finishing operations (grinding of working and base surfaces) are introduced into the technological process. The influence of the gear hole shape is described in [2-5].

In the gears manufacturing the final processing of its central hole, two fixing charts (locating, clamping) are mainly used:

1) along an involute profile through size rolls in a self-centering chuck [6–15];

2) along the outer contour of the gearwheel (without size rolls) in a self-centering chuck;

Initial data: $\alpha = 20^{\circ}$ - profile angle of the initial contour; $r = 22.5$ mm - the radius of the gearwheel pitch circle; $Re = 25$ mm - outer radius of the gearwheel; $m = 2.5$ mm - gearing module; $z = 18$ - number of teeth. The analytical calculation of the device structural elements can be performed according to the method described in [16, p. 183].

The fixing chart calculation by the graphical method is carried out using a simplified way of drawing the tooth profile with the subsequent selection of the size rolls diameter, so that the contact points of the teeth profile with the rolls lie as close as possible to the radius of the gearwheel pitch circle. Let's consider this tracing with an example.

All tracings are performed on a computer in the modern computer-aided design system Compass V19.1 (Any other CAD system can be used for tracing).

Main body

From point O the radiuses of the circles r_{VP}, r_O, r, R_e (Fig. 1) are drawn. From point O at the gear vertical axis, angle φ is postponed. From the point of intersection of the right angle φ with the radius of the pitch circle, are being drawn a chord in both directions equal to $S_H/2$ (the tooth thickness along the chord of the pitch circle). At both ends of the chord, points A and B are being mark. Point A is being connected with center O by a straight line. The center O_1 are being found by dividing the line OA by 2. From the center O_1 the radius of the circle R_1=OA/2=r/2 is being drawn crossing point A, draw it to the intersection with the main circle radius r_O, find the point O_2. Draw an arc CAE from point O2 with radius R2. The resulting curve is a tooth involute. Point B is being connected with center O with a straight line. Dividing line OB by two, we find the center

O_1 '. From the center O_1 'draw the radius of the circle R_1'=OB/2 = r/2, crossing point B, draw to the intersection with the radius of the main circle r_O, find the point O_2 '. Draw an arc KBF from point O_2 'with radius R_2'. The resulting arch is a tooth involute. The CAEFBK points are the tooth profile outline. Using the function (mirror image relative to the gear vertical axis), is being obtained the profile of the second tooth. The teeth root is being constructed using a circle (a circle tangent to three curves - two involutes and the radius of the circle of the root r_{VP}). Circle is being entered d '= 5 mm between two teeth (diameter of the size rolls).

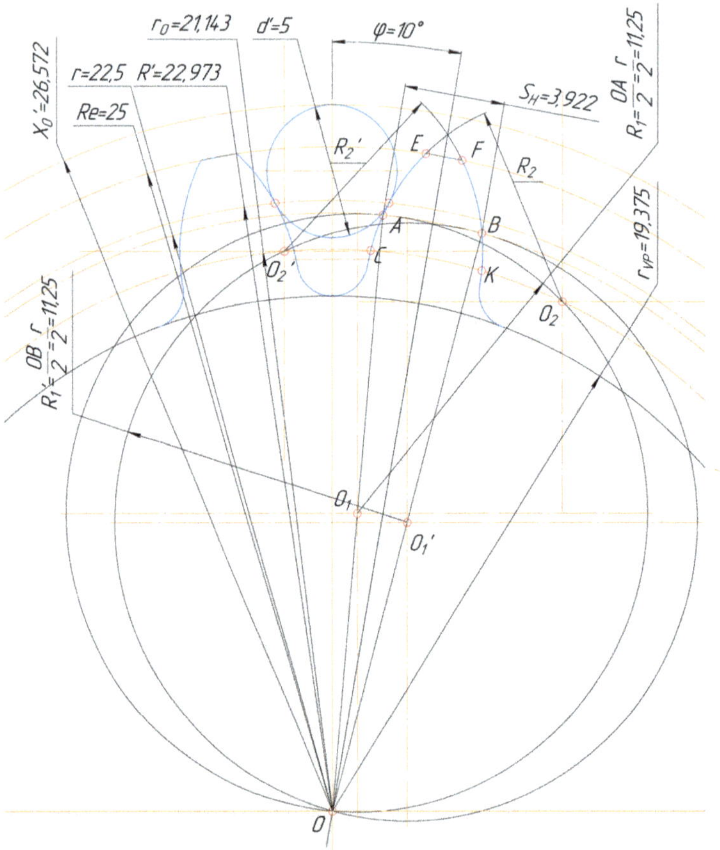

Fig. 1. Calculating chart for the roll diameter and the contact radius with the jaws by the graphical method.

To determine the radius value of the circle on which the contact points of the roll with the teeth are located, using the snapping from the center O, a circle draw through the contact points of the circle d ’with the teeth, graphically defining that R’ = 22.973 mm.

To determine the value of the chuck jaws contact radius with the rolls at the moment of fixing when the gearwheels are being machined, from the center O a circumscribed circle is being drawn around the size rolls. Graphically are being determined that X_O’= 26.572 mm.

Conclusion

The developed graphical calculation method is convenient for use in production, easy to understand, does not require calculating the engagement angle involute or finding its value in tables. It is possible to fully automate the construction of a tooth profile in CAD - Compass V19.1.

Acknowledgments

The reported study was funded by RFBR, project number 20-38-90226.

References

[1] A. S. Serkov, V. B. Masyagin, L. B. Serkova, Optimizatsiya tekhnologii proizvodstva shesterney, zubchatykh kolos, satellitov, Izvestiya Tul'skogo Gosudarstvennogo Universiteta. Tekhnicheskiye Nauki. 14 (2021) 134-139.

[2] O.V. Zhed', V.V. Kopylov, Vliyaniye formy otverstiya zubchatogo ventsa na yego predvaritel'noye napryazhonnoye sostoyaniye pri sborke s bazovoy detal'yu Chast' 1, Gruzovik. 12 (2017) 14 -15.

[3] O.V. Zhed', V.V. Kopylov, Vliyaniye formy otverstiya zubchatogo ventsa na yego predvaritel'noye napryazhonnoye sostoyaniye pri sborke s bazovoy detal'yu Chast' 2, Gruzovik.1 (2018) 18-22.

[4] V.A. Rogov, O.V. Zhed', Vliyaniye predvaritel'nogo napryazhonnogo sostoyaniya zub-chatogo ventsa vsledstviye natyaga s bazovoy detal'yu v raschotakh na izgibnuyu prochnost' zuba, Tekhnologiya mashinostroyeniya.8 (2018) 31–39.

[5] O. Zhed, A.Koshelenko, K. Khishova, The effect of the stressed state of the ring gear due to interference with the base part in calculating the flexural strength of the tooth, IOP Conference Series: Materials Science and Engineering, 709 (2020) 022035-1–022035-8. https://doi.org/10.1088/1757-899X/709/2/022035

[6] A. S. Serkov, I. I. Sayfulin, O.V. Palina, Snizheniye sebestoimosti shlifovaniya posadochnykh otverstiy i tortsov tsilindricheskikh pryamozubykh zubchatykh koles za schot vnedreniya universal'noy osnastki, Innovatsii, kachestvo i servis v tekhnike i tekhnologiyakh, Kursk, (2017) 323–328.

[7] A.S. Serkov, L.B. Antropova, Vnedreniya universal'noy osnastki v operatsiyu shlifovaniya tsentral'nykh otverstiy tsilindricheskikh pryamozubykh zubchatykh kolos, Metrologiya, standartizatsiya i upravleniye kachestvom, Omsk (2018) 105–108.

[8] A.S. Serkov, L.B. Antropova, Shlifovaniye tsentral'nogo otverstiya tsilindricheskogo pryamozubogo zubchatogo kolesa s pomoshch'yu universal'noy osnastki, Problemy effektivnogo ispol'zovaniya nauchnogo potentsiala obshchestva, Sterlitamak, 3 (2019), 178–181.

[9] A.S. Serkov, L.B. Antropova, Obrabotka otverstiya i tortsa tsilindricheskogo pryamozubogo zubchatogo kolesa s pomoshch'yu universal'noy osnastki metodom shlifova-niya, Nauchnyye revolyutsii: sushchnost' i rol' v razvitii nauki i tekhniki, Sterlitamak, 1 (2019) 33–36.

[10] A.S. Serkov, L.B. Antropova, S.F. Yeletskaya, Primeneniye universal'nykh trokhkulachkovykh patronov dlya shlifovaniya otverstiy v tsilindricheskikh pryamozubykh zubchatykh kolos, Metrologiya, standartizatsiya i upravleniye kachestvom, Omsk, (2019) 81–84.

[11] A.S. Serkov, Vnedreniye universal'noy osnastki v protsess shlifovaniya otverstiya i tortsa zagotovki satellita, Izvestiya Tul'skogo Gosudarstvennogo Universiteta, Tekhnicheskiye Nauki, 4 (2020) 316-320.

[12] A.S. Serkov, L.B. Serkova, YU.Ye. Zhdanova, V.B. Masyagin, Vnedreniye universal'noy osnastki v protsess shlifovaniye otverstiya i tortsa zagotovki satellita, Molodozh' i sistemnaya modernizatsiya strany: sbornik nauchnykh statey 5-y Mezhdunarod-noy nauchnoy konferentsii studentov i molodykh uchonykh, Kursk, (2020), 320-323.

[13] Information on http://www.avtomash.ru/gur/2005/20050244.htm

[14] Information on http://www.avtomash.ru/gur/2004/20041246.htm

[15] A.A. Malikov, G.V. Malakhov, A.V. Mikhaylov, Opredeleniye parametrov zubchatogo ventsa zagotovok s predvaritel'no oformlennymi zub'yami, Izvestiya Tul'skogo Gosudarstvennogo Universiteta, Tekhnicheskiye Nauki, 8 (2017) 344-353.

[16] KH. L. Bolotin, F.P. Kostromin, Stanochnyye prisposobleniya, M.: Mashinostroyeniye, (1973)

Investigation of the Orientational Mechanical Properties of Biodegradable Extrusion Films Based on Polyolefins and Beet Pulp

Anton M. Kuzmin[1,a*], Alexander A. Shabarin[1,b] and Sachin S. Raj[2,c]

[1] National Research Mordovia State University, Bolshevistskaya st., 68, Saransk, Republic of Mordovia, Russia 430005

[2]Department of Mechanical Engineering, Gnanamani College of Technology, Namakkal, Tamilnadu, India 637018

[a]kuzmin.a.m@yandex.ru, [b]shab_aa@mail.ru, [c] sachinsraj1991@gmail.com

Keywords: Technological Process, Packaging Films, Cast Film Extrusion, Biodegradable Composite, Polyethylene, Polyethylene Glycol, Beet Pulp

Abstract. The article discusses the possibility of obtaining biodegradable films based on polyolefins and beet pulp by the extrusion method. Biodegradable composites of two mixes with 15% and 25% beet pulp content have been obtained. Compounding was carried out on a twin-screw extruder, and then samples of biodegradable films were obtained by cast film extrusion. The influence of the vegetable filler particles' orientation on the composites mechanical properties has been studied. It has been shown that composites mechanical properties significantly increase in the direction of polymer melt stretching.

Introduction

Filling polymers with vegetable fillers such as wood flour, straw, husk of cereal crops reduces the cost of polymer compositions and can significantly reduce the environmental burden after their use, which is especially important for disposable packaging materials [1-12]. Recently, there have appeared a lot of studies [4-25], which describe the compositions of biodegradable composites based on thermoplastic polymers and natural fillers such as starch, cellulose, rice husks, wood flour, and cereal straw.

The use of film polymer materials in agroindustrial complex is becoming more and more widespread [20-21]. This is due to their low price and excellent operational characteristics. However, after the end of their service life, these materials accumulate in landfills, where they slowly decompose and pollute the environment. Therefore, the development of compositions of biodegradable polymeric materials as well as the development of their production technological processes are currently the priority tasks [24]. The effect of biodegradability can be achieved by introducing vegetable fillers into synthetic thermoplastics [4-17]. In the Russian Federation, the main by-product of sugar beet production is beet pulp. Beet pulp - de-sugar beet shavings remaining after extraction of sugar from it by the diffusion method. The pulp is dried, granulated and pressed into briquettes. Fresh beet pulp contains (in% of dry matter weight): pectin – 24…32, cellulose – 22…30, glycans – 22…30, proteins – 1.5…3.0, ash – 3.0…8.2, lignin – 1.5…3.0, simple sugars – 0.2…0.3. In our opinion, the priority direction of beet pulp recycling can be its use as a filler for polymers.

Studies [11-15] show the use of polyolefins, wood and flax flour to obtain biodegradable compositions. Malaysian scientists [16-19] studied the biocomposites based on various binders with plant waste from sugar palm. These biocomposites have high mechanical properties and thermal stability. However, as the authors note, these biocomposites are limited in widespread use in everyday life.

The purpose of this study was to refine the technological extrusion process for obtaining biodegradable films and study their mechanical properties.

Methods and materials

To produce biodegradable film, the following materials were used:

- low-pressure polyethylene of 273-83 grade, high-pressure polyethylene of 15303-003 grade, ethylene-vinyl acetate copolymer of 12206-007 grade, functionalized by the method of alkaline alcoholysis, were used as a polymer matrix;
- polyethylene glycol of 115 (4000) grade was used as a lubricant;
- crushed beet pulp with a particle size of less than 200 microns and a moisture content of less than 5% was used as a vegetable filler.

Table 1 shows the percentage ratios of polymer composites components used to develop the technological process for biodegradable films production.

Table 1. *Compositions of biodegradable polymer composites*

	Composition 1	Composition 2
HDPE 273-83	37	32
LDPE 15303-003	37	32
Sevilen 12206-007	10	10
Beet pulp	15	25
PEG 115	1	1

At the first stage, the components were mixed in a co-rotating twin-screw extruder to obtain homogeneous granules [2].

At the second stage, a laboratory line was assembled for the biodegradable films production by the cast film extrusion method (Fig. 1), including a drive and measurement unit HAAKE RheoDrive 7 OS, single-screw extruder Rheomex 19/25 PolyLab OS with barrel length L/D = 25, screw 3: 1 Standard Metering, forming extrusion sheet die with 150 mm width and adjustable thickness 0.2 ... 1.2 mm, three-roll sizing device with a pulling unit.

Fig. 1. *Laboratory line for the production of films by cast film extrusion method*

The extrusion temperature was 160° C in all areas. Screw rotation speed was 200 rpm. The torque of the screw shaft during the extrusion process did not exceed 50 Nm. The pulling device speed was 60% higher than the linear speed of rolls rotation of the sizing device, i.e. the longitudinal deformation of the film (stretching) was 1.6 in this case.

The mechanical properties of the composites were obtained in accordance with GOST (State Standard) 11262-2017 and GOST 9550-81. The methods used correspond to international test standards [4]. The tests were carried out on a UAI-7000 M tensile testing machine at a temperature of 23 ± 2° C and a clamping speed of 10 mm/min.

Results and Discussions

The obtained results of mechanical properties are presented in the Table 2.

Table 2. *Physical and mechanical characteristics of biodegradable composite materials with 15% and 25% beet pulp filling.*

Filler content, %	Elastic modulus, MPa	Relative extension, %	Nominal strength limit, MPa	Fracture strength, MPa
15	195 ± 8	141	10,5 ± 0,50	25,3
25	170 ± 11	32	10,4 ± 0,50	13,7

Table 2 shows that the mechanical characteristics of the developed composites are at a high level, which makes it possible to use them for the production of biodegradable packaging films.

Figure 2 shows the appearance of the extruded melt and the entry melt zone from the sheet die into the three-roll sizing device.

Figure 3 shows the films appearance with 15% and 25% filling.

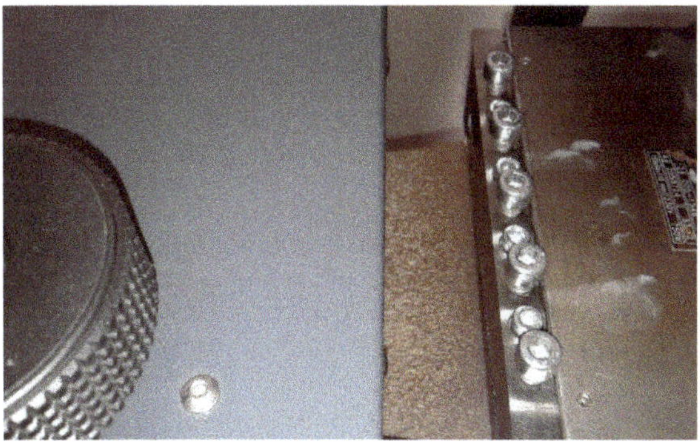

Fig. 2. *Appearance of the extruded melt and the entry melt zone from the sheet die into the three-roll sizing device.*

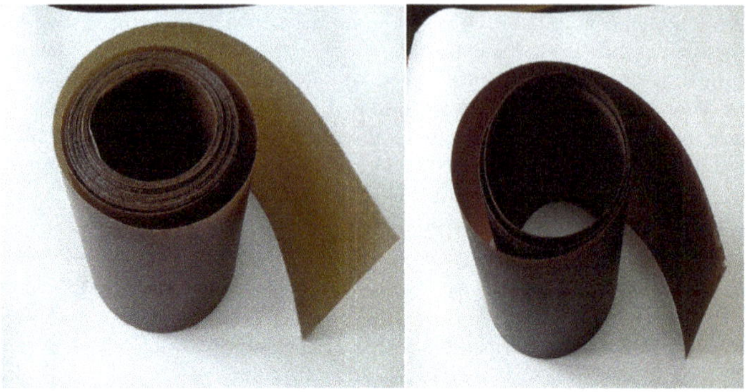

a b

Fig. 3. *Appearance of the obtained films with the content of beet pulp 15% (a) and 25% (b)*

The films obtained were tested for strength along and transverse the direction of stretching according to the above methods. The test results are presented in Table 3.

Table 3. *Stress-strain indicators of films from compositions filled with beet pulp.*

Composition and loading conditions of films		Elastic modulus, MPa	Tensile strength, MPa	Relative extension, %
15 % of pulp	Along stretching	244 ± 21	28,3 ± 3,8	91 ± 19
	Transverse stretching	189 ± 5	10,9 ± 0,12	38 ± 10
25 % of pulp	Along stretching	201 ± 15	16,4 ± 2	92 ± 29
	Transverse stretching	155 ± 4	10 ± 0,4	31 ± 4

According to the results obtained, it follows that during the films stretching a significant increase in strength and elastic modulus in the direction of stretching are achieved. For composites with 15%-filling, the increase in strength was more than 240%, and the increase in elastic modulus was about 30%. For composites with 25%-filling, the increase in strength was more than 160%, and the increase in elastic modulus was about 20%. In the transverse direction, the elastic and strength indicators correspond to the physical and mechanical characteristics of the composites obtained by the compression molding method. The obtained results of biodegradable composites mechanical properties correlate with the results of previous studies [3, 10, 14, 25].

Conclusion

Thus, under laboratory conditions, the possibility of producing films strengthened by uniaxial stretching from the developed biodegradable composites by the method of cast film extrusion has been shown. These characteristics determine the possibility of producing reliable products for packaging agricultural and food products by pneumatic vacuum molding.

References

[1] A.M. Kuzmin, Influence of compatibilizator on the operational and technological properties of thermoplastic composites filled with fine barley straw, Lecture Notes in Civil Engineering, 95, (2021) 325-332. https://doi.org/10.1007/978-3-030-54652-6_49

[2] A.M. Kuzmin, E.A. Radaikina Technology development for the production of thermoplastic composites with agricultural fillers by compounding method on co-directional twin screw extruder, IOP Conference Series: Materials Science and Engineering. 873, 1, (2020) 012022. https://doi.org/10.1088/1757-899X/873/1/012022

[3] S.S. Raj, T.K. Kannan, R. Rajasekar, Effect of wood particulate size on the mechanical properties of PLA biocomposite, Pigment and Resin Technology. 49(6), (2020) 465-472. https://doi.org/10.1108/PRT-12-2019-0117

[4] S.S. Raj, A.M. Kuzmin., K. Subramanian, S. Sathiamoorthyi, K.T. Kandasamy, Philosophy of Selecting ASTM Standards for Mechanical Characterization of Polymers and Polymer Composites, Materiale Plastice, 58(3), (2021) 247-256. https://doi.org/10.37358/MP.21.3.5523

[5] A. Hejna, Potential applications of by-products from the coffee industry in polymer technology – Current state and perspectives, Waste Management. 121, (2021) 296-330. https://doi.org/10.1016/j.wasman.2020.12.018

[6] A. Hejna, M. Barczewski, J. Andrzejewski, P. Kosmela, A. Piasecki, M. Szostak, T. Kuang, Rotational molding of linear low-density polyethylene composites filled with wheat bran, Polymers. 12, 5, (2020) 1004. https://doi.org/10.3390/polym12051004

[7] R.S. Sundaram Pillai, R. Rajamoni, I. Suyambulingam, D. Divakaran, Effect of Fiber Loading Rate on Various Properties of the Fiber Reinforced Polymer Composites. In book: Fracture Failure Analysis of Fiber Reinforced Polymer Matrix Composites (2021)

[8] B.G. Babu, D. Princewinston, S.S. Saravanakumar, A. Khan, P.V. Aravind Bhaskar, S. Indran, D. Divya, Investigation on the Physicochemical and Mechanical Properties of Novel Alkali-treated Phaseolus vulgaris Fibers, Journal of Natural Fibers, 1, 12. (2020). https://doi.org/10.1080/15440478.2020.1761930

[9] J. Neto, H. Queiroz, R. Aguiar, R. Lima, D. Cavalcanti, M.D. Banea, A Review of Recent Advances in Hybrid Natural Fiber Reinforced Polymer Composites, Journal of Renewable Materials, 10(3), (2022) 561–589. https://doi.org/10.32604/jrm.2022.017434

[10] H. Queiroz, M.D., Banea, D. Cavalcanti, J. Neto, The effect of multicale hybridization hybridization on the mechanical properties of natural fiber reinforced composites, Journal of Applied Polymer Science, 138(41), (2021) 51213. https://doi.org/10.1002/app.51213

[11] A.K. Zykova, P.V. Pantyukhov, E.E Mastalygina., C. Chaverri-Ramos, S.G. Nikolaeva, J.J. Saavedra-Arias, M. Poletto, Biocomposites of Low-Density Polyethylene Plus Wood Flour or Flax Straw: Biodegradation Kinetics across Three Environments, Polymers, (2021). 13(13), 2138. https://doi.org/10.3390/polym13132138

[12] M.Schneider, N.Finimundi, M. Podzorova, P.Pantyukhov, M. Poletto, Assessment of Morphological, Physical, Thermal, and Thermal Conductivity Properties of Polypropylene/Lignosulfonate Blends, Materials, 14(3), (2021) 543. https://doi.org/10.3390/ma14030543

[13] A. K Zykova, P. V. Pantyukhov, N.N. Kolesnikova, T. V. Monakhova, A. A. Popov, Influence of Filler Particle Size on Physical Properties and Biodegradation of Biocomposites

Based on Low-Density Polyethylene and Lignocellulosic Fillers, Journal of Polymers and the Environment, 26(4), (2017) 1343–1354. https://doi.org/10.1007/s10924-017-1039-9

[14] A. Zykova, P. Pantyukhov, A. Popov, Ethylene-octene copolymer-wood flour/oil flax straw biocomposites: Effect of filler type and content on mechanical properties, Polymer Engineering & Science, 57(7), (2017) 756–763. https://doi.org/10.1002/pen.24626

[15] M.V. Podzorova, Y.V. Tertyshnaya, P.V. Pantyukhov, L.S. Shibryaeva, A.A. Popov, S. Nikolaeva, Influence of different factors on the destruction of films based on polylactic acid and oxidized polyethylene, AIP Conference Proceedings 1783, (2016) 020185. https://doi.org/10.1063/1.4966479

[16] R.A. Ilyas, S.M., Sapuan, M.M., Harussani, M.Y.A.Y. Hakimi, M.Z.M. Haziq, M.S.N. Atikah, M. Asrofi, Polylactic Acid (PLA) Biocomposite: Processing, Additive Manufacturing and Advanced Applications, Polymers, 13(8), (2021) 1326. https://doi.org/10.3390/polym13081326

[17] N.M. Nurazzi, M.R.M. Asyraf, S.F. Athiyah, S.S. Shazleen, S.A. Rafiqah, M.M. Harussani, A. Khalina, A Review on Mechanical Performance of Hybrid Natural Fiber Polymer Composites for Structural Applications, Polymers, 13(13), (2021) 2170. https://doi.org/10.3390/polym13132170

[18] A. Atiqah, M. Jawaid, S.M. Sapuan, M.R. Ishak, M.N.M. Ansari, R.A. Ilyas, Physical and thermal properties of treated sugar palm/glass fibre reinforced thermoplastic polyurethane hybrid composites, Journal of Materials Research and Technology. 8 (5) (2019). 3726-3732. https://doi.org/10.1016/j.jmrt.2019.06.032

[19] M.A. Azman, M.R.M. Asyraf, A. Khalina, M. Petrů, C.M. Ruzaidi, S.M. Sapuan, M.J. Suriani, Natural Fiber Reinforced Composite Material for Product Design: A Short Review, Polymers, 13(12), (2021) 1917. https://doi.org/10.3390/polym13121917

[20] Aashima, P. Singh, A.S Chatli, A Review on Biodegradable Films, International Journal of Current Microbiology and Applied Sciences 9(5) (2020) 3039-3051. https://doi.org/10.20546/ijcmas.2020.905.362

[21] A. Afzal, Z. Khaliq, S. Ahmad, F. Ahmad, A. Noor, M.B. Qadir, Development and characterization of biodegradable composite film, Environmental Technology & Innovation, 23, (2021) 101664. https://doi.org/10.1016/j.eti.2021.101664

[22] P. Jagadeesh, M., Puttegowda, S. Mavinkere Rangappa, S. Siengchin, Influence of nanofillers on biodegradable composites: A comprehensive review, Polymer Composites. 1-21 (2021). https://doi.org/10.1002/pc.26291

[23] J.T. Varghese, N.S. Sarath Raj, G. Jiji, Development of biodegradable composites and investigation of mechanical behaviour, Materials Today: Proceedings, 38, (2021) 3378–3385. https://doi.org/10.1016/j.matpr.2020.10.478

[24] A. Karimah, M. R. Ridho, S.S. Munawar, D.S. Adi, I. Notowiharjo, R. Damayanti, B. Subiyanto, W. Fatriasari A. Fudholi, A review on natural fibers for development of eco-friendly bio-composite: characteristics, and utilizations, Journal of Materials Research and Technology, 13, (2021) 2442–2458. https://doi.org/10.1016/j.jmrt.2021.06.014

[25] S.R. Theja Reddy, A. V. Ratna Prasad, K. Ramanaiah, Tensile and flexural properties of biodegradable jute fiber reinforced poly lactic acid composites, Materials Today: Proceedings, 44, (2021) 917–921. https://doi.org/10.1016/j.matpr.2020.10.806

Materials Research Proceedings **21** (2022) 150-155 https://doi.org/10.21741/9781644901755-27

The Use of Polymer Materials in the Composition of Asphalt Concrete

K.G. Pugin[1,2,a*], O.V. Yakontseva[2,b], V.K. Salakhova[2,c] and A.M. Burgonutdinov[2]

[1] Perm State Agro-Technological University named after Academician D.N. Pryanishnikov, Perm, Russian Federation

[2] Perm National Research Polytechnic University, Perm, Russian Federation

[a]123zzz@rambler.ru, [b]yakontseva.olga@yandex.ru, [c]veronika815@inbox.ru

Keywords: Asphalt Concrete, Plastic, Waste, Structure Formation, Road Construction, Polymers

Abstract. The research presented in the article shows the possibility of using some polymer materials in the composition of asphalt concrete. The analysis of scientific works in the field of asphalt concrete design shows the efficiency of using plastic waste to obtain asphalt concrete with specified physical, mechanical and operational properties. This technology allows not only to reduce the amount of accumulated plastic waste, but also to improve the condition of highways by increasing the strength of asphalt concrete. Polymers PP and LDPE were selected for the study. As PP samples, we used a nonwoven material used for the production of disposable medical masks, which is suitable for creating linear or dispersed reinforcement of asphalt concrete. LDPE was used as the second polymer sample. The obtained asphalt concrete samples differ from the traditional composition by their increased compressive strength at high temperatures while maintaining plasticity at low temperatures.

For the implementation of a number of state programs of the Russian Federation, in particular "Development of the transport system" and the national project "Safe and high-quality highways", it is necessary to solve a large number of tasks. The main goal of which is aimed at increasing the speed of goods movement, reducing the cost of transportation, improving road safety and improving the quality of highways. To fulfill the tasks set in these programs, large volumes of high-quality road-building materials, in particular asphalt concrete, are required, which would satisfy the growing external loads and have a low cost. It is possible to reduce the cost by using industrial and consumer waste in the structure of asphalt concrete, which makes it possible to obtain asphalt concrete with increased physical and mechanical properties at a low cost. [1-5]

The composition of the asphalt concrete mixture includes a large mineral aggregate (crushed stone), which constitutes the main power frame of the asphalt concrete, small mineral materials (sand, mineral powder) that regulate the mobility of bitumen and fill the voids in the structure. Bitumen of various viscosity is used as a binder material, which combines mineral materials of various sizes into a single monolith. To impart targeted properties, additives are used in the asphalt mixture. Additives can have various purposes, in particular stabilizing (TOPCEL, ITER PPS, KMA), dispersed reinforcement (FORTAFI), adhesive (ITERLENE PE, BN-1401b, APGS-1), plastomers and others. As a rule, additives have a structure and chemical composition different from rocks and bitumen, an increase in the characteristics of asphalt concrete is achieved by changing the structure or physical and mechanical properties of the raw materials used in the composition of the asphalt concrete mixture.

Analysis of scientific literature shows that one of the promising waste for regulating the processes of structure formation in asphalt concrete is polymer waste. They make it possible to obtain asphalt concrete with physical and mechanical properties that meet regulatory requirements and reduce the cost of producing asphalt concrete by saving natural raw materials. [6-10] The use of polymer waste (plastic) in asphalt concrete also solves the issue of waste disposal. The technology of replacing a part of natural raw materials in the composition of asphalt concrete and regulating the physical and mechanical properties with wastes of various nature is known and widely used. In particular, the following are used as mineral components: metallurgical slags; construction waste; waste molding sands; ash from solid fuel combustion; waste from drilling oil and gas wells; chemical production waste and other types of industrial waste with a similar structure to rocks. [11-15] Efficient use of the material resource of waste is possible if the physical-mechanical, chemical and environmental properties of the waste will be close to the properties of the substituted raw material. It is possible to use the specific properties of waste with a structure different from the structure of rocks, which can increase the strength or other characteristics of asphalt concrete. The involvement of waste in the technology of asphalt concrete production requires scientific justification and a large number of preliminary tests of samples of the resulting asphalt concrete, as well as the creation of pilot sites to clarify the operational properties.

Studies published in recent years show the efficiency of using plastic waste to obtain asphalt concrete with specified physical, mechanical and operational properties. [16-17] This technology allows not only to reduce the amount of accumulated plastic waste, but also to improve the condition of roads by increasing the plasticity of asphalt concrete. The optimal content of plastic waste in the composition of asphalt concrete, when part of the bitumen is replaced, is from 5% to 10%. This variation in content is due to the difference in the chemical composition of the plastics. Different types of plastic have different chemical composition, density, and melting point. The properties of the most common groups of plastics are presented in table. 1.

Table 1. *Properties of the most common groups of plastics.*

Plastic type	PET	HDPE	PVC	LDPE	PP	PS
Chemical composition	$(C_{10}H_8O_4)_n$	$(C_2H_4)_n$	$(C_2H_3Cl)_n$	$(C_2H_4)_n$	$(C_3H_6)_n$	$(C_8H_8)_n$
Melting temperature, °C	260	129-145	150-220	125-140	160-170	240
Density g/cm ³	1,38-1,40	0,96	1,35-1,43	0,93-0,97	0,90-0,92	0,90-0,92

The melting point of plastic largely determines the technology of its use in the composition of asphalt concrete. The temperature for preparing the asphalt concrete mixture ranges from 140 to 160 ° C, depending on the type of bitumen used. In this case, two points must be taken into account. The use of higher temperatures in the preparation of asphalt concrete mix leads to rapid aging of bitumen, with a loss of binding, adhesive and plastic properties. Heating plastics increases the risk of formation and release of environmentally hazardous chemicals that are added to plastics to add color, UV resistance, and strength. Studies show that when PET plastics are heated, antimony (a toxic substance of hazard class II) and other carcinogenic compounds are released. Plastics HDPE, LDPE are less hazardous, when they are heated up to 160 ° C, no hazardous substances are released. PVC plastics are one of the most dangerous plastics in existence, since when they are heated to a temperature of 160 ° C and above, carcinogens and lead compounds are released into the atmosphere. PP plastics are considered to be quite safe, however, under some conditions, polypropylene, when heated to more than 160 ° C, can release a biocide into the atmosphere. Polystyrene (PS) is hazardous due to the fact that when it is heated above 100 ° C, styrene is released, which belongs to the III hazard class. In this regard, HDPE, LDPE and PP plastics are the most suitable groups of plastics for the technology of producing asphalt concrete mixtures with ensuring environmental safety. Other groups of plastics can also be effectively used in the technology of producing asphalt concrete mixture, but for its implementation it will be necessary to change the technological scheme and modernize technological equipment to prevent environmental pollution.

The technologies for the use of plastics in the composition of asphalt concrete considered above use only material resource plastics in the form of a melt, replacing part of the volume of bitumen in the composition of an asphalt concrete mixture or acting as a modifier of bitumen, changing its physical and mechanical properties. Currently, there is the formation of polymer waste in the form of: disposable medical masks; a sheet; napkins; overalls; robes, as well as non-woven material used for the needs of agriculture, which are characterized by a homogeneous fine-fiber structure. This makes it possible to consider these wastes as a material for dispersed reinforcement or as a stabilizing component, which is a more efficient technology for using the material resource of the waste. Exposure for 30 minutes at 70 ° C and above effectively destroys the viruses that cause COVID-19.

The study involved medical masks made of spunbond nonwoven material, according to the standard GB 19083-2010 Technical requirements for protective face mask for medical use, manufactured by TESTE (China). The basis of the material is polypropylene. This material is part of the outer and inner layers of the mask. The appearance of the material is shown in Fig. 1. The outer layers of the disposable medical mask are made of larger PP threads, the inner layer is made of thinner PP. A cross-section of a three-layer disposable medical mask is shown in Fig. 2.

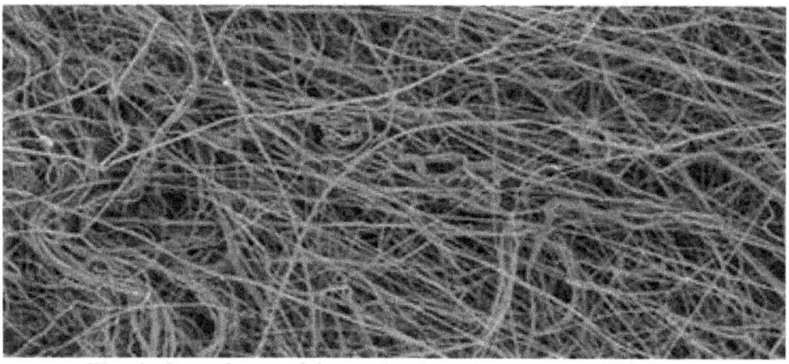

Fig. 1. *Outer layer of the medical mask. Increased 100 times.*

The structure of the nonwoven fabric used for the production of disposable medical masks is suitable for creating linear or dispersed reinforcement of asphalt concrete. Initially, before adding to the asphalt concrete mixture, you can adjust the length of the threads, which will be determined by the size of the pieces into which the material will be cut. Thus, it is possible to create linear or dispersed reinforcement of the asphalt concrete structure.

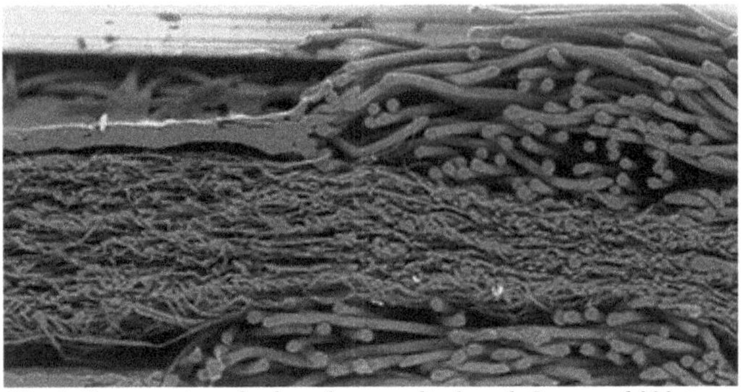

Fig. 2. *Cross section of a three-layer medical mask. Increased 300 times.*

The second plastic was polyethylene, plastic of the LDPE group (Polyethylene film, ST GOST 10354-82), 0.15 mm thick. The choice is justified by the fact that it is used in large volumes, the melting temperature is sufficient for the transition to a plastic state, linear reinforcement of asphalt concrete is possible, as well as partially dispersed. In view of the fact that the melting point is close to the temperature of preparation of the asphalt concrete mixture, it is possible that part of the mass of the film passes into the composition of bitumen, with the formation of modified bitumen.

The most widely used type of asphalt concrete mixture B, the second grade, was chosen as the basis for designing the composition of asphalt concrete samples. The composition of the mixture was used: crushed stone of igneous rocks, fractions 5-20 mm, quarry "Khromisty", Alapaevsk, Sverdlovsk region; sand from screenings of crushing of igneous rocks, fraction 0-5 mm, open pit "Volkovsky mine", Kushva, Sverdlovsk region; activated mineral powder from sedimentary rocks, open pit "Vereskovy", Nevyansk, Sverdlovsk region. BND 100/130 bitumen was used as a binder component of the asphalt concrete mixture. The amount of binder in all formulations was the same - 4.9% over the mineral part. Shredded medical masks and polyethylene were added in an amount of 0.5% by weight of bitumen. Forming and testing of asphalt concrete samples were carried out in the accredited laboratory of LLC Road Service Company, Alapaevsk, Sverdlovsk region in accordance with GOST 12801.

For a preliminary assessment of the performance of bitumen, in which polyethylene was partially dissolved, in particular, strength (shear, bursting), elasticity (determining frost resistance and durability of asphalt concrete), adhesion indicators that determine adhesion properties with mineral particles). With partial dissolution of plastic in bitumen, the obtained indicators of bitumen: the softening temperature of bitumen by KiSh was 58.6 ° C; penetration index 0.94; extensibility at 25 °C - 71 cm, at 0 °C - 4.5 cm.

The data obtained indicate that when plastic is introduced into bitumen, it is partially modified with an improvement in its elasticity at low temperatures and increased resistance to external loads at high temperatures. The modification took place in part due to the low heating temperature of the bitumen of 155 ° C, which was not enough to completely melt the plastic.

Comparative tests were carried out of asphalt concrete samples obtained from traditional raw materials (composition 1), using polyethylene in the form of pieces of 10x10mm film with a thickness of 0.15 mm (composition 2) and using pre-crushed (up to 20x20 mm) disposable medical masks (composition 3) .The main physical and mechanical parameters of the obtained asphalt concrete samples of various compositions are presented in table. 2.

Table 2. *Physical and mechanical properties of asphalt concrete samples*

Indicator name	Requirement of GOST 9123	Content 1	Content 2	Content 3
Average density, g/cm^3	Not standardized	2,53	2,51	2,51
Water saturation, %	from 1.5 to 4.0	2,69	3,12	3,12
Compressive strength: at t = 50°C, MPa	not less than 1.0	1,80	2,18	1,98
at t = 20°C, MPa	not less than 2.2	4,03	4,15	4,10
at t = 0°C, MPa	no more than 12.0	8,85	8,25	7,83
Water resistance	not less than 0.85	0,92	0,90	0,89

The above test results of asphalt concrete samples allow us to conclude that polyethylene (LDPE) and used disposable medical masks (plastic type PP) can be used as a structure-forming element in technologies for producing various types and grades of asphalt concrete. The obtained asphalt concrete samples differ from the traditional one by increased compressive strength at high temperatures (strength limit at 50 ° C) and a reduced strength index at low temperatures (strength limit at 0 ° C). This indicates the best performance properties of the asphalt concrete pavement to external loads in summer and winter. This is achieved through dispersed reinforcement of the asphalt concrete structure and bitumen modification.

It is planned to continue research on the effect of plastics of the LDPE and PP group on the structure formation of asphalt concrete, on the change in physical, mechanical and operational properties. During the study, various results were obtained with different technologies for introducing plastics into the composition of the asphalt concrete mixture. The best results were achieved by heating and mixing bitumen and LDPE together.

References

[1] Pugin, K.G. The structure and properties of asphalt concrete on the basis of waste foundry sand, Materials Science Forum, 2021, 1037 MSF, p. 721–728

[2] Pugin, K.G. Reducing the emission activity of vanadium from building materials, Materials Science Forum, 2021, 1037 MSF, p. 705–712

[3] Vlasov, A.S., Pugin, K.G., Surkov, A.A. Geoecological assessment of the technology for using drilling waste in the composition of asphalt-concrete // Neftyanoe Khozyaystvo - Oil Industrythis link is disabled, 2020, 2020(12), p. 139–142

[4] Vaisman, Y.I., Pugin, K.G., Vlasov, A.S. Using the Resource Potential of Drill Cuttings in Road Construction, IOP Conference Series: Earth and Environmental Science, 2020, 459(2), 022078

[5] Brar, Tejwant S., M. Arif Kamal, and Pinto Emerson. Recycling of Construction and Demolition Waste Material for Energy Savings in India \\ Key Engineering Materials 632 (2014): 107–17. https://doi.org/10.4028/www.scientific.net/kem.632.107

[6] Pugin, K.G. Ensuring environmental safety when using polymer waste in technologies for obtaining building materials // Journal of Physics: Conference Seriest, 2021, 1926(1), 012048

[7] Pugin, K.G., Pugina, V.K. The use of waste in the composition of organic-mineral mixtures used in road construction, Journal of Physics: Conference Series, link is disabled, 2021, 1926(1), 012059

[8] Wang, Ch., WANG, Ni, W., Zhang, S., Wang, S., Gai, G. & Wang W. Preparation and properties of autoclaved aerated concrete using coal gangue and iron ore tailings. Construction and Building Materials, 104 (2016). 109-115. https://doi.org/10.1016/j.conbuildmat.2015.12.041

[9] Arabani M., Mirabdolazimi S.M. Experimental investigation of the fatigue behaviour of asphalt concrete mixtures containing waste iron powder // J. of Materials Science and Engineering. – 2011. – Vol. 528, iss. 10–11. – P. 3866–3870.

[10] Safeer Ullah, Muhammad Raheel, Rawid Khan, Muhammad Tariq Khan, Characterization of physical & mechanical properties of asphalt concrete containing low- & high-density polyethylene waste as aggregates // Construction and Building Materials, Volume 301, 2021, 124127. https://doi.org/10.1016/j.conbuildmat.2021.124127

[11] Angel Mateos, Miguel Angel Millan, John T. Harvey, Fabian Paniagua, Rongzong Wu,

[12] Mechanisms of asphalt cracking and concrete-asphalt debonding in concrete overlay on asphalt pavements, Construction and Building Materials, Volume 301, 2021, 124086,. https://doi.org/10.1016/j.conbuildmat.2021.124086

[13] Weiguang Zhang. Effect of tack coat application on interlayer shear strength of asphalt pavement: A state-of-the-art review based on application in the United States // International Journal of Pavement Research and Technology. – 2017. – Vol. 10, iss. 5. – P. 434–445. https://doi.org/10.1016/j.ijprt.2017.07.003

[14] Effects of coarse aggregate angularity on the microstructure of asphalt mixture / Junfeng Gao, Hainian Wang, Yin Bu, Zhanping You, MohdRosliMohd Hasan, Muhammad Irfan // Construction and Building Materials. – 2018. – Vol. 183. – P. 472–484. https://doi.org/10.1016/j.conbuildmat.2018.06.170

[15] Primary investigation on the relationship between microstructural characteristics and the mechanical performance of asphalt mixtures with different compaction degrees / Pengfei Liu, Jing Hu, Gustavo Canon Falla, Dawei Wang, Sabine Leischner, Markus Oeser // Construction and Building Materials. – 2019. – Vol. 223. – P. 784–793. https://doi.org/10.1016/j.conbuildmat.2019.07.039

[16] Effect of aggregate contact characteristics on densification properties of asphalt mixture / Iange Li, Peilong Li, Jinfei Su, Yu Xue, Wenyu Rao // Construction and Building Materials. – 2019. – Vol. 204. – P. 691–702. https://doi.org/10.1016/j.conbuildmat.2019.01.023

[17] A state-of-the-art review of parameters influencing measurement and modeling of skid resistance of asphalt pavements / Reginald B. Kogbara, Eyad A. Masad, Emad Kassem, A. (Tom) Scarpas, Kumar Anupam // Construction and Building Materials. – 2016. – Vol. 114. – P. 602–617. https://doi.org/10.1016/j.conbuildmat.2016.04.002

[18] Impact of particle morphology on aggregate-asphalt interface behavior / Dongliang Kuang, Ben Zhang, Yuan Jiao, Jianhong Fang, Huaxin Chen, Lu Wang // Construction and Building Materials. – 2017. – Vol. 132. – P. 142–149. https://doi.org/10.1016/j.conbuildmat.2016.11.132

Materials Research Proceedings **21** (2022) 156-160	https://doi.org/10.21741/9781644901755-28

About One of the Approaches for the Research of the Stress-Strain State of a Flange Connection with a Seal Made of an Alloy with Shape Memory

Andrey Boikov[1,a], Lubov Mironova[1,b*] and Sergey Shishkin[1,c]

[1]Moscow Aviation Institute (State National Research University), 4, Volokolamskoe sh., Moscow, 125993, Russia

[a]a.boickov@yandex.ru, [b]mironova_lub@mail.ru, [c]ssshisha@yandex.ru

Keywords: Elastic – Plastic Problem, Materials with Shape Memory, Multilayer Ring, Clutch

Abstract. Design features of a flange connection with a seal made of an alloy with shape memory and the most commonly used methods of researching of the stress-strain state of such connections are estimated. Alternative approach for stress-strain state analysis is proposed, it is based on modeling of the contact zone of the sealing surfaces by means of an equivalent gap between the layers, the value of which changes during axial compression of the multilayer ring and goes into tension. Formulas for determining of contact stresses at the border of layers, which take into consideration the variable physical and mechanical properties of the materials of each layer are presented.

Introduction

Currently, a lot of practical experience has been accumulated in the sphere of design and exploitation of various constructions of flange connections.

Among many constructions, flange connections with a shape memory alloy coupling are promising, which allow to connect and produce pipelines with spatial branches or a large length. Construction of connection of two flanges by means of clutch, made of an alloy with shape memory, is presented in the figure 1 (a). Such connections are called thermomechanical connections.

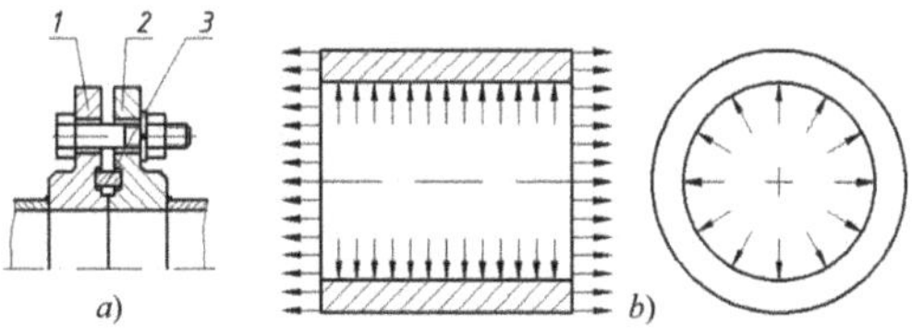

Fig. 1. Flange connection: a – construction; b – analysis model; 1, 2 – flanges, 3 – seal.

Thermomechanical connections of pipelines are a traditional and sufficiently proven field of application of shape memory alloys in terms of production. High operational characteristics and wide technological capabilities have led to the mass use of these connestions in shipbuilding, nuclear and aerospace engineering. The research of the strength of such connections is an urgent problem in ensuring operational reliability and issues of prolonging of the life of technical units.

Materials Research Proceedings **21** (2022) 156-160 https://doi.org/10.21741/9781644901755-28

Problem statement

Flange connection construction with shape memory alloy coupling is estimated.

Flange connection consists of flanges *1* and *2*, connected to the edge of pipes by means of welding. There are grooves, maiden on the end face of flanges. There is seal *3*, mounted in it, made of alloy with shape memory. End faces of the grooves are bearing for the seal.

During the tightening of the bolts, the flanges converge in axial direction. Bearing surfaces of the grooves press on the seal, compressing it. Material of the seal deforms and expands radially under the action of compressive forces, filling the groove cavity. Contact stresses appear at the contact borders of the flange and seal surfaces in the radial and axial directions. The proposed model of the seal is a ring. Therefore, in many cases, well-known equations of the theory of axisymmetric deformation of rings are used for analyze it's stress-strain state, taking into account plastic deformation at small displacements, Fig 1 (b), [1-3].

The proposed formulas for determining the main stresses in the elastic and plastic zones of the analysis model are functions of the radius only [2], and do not depend on the physical and mechanical properties of materials. For our case, the elastic modulus of the flange and seal materials is different. In addition, the material of the seal has a shape memory effect. The phenomenon of pseudoelasticity is of the greatest interest here, in which plastic deformation, reaching tens of percent, is restored after the load is removed [4].

The engineering analysis of contact stresses in the structures of thermomechanical connections is based on modeling of the thermodeformation loading of a clutch, made of an alloy with shape memory by means of thermomechanical diagram [5]. It is a nonlinear dependence of the reactive return stresses on the magnitude of the under-recovery deformation.

However, the resistance of the pipe during forming is the cause of incomplete restoration of the original dimensions of the clutch during manufacturing, which leads to a contact load in the sealed joint.

Here, modeling of the stress-strain state based on the representation of the contact zone of the sealing surfaces by means of equivalent gap can be quite an acceptable way. We suppose, that the bolts are tightened and material of the seal has filled the groove cavities under the action of a compressive normal force. Then a multilayer ring can be used as analysis model, figure 2.

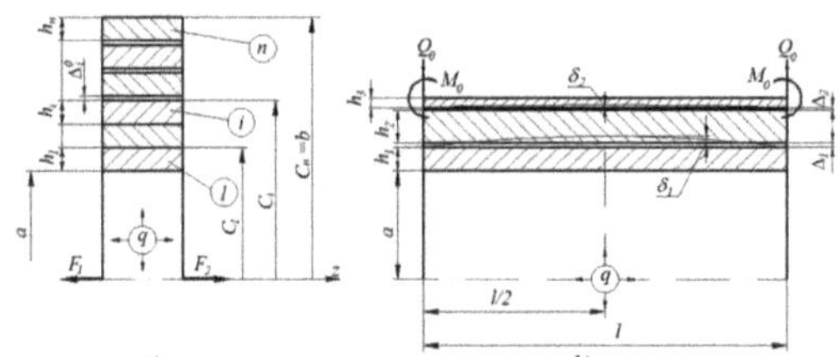

a) *b)*

Fig. 2. *Multilayer ring: a – number of layers is n; b – number of layers is 3.*

The problem statement is formulated as follows. We suppose, that there is a complete multilayer ring in the disk with *n* number of layers, and the number of layers is not limited. The ring is under the action of internal pressure q, axial force $P_z = q\pi a^2$ and under heating conditions $T = T(r)$ provided by the thermomechanical connection technology, Fig. 2 (a). There are gaps or tights between the layers before loading, which are small compared to the thickness of the minimum layer of the construction.

Materials Research Proceedings 21 (2022) 156-160
https://doi.org/10.21741/9781644901755-28

Following the work [6], it is required to determine the stresses and contact pressures at the boundary of the ring layers, taking into consideration micro-gaps and micro-dimensions of surfaces under given initial conditions.

From the point of view of the engineering approach, all the considered layers in the form of mating surfaces refer to smooth cylindrical conjugations consisting of enclosing and encompassing smooth cylindrical surfaces. These interfaces are provided by the presence of gaps or tightnesses, which determines the freedom of their relative movement or the degree of resistance to mutual displacement during the thermomechanical connection.

Basic approach and method description

Let's denote the gaps or tightnesses which have place before loading in the multilayer structure Δ_i^0, and we assume that the gap corresponds to a positive value of Δ_i^0, the tightness – to negative one. The thickness of the layer is determined by the value h_i, and the layer itself is formed as a segment c_i - c_{i-1}, $i = (1, 2, 3)$.

Under the influence of internal pressure, the layers work in conditions of mutual contact, as a result of which, a contact pressure p is formed between them. In this case, the micro-dimensions of the surface layer deform and move in the radial direction. The magnitude of such a displacement Δ_p is a function of the contact pressure and it is expressed by the dependence

$$\Delta_p = \Delta_p(p). \tag{1}$$

The surface layer has roughness, the magnitude of which depends on the manufacturing technology of the product. When rough surfaces come into contact, contact is formed between them at separate points. If there is a temperature effect during the manufacturing process, then a thermal contact resistance R_K arises, the value of which is determined by the ratio

$$R_K = R_K(p). \tag{2}$$

The values (1) and (2) are determined experimentally, so we consider them known.

The stress-strain state of the ring is described by the equilibrium equation:
for the highlighted element of the ring

$$r\frac{d\sigma_r}{dr} = \sigma_t - \sigma_r; \tag{3}$$

for the highlighted segment of the ring:

$$\int_a^b \sigma_z r dr = \frac{qa^2}{2}, \tag{4}$$

where σ_t, σ_r, σ_z are circular, radial and axial stresses, correspondingly.
In the case of thermal action, the equation of state of any point on the surface of the ring can be expressed by the following relation:

$$\varepsilon_r = \frac{1}{E(c_i)}\left[\sigma_r - \nu(\sigma_t + \sigma_z)\right] + \alpha(c_i)T, \tag{5}$$

Materials Research Proceedings 21 (2022) 156-160 https://doi.org/10.21741/9781644901755-28

where $E(c_i)$ is the modulus of elasticity; ν is the Poisson's ratio; $\alpha(c_i)$ is the coefficient of thermal expansion of each layer of the material. We consider the case of a plane deformation when ε_z = const. Let 's write down the equation of thermal conductivity for a multilayer cylinder with stationary heat exchange in the following form

$$\frac{dT}{dr} = -\frac{\overline{Q}}{2\pi r}\left[\frac{1}{\lambda(c_i)} + \sum_i R_K(p_i)c_i \cdot \delta\left(1 - \frac{c_i}{r}\right)\right], \tag{6}$$

where Q is the heat flux, acting in a unit of cylinder length; $\lambda(c_i)$ is the thermal conductivity coefficient of each layer of material; δ is the Dirac delta function.

If the deformations are compatible, radial displacement and relative radial deformation are expressed by the equalities [7]

$$\frac{du}{dr} = \varepsilon_r - \sum_i \left(\Delta_i^0 + \Delta_p(p_i)\right) \cdot \delta\left(1 - \frac{c_i}{r}\right). \tag{7}$$

$$\frac{d\varepsilon_\theta}{dr} = \frac{\varepsilon_r - \varepsilon_\theta}{r} - \frac{1}{r}\sum_i \left(\Delta_i^0 + \Delta_p(p_i)\right) \cdot \delta\left(1 - \frac{c_i}{r}\right). \tag{8}$$

Here ε_r, ε_θ are radial and circumferential deformations; $p_i = |\sigma_r(c_i)|$; $\sigma_r(c_i) \le 0$.

By integrating the system of equations (3), (6) – (8) we obtain

$$\sigma_r = -q\psi_1\left(\frac{a}{r}\right) + \sigma_\theta^0\psi_2\left(\frac{a}{r}\right) + \frac{\overline{Q}E(c_i)\alpha(c_i)}{(1-\nu)4\pi\lambda(c_i)} f_{rT}\left(r, \overline{p}\right) - \frac{E(c_i)}{1-\nu^2} f_{r\Delta}\left(r, \overline{p}, \overline{\Delta}^0\right); \tag{9}$$

$$\sigma_t = -q\psi_2\left(\frac{a}{r}\right) + \sigma_t^0\psi_1\left(\frac{a}{r}\right) + \frac{\overline{Q}E(c_i)\alpha(c_i)}{(1-\nu)4\pi\lambda(c_i)} f_{tT}\left(r, \overline{p}\right) - \frac{E(c_i)}{1-\nu^2} f_{t\Delta}\left(r, \overline{p}, \overline{\Delta}^0\right); \tag{10}$$

$$\sigma_z = \sigma_z^0 + \frac{\overline{Q}E(c_i)\alpha(c_i)}{(1-\nu)4\pi\lambda(c_i)} f_{zT}\left(r, \overline{p}\right) - \frac{\nu E(c_i)}{1-\nu^2} f_{z\Delta}\left(r, \overline{p}, \overline{\Delta}^0\right); \tag{11}$$

$$T = T_0 - \frac{\overline{Q}}{2\pi\lambda(c_i)} f_{zT}\left(r, \overline{p}\right). \tag{12}$$

where f_{rT}, $f_{\theta T}$, f_{zT} are radial, circular and axial functions depending on the temperature distribution, the contact pressure between the layers and the thermal resistance, determine the name from article [7].

Expressions (9) – (12) use the following notation: vector of the contact pressure value is $\overline{p} = (p_1,\ p_2, \dots, p_{n-1})$; vector of initial gaps between cylinder layers is $\overline{\Delta} = (\Delta_1,\ \Delta_2, \dots, \Delta_{n-1})$; σ_t^0, T_0, are stress and temperature at points on the inner surface of the cylinder.

The boundary conditions on the inner and outer surfaces of the cylinder are formulated as follows, when $r = a$; $\sigma_r = -q$; $r = b$; $\sigma_r = 0$, then

$$\sigma_i^0\psi_2\left(\frac{a}{b}\right)-q\psi_1\left(\frac{a}{b}\right)+\frac{\overline{Q}E(c_i)\alpha(c_i)}{(1-\nu)4\pi\lambda(c_i)}f_{rT}\left(b,\ \overline{p}\right)-\frac{E(c_i)}{1-\nu^2}f_{r\Delta}\left(b,\ \overline{p},\ \overline{\Delta}\right)=0. \tag{13}$$

Substituting the values of $r = c_i$, $(i = 1,2,3)$ into equation (9), we obtain a system of nonlinear equations regarding to the values of the contact pressure between the layers

$$p_i=q\psi_1\left(\frac{a}{b}\right)-\sigma_i^0\psi_2\left(\frac{a}{b}\right)-\frac{\overline{Q}E(c_i)\alpha(c_i)}{(1-\nu)4\pi\lambda(c_i)}f_{rT}\left(b,\ \overline{p}\right)+\frac{E(c_i)}{1-\nu^2}f_{r\Delta}\left(b,\ \overline{p},\ \overline{\Delta}\right)=0. \tag{14}$$

The solution of the system of equations (13) and (14) has mathematical difficulties. Therefore, to determine the contact pressure between the layers, we use the well-known formula for elongating of the radius of the cylinder δ under the action of internal pressure from the work [8], which forms the contact pressure between layers 1, 2, 3 at the border c_i, Fig 2 (b) Taking into consideration the thermal expansion of the material under thermal action, we believe that [6]

$$\delta_i=\frac{qa^2}{E_i\sum\limits_{i=1}^{n}h_i}\left(1-\frac{\nu}{2}\right)+\alpha_{i(T)}Th_i;\ p_i=\frac{(\delta_i-\Delta_i)E_ih_i}{a^2\left(1-\frac{\nu}{2}\right)}. \tag{15}$$

Conclusion

The proposed approach makes it possible to evaluate the stress-strain state of a flange connection with a seal made of shape memory alloy and determine the maximum contact pressure between layers with acceptable accuracy in engineering practice. According to the strength criterion, the calculated values of the contact pressure should not exceed the permissible values of the contact stress.

References

[1] N.I. Bezuhov, Bases of the theory of elasticity, plasticity and creep, Vysshaya shkola, Moscow, 1968.

[2] N.N. Malinin, Applied theory of plasticity and creep, Mashinostroenie, Moscow, 1975.

[3] R. Hill, The Mathematical theory of plasticity, at the Clarendon press, Oxford, 1950.

[4] V.A. Lihachev, S.L. Kuz'min, Z.P. Kamenceva, Shape memory effect, Leningrad, LGU, 1987.

[5] S.V. Shishkin, Experimental determination of the generalized thermomechanical diagram of shape memory alloys under axisymmetric bending, J. Industrial Laboratory 60 (2) (1994) 124-125.

[6] L.I. Mironova, A.S. Ivanov, Effect of defects on technological thermal stress state of multilayer cylindrical structure, J. Engineering & Automation Problems, 2 (2013), 121-125.

[7] P.G. Pimshtejn, V.N. Zhukova, Calculation of stresses in a multilayer cylinder taking into account the peculiarities of the contact of the layers, J. Strength problems 5 (1977) 71-75.

[8] Timoshenko S, Vojnovskij – Kriger S. Theory of plates and shells.MeGraw-Hill Boor Company, INC, New-York, Toronto, London, 1959.

Theoretical Justification of Coating Adhesion during Gas-Thermal Spraying

Yuri A. Kuznetsov[1,a], Michael F. Selemenev[2,b*], Igor N. Kravchenko[3,c], Alexander V. Spasibin[2,d] and Grigory B. Pankov[1,e]

[1]Orel State Agrarian University named after N.V. Parakhin, Orel, Russian Federation

[2]Orel State University named after I.S. Turgenev, Komsomolskaya st. 95, Orel, Russian Federation

[3]Russian State Agrarian University-Moscow Agricultural Academy named after K.A. Timiryazev, Moscow, Russian Federation

[a]kentury@yandex.ru, [b]selemenev2007@yandex.ru, [c]kpavchenko-in71@yandex.ru, [d]spasibin99@inbox.ru, [e]grigo-pankov@mail.ru

Keywords: Gas-Thermal, High-Speed Spraying, Coating, Adhesion Energy, Particle

Abstract. The article provides a brief overview of modern methods of thermal spraying. Particular attention is paid to high-speed flame spraying. The theoretical substantiation of the adhesion of coatings formed on machine parts using the methods of thermal spraying is presented.

Introduction

The term "thermal spray" refers to all coating processes with wire or powder materials that do not degrade at high temperatures. They are introduced into the high-temperature zone and sprayed with a gas jet or compressed air, thus forming small particles that move at high speed and fall on a prepared surface, where they form a coating with desired properties [1, 2].

The Swiss inventor Max Ulrich Schoop (1870-1956) is considered the founder of the thermal spray coating method, in which heated metal particles are sprayed onto the surface of the part.

Depending on the source of heat, the methods of thermal spraying are divided into the following types [3-7]: flame spraying, plasma spraying, plasma arc spraying, high-frequency plasma spraying, detonation spraying, spraying in a controlled atmosphere, spraying in a dynamic vacuum and crucible spraying.

[Crucible metallization, plasma spraying, electric arc metallization, PN-21 plasmatron, supersonic gas-dynamic spraying, supersonic gas-flame spraying, detail spraying, supersonic plasma-jet spraying, supersonic plasma-channel plasma spraying, oxygen-kerosene-suction spraying]

Flame spraying is a process in which the sprayed powder material is fed into the flame of an acetylene-oxygen or propane-oxygen burner, melted and transferred by compressed air to the surface of the part, where, while cooling, it forms a coating [2].

Flame sprayed coatings can harden the surfaces of parts and protect them from wear, corrosion and thermal stress.

Methods of flame spraying can be divided into two groups [6-11]: conventional (subsonic) spraying and high-speed (supersonic) spraying. The main disadvantages of gas-flame coatings obtained by subsonic spraying are their low adhesion strength to the surface of the part and low mechanical properties. The most interesting is high-speed spraying. When used, coatings can be thicker and stronger, and therefore can better protect operating parts from wear, corrosion or high temperatures.

In supersonic gas-flame spraying, a sprayed metal powder is introduced into a gas stream, with which it is heated, accelerated (usually more than 5 speeds of sound) and directed to the surface of the part.

There are two types of high-speed gas-flame spraying: high-speed air-fuel spraying and high-speed oxy-fuel spraying.

High-speed air-fuel spaying (HVAF-spaying) is based on acceleration and heating of sprayed particles in a high-speed flow of hydrocarbon fuel combustion products in compressed air (Figure 1) [2, 7-10]. Combustible gases (methane, propane, hydrogen) or liquid fuel (kerosene, diesel fuel) are used as fuel. The maximum temperature attainable in the combustion chamber of the burner is about 2200K, the maximum gas flow rate is up to 1800 m / s. The speed of the sprayed particles can be 300-600 m / s. The advantages of this type of spraying are efficiency, productivity and high quality of the resulting coatings. The coatings of aluminum, zinc, copper, their alloys and mixtures applied by this method are of high quality.

High-speed oxy-fuel spraying (HVOF-spraying) is based on acceleration and heating of sprayed material particles in a high-temperature high-speed flow of hydrocarbon fuel combustion products in oxygen [2, 6-8]. With a temperature in the combustion chamber reaching 3300K and a gas flow rate exceeding 2300 m / s, the method makes it possible to apply coatings from most existing refractory materials. The speed of the sprayed particles can be 400-900 m / s. This type of spraying is suitable for creating very hard wear-resistant coatings on expensive parts of machines and mechanisms.

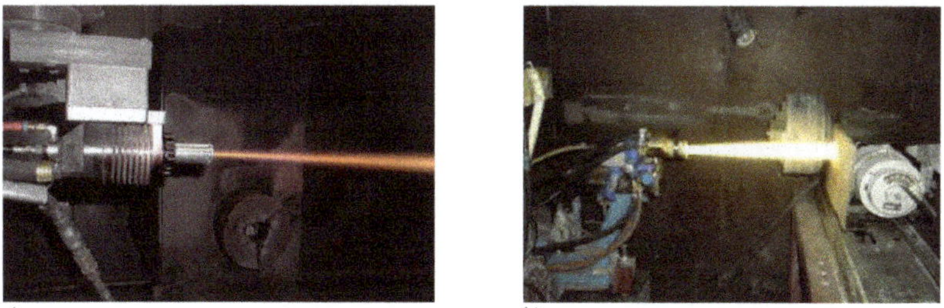

a) b)

Fig. 1. *Spraying the coating with high-speed air-fuel spraying using an HVAF-burner (a) and high-speed oxy-fuel spraying using an HVOF burner (b).*

In gas-thermal spraying, the physical characteristics of the sprayed particles are largely determined by the adhesion energy of the sprayed materials to the base [12].

Analysis

The concept of adhesion energy is closely related to the concept of surface energy [13]. Surface energy is energy concentrated under a unit of free surface area of a body. If we assume that the surface energy is distributed continuously over the volume, concentrating mainly near its boundary S_a, then for a body occupying a semi-infinite space, it is determined by the equality [13]:

$$W_p = \int_0^{\infty} w(h)dh, \quad (1)$$

where: W_p – surface energy, J / m^2;

$w(h)$ – bulk density of the potential of internal forces at a distance h from the area S_a, J / m^3.

The layer of material in which the surface energy is concentrated under real conditions is very small in comparison with the dimensions of most of the investigated bodies - on the order of ten interatomic distances. Therefore, in many cases, a real body in relation to its surface layer can be considered semi-infinite, and the value W_p distributed along S_a evenly. The consequence of this is the formula (1).

Adhesion is caused by various types of interactions between molecules and atoms of contacting bodies [14]. Their long-range character [15] gives grounds to use nonlocal models to describe the state of adhesion within the framework of continuum mechanics. For example, it was shown in [16] that a nonlocally elastic medium can be approximated by a model of a medium of class N (the energy density depends on the displacement gradients up to the Nth order inclusive). In this paper, it is proposed to use a linear medium of the second class [16]. In [17], it was generalized by introducing the assumption of the presence of a nonclassical initial stress state in each of the bodies. This is what made it possible to describe the effects of surface energy, the energy of adhesion. The bottom line is as follows.

It is assumed that any body *B*, limited by the surface *S*, is obtained by instantaneous separation from the infinite medium *Q*, not subject to any external influences. State *B* before selection is taken as reference. Particles *B* in this case are in a force field created by a set of particles that form the exterior of *B* in *Q*. Its configuration is determined by the shape of the boundary *S*, the lines of force have a unit vector as a tangent $\bar{e}$. The action of the external field is balanced by the action of internal forces. The theory proposed in [16, 17] uses not one, as the classical model of the medium, but two stress tensors. First $-\overset{1}{P}$ performs work on the generalized displacements that form their first gradient. Second $-\overset{2}{P}$ does work on the components of the second displacement gradient. The stress state in the body B, mentally isolated in an infinite medium, caused by the above-mentioned field, is described by a tensor of the second type, determined by the equality [17]

$$\overset{2}{P}_0 = T_0 \cdot \bar{e}, \tag{2}$$

where: $T_0 = \pi_0 \delta_{ij} \delta_{rk} \overline{e_i e_j e_k e_l}$ – tensor of the fourth rank, characterizing the ability of an isotropic medium to develop the described stress state, N / m [18]; π_0 – newly introduced physical constant; $\overline{e_k}$ – orthonormal basis; k =1, 2, 3.

With a real selection *B*, the appearance disappears, the uniformity of the distribution of internal forces is disturbed. The consequence is a change in the state of affairs in the border area.

This qualitative description is quantitatively expressed by the dependence of the potential on the characteristics of the deformed state, determined by the relation [19]

$$w = \frac{2A_1 - A_2}{2} g_{ij} g_{ij} + \frac{A_2}{2} g_{kk} g_{ll} + 2A_1 b^2 Z_{ijk} Z_{ijk} + \pi_0 \delta_{ij} e_k Z_{ijk}, \tag{3}$$

$$2A_1 = E\frac{1-\nu}{(1+\nu)(1-2\nu)}; k = \frac{\nu}{1-\nu}; A_2 = \frac{E\nu}{(1+\nu)(1-2\nu)}; \tag{4}$$

where: E – Young's modulus, H/м^2;

ν – Poisson's ratio;

In formula (3), the rule for repeated indices is used, the expression $g_{ij} \cdot g_{ij}$ means the amount:

$$\sum_{i,j=1}^{3} g_{ij} \cdot g_{ij} = g_{11} \cdot g_{11} + g_{12} \cdot g_{12} + g_{13} \cdot g_{13} + g_{21} \cdot g_{21} + g_{22} \cdot g_{22} + g_{23} \cdot g_{23} + g_{31} \cdot g_{31} +$$

$$+ g_{32} \cdot g_{32} + g_{33} \cdot g_{33}$$

where: $g_{ij} = \frac{1}{2}\left(u_{i},_{j} + u_{j},_{i}\right)$ components of the classical strain tensor in the basis $\bar{e}_k$;

$Z_{ijk} = u_{i},_{jk}$ – components of the second displacement gradient;

δ_{ij} – Kronecker symbol;

e_k – vector components $\bar{e}$;

$$\pi_0 = \frac{W_p}{k}; b = \frac{3W_p}{4k^2 A_1}, \tag{5}$$

where: b, π_0 – additional constants introduced in the theory [16] to take into account the properties of an elastic medium.

Components $\overset{1}{P}$ и $\overset{2}{P}$ are determined by differentiating the right-hand side of formula (3) with respect to g_{ij} и Z_{ijk} respectively.

Energy of adhesion of bodies $B^{(1)}$ and $B^{(2)}$ is usually determined by the loss per unit area of the formed adhesive seam of their free energy [13]. This definition is realized in the equality

$$F_a = W_p^{(1,2)} - W_p^{(1)} - W_p^{(2)}; \tag{6}$$

The theory makes it possible to take into account the volumetric distribution of the mentioned energy changes. Therefore, to determine F_a In the general case, we proposed to use the value S_a averaged over the contact surface determined by the equality

$$F_a = \langle F_a \rangle = \frac{1}{S_a} \sum_{j=1}^{2} \int_{V^{(j)}} \left(w^{(j)} - w_1^{(j)}\right) dV^{(j)}, \tag{7}$$

where: $w_1^{(j)}$ – the density of the potential of the body with number (j) before it enters the adhesive contact;

$\langle F_a \rangle$ – average F_a on the surface S_a.

Density $w^{(j)}$ are determined when solving the problem of the stress-strain state of a system of two bodies stuck together in the absence of external influences [20]. For a system of two plates with thicknesses $h^{(j)}$, this formula takes the form (6), in which the terms are determined by the equalities

Materials Research Proceedings **21** (2022) 161-168 https://doi.org/10.21741/9781644901755-29

$$W_p^{(1,2)} = \int_{-h^{(1)}}^{0} w^{(1)}(x)dx + \int_{0}^{h^{(2)}} w^{(2)}(x)dx; \tag{8}$$

$$W_p^{(1)} = \int_{-h^{(1)}}^{0} w^{(1)}(x)dx; \quad W_p^{(2)} = \int_{0}^{h^{(2)}} w_1^{(2)}(x)dx, \tag{9}$$

where: x - coordinate measured from the contact plane in the direction perpendicular to it.

The result of the calculations was a rather cumbersome expression that determines the dependence of the adhesion energy on the constants characterizing the mechanical properties of the materials of parts, and the ratios $h^{(j)} / b^{(j)}$; $j = 1,2$. Their order for coatings with a thickness of $\approx 10^{-6}$ *m* is such that one can neglect the dependence of the adhesion energy on the thickness of the coatings. For multilayer structures, this would mean neglecting the mutual influence of states in the vicinity of adjacent adhesive joints. As a result, to calculate the adhesion energy, we obtained the formula:

$$F_a = \frac{W_p^{(1)} W_p^{(2)} (k^{(1)} + k^{(2)})^2}{W_p^{(1)} k^{(2)2} + W_p^{(2)} k^{(1)2}}; \tag{10}$$

Let us assume that the contacting bodies consist of the same material. In other words, these are two parts of the same body, consisting of only one material with characteristics constant throughout the body. Then

$$W_p^{(1)} = W_p^{(2)}; \; k^{(1)} = k^{(2)}; \tag{11}$$

Both expression (9) and expression (10) in this case indicate that

$$F = 2W_p^{(2)}, \tag{12}$$

That is, the adhesion energy is equal to twice the surface energy of the body - the work required to divide a body of a unit cross-sectional area into two parts. This result qualitatively confirms the correctness of the stated views. In addition, they were confirmed quantitatively by comparing the calculations of the F value with the data known in the literature [21].

Introducing the notation:

$$x = \frac{k^{(2)}}{k^{(1)}}; \; f = \frac{F}{W_p^{(1)}}; \; \beta = (W_p^{(2)} / W_p^{(1)}) > 1, \tag{13}$$

based on formula (10), we obtain the dependence

$$f = \frac{\beta(1+x)^2}{x^2 + \beta}, \tag{14}$$

It is seen that at $x \to \infty : f \to \beta$, at $x \to 0 : f \to 1$. That is, if $k^{(1)} \to 0$, then the adhesion energy tends to the largest of the surface energies – $W_p^{(2)}$.

If $k^{(2)} \to 0$, then $F \to W_p^{(1)}$. At $x_m = \beta$ function (14) has a maximum equal to $f_m = 1 + \beta$.

If $x = \dfrac{k^{(2)}}{k^{(1)}}$ coincides with the ratio of the surface energies of the contacting bodies, then the energy of their adhesion reaches the maximum (given $W_p^{(1)}$ and $W_p^{(2)}$) value equal to

$$F = W_p^{(1)} + W_p^{(2)} = \frac{2W_p^{(1)} + 2W_p^{(2)}}{2}, \tag{15}$$

For real materials, the determination of surface energy W_p is difficult. Therefore, it is proposed to determine it approximately on the basis of the following assumptions. W_p

In [22, 23], based on the theory of Griffiths cracks, it was assumed that

$$\sigma_f = \sqrt{EW_p / 4c}, \tag{16}$$

where: *2c* – length of cracks observed in the fractured layer;

σ_f – tensile strength, N/m^2.

It is also noted there that the value of the ultimate strength usually observed during the destruction of a body is determined by the formula

$$\sigma_f = \frac{E}{500}, \tag{17}$$

This corresponds to the presence of cracks with a length of 2s = 10-6 m.Taking into account the above estimates, we can obtain the expression of interest to us for determining the value of the surface energy

$$W_p = 8E \cdot 10^{-12}, \tag{18}$$

Conclusion

Thus, knowing Young's moduli and Poisson's ratios of materials of parts and coatings in fixed contact conditions, it is possible to quantify the maximum degree of their possible adhesion.

References

[1] Puchin E.A. Machine repair technology: textbook for universities / E.A. Puchin, N.A. Ochkovsky, V.S. Novikov. - M .: KolosS, 2011 .-- 488 p.

[2] Baldaev L.Kh. Thermal spraying: textbook for universities / L.Kh. Baldaeva, V.N. Borisov, V.A. Vakhalin. - M .: Market DS, 2007 .-- 344 p.

[3] Goncharov V.S., Guryanov A.N., Domashenko Yu.V. Formation of supersonic flows in gas-flame burners. 6th International Conference "Films and Coatings 2001". - SPb .: Publishing house of SPbSU, 2001.S. 48-52.

[4] Alkhimov A.P., Klinkov S.V., Kosarev V.F. and others. Cold gas-dynamic spraying. Theory and Practice / Ed. V.M. Fomin. - M .: FIZMATLIT, 2010 .-- 536 p.

[5] Klinkov S.V., Kosarev V.F. Measurements of cold spray deposition efficiency // J. of Thermal Spray Technology. 2006. Vol. 15.No. 3. P. 364-371. https://doi.org/10.1361/105996306X124365

[6] Dykhuizen R. C., Smith M. F. Gas dynamic principles of spray // J. of Thermal Spray Technology. 1998. Vol. 7.No 2.P. 205-212. https://doi.org/10.1361/105996398770350945

[7] Microstructural characteristics of cold-sprayed nanostructured WC – Co coatings / R.S. Lima. J. Karthikeyan, C.M. Kay et. al. Thin Solid Films. 2002. Vol. 416. P. 129-135. https://doi.org/10.1016/S0040-6090(02)00631-4

[8] Van Steenkiste T.N. Smith J.R. Evaluation of coatings produced via kinetic and cold spray processes // J. of Thermal Spray Technology. 2004. Vol. 13.No. 2.P. 274-282. https://doi.org/10.1361/10599630419427

[9] Helfritch D. Champagne V. Optimal panicle size for the cold spray process / U.S. Army Research Lab. Aberdeen Proving Ground, MD // Building on 100 Years of Success: proc. of the Intern. Thermal Spray Conf. (Seattle, WA, USA). ASM International. 2006.

[10] Eden T., Wolfe D. E., Potter J., et al. Cr1C2-Based coatings applied by high velocity particle consolidation (HVPC) for wear resistant applications / Applied Research lab. The Pennsylvania State University: proc. of ASM Cold Spray Conf. (Akron, Ohio, USA). ASM International. 2004.

[11] Zhang D., Shipway P.H., McCartney D.G. Panicle-substrate interactions in cold gas dynamic spraying. Advancing the Science and Applying the Technology: proc. Of the Intern. Thermal Spray Conf. (Orlando, FL, USA). ASM International. 2003.

[12] Shorkin VS, Kuznetsov Yu.A., Akimochkin AV On the possibility of taking into account the composition of the coating and the nature of its interaction with the base when calculating the energy of adhesion // In the collection: Progressive technologies in transport systems. Collection of reports of the 5th Russian scientific and technical conference. Part 1. - Orenburg. IPK OSU, 2002 .-- S. 171-173.

[13] Shorkin V.S., Batishchev A.N., Kuznetsov Yu.A. Analysis of the stress state of wear-resistant coatings of reconditioned parts // Mechanization and electrification of agriculture. 2003. No. 3. S. 28-30.

[14] Simon A.D. Adhesion of films and coatings. - M .: Chemistry, 1977 .-- 352 p.

[15] Leibfried G. Microscopic theory of mechanical and thermal properties of crystals. - M.-L .: State. ed.- in f.-m. lit. - 312s.

[16] Small V.I. On the nonlocal theory of elasticity. // Strength and plasticity.- Moscow: Nauka, 1971.- P. 74 - 78.

[17] Tupin R.A. The theory of elasticity, taking into account moment stresses. // Mechanics / Coll. translations, 1965, No. 3.- S. 113 - 140.

[18] Shorkin V.S. Theory of elasticity of surface layers of solids. 2. Mechanics.- 1995.- S. 169-179.

[19] Gordon V.A., Tinyakova E.V., Shorkin V.S. Description of the stressed state of bodies caused by their adhesion. // Scientific and technical problems of predicting the reliability and durability of structures and methods for their solution. / 4th Intern. conf. June 26 - 28, 2001 - SPb .: "Nestor", 2001 - S. 87 - 88

[20] Shorkin V.S. Features of elasticity of surface layers of solids. // Elasticity and inelasticity. / Materials Int. n.symp. by probl. fur. deform. tel. / Ed. Kiyko I.A., Israilova M.Sh., Brovko G.L.- M .: Moscow State University, 2001-S. 453-454.

[21] Tinyakova E.V., Shorkin V.S. Theoretical calculation of the energy of adhesion. // Films and coatings. / 6th Intern. conf. "Films and Coatings' 2001". Proceedings. / Ed. Klubnikina V.S.- SPb .: Publishing house of SPbSTU, 2001- S. 611- 612.

[22] Averbakh B.L. Some physical aspects of destruction: / Ed. Libovitz G. Destruction. - v.1.- M .: Mir, 1973- S. 471 - 504.

[23] Petch N. Metallographic aspects of destruction./ Libovitz G. Destruction. - v. 1- M .: Mir, 1973- S. 377 - 420.

Development of Abrasive Grain Coating Technology Protective Polymer Casing for Water-Abrasive Cutting

Olga G. Kozhus[1,a*], Gennady V. Barsukov[1,b], Larisa Yu. Frolenkova[1,c], Evgeny R. Arkhipov[1,d] and Alexander A. Kazarin[1,e]

[1]Orel State University named after I.S. Turgenev, Komsomolskaya st. 95, Orel, Russian Federation

[a]okozhus@mail.ru, [b]awj@list.ru, [c]larafrolenkova@yandex.ru, [d]arhipov199806@yandex.ru, [e]kazarin1999@gmail.com

Keywords: Abrasive, Polymer, Waterjet Cutting, Fluidized Bed, Film Coating

Abstract. The article presents the technology of covering abrasive grains with a protective polymer shell, which allows to reduce the wear of the focusing tube of the nozzle of a hydroabrasive installation. A description of the manufacturing process of a prototype abrasive with a polymer coating in a fluidized bed is presented in order to determine the critical processing modes at which a high-quality product will be obtained in a minimum operating time of the installation.

Introduction

The productivity of applying a polymer shell to the surface of an abrasive grain is estimated as the mass of the processed material per hour of operation of the installation. At the same time, important processes must be ensured that take place during the encapsulation of the abrasive in a fluidized bed. The speed of their passage affects the processing performance. These processes include the adhesion of the polymer to the abrasive surface followed by drying. In this case, it is necessary to exclude the agglomeration of particles and their deposition on the gratings, meshes and walls of the equipment.

The performance of fluidized bed encapsulation of abrasives is achieved through selection of equipment, polymer type, spray parameters and abrasive properties.

The choice of polymer affects the flow rate of the spray solution during the coating process on the abrasive. The amount of polymer sprayed determines the processing efficiency and the performance of the encapsulation process. However, it is necessary to control the volume of consumption of the polymer solution and the increment in the mass of the processed abrasive. An increase in processing productivity requires an increase in the consumption of the sprayed polymer and the acceleration of the processes occurring in the coating zone. The processing modes should ensure the rapid passage of the processes of polymer spraying, its deposition and adhesion on the abrasive surface, followed by drying. Such modes will be critical and will provide the highest possible processing performance [3].

Main part

The process of encapsulation of a polymer shell on an abrasive under critical technological conditions depends on the following characteristics:

1. Construction and size of the coating booth:
- air flow;
- the temperature of the supplied air.
2. Characteristics of the polymer solution:
- type of polymer and composition of the coating solution;

- consumption of polymer solution;
- the initial temperature of the polymer solution;
- humidity.

3. Parameters of polymer solution spraying:

- characteristics of the spray nozzle (spray angle, nozzle flow rate, orifice diameter);
- the position of the sprayer (above or below the abrasive layer);
- distance from the nozzle edge to the abrasive layer;
- the concentration of the polymer solution;
- spray pressure.

4. Parameters of abrasive particles:

- the size of the abrasive;
- the mass of the processed material in the installation chamber;
- roughness of the abrasive surface;
- wettability of the abrasive surface.

The intensification of the coating process on the abrasive surface is limited by the capabilities of the equipment used. The parameters of the fluidized bed are determined by the volume of the loaded abrasive and the dimensions of the treatment chamber. The speed and productivity of processing in this case is defined as the ratio of the consumption of abrasive to the surface area in the processing zone[4,5,6].

This is due to the fact that when passing air through the loaded abrasive, it is necessary to ensure the flow rate ω, when the particles become fluidized and begin to move. Further increasing the air flow rate to a critical value $\omega_{кр2}$ will increase the height of the fluidized abrasive bed and, as a result, the distance between the particles. Subsequent increase in speed above the critical value $\omega_{кр2}$ will lead to the carryover of the abrasive from the processing area (Figure 1).

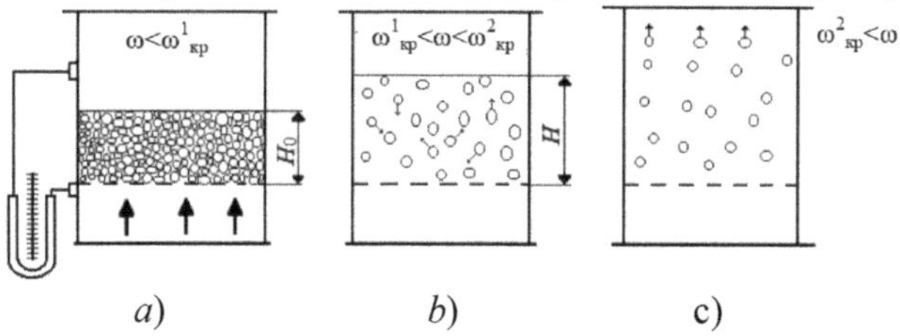

Fig. 1. *Influence of the air flow rate on the process liquefaction of abrasive a) - abrasive layer without movement; b) - fluidized abrasive layer; c) - removal of abrasive from the processing area*

Thus, an increase in the productivity of applying a polymer coating to an abrasive due to a larger loaded volume of abrasive leads to an increase in the size of the installation and the need to use more powerful air injection equipment. Therefore, the critical processing conditions will be determined by the choice of the type and characteristics of the corresponding equipment.

The parameters and characteristics of the equipment will determine the required air flow rate and air flow rate to ensure and maintain a stable fluidized bed.

Another way to increase the productivity of processing on the existing equipment can be an increase in the speed of application of the coating and its subsequent rapid drying. This can be achieved through the use of a fast-drying polymer and an increase in mortar consumption.

However, an increase in the volume of the applied polymer per unit time will lead to an increase in the mass of the abrasive particle to a critical value, when it cannot be and be maintained in suspension. This will lead to agglomeration of particles and their deposition on the lattice and walls of the installation and to stop the encapsulation process.

Unvaporated droplets of sprayed coating solution are more likely than in the central flow area to form agglomerates with a liquid, uncured coating.

The appearance of liquid drops in the near-wall zone may be due to the fact that the flow rate of the sprayed polymer solution is increased [1,2].

In the upper part of the layer, where liquid drops of a polymer coating solution appear, as well as along the walls of the chamber, where these drops can be carried by turbulent air pulsations, both independently and together with abrasive particles, agglomerates of abrasive grains with an unhardened coating may appear. Their large weight, as well as vertical pulsations of the speed of air and abrasive particles, contribute to the fall of agglomerates to the bottom.

The reason for the ingress of wet abrasive particles to the bottom of the chamber and the formation of agglomerates caused by this process can be the slow evaporation of the solvent - water, not only due to the lowered temperature in some areas of the chamber, but also due to the excess of not evaporated polymer solution.

To intensify the process of applying a polymer coating to an abrasive, it is necessary to improve the conditions for blowing the polymer solution through the fluidized bed, to reduce the likelihood of adhesion of abrasive agglomerates on the walls of the chamber, to reduce the vortex movement of particles from the bottom of the chamber to the fluidized layer and back. This is largely due to the difference in the movement of the abrasive flow in the center of the chamber and on its walls, where the operating temperature decreases due to heat losses and the speed of the moving flow decreases.

It is possible to ensure the uniformity of drying temperature and flow rate from the center of the chamber to the boundaries of the liquefied bed by constructive modernization of the liquefaction chamber. A decrease in porosity is equivalent to an increase in the frequency of collision of abrasive particles with a wall at low kinetic energy. You can get rid of this by reducing the wall resistance to air flow. This can be achieved by blowing additional air into it through the walls.

To do this, we will install a mesh cylinder in the liquefaction chamber. The abrasive is loaded not into the installation chamber, but inside the cylinder (Figure 2).

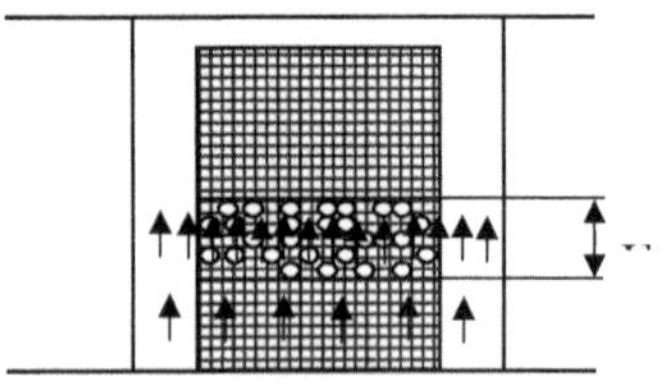

Fig. 2. *The design of the chamber for creating a liquefied abrasive layer with a mesh cylinder for operation in critical modes of polymer coating*

There are no abrasive adhesion conditions on the walls of the mesh cylinder of the chamber. There is no stable boundary layer near the mesh wall, in which the air flow speed is equal to the operating speed. The likelihood of the deposition of agglomerates or wet abrasive particles on the gas distribution grid is reduced due to the uneven distribution of the working speed of the gas flow over the section of the working chamber, the uneven distribution of the air temperature over this section, the near-wall feature of reducing the flow rate in the chamber for applying a protective coating to abrasive particles.

Consequently, the introduced design changes make it possible to increase the absolute velocity of the fluidizing agent near the walls of the mesh cylindrical chamber and equalize its value with other parts of the value. This means that the concentration of abrasive particles near the walls will be commensurate with other zones of the processing chamber. Increasing the relative particle velocity significantly reduces the likelihood of agglomeration.

In the course of the study, it was found that the agglomeration of adhering pairs of particles will proceed faster where the probability of their collision is more likely, that is, where their concentration is greater. Maintaining a liquid state, the viscosity of a polymer liquid significantly depends on the local temperature of the process, the length and shape of macromolecules. Further, the influence of each of the reasons is considered separately [7,8].

The homogeneity of the coating was studied (table 1). At the first stage, the abrasive surface was examined at low magnification (Figure 3). If, during the study, defects in the uniformity of the coating were found, the surface area of interest is considered at the highest possible resolution.

***Fig.** 3. Photos of a prototype abrasive inpolymer shell (LeicoMZ 16 microscope, × 150 magnification)*

***Table 1.** Assessment of coating uniformity*

№ p/p	Coating uniformity index	Result	Grade K_n
1	Holes	Not detected	1
2	Integrity	Not broken	1
3	Chips, delamination	Not detected	1
4	Dust splashes	Not detected	1
5	Coverage uniformity	Yes	1

The value of the uniformity of the coating for each selected grain is determined by the formula:

$$O_a = \frac{\sum K_n}{n} * 100\%, \quad (1)$$

where K_n - indicator score;

n – the number of indicators of uniformity of the coating, is defined.

Based on the results of the assessment, it was concluded that the uniformity of the coating of the abrasive in the polymer shell is 100%.

Next, the thickness of the polymer coating on the abrasive was determined by determining the difference between the linear size of the outer surface of the abrasive with polymer coatings and the size before applying the shell (after dissolving the polymer shell in water) (Figure 4).

a) *b)*

***Fig. 4**. Measurement of abrasive grain with polymer coating a - abrasive in a polymer sheath, b - abrasive without a polymer sheath*

Measurements at selected points on the grain surface were recorded in the test report (Table 2). Then the thickness of the polymer coating for the studied abrasives was determined by the formula:

$$T_a = \frac{(a_n - a_a)}{2}, \text{mkm} \quad (2)$$

where a_n – coated abrasive grain size;

a_a – abrasive grain size uncoated.

***Table 2.** Results of measurements of abrasive grain*

№ p/p	Test object	a_{n}, mm			a_{a}, mm			T_{a}, mkm			T_{cp}, mkm
		1	2	3	1	2	3	1	2	3	
1	Abrasive grain with polymer coating No №1	0,273	0,295	0,310	0,263	0,274	0,260	5	10,5	25	13,5
2	Abrasive grain with polymer coating No №2	0,281	0,304	0,288	0,265	0,286	0,269	8	9	9,5	8,8
3	Abrasive grain with polymer coating No №3	0,320	0,298	0,310	0,301	0,275	0,294	9,5	11,5	8	9,6

The final result is the arithmetic mean value of the coating thickness on the test sample of abrasive:

$$T_{cp} = \frac{\sum T_a}{m}, \text{ mkm} \qquad (3)$$

where m – number of measurements.

According to the measurement results, it was found that the thickness of the polymer coating does not exceed 50 mkm.

The sphericity of the abrasive polymer was estimated as the ratio of the horizontal projection area of the abrasive with a polymer shell to the area of the circle approximating the grain contour. To assess the spherical shape of the abrasive in the polymer sheath, a microscope with a magnification factor of 50 to 1500 was used with the ability to shoot with a digital camera, a computer and the KOMPAS-3D V17 software (Table 3).

The evaluation of the spherical shape of the abrasive grain in the polymer shell was carried out according to the formula:

$$O_{ш} = (1 - S_a / S_{кр})100\ \%, \qquad (4)$$

where S_a – area of the abrasive contour with a polymer sheath;

$S_{кр}$ – area of a circle.

Table 3. *Evaluation of the spherical shape of the abrasive in the polymer shell*

№ abrasive	Picture	Abrasive area, mm^2	Circling area, mm^2	Meaning $O_{ш}$, %
1	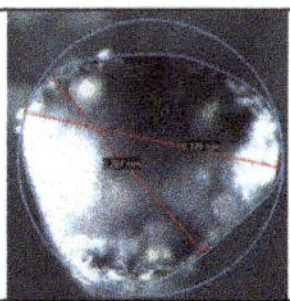	0,078761	0,089404	11 %
2	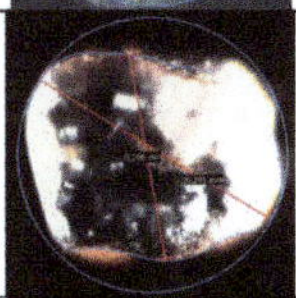	0,049881	0,058215	15 %
3	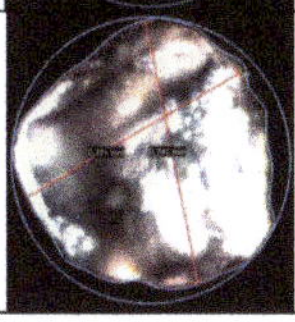	0,091652	0,106631	14 %

Based on the results of the assessment, it was found that the deviation from the spherical shape of the abrasive in the polymer shell does not exceed 15% [9].

Conclusion

Based on the data obtained for the manufacture of a prototype abrasive in a polymer sheath, the following conclusion was made:

1. The bulk density is determined by the original abrasive and does not depend on the processing modes.

2. The spray pressure and the concentration of the polymer in the solution play an important role in the formation of the mass and size characteristics of the abrasive polymer. According to the results obtained, with an increase in the spraying pressure to 100 kPa, the uniformity of the coating increases to 100%, and a coating thickness of about 30 μm is formed.

3. The moisture content of the abrasive in the polymer shell is not affected by all factors, so the processing modes are selected taking into account the work on the critical values of the input factors.

4. The value of the concentration of the polymer in the solution does not significantly affect the quality of the surface. The greatest influence is observed on the value of the air flow rate. Its value should be chosen based on the maximum installation possibility. In our case, this is 40 m2 / h.

5. The deviation of the specific gravity of the abrasive in the shell practically did not exceed 10%. However, for some values with the maximum concentration of the polymer in the solution, the values of the increase in specific gravity up to 12% are recorded.

Acknowledgement
This research was supported by the Russian Science Foundation, grant no. 22-29-01599.

References

[1] Perec A (2018) Experimental research into alternative abrasive material for the abrasive water-jet cutting of titanium. Int J Adv Manuf Technol 97 (1): 1529-1540. https://doi.org/10.1007/s00170-018-1957-2

[2] Miao X, Qiang Z, Wu M, Song L, Ye F (2018) The optimal cutting times of multipass abrasive water jet cutting. Int J Adv Manuf Technol 97 (5): 1779-1786. https://doi.org/10.1007/s00170-018-2011-0

[3] Barsukov GV, Zhuravleva TA, Kozhus OG (2017) Increasing of efficiency of environmentally friendly technology of AWJ of a glass fiber plastic. IOP Conf Ser Earth Environ Sci 50 (1): 012001. https://doi.org/10.1088/1755-1315/50/1/012001

[4] Sutowska M, Kapłonek W, Pimenov DY, Gupta MK, Mia M, Sharma S (2020) Influence of variable radius of cutting head trajectory on quality of cutting kerf in the abrasive water jet process for soda – lime glass. Materials 13 (19): 4277 14. https://doi.org/10.3390/ma13194277

[5] Momber A (2011) Hydrodemolition of concrete surfaces and reinforced concrete. Elsevier

[6] Perec A, Pude F, Grigoryev A, Kaufeld M, Wegener K (2019) A study of wear on focusing tubes exposed to corundum-based abrasives in the waterjet cutting process. Int J Adv Manuf Technol 104 (5): 2415-2427. https://doi.org/10.1007/s00170-019-03971-0

[7] Kozhus OG, Barsukov GV, Shorkin VS, Frolenkova L.Yu. Peculiarities of polymer deposition on the surface of abrasive grain to increase the efficiency of waterjet cutting // Fundamental and applied problems of technology and technology. 2021. No. 2 (346). S. 86-92. https://doi.org/10.33979/2073-7408-2021-346-2-86-92

[8] Kozhus OG, Barsukov GV, Shorkin VS, Stepanov Yu.S., Frolenkova L.Yu. Improving the efficiency of the installation for applying a polymer shell to an abrasive under critical technological conditions // Fundamental and applied problems of technology and technology. 2021. No. 3 (347). S. 91-99. https://doi.org/10.33979/2073-7408-2021-347-3-91-99

[9] Barsukov G.V., Kozhus O.G., Kozhus S.V., Pikalov D.N., Gukov A.A. Calculation of the fluidization rate of abrasive particles in the apparatus for applying a polymer coating // Certificate of registration of the computer program 2020617325, 03.07. Application No. 2020616334 dated 16.06.2020.

Analysis of Magnetic Forces in the Working Clearance with Magnetic-Abrasive Treatment of Inductors on Standing Magnets

A.M. Ikonnikov[1,a*], S.L. Leonov[1,b], D.E. Solomin[1,c] and A.A. Kulakov[2,d]

[1]Altai State Technical University. I.I. Polzunova, Barnaul, Russian Federation

[2]Chenstochiv Polytechnic University, Czestochowa, Republic of Poland

[a]iamagtu@mail.ru, [b]sergey_and_nady@mail.ru, [c]d147s@mail.ru, [d]akulawik@icis.pcz.pl

Keywords: Magnetic Abrasive Treatment, Working Gap, Magnetic Abrasive Powder, Magnetic Force

Abstract. The authors describe the method of calculating the magnetic forces in the working gap in the case of magnetically abrasive machining of flat surfaces of billets from magnetic materials by the periphery of a circular inductor on permanent magnets. The application of the software package ANSIS Maxwell for the calculation of the magnetic induction method in the working gap and the magnetic forces of the magnetically abrasive powder acting on the grain is shown. As a result of the work, the magnetic induction in the working gap was calculated for magnetically abrasive machining of flat surfaces of billets from magnetic materials by an inducer on permanent magnets. Also, calculations showed the distribution of the magnetic abrasive powder in the working gap, depending on the material of the workpiece being processed. In the case of magnetically abrasive machining of a magnetic workpiece, the powder in the working gap is concentrated in the zones with the greatest density of force lines - under the inductor poles. An analysis is made of the distribution of magnetic forces in the working gap during magnetic abrasive machining.

Introduction

Magnetic abrasive machining is one of the most effective finishing operations [1]. The essence of magnetic abrasive processing is the effect on the workpiece of a powder ferromagnetic mass, compacted by the forces of a magnetic field. It is from the magnitude and distribution of the magnetic induction that the ferromagnetic mass will be distributed in one way or another in the working gap between the magnetic inductor by the workpiece surface being processed.

In magnetic abrasive processing, the role of the cutting tool is performed by a portion of the ferromagnetic powder, and the workpiece for processing is placed at a certain distance from the magnetic inductor. This space is partially or completely filled with a magnetic abrasive powder. Under the influence of the magnetic field, the grains of the ferromagnetic powder are formed into chains, forming a kind of "brush". This brush is held by the forces of the magnetic field in the working space, and when the workpiece moves relative to the inductor, it removes from the surface of the workpiece [2]. Knowing the magnitude of the magnetic forces in the working gap, it is possible to determine the location of the grains of the magnetic-abrasive powder in the working gap.

Calculation of magnetic forces in the working gap during magnetic abrasive processing

It is known that the complexity of calculating the magnetic forces acting on the grains inside the working gap is aggravated by the fact that they have different shapes and sizes. It is not possible to take this manifold into account and it is advisable to assume that the working gap is filled with a continuous medium homogeneous in density and magnetic properties [1].

The magnetic forces acting on the grains of the ferromagnetic powder in the process of magnetic abrasive processing are calculated by the finite element method. The Ansys environment [3] will be used for this.

In this study, a planar problem is solved, and at the same time, a continuous medium homogeneous in density and magnetic properties is divided into individual grains. In this case, it is assumed that the grains of the ferromagnetic powder are ordered and have the shape of a ball.

During magnetic abrasive processing, permanent magnets are located in the inductor so that the magnetic system of the inductor consists of magnetic cells, each of which consists of a magnet and two steel magnetic circuits located at the pole side surfaces of the magnet (Fig. 1) [4].

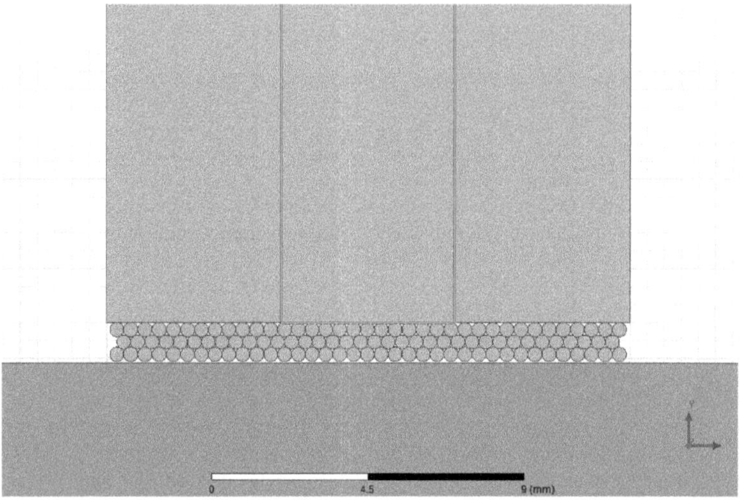

Fig. 1. *Magnetic system*

In this study, the calculation of the magnetic forces in the process of magnetic-abrasive processing of a workpiece from a non-magnetic material, for example, aluminum was made. It was a non-magnetic material that was taken as a sample, since it does not affect the acting magnetic forces in the working gap. When polishing a non-magnetic workpiece, the lines of force pass from one pole of the inductor to another mainly along the working gap under the magnet, concentrating at the side edges of the steel magnetic circuits (Fig. 2) [5]

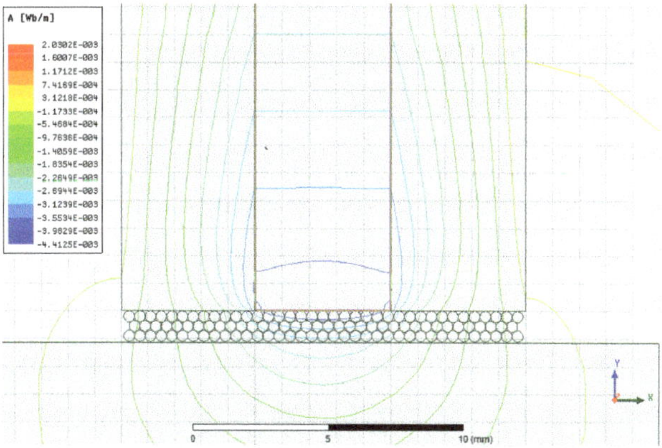

Fig. 2. *Pattern of magnetic field lines*

Because of this, the magnetic-abrasive powder in the working gap is concentrated under the magnets (Fig. 3).

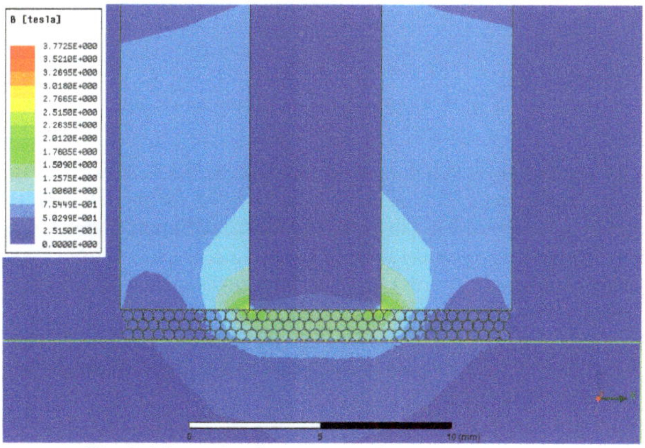

Fig. 3. *Magnetic abrasive powder concentration*

The next step is to calculate the magnetic forces acting on the grains of the ferromagnetic powder from the side of the magnetic field. These magnetic forces keep the powder on the working surface of the magnetic inductor [6]. Depending on the location of each specific grain in the working gap, the corresponding magnetic field force acts on the grain in a state of equilibrium.

As a result, the values of all magnetic forces acting on all grains of the ferromagnetic powder located in the working gap in a state of equilibrium were obtained [7-10]. Each grain has its own individual number, and its own value of the magnetic force.

Materials Research Proceedings **21** (2022) 176-182 https://doi.org/10.21741/9781644901755-31

Analysis of the magnetic forces acting in the working gap

The graphs show the magnetic forces applied to the grains (Fig. 4).

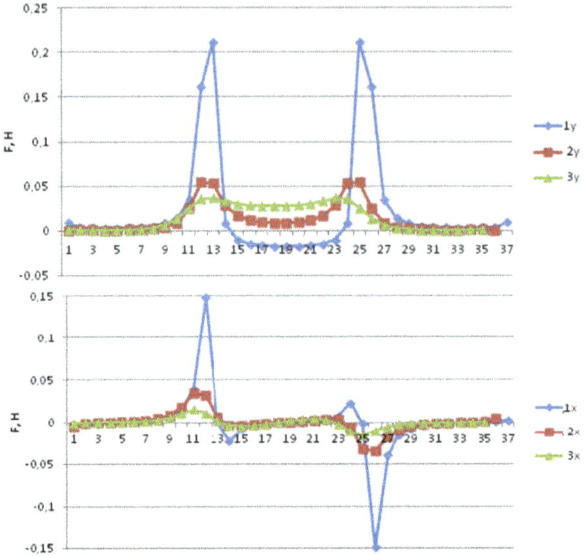

Fig. 4. *Magnetic forces acting on grains*

Analysis of the graphs shows that horizontal forces compress the rows of grains, ensuring their horizontal stability in the rows. This is due to the fact that for the leftmost grains the force is positive and tries to move the row to the right, and for the rightmost grains it is negative and tries to move the row to the left. The vertical force P_y presses the first row of grains against the inductor (the force is positive - directed upwards), and the lower row - against the workpiece (the forces are negative - directed downward).

If we consider only 1 row of grains (at the inductor), then it is, in fact, in equilibrium: (Fig. 5, 6) shows the forces acting on the inductor and between the grains of the 1st row. All forces are positive - the grains are in balance.

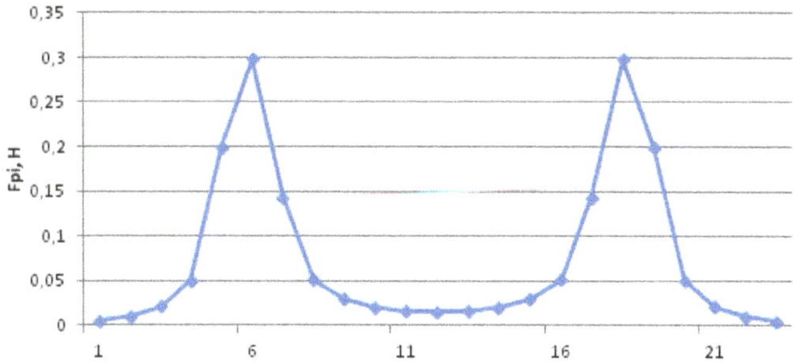

Fig. 5. *The graph of forces between the grains of the first row and the inductor* F_{pi}

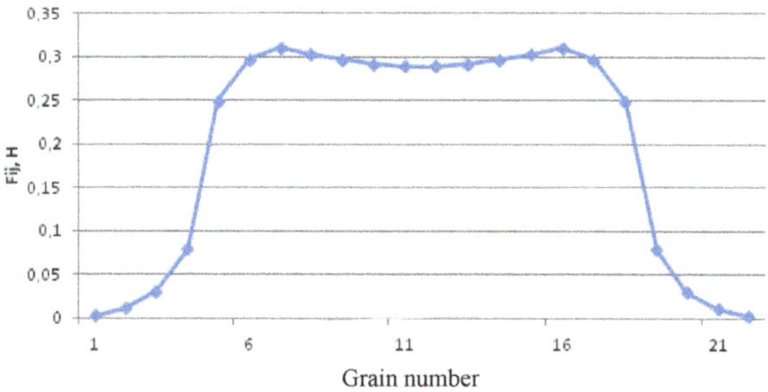

Fig. 6. *Plot of forces between grains of the first row* F_{ij}

However, if we consider the second row of grains, the equilibrium is disturbed. For the first row, equilibrium is maintained. This is due to the fact that for the extreme grains, some forces between the first and second rows have negative values. For the first row, equilibrium is maintained. If we consider all three rows, the instability becomes even greater. The instability is enhanced by the grains of the third row, which "break away" from the grains of the second row. It should be noted that the equilibrium of grains is considered without taking into account their interaction with the workpiece - cutting forces are not taken into account. With the addition of the normal components of the cutting forces acting on the grains of the third row, the stability of the grain system can be restored.

To take into account and determine the cutting forces, they must be applied to the grains of the third row, adding them with magnetic forces directed along the Y-axis - perpendicular to the workpiece surface (normal components). In this case, the maximum values of the cutting forces are determined from the condition of equilibrium of the grain system - the absence of negative forces between the grains. In fact, the cutting forces should eliminate the negative values of the contact forces between the grains of the first and second rows and the second and third rows, but not cause the appearance of negative forces between the grains of the first row. The calculation of the values of the cutting forces is reduced to their selection by the method of least squares to ensure that the forces between the grains of the first row are close to zero. The limitation is the absence of negative forces both in the first Rada and between the rows. Figure 7 shows the cutting forces calculated using this criterion. The curve is well approximated by the expression

$$Pi = A[1-\exp(a_i+b_i^2)]. \quad (1)$$

where i - is the number of the third row grain.

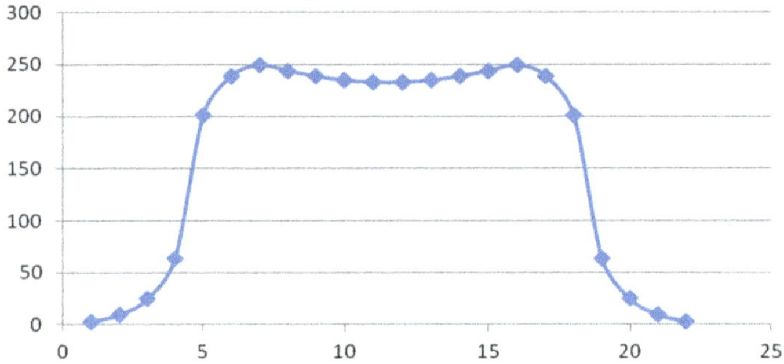

***Fig.** 7. Cutting forces, N*

Due to the symmetry of the curves, the approximation was carried out for grains with numbers 1...18. The right half of the dependencies is symmetrical. Therefore, for grains 19...36, the dependence has the same form, but the numbers of grains are calculated by the formula $i = 37-i_1$, where $i_1 = 19...36$.

Summary

As a result of the work done, the state of equilibrium of the grains of the magnetic-abrasive powder in the process of magnetic-abrasive processing was investigated. For this, the values of the magnetic forces in the working gap have been calculated. This calculation is carried out by the finite element method, and its result is the values of all magnetic forces acting on all grains of the ferromagnetic powder located in the working gap. Knowing the magnitude of the magnetic forces, it is possible to determine the location of the grains of the magnetic-abrasive powder in a state of equilibrium. The next step is to calculate all mechanical forces in the working gap by solving the problem of static equilibrium. In contrast to the finite element method, this calculation method allows one to neglect the stresses arising from point contact of powder grains in a state of equilibrium. The result of solving this problem will be the identification of the configuration of the grains of the ferromagnetic powder in the state of static equilibrium.

References

[1] Baron, Yu. M. Magnetic-abrasive and magnetic processing of articles and cutting tools / Yu. M. Baron. - L.: Mechanical Engineering, 1986. - 176 p.

[2] Sakulevich, F.Yu. Fundamentals of magnetic abrasive processing / F. Yu. Sakulevich. - Minsk: Science and Technology, 1981. - 328 p.

[3] Prihodko, SP Simulation of the process of magnetically abrasive machining of machine parts on computers / SP Prikhod'ko // Finishing-finishing machining methods and tools in machine-building technology. - Barnaul, 1987. - P. 115-119.

[4] Kulavik, AA Calculation of the magnetic forces acting on the grains of a ferromagnetic powder in the process of magnetically abrasive processing / AA Kulavik, SL Leonov, AM Ikonnikov, RV Gre - benkov // Collected Works of the III All-Russian Youth Scientific and Practical School November 29- 30, 2017 "Strengthening Technologies and Functional Coatings in Mechanical Engineering", KuzGTU, Kemerovo, 2017. Access mode: http: // science .kuzstu.ru / wp-content / Events / School / utec / 2017 / school / pages / Articles / 108.pdf

[5] Prikhod'ko, SP Magnetic-abrasive polishing by inducers on permanent magnets / SP Prikhod'ko, Yu. M. Baron // Avtotorktornoeelektrooborudovanie. - 1983. - No. 5. - P. 11-14.

[6] Oliker, VE Powders for magnetically abrasive treatment of wear-resistant coatings / V.E. Oliker. - Moscow: Metallurgy, 1990. - 175 p.

[7] Leonov, SL Automatic regulation of the working gap in the case of magnetically abrasive machining of three-dimensional complex surfaces / SL Leonov, AM Ikonnikov, RV Grebenkov // Materials of the 1st International scientific-practical conference "Actual problems in machine-building", Novosibirsk, 2014. - P. 162-166.

[8] Prikhodko, SP The role of eddy currents in the process of magnetically abrasive processing / SP Prikhod'ko // Finishing-finishing processing methods and tools in engineering technology. - Barnaul, 1989. - P. 104-107.

[9] Krymsky, M.Distribution and consolidation of a magnetic-abrasive powder in the working gap of the machine tool / MD Krymsky // Magnetic abrasive materials and methods for their testing. - Kiev, 1980. - P. 92-97.

[10]Konovalov, Ye. G. Pure machining of parts in a magnetic field by ferromagnetic powders / EG Konovalov, GS Shulev. - Minsk: Science and Technology, 1967. - 125 p.

Change in Properties of Bitumen Used for Road Construction in Bitumineral Mixtures

K.G. Pugin[1,2,a*], O.V. Yakontseva[2,b], V.K. Salakhova[2,c] and K. Yu. Tyuryukhanov[2,d]

[1]Perm State Agro-Technological University named after Academician D.N. Pryanishnikov, Perm, Russian Federation

[2]Perm National Research Polytechnic University, Perm, Russian Federation

[a]123zzz@rambler.ru, [b]yakontseva.olga@yandex.ru, [c]veronika815@inbox.ru, [d]turuchanov.k.u@list.ru

Keywords: Bitumen, Asphalt Concrete, Bitumen Aging, Crushed Stone, Mineral Aggregate

Abstract. Premature destruction of asphalt concrete can be caused by the aging of bitumen, which is associated with a change in the physical and chemical properties of bitumen. The article shows that in most cases, the improvement of the characteristics of asphalt concrete is achieved through the introduction of additives that affect the basic properties of the bituminous binder, such as penetration, softening temperature, viscosity. The influence of the chemical composition of the mineral filler on the rate of bitumen aging has been experimentally proved. The obtained research data show that the increase in the rate of aging of bitumen is influenced by the compounds of silicon and aluminum. On the other hand, metals such as iron and titanium contribute to the preservation of the initial plasticity of bitumen, and the accumulation of asphaltenes in the structure of the bitumen-mineral mixture slows down.

Introduction

Currently, asphalt concrete remains the most common material for road construction. The construction of a modern transport infrastructure requires a large amount of asphalt concrete with increased performance indicators. This is due to an increase in the load on the pavement structure and the requirement to increase the operating time of the pavement before major repairs, to reduce the cost of work and used road materials, including asphalt concrete. Currently developed new technologies for producing asphalt concrete make it possible to obtain asphalt concrete with higher performance, but they imply the use of expensive polymer or composite materials, new technological equipment, expensive raw materials in the compositions and are not always feasible due to the climatic features of the construction area. [1-7] This leads to a significant rise in the cost of road construction. Premature destruction of asphalt concrete (wear, destruction of the surface layer of the road surface) can be caused by the aging of bitumen, which is associated with a change in the physical and chemical properties of bitumen. [8-10] The aging process of bitumen can occur during the preparation of an asphalt mixture at high temperatures or during operation asphalt concrete. The last aging process is of great importance for the initial physical and mechanical properties of bitumen, the presence of polymer and stabilizing additives in the asphalt concrete mixture, and the chemical composition of stone materials used as coarse and fine aggregates. [11-13] Analysis of scientific literature showed that the chemical composition of stone materials used in the composition is not fully taken into account when choosing raw materials for the production of asphalt concrete. As a rule, the content of silicon oxides in the composition of the rock is taken

Materials Research Proceedings 21 (2022) 183-188 https://doi.org/10.21741/9781644901755-32

into account, which determines the adhesion of bitumen to the surface of stone materials. [14-17] There are practically no studies on the effect of other chemical elements on the physical-mechanical, technological and operational properties of asphalt concrete. The relevance of such studies is due to the need to use local rocks for road construction, which have generally similar physicochemical properties, but differ in the presence of one or two chemical compounds characteristic of a given area. The use of local raw materials allows to reduce the cost of road construction by reducing transport costs. This is especially true for areas where there are no quarries for the extraction of stone materials suitable for use in the composition of asphalt concrete. In this regard, the study of the influence of the chemical composition of rocks on the aging process of bitumen is an urgent topic of research.

Bitumen for asphalt pavements does not always have the required parameters, such as adhesion, increased elasticity, resistance to aging, heat, workability of the mixture. To improve the quality properties of bitumen and asphalt concrete mixtures, additives and modifiers are used, for example: to increase elasticity, improve adhesion to mineral aggregates, reduce viscosity, increase resistance to the aging process, and prevent the binder from flowing off the aggregate surface. In some cases, the required properties can be changed by changing the properties of the mineral filler. [18-21]

Adhesion improvers are surfactants that lower the surface tension in the bitumen-aggregate phase and thus improve wetting. A number of scientific articles present the results of testing the use of additives and indicate their positive effect on the adhesion of bitumen to aggregate to acidic and neutral aggregates (granite, quartz). However, a number of researchers show that these additives also affect the basic properties of the bitumen binder, such as penetration, softening point, viscosity. [22, 23] In the production of low-temperature asphalt mixtures (WMA), organic compounds such as amide waxes are used as additives. mountain waxes, low Fischer-Tropsch (FT) paraffin wax, zeolites. Each of them has different characteristics, while others affect the asphalt production process. Waxes and paraffin's also affect the physical and mechanical properties of asphalt concrete, and zeolites improve the workability of the mixture. The type of additive should be chosen carefully so that the melting point of the additive is higher than the expected operating temperatures. This reduces the risk of permanent deformation and minimizes cracking of the asphalt at low temperatures. Surfactants reduce the surface tension when bitumen comes into contact with the aggregate and improve wetting. Modifiers, unlike additives, are part of the bitumen structure and create a three-dimensional network. Bitumen modified by the addition of high molecular weight substances, such as synthetic polymers, changes the thermo viscous and elastic-viscous properties of bitumen. The quality level depends on the type of modifier and its amount in the bitumen, the type of bitumen and chemical composition, as well as on the technology, mainly the homogenization method.

All of the above additives and techniques can improve the characteristics of asphalt concrete at the time of its production and the initial period of operation. Little attention is paid to such an indicator as thermal aging of bitumen, which determines the service life of asphalt concrete.

Methods and materials

At the Perm National Research Polytechnic University, at the Department of Roads and Bridges, together with the Perm Agrarian-Technological University, studies were carried out on the effect of the chemical composition of the mineral filler on the aging of bitumen. The following materials were used for the study: crushed stone of the Alapaevskoye deposit, the Khromisty open-pit mine; crushed stone of the Druzhininsky deposit, the Druzhininsky quarry; crushed stone of the Gusevogorsk deposit, open pit "KGOK". The chemical composition of crushed stone from different quarries is presented in Table 1.

***Table 1**. Chemical composition of crushed stone from different quarries*

Mass fraction	Quarry "Chromisty"	Quarry "Druzhininsky"	Open pit "KGOK"
Calcium oxide, %	29,88	63,30	30,64
Magnesium oxide,%	11,58	10,32	12,58
Silicon dioxide, %	46,22	26,22	39,74
Sulfur, %	-	0,02	0,02
Phosphorus, %	-	0,03	0,02
Iron oxide, %	2,94	0,06	14,55
Aluminum oxide, %	9,50	0,16	-
Manganese oxide,%	0,06	0,02	0,14
K_2O+Na_2O, %	0,28+4,94	0,04+0,15	-
TiO_2, %	0,24	0,005	2,26
CrO_5, %	-	-	0,05

Crushed stone before mixing with bitumen was crushed to a fraction of 2.5 - 5.0 mm. The research used bitumen BND 100/130 produced by PJSC "Lukoil". To determine the indicators of bitumen, the methodology GOST 18180 Petroleum bitumen was used. Method for determining the change in mass after heating. During the experiment, heated crushed stone and bitumen were mixed and kept in a heating cabinet. The crushed stone / bitumen ratio was ¼. The study was carried out using the equipment: glass cups, type ЧБН in accordance with GOST 25336 with an outer diameter of (100 ± 1) mm; drying cabinet with a volume of 67 dm^3, equipped with a thermostat that maintains the temperature with a deviation of no more than ± 1 °C, with a ventilation hole with a diameter of 10 mm; metal sieve with mesh N 07 in accordance with GOST 6613; glass thermometer technical performance P with a graduation value of 1 °C, up to 200 °C in accordance with GOST 28498; desiccators according to GOST 25336; technical laboratory scales with a weighing error of no more than 0.01 g.

Results and discussion

The data obtained on the study of the indicators of bitumen after aging and contact with mineral filler from different quarries are presented in Table 2.

***Table 2.** Indicators of bitumen after testing.*

	Indicator	Initial bitumen	Initial bitumen , after testing	Bitumen with crushed stone quarry "Chromisty"	Bitumen with crushed stone quarry "Druzhininsky"	Bitumen with crushed stone "KGOK"
1	KiSh	46,8	52,8	52,5	52,5	51,7
2	Depth of needle penetration, at 25 ° C	100	66	58	64	77
3	Penetration index	-0,23	0,09	-0,23	0,02	0,33
4	Mass loss after testing		0,06	0,24	0,20	0,17
5	Extensibility at 25 ° C, cm	149,3	117,1	66,0	99,3	118,6
6	Extensibility at 0 ° C, cm	5,5	4,2	4,4	4,0	4,1

Aging of bitumen can occur under the influence of many factors: temperature, light, chemical reactions. The process of bitumen aging itself can be represented as the transition of the constituent parts of bitumen from one form to another. Oils turn into resins, resins into asphaltenes. As a number of researchers show, the process of conversion of oils to resins is much slower than the conversion to asphaltenes. Over time, the asphaltene content increases in bitumen, which leads to the loss of such important properties of bitumen as extensibility and plasticity, while brittleness increases. Aging of bitumen causes irreversible structural changes in asphalt concrete, leading to the destruction of road surfaces. Asphalt concrete largely loses its ability to absorb tensile stresses. With a sharp drop in temperature in winter, the formation of cracks on the asphalt concrete pavement is observed and the process of destruction of the surface layer intensifies. Joint heating with mineral materials can increase the rate of bitumen aging. In this case, some mineral materials act as catalysts, in the presence of which the transition of the bitumen constituents occurs more intensively. Aging of bitumen-mineral mixtures can be divided into two periods. Until a certain time, the processes of aging of bitumen can be characterized only as processes of strengthening the structure of asphalt concrete, due to improved adhesion to the mineral surface (the acid number increases in bitumen), as a result of which the water resistance and frost resistance of asphalt concrete increases. An increase in the viscosity of bitumen during this period leads to an optimal increase in the strength and deformation resistance of asphalt concrete. Further, due to an increase in the brittleness of bitumen, adhesion decreases, leading to a decrease in the corrosion resistance of the road surface, and due to excessive rigidity, cracks appear in it more often.

Conclusions

Aging of bitumen and bitumen-mineral mixtures differs in the ratio of the duration of these periods. For the most resistant to aging bitumen, the first period lasts a long time, and is practically commensurate with the service life of asphalt concrete pavements. For coatings with bitumen that are not resistant to aging, after a year or two, a period of intense aging may occur, which is due to the presence of certain chemical elements. Additionally, it can be assumed that other chemical elements can, on the contrary, hinder the aging process of bitumen. The obtained research data show that the increase in the rate of aging of bitumen is influenced by the compounds of silicon and aluminum. On the other hand, metals such as iron and titanium contribute to the preservation of the initial plasticity of bitumen, and the accumulation of asphaltenes in the structure of the bitumen-mineral mixture slows down. In this regard, it can be concluded that when selecting materials for the design of an asphalt concrete mixture, it is necessary to take into account the chemical composition of rocks and their ability to influence not only the adhesion of bitumen, but also its ability to thermal aging upon direct contact of bitumen and mineral particles.

References

[1] Benosman A S, Senhadji Y, Mouli M 2015 Application of Polymer-Mortar Composites as a Sustainable Building Material *Key Engineering Materials* **650** 21–28. https://doi.org/10.4028/www.scientific.net/KEM.650.21

[2] Yang E H, Liu Y Q, Chen Z T 2015 Environmental Sustainability through Recycling Incineration Bottom Ash for the Production of Autoclaved Aerated Concrete *Key Engineering Materials*, **650** 51–70. https://doi.org/10.4028/www.scientific.net/KEM.650.51

[3] Kannamma D, Meenatchi Sundaram A 2015 Implications of Building Material Choice on Outdoor Microclimate for Sustainable Built Environment *Key Engineering Materials* **650** 82–90. https://doi.org/10.4028/www.scientific.net/KEM.650.82

[4] Hussain A, Arif Kamal M 2015 Energy Efficient Sustainable Building Materials: An Overview *Key Engineering Materials* **650** 38–50. https://doi.org/10.4028/www.scientific.net/KEM.650.38

[5] Vaisman Y I, Pugin K G, Vlasov A S 2020 Using the Resource Potential of Drill Cuttings in Road Construction *IOP Conference Series: Earth and Environmental Science* **459(2)** 022078. https://doi.org/10.1088/1755-1315/459/2/022078

[6] Xu, S.; Yu, J.; Zhang, C.; Sun, Y. Effect of ultraviolet aging on rheological properties of organic intercalated layered double hydroxides modified asphalt. Constr. Build. Mater. 2015, 75, 421–428. https://doi.org/10.1016/j.conbuildmat.2014.11.046

[7] Tyuryukhanov K Y, Pugin K G 2020 An Impact of Waste Foundry Sand on Asphalt Concrete Mixture *IOP Conference Series: Materials Science and Engineering* **753(2)** 022079. https://doi.org/10.1088/1757-899X/753/2/022079

[8] Wu, S.; Zhao, Z.; Xiao, Y.; Yi, M.; Chen, Z.; Li, M. Evaluation of mechanical properties and aging index of 10-year field aged asphalt materials. Constr. Build. Mater. 2017, 155, 1158–1167. https://doi.org/10.1016/j.conbuildmat.2017.08.102

[9] Abu Qtaish, L.; Nazzal, M.D.; Abbas, A.; Kaya, S.; Akinbowale, S.; Arefin, M.S.; Kim, S.S. Micromechanical and chemical characterization of foamed warm-mix asphalt aging. J. Mater. Civ. Eng. 2018, 30, 04018213. https://doi.org/10.1061/(ASCE)MT.1943-5533.0002430

[10] Bowers, B.F.; Huang, B.; Shu, X.; Miller, B.C. Investigation of reclaimed asphalt pavement blending efficiency through GPC and FTIR. Constr. Build. Mater. 2014, 50, 517–523. https://doi.org/10.1016/j.conbuildmat.2013.10.003

[11] Pugin K G 2020 Influence of quality indicators of the surface of mineral particles on the properties of asphalt concrete *IOP Conference Series: Materials Science and Engineering* **971(3)** 032043. https://doi.org/10.1088/1757-899X/971/3/032043

[12] Lee, S.J.; Hu, J.; Kim, H.; Amirkhanian, S.N.; Jeong, K.D. Aging analysis of rubberized asphalt binders and mixes using gel permeation chromatography. Constr. Build. Mater. 2011, 25, 1485–1490. https://doi.org/10.1016/j.conbuildmat.2010.08.004

[13] Wang, P.Y.; Zhao, K.; Glover, C.; Chen, L.; Wen, Y.; Chong, D.; Hu, C. Effects of aging on the properties of asphalt at the nanoscale. Constr. Build. Mater. 2015, 80, 244–254. https://doi.org/10.1016/j.conbuildmat.2015.01.059

[14] Chen, A.; Liu, G.; Zhao, Y.; Li, J.; Pan, Y.; Zhou, J. Research on the aging and rejuvenation mechanisms of asphalt using atomic force microscopy. Constr. Build. Mater. 2018, 167, 177–184. https://doi.org/10.1016/j.conbuildmat.2018.02.008

[15] Zhang, D.; Zhang, H.; Shi, C. Investigation of aging performance of SBS modified asphalt with various aging methods. Constr. Build. Mater. 2017, 145, 445–451. https://doi.org/10.1016/j.conbuildmat.2017.04.055

[16] Yu, J.; Dai, Z.; Shen, J.; Zhu, H.; Shi, P. Aging of asphalt binders from weathered asphalt mixtures compared with a SHRP process. Constr. Build. Mater. 2018, 160, 475–486. https://doi.org/10.1016/j.conbuildmat.2017.11.060

[17] Yan, C.; Huang, W.; Tang, N. Evaluation of the temperature effect on Rolling Thin Film Oven aging for polymer modified asphalt. Constr. Build. Mater. 2017, 137, 485–493. https://doi.org/10.1016/j.conbuildmat.2017.01.135

Materials Research Proceedings 21 (2022) 183-188 https://doi.org/10.21741/9781644901755-32

[18] Dondi, G.; Mazzotta, F.; Simone, A.; Vignali, V.; Sangiorgi, C.; Lantieri, C. Evaluation of different short term aging procedures with neat, warm and modified binders. Constr. Build. Mater. 2016, 106, 282–289. https://doi.org/10.1016/j.conbuildmat.2015.12.122

[19] Xu S., Fan Y., Feng Z., Ke Y., Zhang C., Huang H. Comparison of quantitative determination for SBS content in SBS modified asphalt Construction and Building Materials, Volume 282, 2021. https://doi.org/10.1016/j.conbuildmat.2021.122733

[20] Chomicz-Kowalska, A., Mrugala, J., Maciejewski, K.: Evaluation of Foaming Performance of Bitumen Modified with the Addition of Surface Active Agent, WMCAUS, 245 (2017), pp.1-11. https://doi.org/10.1088/1757-899X/245/3/032086

[21] 21.Iwanski, M., Chomicz-Kowalska, A., Maciejewski, K.: Application of synthetic wax for improvement of foamed bitumen parameters, Construction and Building Materials 83 (2015), pp.62-69, 2015. https://doi.org/10.1016/j.conbuildmat.2015.02.060

[22] A. Vaitkus, D. Cygas, A. Laurinavicius, Z. Perveneckas, Warm mix asphalt research, analysis and evaluation, The Baltic Journal of Road and Bridge Engineering, 4 (2), 2009, pp. 80-86. https://doi.org/10.3846/1822-427X.2009.4.80-86

[23] P. Radziszewski, K. Kowalski, J. Król, M. Sarnowski, J. Piłat: Quality assessment of bituminous binders based on the viscoelastic properties: polish experience.Journal of Civil Engineering and Management, vol. 1(20), 2014, pp. 111-120. https://doi.org/10.3846/13923730.2013.843586

Materials Research Proceedings **21** (2022) 189-193 https://doi.org/10.21741/9781644901755-33

Tool Bits with Elastic Damping Inserted Elements Formed by Semi-Inserts with Different Rigidity

Sergey G. Novikov[1,a], Vitaliy V. Malykhin[2,b], Sergey A. Chevychelov[2,c*] and Fedor V. Novikov[3,d]

[1] Regional Open Social Institute, 85, Mayakovskogo ul. Kursk, 305009, Russia

[2] Southwest State University, 94, 50 Let Octyabrya ul., Kursk, 305040, Russia

[3]Kharkiv National Economic University named after Semyon Kuznets, 9A, Nauli ul., Kharkiv, 61166, Ukraine

[a]novikov.s.46@mail.ru, [b]mtio29@yandex.ru, [c]tschsa@yandex.ru, [d]fokusnik1@rambler.ru

Keywords: Tool Bit, Holder, Vibration, Impact Load, Damping Semi-Inserts, Semi-Inserts Rigidity

Abstract. Vibration during turning is a consequence of premature tool wear and an increase in the roughness of parts. The designs of tool bits that increase their durability, reduce vibrations arising during the cutting process, and improve the quality of processing are of practical and scientific interest. Existing developments of vibration-damping bits with elastic damping inserts of constant rigidity, in which a holder or mandrel with a cut-off insert is installed, do not meet the requirements for the efficiency of vibration reduction due to the impossibility of the necessary vibration damping; the consumption of insert materials is high. Innovative designs of tool bits with inserts formed by semi-inserts with different rigidity are proposed, which allows improving the quality of processing by increasing the efficiency of vibration damping, reducing material consumption, because the length of the semi-inserts is less than half the length of the full insert. The tool bits can be used in the area of cutting materials, mechanical engineering and tool manufacturing.

Introduction

Reducing the vibration level during cutting is the most important problem, especially taking into account the increasing requirements for the quality of processing [1-4]. One of the ways to solve this problem is to improve the designs of cutting tools in order to increase the stability of the cutting process [5-10]. In particular, the designs of damping tool bits which make it possible to address the most important task of mechanical engineering to increase the efficiency of processing structural materials are being developed; this task involves achieving the highest productivity while ensuring a given level of quality of the surface layer of parts [11-15].

However, at present, the development of damping tool bits does not fully meet the imposed requirements. A significant drawback of most of the existing designs of tool bits is that they are rigidly attached to the machine [11-14], which does not allow achieving effective vibration reduction during processing. Many damping bits also have complex designs [11,13,14], technologies of their manufacturing are very sophisticated, and normal operation of some machines is highly doubtful.

New designs of bits to increase their durability, reduce vibrations arising in the cutting process, and improve the quality of processing are of practical and scientific interest.

Experimental procedure

Materials and methods for solving the problem. There is a damping tool bit containing a cut-off insert with a unit for its attachment in a holder which has a rabbet and an inserted element made of a material with a high damping property; the rabbet is made evenly along the perimeter of the end of the holder along the length from its end to the projection with the cut-off insert; the end of the holder with the rabbet is placed in a metal body made in the form of a rectangular parallelepiped with the same gaps along its base and walls and the gaps are filled with the material of the mentioned inserted element, and the holder is installed without the possibility of contact with the body [15]. The tool bit has the following drawbacks:

1. When turning a product, the perturbation cutting force causes vibrations of the tool bit, however, a material with a high damping property has constant rigidity for the solid inserted element, therefore, the pair of restoring forces arising in it, although creating a moment that makes the tool bit return to the position of static equilibrium and dampens vibrations, is not sufficient to minimize vibrations to the desired level, and does not allow achieving the required processing quality.

2. Excessive consumption of the material used for inserted element due to the fact that it fills the gaps along the entire length of the rabbet of the holder; the gaps between the surfaces of the holder with the rabbet and the inner cavity of the metal body have the form of rectangular parallelepipeds and are rather large.

The technical purpose of the use of the proposed damping tool bit is to improve the quality of processing and reduce the consumption of the material used for insert element.

Figure 1 shows a general view of the tool bit.

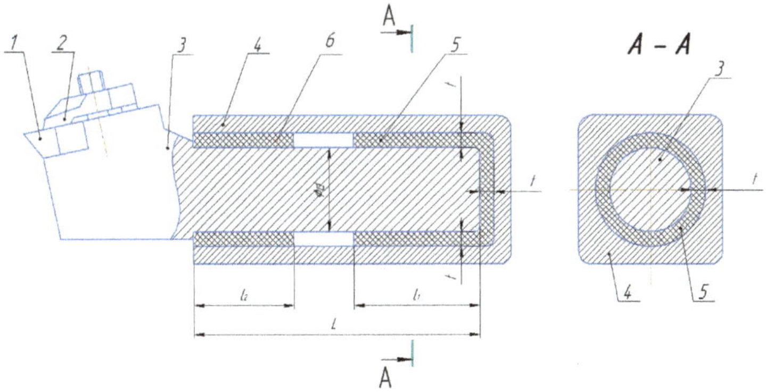

Fig. 1. *A tool bit with damping semi-inserts of different rigidity:*
1 – cut-off insert; 2 – insert fastening unit; 3 – holder; 4 – body; 5, 6 – damping semi-inserts.

The damping tool bit operates as follows. The required cutting conditions for structural material are specified. During workpiece processing, the perturbation cutting force causes the tool to vibrate. In the case of making the inserted element along the entire length L of the cylindrical rabbet of holder 3 of diameter d or consisting of semi-inserts 5, 6 of materials with high damping properties of the same rigidity ($j_1 = j_2$), a pair of restoring forces arise in the solid inserted element, and two identical pairs of restoring forces arise in semi-inserts 5, 6 creating moments that tend to return the bit to a static equilibrium position and damping vibrations, which are insufficient to minimize vibrations to the desired level.

In the proposed tool bit, the inserted element is formed of cylindrical semi-inserts 5, 6 of materials with high damping properties of various rigidity ($j_1 > j_2$), and the lengths l_1, l_2 of semi-inserts 5, 6 are less than half the length L of the rabbet ($l_1<L/2$, $l_2<L/2$). Accordingly, in the semi-inserts, there arise restoring parallel forces directed in opposite directions and not equal in modulus, while the moduli of forces in semi-insert 5 are more than the moduli of forces in semi-insertion 6. Therefore, in addition to the moment from a pair of forces equal in modulus to the forces in semi-inserts 5 and 6, in semi-insert 5 there are still forces equal to the difference in force moduli in the semi-inserts and creating an additional restoring moment, which increases the efficiency of damping tool bit vibrations. (The additional moment is equal to the product of the difference between the moduli of the indicated forces by the arm). So, we get highly effective damping of longitudinal, transverse vibrations and impact loads arising in the cutting process; the reliability of this process increases in both turning and planing technological operations. Thus, by selecting different rigidity materials with the of high damping properties for semi-inserts 5, 6, it is possible to minimize tool bit vibrations to the required level and achieve the required processing quality.

The design allows:

1. To improve the quality of processing by increasing the efficiency of tool bit vibrations damping due to the use of the inserted element formed of two semi-inserts made of materials with high damping properties and having different rigidity; in this case, the regidity of the semi-insert located at the end of the holder at the bottom of the body base is higher than the rigidity of the semi-insert placed on opposite end of the holder with rabbet.

2. Reduce the consumption of the material used for inserted damping element, since it consists of two semi-inserts placed at a distance from each other; moreover, the length of both semi-inserts is less than half the length of the rabbet of the holder end; in addition, the insert fills the annular gap between the cylindrical rabbet of the holder and the cylindrical rabbet of the body, minimizing the consumption of materials used for semi-inserts.

This leads to the technical result of improving the quality of processing and reducing the consumption of the material used for inserts.

Results and discussion. The originality of the proposed damping tool bit is that in the rabbet made in the shape of mandrel 3, the lower end of mandrel 3 is placed with gaps *t* uniform along the bottom and walls of the rabbet filled with an inserted element made of material with high damping properties, the inserted element is formed of two spaced apart from each other semi-inserts made of materials with high damping properties and having different rigidity; the first of the semi-inserts – centre bearing 5 is placed at the end of mandrel 3 at the bottom of the rabbet, and the second semi-insert 6 is placed at the upper part of mandrel 3 placed in the rabbet; in this case, the length h_1 of centre bearing 5 and h_2 of the second semi-insert 6 is less than half the depth h/2 of the rabbet, and the rigidity of centre bearing 5 is higher than the rigidity of the second semi-insert 6, which makes it possible:

1. To improve the quality of processing by increasing the efficiency of damping vibrations of the cut-off insert rigidly mounted on the upper surface of the mandrel by forming an insert element from a centre bearing and a semi-insert, made of materials with high damping properties and having different rigidity, and the rigidity of the centre bearing is higher than the rigidity of the semi-insert.

2. To simplify the design of the tool bit in such a way that the lower end of the mandrel is placed directly in the rabbet, made in the shape of the mandrel, with uniform gaps along the bottom and walls of the rabbet, which does not require additional structural elements to be placed.

3. Reduce the consumption of the inserted element material, since it consists of a centre bearing and a second semi-insert placed at the lower end of the mandrel at a distance from each other, and the length of the centre bearing and the semi-insert is less than half the depth of the rabbet; in addition, material consumption for manufacturing additional structural elements for placing the mandrel in the rabbet is not necessary.

Summary

Thus, the proposed damping tool bits make it possible to achieve a technical result to improve the quality of processing, simplify the design of tool bits and reduce material consumption for their manufacturing. The tool bits can be used in the field of materials cutting, mechanical engineering and tool production.

References

[1] Maslennikov A.V., Chevychelov S.A., Merzhoeva M.S., Gatiev M.S., Sidorova V.V. Influence of axial harmonic oscillations on chip formation when drilling holes in ductile metals. Russian Engineering Research. 2014. T. 34. № 11. C. 722-724. https://doi.org/10.3103/S1068798X14110136

[2] Grechukhin, A.N., Kuts, V.V., Razumov, M.S. Solving Problem of Curved Surface Approximation by Layers with Constant and Variable Sections During Forming by Additive Methods (2020) Lecture Notes in Mechanical Engineering, pp. 239-248. https://doi.org/10.1007/978-3-030-22041-9_28

[3] Kuts, V.V., Razumov, M.S., Sidorova, M.A. Development of the Internally Cutting Milling Cutter Model with Radial Advance for Treatment of Profiling Rolls (2017) Procedia Engineering, 206, pp. 1497-1502. https://doi.org/10.1016/j.proeng.2017.10.667

[4] Emelyanov S.G., Chevychelov S.A. Improving the quality of the processed surface by choosing a rational shaping scheme. Mechanical engineering and automation issues. 2004. No. 3. P. 86-88.

[5] Emelyanov S.G., Chevychelov S.A. Modelling the error of the profile formed by hyperbolic cutters. Bulletin of mechanical engineering. 2008. No. 1. P. 38-40.

[6] Maslennikov A.V., Chevychelov S.A., Golubev I.G. Rear angle of a helical bit in hole shaping with axial vibration. Russian Engineering Research. 2013. T. 33. № 5. C. 295-298. https://doi.org/10.3103/S1068798X13050092

[7] Razumov, M., Grechukhin, A., Maslennikov, A. Geometric Parameters of Cutting Tools that Can be Used for Forming Sided Surfaces with Variable Profile. Archives of Metallurgy and Materials, 2017, 62(1), стр. 33–40. https://doi.org/10.1515/amm-2017-0005

[8] Kuts, V.V., Gridin, D.S. Comprehensive study of the process of cutting screw grooves on the inner surface of a cylindrical steel sleeve with rotary cutting mandrels (2021) Chernye Metally, 2021 (2), pp. 63-68. https://doi.org/10.17580/chm.2021.02.11

[9] Kuts, V., Razumov, M., Kalutskikh, N. Modelling of cutting block cut surface at faceted surfaces machining using planetary gear set (2017) MATEC Web of Conferences, 129, статья № 01036. https://doi.org/10.1051/matecconf/201712901036

[10] Grechishnikov, V.A., Kuts, V.V., Razumov, M.S., Romanov, V.B., Pivkin, P.M., Grechukhin, A.N., Yurasov, S.Y. Errors in shaping by a planetary mechanism (2017) Russian Engineering Research, 37 (9), pp. 824-826. https://doi.org/10.3103/S1068798X17090118

[11] Tool bit: RF Patent 2009768: MPK[5] B23B27/00/ V.A. Rogov; applicant patent holder - Peoples' Friendship University of Russia. – No. 5014352/08; appl. 02.12.91; publ. 03.30.94.-3 p.

[12] Tool bit: RF Patent 2280542: MPK[7] B23B27/00/ L.A. Vasin, N.N. Borodkin; applicant patent holder Tula State University. -25.04.2005; publ. 27.07.2006. 4 p.

[13] Tool bit: RF Patent 2287405: MPK B23B27/10/ G.M. Polezhaikin, N.A. Ladoshkin, A.G. Fomin L.A., I.G. Polezhaikin, T.G. Efanyeva; applicant - patent holder Mordovia State University named after N.P. Ogarev. No. 2005 124 690/02; appl. 02.08.2005; publ. 20.11.2006. 3p.

[14] Tool bit: RF Patent 2107588: MPK B23B27/16/ V.G. Voitenko, E.Yu. Tatarkin, A.A. Sitnikov; applicant-holder Altai State University. No96123454/02; appl. 15.12.96; publ. 27.03.98. 3p.

[15] Yatsun E.I., Malykhin V.V., Zubkova O.S. Novikov S.G. Instrumental support of machining processes with hard alloys and composites: monograph. Kursk: "Universitetskaya kniga". 2016. Pp.160-124

Analytical and Experimental Study of Grinding Forces with Flap Wheels

Alexander N. Unyanin[1,a*] and Wladimir S. Chistjakov[1,b]

[1]Faculty of Mechanical Engineering, Ulyanovsk State Technical University, 32, Severny Venetz str., 432027, Ulyanovsk, Russia

[a]a_un@mail.ru, [b]v_s_chistyakov@mail.ru

Keywords: Grinding, Abrasive Grain, Flap Wheel, Force, Workpiece

Abstract. Analytically, the dependences for calculating the grinding forces with flap wheels are obtained. The forces were defined as the sum of the forces associated with dispersion and friction of the cutting and plastically deforming petal grains against the workpiece. The dependencies take into account the change in the depth of penetration of the grain into the workpiece material along the length of the arc of the contact of the circle with the workpiece. Numerical modeling and experimental study of forces have been carried out. The discrepancy between the calculated and experimental values of the forces does not exceed 20%.

Introduction

Abrasion with flap wheels made of abrasive paper is used to polish and remove the defective layer from the surfaces of the workpieces.

In the process of processing, the petal circle is positioned relative to the workpiece surface to be processed at a distance at which the petals are deformed. The radial force with which the flap wheel acts on the workpiece during processing is made up of centrifugal and elastic components [1]. This force can be varied by changing the speed of the circle and deformation of the petals. The radial force with which the flap wheel acts on the workpiece must be equal to the radial component of the force of dispersion of the workpiece material by the circle. Otherwise, the required rate of material removal from the workpiece will not be provided. Therefore, it is necessary to know the forces arising in the process of grinding with flap wheels.

Mathematical models for calculating grinding forces analytically were developed by S.N. Korchak [2], L.N. Filimonov [3], V.I. Ostrovsky [4], V.V. Efimov [5], J. Peklenik, A. Pahlitzsch and other researchers.

S.N. Korchak [2] established the relationship between the cutting forces of a single abrasive grain (AG) and shear and compression stresses, which depend on the strength of the workpiece material at the deformation rate and temperature in the grinding zone. L.N. Filimonov [3] obtained equations for calculating the cutting forces by a single AG, taking into account the inertial component. This is important for grinding at high wheel speeds, when the inertial component is of great importance. V.V. Efimov [8] switched from calculating cutting forces with a single grain to cutting forces with a grinding wheel. Not only grains producing microcutting are taken into account, but also grains performing plastic deformation. Models of grinding forces, including those with superimposed vibrations, are presented in [6], [7], [8].

The dependencies for calculating the grinding forces with flap wheels are given in [9]. However, they do not take into account the change in the depth of penetration of abrasive grains on the trajectory of their contact with the workpiece and the dispersion of the workpiece by plastically deforming grains.

Analytical research

The components of the grinding force were presented as the sum of the forces associated with the dispersion of the workpiece material and the friction of the petal grains against the workpiece, which carry out cutting and plastic deformation. We took into account the peculiarities of the location of abrasive grains along the depth of the working layer of the petals made of a grinding belt [10]. The geometric shape of the grains was presented in the form of a truncated cone, the diameter of the small circle of which is equal to the size of the blunt area formed on the grain during wear.

The radial P_y and tangent P_z components of the grinding force were determined as:

$$P_y = P_{yr1} + P_{yr2} + P_{yd1} + P_{yd2}; \quad P_z = P_{zr1} + P_{zr2} + P_{zd1} + P_{zd2}, \tag{1}$$

where P_{yr1} and P_{zr1}, P_{yr2} and P_{zr2} – radial (index y) and tangential (z) components of the force due to microcutting of the workpiece material and friction of the cutting grains of the petals on the workpiece, N; P_{yd1} and P_{zd1}, P_{yd2} and P_{zd2} – components of the force due to plastic deformation of the workpiece material and friction of plastically deforming grains on the workpiece, N.

Cutting grains are located at a distance y from the conditional outer surface of the petal: $0 \le y \le y(l)$, where $y(l)$ – a function describing the dependence of the depth y of occurrence of AG on the length l of its contact with the workpiece. In determining P_{yr1} multiple integration is used to sum the microcutting forces from the individual AG.

$$P_{yr1} = \int_0^{l_k} \left[\int_0^{y(l)} P_{yr1i} \cdot n_{zk1} \cdot c_k \cdot dy \right] \cdot dl, \tag{2}$$

where l_k – arc length of contact of the cutting grain with the workpiece, m; P_{yr1i} – microcutting force by individual AG, N; n_{zk1} – the number of AG on the surface of the petal in the section by a plane parallel to the axis of the circle, 1 / m; ($n_{zk1} = Z_0 \cdot H$, where Z_0 – number of AG on a single lobe area, 1 / m^2; H – the size of the workpiece surface to be machined in a direction parallel to the axis of the circle, m); c_k – coefficient characterizing the distribution of grain depths in the working volume of the petal, 1 / m.

$$y(l) = h_{ur} + a_z(l), \tag{3}$$

where h_{ur} – deterioration of grain performing microcutting, м; $a_z(l)$ – function describing the change in the depth of penetration of AG into the workpiece material along the contact length l.

Forces to cut the workpiece material with a single grain:

$$P_{yr1i} = \tau_s \cdot F_m \cdot k_y, \tag{4}$$

where τ_s – shear strain during microcutting by grain of the workpiece material, Pa; F_m – area of the workpiece metal removed by a single grain, м^2; k_y – coefficient [5].

$$F_m = (2h_{ur} + a_{ir}) \cdot a_{ir} \cdot tg\gamma, \tag{5}$$

where a_{ir} – the depth of penetration into the workpiece of the i-burning AG, m [5]; γ – half of the angle at the apex AG, degree.

To calculate $a_z(l)$ used the dependence [11]:

$$a_z(l) = \frac{V_s \cdot sin\alpha_{max} \cdot l_m \cdot l}{\pi \cdot D_k \cdot n_k \cdot l_k}, \tag{6}$$

where l_m – average distance between the cutting grains of the petal, m; V_s – flap wheel longitudinal feed speed, m / min; D_k – circle diameter, m; n_k – wheel speed, 1 / min; α_{max} – angle, degree [11].

The result is the dependence:

$$P_{yr1} = k_y \cdot \tau_s \cdot B_p \cdot (2 \cdot h_{ur}^2 \cdot a_{kr} \cdot l_k + (6 \cdot h_{ur} \cdot a_{kr} + 7 \cdot h_u^2 + a_{kr}{}^2) A \cdot \frac{l_k^2}{2} + h_{ur} \cdot a_{kr}{}^2 \cdot l_k +$$
$$+(h_{ur} + a_{kr}) \cdot h_{ur}^2 \cdot l_k + \frac{h_{ur}^3 \cdot l_k}{3}), \tag{7}$$

where a_{kr} – critical depth of microcutting, when exceeding which the grain performs microcutting, m;

$$A = \frac{V_s \cdot sin\alpha_{max} \cdot l_m}{\pi \cdot D_k \cdot n_k \cdot l_k}; \quad B_p = tg\gamma \cdot n_{zk1} \cdot c_k.$$

To calculate P_{yp2} dependences were obtained:

$$P_{yr2} = \frac{\pi \cdot l_{2r}^2 \cdot k_y \cdot n_{zk1} \cdot c_k}{12 \cdot \mu_s} \cdot \tau_s \cdot (h_{ur} \cdot l_k + \frac{A \cdot l_k^2}{2}),$$

where l_{2r} – the size of the blunt area on the cutting grain, m; μ_s – internal friction coefficient.

To calculate the radial component of the force from plastically deforming grains associated with deformation of the workpiece material (P_{yd1}) and friction of plastically deforming grains against the workpiece (P_{yd2}), dependences were obtained.

$$P_{yd1} = \frac{c \cdot \sigma_{ts} \cdot n_{zk1} \cdot c_k \cdot tg\gamma}{2} (sin\gamma + \mu_0 \cdot cos\gamma) \cdot l_{k1} \cdot a_{kr} \cdot (\frac{a_{kr}^2}{3} - h_{ud}^2); \tag{8}$$

$$P_{yd2} = c \cdot \sigma_{ts} \cdot \frac{k_y \cdot \pi \cdot l_{2d}^2 \cdot n_{zk1} \cdot c_k}{12} \cdot a_{kr} \cdot l_{k1}, \tag{9}$$

where l_{k1} – length of the trajectory of contact of the plastically deforming grain of the tape with the workpiece, m; c – coefficient [5, 11]; σ_{ts} – yield strength of the workpiece material at the temperature and speed of grinding, Pa; l_{2d} – the size of the blunting area on the plastically deforming grain, m; h_{ud} – deterioration of grains carrying out plastic deformation, m; μ_0 – coefficient of friction of grain on the workpiece.

To calculate the tangential component of the force P_z dependences were obtained:

$$P_{zr1} = \frac{k_z}{k_y} P_{yr1}; \quad P_{zr2} = \mu_0 \cdot P_{yr2}; \tag{10}$$

$$P_{zd1} = B_d \cdot (\mu_0 \cdot sin\gamma + cos\gamma) \cdot l_{k1} \cdot a_{kr} \cdot (\frac{a_{kr}^2}{3} - h_{ud}^2); P_{zd2} = \mu_0 \cdot c \cdot \sigma_T \cdot D_d \cdot a_{kr} \cdot l_{k1}, \tag{11}$$

where k_z – coefficient [5].

When calculating the average distance between the cutting grains of the petal, it was assumed that the number of AG in contact with the workpiece increases linearly with increasing depth of their occurrence in the working volume of the petal. The number of AG on a single area of a petal made of a grinding belt was calculated from the dependence [10].

When calculating the shear stresses averaged over the shear plane, which determine the microcutting force by the petal grains, we used the data of the work [2].

A numerical simulation of the grinding force and their components during the processing of workpieces made of aluminum alloy D16 with a flap wheel with an outer diameter of D_k = 150 mm, abrasive grain material – 25A; grit P60; wheel height H = 0,02 m. Grinding mode: working speed of the petal circle V_k = 26 m / s; longitudinal feed speed V_s = 1,2 and 3 m / min; grinding depth t = 0,0048 mm.

Experimental research

Experimental studies were carried out on a surface grinding machine under the same conditions under which the numerical simulation of forces was performed. The components of the grinding force were measured with an UDM-100 dynamometer, the signal from which was fed to the TDA amplifier and then to the 16/16 – SIGMA / USB ADC. Used the ZetLabStudio software.

The results of numerical modeling and experimental research are presented in Table 1.

Table 1. *Results of numerical modeling and experimental study of the grinding force component* P_y

Longitudinal feed speed V_s, m / min	1,2	3
Calculated value P_y, N	24,8	26,3
Experimental value P_y, N	26,9	30,0

With an increase in the longitudinal feed rate from 1,2 to 3 m / min, the calculated and experimental values of the force Py increased by 6 and 12%, respectively.

The maximum discrepancy between the calculated and experimental values does not exceed 14%, which indicates the possibility of using the obtained analytical dependences for calculating the grinding forces with flap wheels.

Key findings

Summing up the forces associated with the dispersion and friction of the cutting and plastically deforming petal grains on the workpiece, we obtained analytical dependences for calculating the grinding forces with flap wheels,

Numerical modeling and experimental studies have confirmed the possibility of using the obtained analytical dependences for calculating the grinding forces with flap wheels.

With an increase in the longitudinal feed rate 2,5 times, the component P_y of the grinding force increases by 12%.

Acknowledgments

This work was supported by the Russian Foundation for Basic Research and the Government of the Ulyanovsk Region within the framework of the scientific project No. 19-48-730002.

References

[1] Gdalevich A.I., Zhitnitskiy S.I., Khrychev V.I. etc. Polishing of parts with flap wheels. - M .: Mechanical Engineering, 1980. - 80 p.

[2] Korchak S.N. Productivity of the process of grinding steel parts. - M.: Mechanical Engineering, 1974. - 280 p.

[3] Filimonov L.N. Resistance of grinding wheels. - L .: Mechanical engineering. Leningrad branch, 1973. - 136 p.

[4] Ostrovsky V.I. Theoretical foundations of the grinding process. - L.: Publishing house of Leningrad State University, 1981. - 144 p.

[5] Efimov V.V. Model of the grinding process using lubricating coolant. – Saratov: Publishing house Sarat. un-that. - 1992. - 132 p.

[6] T. Yu, H. Su, J. Dai, W. Zhou Grinding Force and Specific Grinding Energy in Process of Grinding SiC with Single Grit. Nanjing Hangkong Hangtian Daxue Xuebao. Journal of Nanjing University of Aeronautics and Astronautics. – 2018. – 50(1):120 -125.

[7] Deguo Li, Jinyuan Tang, Haifeng Chen, Wen Shao. Study on grinding force model in ultrasonic vibration-assisted grinding of alloy structural steel. The International Journal of Advanced Manufacturing Technology. - 2019. - 101(9–12)

[8] Unyanin A.N., Khusainov A.S. The ultrasonic grinding process temperature field study // MATEC Web of Conferences. - 2017. C. 01011. SJR - 0,166. https://doi.org/10.1051/matecconf/201712901011

[9] Dubrovsky P. V. Grinding of titanium alloys with flap wheels. - Ulyanovsk: UlSTU, 2000. - 100 p.

[10] Verezub V.N. Sanding with abrasive belts. - M.: Mechanical Engineering, 1972. - 103 p.

[11] Khudobin L.V., Unyanin A.N. Minimization of contamination of grinding wheels. - Ulyanovsk: UlSTU, 2007. - 298 p.

Forming of Composite Cutting Layer on Tool Steel by Heat Treatment of CVD Coatings

Irina Belashova[1], Larisa Petrova[1], Peter Demin[1,a*] and Mikhail Prokofyev[2]

[1]MADI University, Technology of structural materials department, 125319 Leningradskiy prospect 64 Moscow, Russia

[2]Moscow Aviation Institute (National Research University) 125993, Volokolamskoe shosse, 4 Moscow, Russia

[a]petr-demin@yandex.ru

Keywords: CVD, Coatings, Composite Layer, Heat Treatment, Chromium Carbides

Abstract. CVD chromium coatings are evaporated on steel substrate from chrome-organic compounds. For crystallization with forming of nano-particles of chromium carbides, subsequent heating (annealing) of tool steel with hybrid coatings is carrying out. Significant increase of micro-hardness of the coating up to 27000 MPa is observed due to the dispersion strengthening. Optimal annealing parameters (temperature and duration) are determined, which maximally strengthen the coatings and increase their adhesion to the steel substrate.

Introduction

Different techniques of surface strengthening are actively used for increase of service life of cutting tools. Among these techniques, the methods of coatings deposition are of special interest [1-3]. CVD method (Chemical Vapour Deposition) consists in condensation of a metal at a substrate after its reduction from decomposed chemical compound [4-5]. The technique of vapor deposition of chromium coatings from so-called MOC (metal-organic compounds) is an effective method for tool steels strengthening [3, 5]. CVD chromium coatings have high physics-mechanical properties; however, the problem of peeling exists for coatings receiving without noticeable diffusion penetration of deposited element [6]. Precipitation of new phases may cause unfavorable stresses in the surface layer, which also lead to decrease in adhesion.

For coatings structure modification, subsequent heat treatment is used after the deposition [7-8]. The effect of additional heating of a coating on increase its adhesion to the steel substrate appears due to the forming of transition diffusion zone under the coating [9]. Heat treatment of amorphous chromium coatings gives the opportunity of carrying out controlled crystallization with precipitation of nano-dispersed chromium carbides and forming of strengthened composite layer [10]. In this paper, the influence of thermal effects of annealing on structure and properties of CVD chromium coatings on a tool steel is studied. The purpose of the study consists in definition of optimal deposition mode and subsequent annealing regimes for forming the structure for better adhesion and high strengthening level.

Methods of experimental study

CVD chromium coatings were received on substrates of low-alloyed tool steel grade T31507 by vapor deposition from chrome-organic compounds. In the reactor, condensation of chromium on a steel substrate occurs with forming of coatings with the thickness depending on the deposition time. The main parameters influencing on structure and properties of the coatings are: temperature of deposition, partial pressure in the reactor, velocity of vapors delivery to the reactor, and duration of the process. Samples of steel with chromium coatings were annealed directly in the reactor in vacuum with addition of argon at temperatures 510…560°C during the time up to 180 min.

Microstructure of coatings was investigated by standard methods of metallographic analysis and scanned electronic microscopy. Strengthening of coatings was evaluated by measuring of their microhardness with loading of 50 grams.

Residual stresses parameters of samples with coatings ($-\sigma^{\phi}_{1}$) were determined by measuring of their deflection. Adhesion strength of chromium coatings to the steel substrate was assessed during tensile tests with measuring the relative adhesion index RAI = $(m_C/m_0)\cdot 100\%$, where m_C is the mass of the specimen after compression tests; and m_0 is the initial mass of the specimen before tests.

Results and discussion

The lowest temperature of deposition (420…430°C) is limited by the intensity of evaporation process. At these temperatures, the kinetic mode of deposition is realized; and the coatings are fully amorphous. At high temperatures (510…520°C), the diffusion mode of deposition provides intensive mass transfer of initial chemical compound to the surface of a substrate forming fully crystallized coating morphology. At intermediate temperatures (450…480°C), growth of the coatings occurs according to the hybrid diffusion-kinetic mode. Hybrid coating have layered structure (Fig. 1 a) consisting of alternating amorphous and crystalline interlayers that is the consequence of cyclic delivery of MOC.

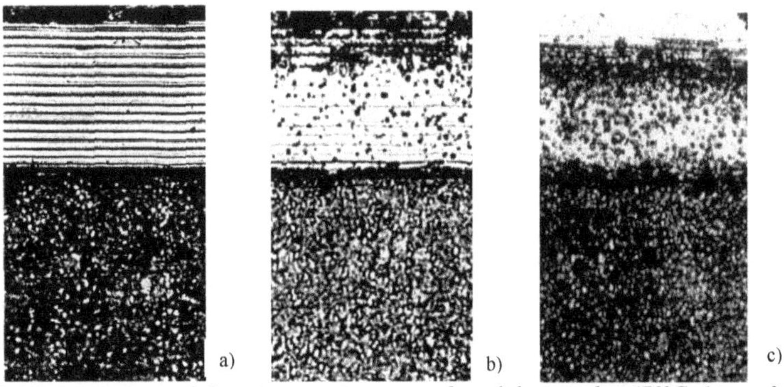

Fig. 1. Microstructures of chromium coatings on tool steel deposited at 470°C (time of deposition is 15 min) , x500: a) – as-received; b) – annealed at 550°C 0.5h; c) – annealed at 550°C 1.5h

The total thickness of hybrid coatings after 15 min of deposition is about 6 microns. The light layers are amorphous solid solution of carbon in chromium, whereas the dark layers have crystal structure composed of chromium and its carbides Cr_7C_3, $Cr_{23}C_6$. Changes of MOC concentration near the substrate leads to redistribution of carbon between the layers: crystal phase contains more carbon than the amorphous one.

The mode of deposition affects the residual stresses in the coating and, consequently, influence on its adhesion. The minimum stress level corresponds to the kinetic mode of deposition and formation of amorphous coatings (Fig. 2). With transition to the hybrid deposition mode an increase in stresses is observed. This effect is associated with the appearance of crystalline carbides interlayers in the amorphous phase, which is accompanied by an increase in volume. In the coatings received in the diffusion mode of deposition (510°C), the stresses reach their maximum.

As the diffusion deposition temperature increases, carbide inclusions become larger, which leads to a decrease in internal stresses.

The level of residual stresses depends not only on the deposition temperature, but also on the coating thickness, which is determined by the time of deposition: the value of residual stresses is higher with increasing coating thickness (Fig. 2).

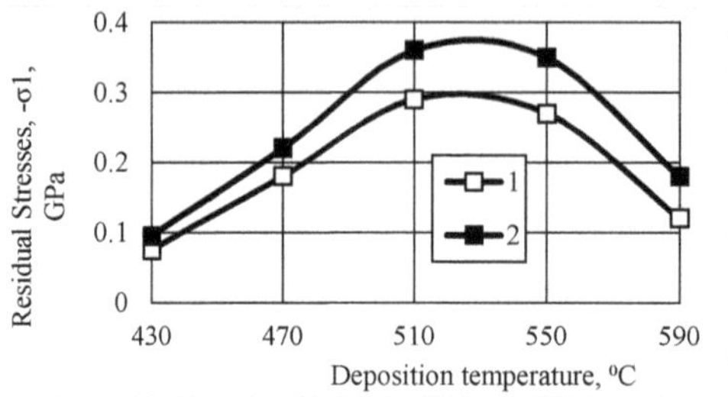

Fig. 2. *Residual stresses in chromium coatings with the thickness of 3 μm (1) and 6 μm (2) deposited at different temperatures.*

Annealing of coatings deposited in the kinetic and in the diffusion modes does not results in noticeable modification of their structure. Amorphous coatings formed in the kinetic mode of deposition have no conditions for nucleation of a crystal phase. Coatings received in the diffusion mode of deposition are sustainable to heating due to the fully formed crystal structure.

Heat treatment of hybrid coatings allows implementing the crystallization of amorphous layers. Dispersed precipitations of chromium carbides in crystal interlayers may serve as the nucleuses for crystallization process. As far as the amorphous phase is unstable, the system drives to the equilibrium by forming crystal structure based on chromium carbides.

Annealing at temperatures below 500°C does not result in any changes of coatings' structure. At temperatures higher than 600°C, the initial layered structure of the coating disappears; and they receive crystal-grained structure.

For precipitation of crystal nucleuses in amorphous phase, the annealing at intermediate temperatures 540-560°C should be used. After soaking at these temperatures, precipitations of carbide particles are observed in hybrid coatings (Fig. 1 b). Increase of the annealing time leads to further activation of crystallization process; as a result, crystal structure is spread progressively to the whole coating, and amorphous phase disappears (Fig. 1 c). A transitional zone is formed under the coating with the thickness up to 20 microns that makes micrihardness profiles smoother (Fig. 3).

Strengthening achieving in chromium coatings just after deposition depends on the deposition mode (deposition temperature). Microhardness of amorphous coatings received in the kinetic mode of deposition is up to 14000 MPa. Maximal microharness (up to 19500 MPa) is achieved after deposition at temperatures of initial diffusion regime (500...510°C) when fine-grained structure of coatings is formed. Chromium coatings of hybrid deposition mode (450...480°C) have microhardness value about 16000 MPa.

Annealing of hybrid coatings may increase their microhardness additionally (Fig. 3). Strengthening of heat-treated coatings depends on the temperature of annealing (Fig. 4 a). Annealing at the temperatures below 500oC does not make changes of microhardness. Increase of the temperature enough for starting the crystallization (540...550°C) results in the explosive growth of microhardness up to 22000...27000 MPa. This is the effect of dispersion hardening by nanoparticles of chromium carbides coherent with solid solution. Increase of the annealing temperature leads to slight decrease of microhardness due to the coagulation of chromium carbides.

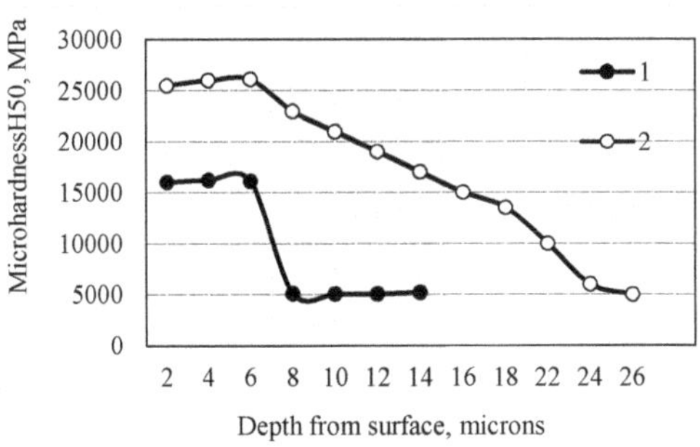

Fig. 3. *Microhardness profiles of coatings on tool steel after deposition in hybrid mode (1), and after subsequent annealing at 550°C, 1 hour (2)*

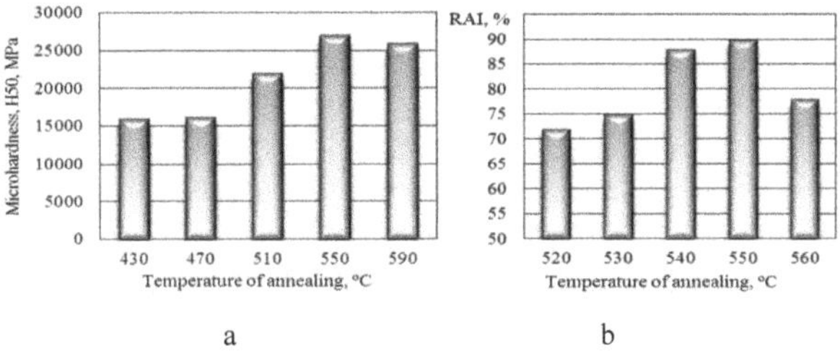

Fig. 4. *Influence of annealing temperature on microhardness of CVD coatings (a) and on their adhesion (b)*

Adhesion resistance of coatings depends on the degree of crystallization: crystallized carbides appearing during the annealing increases the level of internal stresses. Completion of crystallization at higher temperatures decreases distortions and weakens cohesive interaction in the coating. The level of adhesion resistance slightly decrease (by 7-10%) at combined crystallizing treatment (Fig. 4 b).

Conclusion

The formation of strengthened composite chromium-carbide layers on the surface of tool steel has been experimentally confirmed by annealing of chromium CVD coatings deposited in a hybrid (kinetic-diffusion) temperature mode. The microstructure of the composite coatings is a result of crystallization of amorphous interlayers and formation of nano-particles of chromium carbides. This increases the microhardness of the coating due to dispersion strengthening.

Heating the coating at annealing temperatures of 540 ... 560°C activates diffusion processes. A transition layer is formed between the coatings and the substrate, which increases their adhesion. This correlates with the increase of the compressive residual stresses after the crystallization and precipitation of coherent carbides.

There are optimal annealing parameters (temperature and duration), which maximally strengthen the coatings and increase their adhesion to the steel substrate. The lower limits of these parameters are associated with the crystallization initiation, and the upper limits are determined by the coagulation of carbide particles, which causes stress relaxation.

Acknowledgments
The material was prepared within the framework of scientific research under the project №FSFM-2020-0011 (2019-1342); experimental studies were carried out using the equipment of the MADI Centre of collective usage.

References

[1] M. Diesselberg, H.R. Stock, P. Mayr Friction and wear behaviour of PVD chromium nitride supported carbon coatings/ Surface and Coatings Technology. 2004. T. 188-189. № 1-3 SPEC.ISS. C. pp. 612-616. https://doi.org/10.1016/j.surfcoat.2004.07.023

[2] L.S. Kremnev, A.K. Onegina, L.A. Vinogradova, I.Yu Sapronov. Special features of the composition, structure and properties of high-speed steels for cutting tools with ion-plasma coatings based on titanium nitride, Metal Science and Heat Treatment, 2012. T. 54. № 1-2. pp. 3-7. https://doi.org/10.1007/s11041-012-9446-y

[3] V.I. Yurshev, R.I. Mukatdarov, I.V. Yurshev Surface hardening of tools by depositing a pyrolytic chromium carbide coating, Metal Science and Heat Treatment. 2015. T. 57. № 1. pp. 107-111. https://doi.org/10.1007/s11041-015-9845-y

[4] T.A. Tabaza, O.T. Tabaza, A. Al-Sakarneh CVD technology for preparing chromium oxide coatings, study of the kinetics of growth of coatings, Key Engineering Materials. 2018. T. 765 KEM. C. pp.193-198. https://doi.org/10.4028/www.scientific.net/KEM.765.193

[5] L.L. Ivanov Development of strengthening technology for cutting tools by deposition of pyrolitic chromium coatings, PhD Theses, Moscow, MADI, 1988. 242

[6] A. Vereschaka, A. Aksenenko, S. Shevchenko, et al. Effect of adhesion and tribological properties of modified composite nano-structured multi-layer nitride coatings on WC-Co tools life/Tribology International. 2018. T. 128. pp. 313-327. https://doi.org/10.1016/j.triboint.2018.07.039

[7] X. Pang, K. Gao, F. Luo, H. Yang, L. Qiao, Y. Wang, A.A. Volinsky Annealing effects on microstructure and mechanical properties of chromium oxide coatings, Thin Solid Films. 2008. T. 516. № 15. C. pp. 4685-4689. https://doi.org/10.1016/j.tsf.2007.08.083

[8] O.V. Somov, V.A. Pashkin, S.V. Savushkina, B.L. Krit Pyrolytic chromium carbide coatings on steel: effect of annealing on morphology and elemental composition,Surface Engineering and Applied Electrochemistry. 2020. T. 56. № 2. C. pp.140-146. https://doi.org/10.3103/S1068375520020179

[9] L.G. Petrova, P.E. Demin, G.Y. Timofeeva, A.V. Kosachev, V.D. Aleksandrov Influence of nitriding on the wear resistance and adhesion of zinc coatings on steel, Russian Engineering Research. 2019. T. 39. № 8. C. pp.696-698. https://doi.org/10.3103/S1068798X1908015X

[10] I.S. Belashova, L.G. Petrova Dispersion Hardening of Amorphous Chromium Coatings by Crystallized Nano-Particles of Carbides , European Conference of Heat Treatment and 21st IFHTSE Congress, May 2014, Munich, Germany, pp..505-510

Materials Research Proceedings **21** (2022) 204-208 https://doi.org/10.21741/9781644901755-36

Nitriding of High Speed Steel for Improvement of Tools Resistance

Larisa Petrova[1], Victor Vdovin[1], Peter Demin[1,a*] and Alexandra Sergeeva[1]

[1]MADI University, Technology of structural materials department, 125319 Leningradskiy prospect 64 Moscow, Russia

[a]petr-demin@yandex.ru

Keywords: Nitriding, High Speed Steel, Tool Resistance, Phase Composition, Hardening

Abstract. The article is devoted to the study of the gas nitriding process, which makes it possible to obtain high-quality diffusion layers in high-speed steel M2 on the basis of an internal nitriding zone without a brittle nitride zone. The results of studies of the nitrided steel phase composition with a change of the saturating atmosphere during dilution of ammonia by hydrogen are presented. An increase in the resistance of the nitrided tool when drilling structural steel is shown, which is associated with the dispersion hardening of the internal nitriding zone with tungsten nitrides.

Introduction

Today, the constantly developing industry is tightening the requirements for the service life of tools designed for processing various machine parts and devices. The most stringent requirements are imposed on tools made of high-speed steels (HSS) [1-2], working in complex conditions where high wear resistance and heat resistance are required along with resistance to shock loads. Increasing the durability of a high-speed tool contributes to resource saving, since it provides savings in metals used for alloying steels, such as tungsten and molybdenum, which makes it relevant to research on improving technologies for hardening the cutting parts of the tool. The advantages of surface hardening are not only in reducing the consumption of alloying elements, but also in maintaining a ductile core, which provides resistance to shock loads.

Currently, many different methods of surface hardening of the tool are used, which have their advantages and disadvantages [3-5]. An unique position among such processes is occupied by chemical-thermal treatment (CTT) [6-7]. For example, nitriding is widely used to increase the reliability and durability of a wide range of steels, both structural and tool [8-9]. One of the main advantages of the nitriding process is the ability to control the nitrided layers phase composition in various steels, depending on the required surface properties. The structure of the diffusion layer is regulated by controlling the nitrogen potential of the saturating atmosphere. The advantage of the process in relation to high-speed steels is a relatively low saturation temperature, which makes it possible to preserve the core structure formed by the basic heat treatment.

The optimal phase composition of the nitrided cutting edge, which provides the necessary tool durability, differs depending on the cutting conditions, the type and size of the tool, as well as on the material being processed. To prevent the compact nitrides zone formation at the surface, technological techniques based on changes in the nitrogen potential of the atmosphere are used: dilution of ammonia, for example, with an inert gas, hydrogen or with products of ammonia dissociation.

The purpose of this work is to study the structure of the nitrided layer in high-speed steel during nitriding in ammonia and hydrogen, as well as to determine the composition of the saturating atmosphere to increase the drills durability.

Methods of experimental study

The studies were carried out on samples and drills with a diameter of 5 mm made of high-speed tool steel HSS M2, the chemical composition of which according to ASTM A600 Standard is given in the Table 1.

Table 1. *Chemical composition of M2 steel*

Element	C	Cr	V	Mo	W	P	S
Weight %	0.78-0.88	3.75-4.50	1.75-2.20	4.50-5.50	5.50-6.75	0.03	0.03

Steel samples and drills were used in the condition after standard heat treatment. Quenching from 1250°C and triple tempering at 560°C forms the structure of the steel cutting edge with dispersed tungsten carbides in solid solution (Fig. 1a).

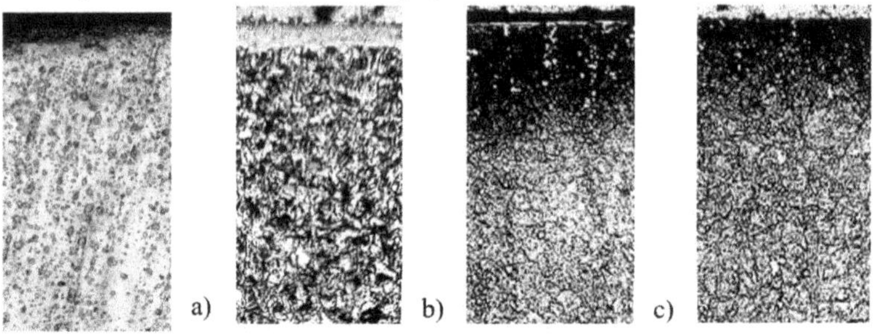

Fig. 1. *Microstructure of M2 steel after standard heat treatment (a) and after nitriding in pure ammonia (b), in a mixture of 75%NH_3+25%H_2 (c), in a mixture of 50%NH_3+50%H_2 (d)*

Nitriding was carried out on a laboratory equipment in a muffle furnace comprising a nitriding reactor. Process temperature of 540°C was selected below the tempering temperature to keep the matrix hardness at the initial level. Saturation was carried out in mixtures of ammonia and hydrogen; variation in the degree of ammonia dissociation was achieved by changing the hydrogen content in the atmosphere (25%H_2, 50%H_2, and 75% H_2). The regulation of the gas mixtures composition in the furnace was carried out using a control unit for the dosing of gas components. The atmosphere composition at the entrance and the temperature in the furnace were set using software and maintained automatically using the control of the process technological parameters.

The nitrided steel microstructure was studied by metallographic analysis using an Axiovert 25 CA optical microscope after electrolytic etching of microsections in an aqueous solution of oxalic acid. Studies of the substructure and layer phase composition were carried out by electron microscopy (SMM 2000), scanning probe microscopy (SPM NanoEducator) after electrolytic etching and by X-ray diffraction analysis (DRON-3). Identification of precipitations, determination of their size and volume fraction were carried out using images on micro-diffraction patterns and dark-field images obtained in the corresponding reflexes. X-ray examinations were taken in monochromatic $Fe_{K\alpha}$ radiation.

A comparative assessment of the drills service life with different nitrogen potential was carried out by the method of studying the degree of bluntness during cutting holes by recording sound using a microphone fixed to the workpiece. During the mechanical processing of carbon steel sheet (0.2%C) with a thickness of 2 mm on a drilling machine, sound waves were recorded on a personal

Materials Research Proceedings **21** (2022) 204-208
https://doi.org/10.21741/9781644901755-36

computer. The number of holes made by the drill until blunting (N) was determined before the appearance of a specific sound when the critical value of the amplitude of sound vibrations was exceeded.

Results and discussion

Nitriding of high-speed steel in pure ammonia forms a brittle nitride zone on the surface (Fig. 1b). This layer easily crumbles during cutting, which leads to catastrophic destruction of the cutting edge. Therefore, nitriding processes were carried out in diluted ammonia to determine the conditions under which a nitride zone does not form.

To determine the saturation time to obtain a layer of a given thickness, studies of the nitriding process kinetics in atmospheres of various compositions were carried out, and corresponding nomograms were obtained. For example, it takes 1 hour to form a diffusion layer with a thickness of 50 μm when nitriding in a mixture of 50%NH_3+50% H_2 (Fig. 2).

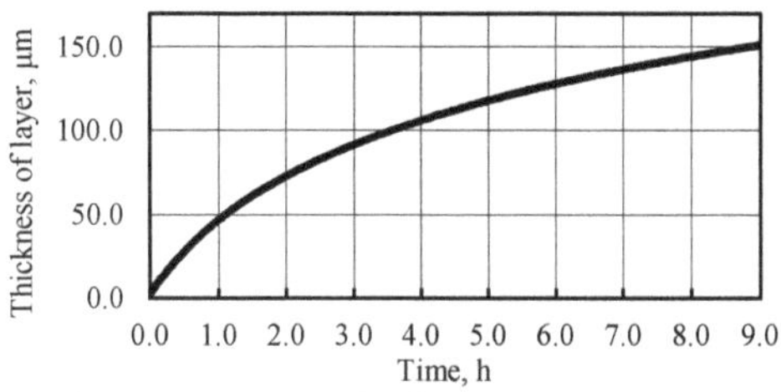

Fig. 2. *Nomogram for determination of diffusion layer thickness during nitriding in 50%NH_3+50%H_2 depending on saturating time.*

It is known that the saturating atmosphere composition during nitriding is related to the nitrogen potential of the atmosphere π_N, which is determined by the partial pressures ratio of hydrogen and ammonia in the system. Formation of phases in the nitrided layer depends on the nitrogen potential and the process temperature. According to the known regularities, nitriding with a low nitrogen potential (in highly dilute ammonia) may form nitrided layers of α-solid solution without a surface nitride zone. With an increase in the nitrogen potential, ferrous nitrides are formed in order of increasing nitrogen content: γ'-Fe_4N→ε-$Fe_{2-3}N$. Table 2 shows the phases identified by X-ray analysis on the surface of M2 steel nitrided in various mixtures of ammonia with hydrogen. In Fig. 1c the nitride zone of ε+γ' is observed, formed in the steel saturated in a mixture of 75%NH_3+25%H_2. When nitriding in a mixture of 50%NH_3+50%H_2, a nitride zone is not discovered in the microstructure (Fig. 1d).

Due to the high concentration of carbon and alloying elements in the M2 steel, the area of γ'–phase existence significantly narrows, and the area of α-phase expands, while the thermodynamic possibility of formation of alloying elements nitrides appears. No alloying elements' nitrides were detected directly on the surface, whereas in the internal nitriding zone at a distance of 10 microns from the surface, nanoparticles of tungsten nitrides (no more than 10 nm) are determined in solid solution (Fig. 3). Reflexes of α–phase (335) and of WN nitride (001) are indicated on X-ray pattern. Such a structure of the internal nitriding zone is a prerequisite for the internal nitriding zone dispersed strengthening.

Table 2. *Phase composition of the nitrided surface on M2 steel*

Saturation environment	Phases on the nitrided surface
Without nitriding	-
100% NH_3	ε-$Fe_{2-3}N$
75% NH_3+25% H_2	ε-$Fe_{2-3}N$+γ'-Fe_4N +$Fe_\alpha(N)$
50% NH_3+50% H_2	γ'-Fe_4N +$Fe_\alpha(N)$
25% NH_3+75% H_2	$Fe_\alpha(N)$

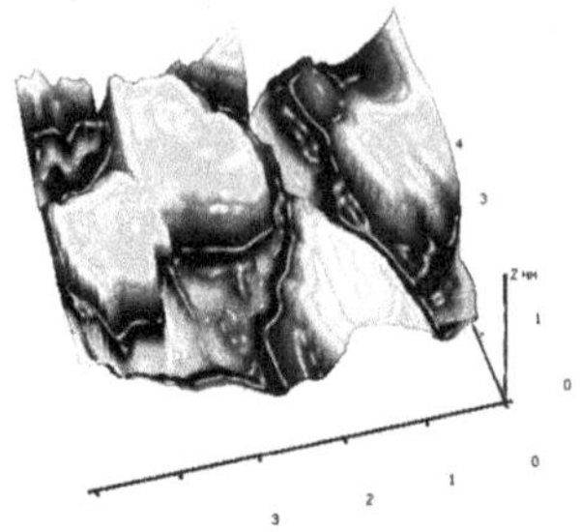

Fig. 3. *Detection of tungsten nitride particles in the nitrided layer of M2 steel by SEM*

Fig. 4 shows a comparison of drill resistance by the number of drilled holes measured during drilling tests. In general, the service life of drills, estimated by the sound change when drilling holes, increases after nitriding in ammonia-hydrogen mixtures. The operating time of a drill until the specific sound of blunting depends on the composition of the saturating atmosphere, i.e. it is determined by the nitrided layer phase composition.

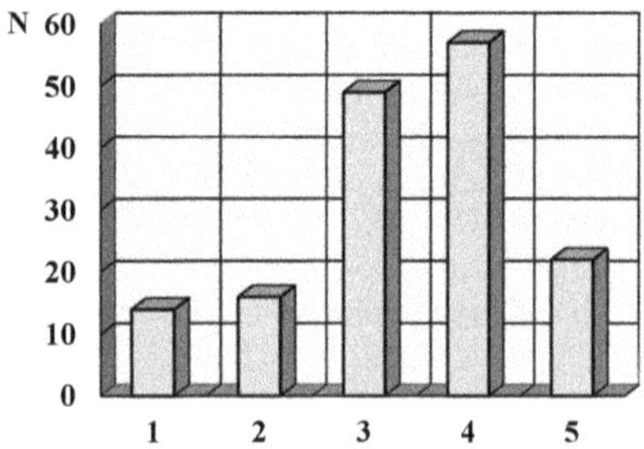

Fig. 4. *Comparison of drills resistance (numbers of drilled holes until blunting): 1 – non-nitrided drill; nitrided in: 25%NH_3+75%H_2 (2), 50%NH_3+50%H_2 (3), 75%NH_3+25%H_2 (4), 100%NH_3 (5)*

The sound amplitude during drilling reaches a critical value faster when using a non-nitrided drill, which has a minimum resistance index. Drills with the nitrided layer without a surface nitride zone formed in an atmosphere of 25%NH_3+ 75%H_2 increase the durability of the drill slightly. Drills nitrided in pure ammonia also do not provide a large increase in durability, since they have a brittle ε-phase zone on the surface, which quickly cleaves off during drilling. Drills nitrided in

$75\%NH_3$ $+25\%H_2$, having the nitride zone of $\varepsilon+\gamma'$, work the longest before blunting; their resistance is 4 times higher comparing to un-nitrided tool.

Conclusion

Nitriding process of high-speed steel at t=540°C is investigated in mixtures of ammonia and hydrogen. The nitriding modes are determined that from diffusion layers of an internal nitriding zone with precipitation of dispersed tungsten nitrides and without brittle ε-phase zone at the surface. Nitriding in this mode significantly increases the durability of drills, due to the dispersion hardening of the internal nitriding zone. The tools nitrided in ammonia diluted with $25\%H_2$ and with $50\%H_2$ has the best resistance. Strong dilution of ammonia ($25\%NH_3+75\%H_2$) significantly reduces the nitriding effect, which is associated with a small layer thickness and a low level of solid solution hardening.

Acknowledgments

The material was prepared within the framework of scientific research under the project №FSFM-2020-0011 (2019-1342); experimental studies were carried out using the equipment of the MADI Centre of collective usage.

References

[1] L.S. Kremnev From high-speed tungsten steel to high-temperature molybdenum steel: a century of high-speed steel/Steel in Translation. 2009. T. 39. № 12. pp. 1111-1118. https://doi.org/10.3103/S0967091209120195

[2] A.S. Chaus, M. Dománková Unknown high-speed steel/Materials Letters. 2021. T. 292. pp. 129653. https://doi.org/10.1016/j.matlet.2021.129653

[3] Y. Li, G. Zhang, J. Yao, F. Xiao,Y. Wang ,Y. Liu M2 High speed steel irradiated by high-intensity pulsed ion beam/Cailiao Yanjiu Xuebao. 2005. T. 19. № 6. pp. 613-618.

[4] V.I. Yurshev, R.I. Mukatdarov, I.V. Yurshev Surface hardening of tools by depositing a pyrolytic chromium carbide coating, Metal Science and Heat Treatment. 2015. T. 57. № 1. pp. 107-111. https://doi.org/10.1007/s11041-015-9845-y

[5] A. Vereschaka, A. Aksenenko, S. Shevchenko, et al. Effect of adhesion and tribological properties of modified composite nano-structured multi-layer nitride coatings on WC-Co tools life/Tribology International. 2018. T. 128. pp. 313-327. https://doi.org/10.1016/j.triboint.2018.07.039

[6] U. Tolibjon Hardening high-speed steel R6M5 using a combined heat treatment technology/Lecture Notes in Networks and Systems. 2022. T. 305. pp. 153-161. https://doi.org/10.1007/978-3-030-83368-8_15

[7] V.A. Aleksandrov, L.G. Petrova, A.S. Sergeeva, V.D. Aleksandrov, E.U. Akhmetzhanova Production of tool coatings by chemicothermal plasma methods, Russian Engineering Research. 2019. T. 39. № 8. pp. 693-695. https://doi.org/10.3103/S1068798X19080033

[8] I.S. Belashova, L.G. Petrova, V.D. Aleksandrov, P.E. Demin. Improving the properties of low-alloy and carbon steel tools by cyclic nitriding, Russian Engineering Research. 2018. T. 38. № 1. pp. 53-56. https://doi.org/10.3103/S1068798X18010057

[9] P.L.C. Serra, A.S.A. Furtado, R.R.M. de Sousa, J.R. de Barros Neto, W.R.V. Sampaio, M.C. Feitor, T.H. de Carvalho, Costa Study of plasma nitriding and duplex treatment on high speed steel drills/Revista Materia. 2020. T. 25. № 2. pp. 1-15. https://doi.org/10.1590/s1517-707620200002.1072

Materials Research Proceedings **21** (2022) 209-213 https://doi.org/10.21741/9781644901755-37

Dynamic Characteristics of Reinforced Concrete Beams Made of Carbonate Concrete

Vladimir Belov[1,a*], Pavel Kuliaev[2] and Temur Barkaya[2]

[1] Chair of Building Products and Structures Manufacture, Russian academy of architecture and building sciences, Tver State Technical University, Tver, Russia

[2]Department of the Construction materials and technologies, Tver State Technical University, Tver, Russia

[a]vladim-bel@yandex.ru

Keywords: Reinforced Concrete, Carbonate Concrete, Strength, Dynamic Parameters, Matrix

Abstract. The ability of solid structures to absorb a certain part of the energy of dynamic impacts has not been properly reflected in impact theories. Meanwhile the effect of material properties on the various structures in the impact is so much that ignoring it when the solution of a large number dynamic problems makes it impossible to explain without distorting quantitatively and qualitatively, many of the actually observed phenomena, for example, equalization of dynamic stresses in places of their concentration and fluctuation of other parameters. In the article, two independent parameters for conventional reinforced concrete beams and those made of limestone concrete are compared, namely dynamic coefficient and the values of elastic rebound in impact. The effect of the reinforcement is not discussed in the paper.

Introduction

Research were carried out in the field of dissipative properties under different aspects of concrete and the use of screenings for crushed carbonate rocks and other staff for use as a filler in the preparation of fine-grained concrete to be the solid staff of reinforced concrete beams at different treatment as well as in view of crack resistance [1,2,3]. The main criterion of concrete properties considered in there were dissipative properties, enclosed by the area of crack resistance and critical stress intensity factors. In addition to internal forces, dissipative forces also resist the action of loads, which cause energy dissipation in structures during vibrations. These forces, which are non-conservative, are divided into external and internal. External dissipative forces include aerodynamic and bond-drag forces that occur during deformations of the base of the structure. Internal forces (or internal friction) are caused by microplastic deformations of the particles of concrete and bonds with reinforcement and friction in the joints of the elements (structural hysteresis). The influence of all these contra-forces on the work of structures is different and depends on many factors [4,5,6]. For example, at relatively small impact (cyclic) loads, weak bonds between elements carried out by dry friction intensively affect the energy dissipation. Under certain conditions, friction in the joints can dissipate energy several times greater than the internal friction in the construction material [7,8]. The aerodynamic drag for conventional structures is insignificant and it can be ignored. It was also shown in papers that limestone aggregate and filler together with a superplasticizer increases crack resistance and reduces creep deformations, which makes fine-grained carbonate concrete durable material [9]. Knowledge of the stress-strain distribution in concrete through the crack-bound energy dissipation with limestone aggregate and filler in impact is necessary when designing certain types of reinforced concrete structures. These structures are made with the addition of limestone filler which performs elastic energy absorbing effect, critical in impacts. The last consists in increasing the level of crack-forming-stress extinguishing areas [10], plastic by nature and creepy, contributing to dynamic features of reinforced concrete structures.

Experiment

In the process of deformation in impact, the beam passes, as in case of static loading, three stages, prior to the formation of cracks, after the formation of cracks, until the yield is reached in the longitudinal stretched reinforcement; after the yield is reached until destruction. At the same time, a number of significant features occur during shock loading. In general, they depend on the dynamic properties of the structure (usually characterized by the lowest natural frequency ω_1), the ratio of the masses of the impactor and the beam, the impact velocity v_0, the impact time and a number of other factors. For a complete understanding of the mechanisms of failure in impact, data on mechanical properties are combined with some parameters describing the patterns of destruction. Such are dynamic coefficient, the values of elastic rebound. All this data can be obtained in dynamic and, partly, static tests. For the experiment, a class A300 reinforcement was taken, longitudinal with a diameter of 12 mm, transverse-8 mm. The coupling of the frame was performed using electrodes that are designed to connect low-carbon and ordinary steels by manual arc welding. They are metal rods with a length of 350 mm and a diameter of 3 mm. Longitudinal rods were 1.45 m. The rutile coating ensures stable arc burning regardless of the type of current. Before welding, dirt and dust are removed from the reinforcement. With a small amount of work, they are cleaned to a metallic shine, which has a positive effect on the quality of the seam. The preparation of the concrete mixture of components was carried out by semi-dry mixing. At first, cement and sand were filled into a mixing container with subsequent water filling. Everything was mixed until there were no lumps and dry sediment on the bottom visually. After that, granite rubble was slowly dosed down. The whole mixture was mixed for about 10-15 minutes, with one pause of 5-6 minutes. Compositions per 1 m^3 consisted of cement – 414 kg., water-201.4 kg., sand (limestone filler – 515-762 kg., aggregate (granite and limestone) -968-1170 kg. Sledgehammer tip weighed 5 kg. The following are the 6 beams (Fig. 1), three – conventional (to obtain a medium value), three – made of carbonate concrete and impact device (Fig. 2).

Fig. 1. *Formwork with reinforcement (left) and test 6 beams (right)*

Materials Research Proceedings **21** (2022) 209-213 https://doi.org/10.21741/9781644901755-37

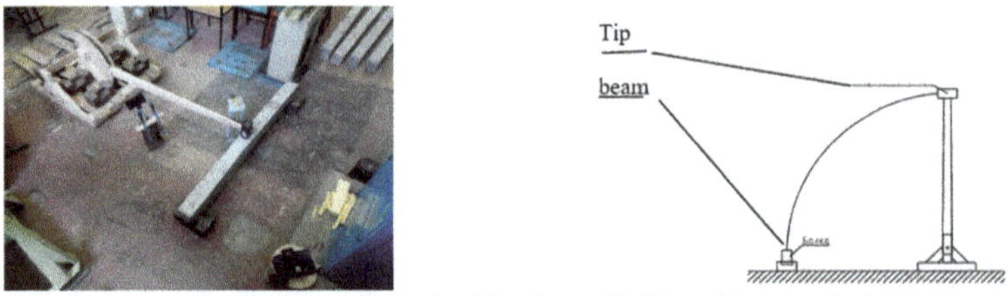

Fig. 2. *Impact device (overall view) with a beam (left)) and impact device scheme (right)*

Photo and video documentation were made using a camera aimed at a measuring scale, which is a measuring tape measure, tied with glue tape to a concrete beam. With the help of the recording on the camera, a storyboard was made in order to determine the rebound of the tip from the beam after a dynamic impact. Rebound values were obtained from direct measurement (Fig. 3).

Fig. 3. *Sledgehammer tip rebound measurement technique*

The table 1 represents Medium sledgehammer tip impact rebounds.

Table 1. *Medium sledgehammer tip impact rebound, mm*

N b/o	Beam	Medium impact rebound, mm
1	Of conventional concrete	72,5
2	Of carbonate concrete	38,5

According to the obtained data of the impact experiment, it can be concluded that the rebound of ordinary concrete is almost twice as much as that of carbonate concrete, extinguishing the elastic part of the total inner energy of microcracking process.

The following is a procedure to determine dynamic coefficient. Let's determine the impact force by the formula

$$F_d \cdot \Delta t = m \cdot V; \quad (1)$$

Hence,

$$F_d = m \cdot V / \Delta t$$

here, F_d - is an impact force, m – tip weight, V – tip impact velocity, Δt - contact time in seconds. It is the period within one rebound (taken from video frames interim time) of the impact part of the impact machine, we can take this time equal to half the time of one video frame, i.e. 0.01 seconds.

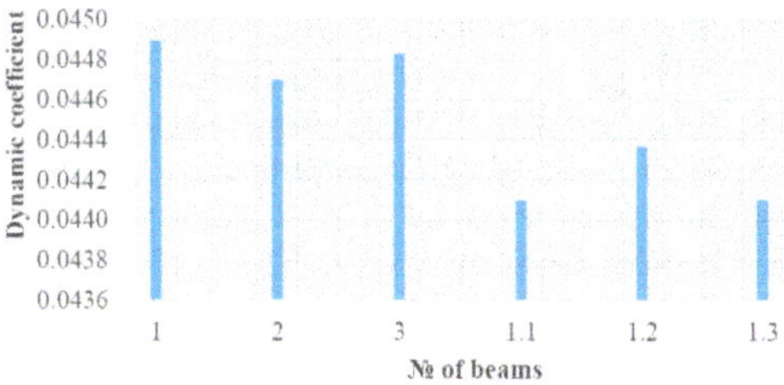

Fig. 4. *Dynamic coefficient values for conventional concrete beams (1,2,3) and carbonate concrete beams (1.1,1.2,1.3)*

Hence, the dynamic coefficient (Fig. 4) medium values in relation to the impact force according to the results obtained from the formula above:

Table 2. *Dynamic coefficient medium values*

N b/o	Beam	dynamic coefficient medium value
1	Of conventional concrete	0.0448
2	Of carbonate concrete	0.0441

From the table 2 it is obvious that dynamic coefficient of carbonate concrete is less than that of conventional concrete.

This means that the “dynamic versus static” mechanical parameters (failure-forces, deflections) of carbonate concrete in general are less and therefore better than those of an ordinary concrete. This also indicates that carbonate-cement matrix of carbonate beams, even apart from steel reinforcement effects, is a good external impact energy absorber.

Conclusions

The carbonate filler helps to seal the cement stone and strengthen the contact zone between the cement stone and the filler, accumulates part of the water, increases the degree of hydration of cement and, on the other hand, contributes to the formation of microplastic centers, absorbing the impact energy, thereby reducing elastic rebounds and the values of dynamic coefficient, making the beams on carbonate concrete more resistible to impact loads. It is established that the introduction of a fine limestone filler into the mixture about 50% of cement contributes to the extinguishing the local stress concentrators effect on internal micro crack formation, a great deal due to internal friction. In this case, a more uniform and durable composite structure is formed between the filler and the binder. This increases the crack resistance of carbonate concrete and the dynamic strength and reduces the level of elastic canvas failure, which leads to an increase in the efficiency and improvement of the physical and mechanical properties of carbonate concrete. It is shown that the use of carbonate concrete allows achieving more uniform inner matrix and crack good resistance by increasing the operational and technological properties of carbonate concrete and at the same time reducing the energy and material costs of its production and contributing to an increase in the efficiency of carbonate concrete. The limestone filler has a wide range of granule sizes, which contributes to filling the voids between the cement particles during processing, which leads to a reduction in the need for water, improving the uniformity of the mixture and unifying the microstructure of the composite. Due to the short-term volumetric plastic deformation of the matrix, the volumetric alignment of the composite microstructure improves the dynamic characteristics of carbonate concrete, restraining the growth of deformations, especially in the long term, without significant losses of technological properties at the initial stage. Higher dynamic parameters of carbonate concrete also occur due to better packing of particles inside the concrete matrix and a decrease in porosity caused by the transfer of microcracks from air and capillary regions to micro-subzones and compression of microcracks to internal microstructural nuclei, depriving the composite structure of microplastic energy-consuming regions. This makes the base composite (ordinary heavy concrete) more brittle and unconnected at the micro level than the carbonate composite.

References

[1] Ansari F. Analysis of Micro-Cracked Zone in Concrete. Fract. Toughness and Fract. Energy Concr. Proc. Int. Conf., Lausanne, Oct.1-3, Amsterdam pp.229-240 (1986).

[2] Beaudoin J.J. Mechanisms of Subcritical Crack Growth in Portland Cement Paste. Fract. Toughness and Fract. Energy Concr. Proc. Int. Conf., Lausanne, Oct.1-3, Amsterdam pp.11-19 (1986).

[3] Derucher K.N. Failure Mechanism of Concrete. Compos. Mater.: Test. and Des. 5th Conf., New Orleans, La, Philadelphia. pp.664-679 (1999).

[4] Desov A.E. Basic principles of high-strength concrete. Transpr. Res. Rec. №**504** pp.37-42 (2001).

[5] Detriche Ch.H., Ramoda S.A. Effect of the Composition of Mortars and Testing Procedures on Fracture Toughness. Fract. Toughness and Fract. Energy Concr.Proc. Int. Conf., Lausanne, Oct.1-3, Amsterdae.a. pp.291-298 (1995).

[6] Diamond Sidney, Bentur Arnon. On the Cracking in Concrete and Fiber-Reinforced Cements. Appl. Fract. Mech. Cementitious Composites. Proc. NATO Adv. Res. Workshop, Evanston, Sept.4-7, Dordrecht pp.87-140 (2005). https://doi.org/10.1007/978-94-009-5121-1_4

[7] Fanella David, Krajcinovic Dusan. Size Effect in Concrete. J. Eng. Mech. №**4**. pp.704-715 (1998). https://doi.org/10.1061/(ASCE)0733-9399(1988)114:4(704)

[8] Kishitani Koichi, Moeda Koichi. Compressive fracture of cracked Mortar. Cem. Assoc. Jap. Rev. 13th Gen. Meet. Techn. Sess., Tokyo, Sem. Gijutsu Nempo, pp.216-217 (1985).

[9] Okada Kiyoshi, Koyanagi Wataru, Rokugo Keitetsu. Fracture Process of Concrete in Compression. Proc.2nd Int. Conf. Mech. Behav. Mater., Boston, Masspp.1358-1362 (1988)

[10] Struble L.J., Stutzman P.E., Fuller E.R. Microstructural Aspects of The Fracture of Hardened Cement Paste: [Pap.] 1st Symp. Rel. Multilayer Ceram. Capacitors, University Park, Pa, May 11-12, J. Amer. Ceram. Soc12. -pp.2295-2299 (1989). https://doi.org/10.1111/j.1151-2916.1989.tb06078.x

Studies of Highly Dispersed Titanium Carbide Powder Obtained from Scrap Tungstenless Cemented Carbide Alloys

V. Gavrish[1,a], T. Chayka[1,b*], N. Cherkashina[2,c] and O. Gavrish[1,d]

[1]Sevastopol State University, 33 Universitetskaya Street, Sevastopol, 299053, Russia

[2]Belgorod State Technological University named after V.G. Shoukhov, Kostyukovstr., Belgorod,308012, Russia

[a]vmgavrish@sevsu.ru, [b]KimChayka@yandex.ru, [c]cherkashina.ni@bstu.ru, [d]olga_gavrish_53@mail.ru

Keywords: Titanium Carbide, Highly Dispersed Powder, Carbide Waste

Abstract. The results of research of characteristics of highly dispersed titanium carbide powder obtained from carbide waste of TN-20, TN-25, TN-30 types are presented. The powder particles were studied using analytical methods including scanning microscopy, X-ray diffraction, differential thermal analysis. The obtained results confirm the formation of nanosized particles of titanium carbide of monocrystalline form.

Introduction

Titanium carbide is one of the most demanded materials widely used in industry, modern engineering and medicine [1-3]. The wide range of applications is due to the unique combination of chemical and physical properties of titanium carbide (high wear resistance, corrosion resistance, chemical inertness, high melting point, flexural modulus, etc.) [4-7]. The use of high-dispersity titanium carbide powders (micro- and nanopowders), which provide a significant improvement of the properties of materials based on titanium carbide, is becoming increasingly important [8-11]. The most studied methods of titanium carbide nanopowders production are chemical vapor deposition, self-propagating high-temperature synthesis with laser ignition, sol-gel method, self-propagating high-temperature synthesis, etc. [12-16]. All these methods are characterized by very low productivity and rather high cost of the final product. Increasing industrial use makes it necessary to reconsider the attitude to the methods of production of titanium carbide nanopowder both in terms of ensuring its high dispersity and purity and in terms of productivity, energy intensity and economy of its production technology.

In this regard, an alternative technological approach is the production of highly dispersed TiC powders from titanium-containing carbide waste by the technology [17]. This method offers ample opportunities for the production of nanopowders on an industrial scale, as it is characterized by minimal technological equipment, environmental safety, rather low operating costs and competitive prices.

Experimental part

The powder obtained from the scrap of titanium hard alloys of TN-20, TN-25, TN-30 grades was investigated. The chemical composition is presented in Table 1.

Table 1. *Chemical composition of initial raw materials for obtaining titanium carbide powder*

Alloy grade	Component content in titanium alloy scrap, mass %				
	TiC	Ni	Mo	Nb	WC
TN-20	79.0	14,0	6.0	0,05	0,95
TN-25	74.0	19,3	6.0	0.01	0.69
TN-30	70.0	22,5	7.0	-	0.50

The main component of the slab is titanium carbide TiC cemented by nickel-molybdenum bond (Ni+Mo), there are also small impurities of niobium (Nb) and tungsten carbide (WC). As a result of treatment of alloys by technology [17] specially prepared dispersing solutions, there is a destruction of the alloy matrix and particle refinement due to the improvement of the machining process.

The crystalline structure of the fillers was investigated by X-ray diffraction (ARL X'TRA, ThermoTechno) with CuKα source in the 2θ angle range from 4 to 56° in asymmetric coplanar mode with sliding angle of incidence α=3o (θ-scan). Phase identification and peak indexing were performed using the ICDD (International Centre for Diffraction Data) PDF-2 database.

The microstructure of the obtained powder was studied using a scanning electron microscope "TESCAN MIRA 3 LMU" (Poland).

For the analysis of the particle size we used the Particle Metric software of the scanning electron microscope PHENOM proX (Phenom-World B.V. (Netherlands).

Differential thermal analysis of the powder was carried out on a NETZSCH STA 449F1. Thermal analysis of the powder was carried out in a platinum crucible in an argon atmosphere in the temperature range of 25...1000 °C at heating of 10 °C per minute.

Analysis of experimental results

Fig. 1 shows a diffractogram of the powder obtained from carbide scrap.

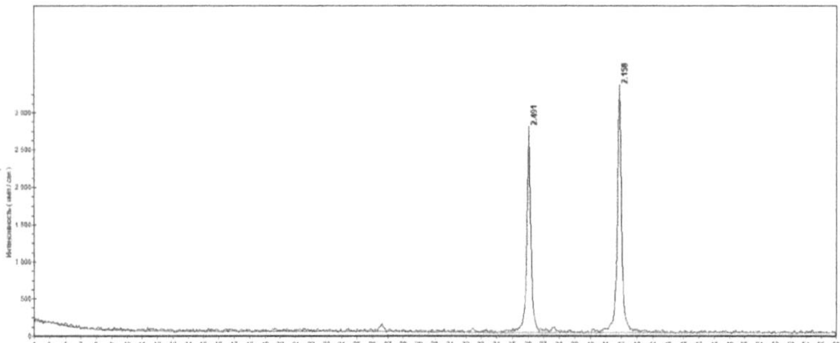

Fig. 1. *Powder diffractogram*

Two characteristic reflexes at angles 2 θ 36 and 42°, which correspond to pure titanium carbide, are present on the X-ray image (PDF card No. 31-1400).

The microstructure of the powder is shown in Fig. 2

Fig. 2. *Microstructure of titanium carbide (TiC) powder*

As can be seen from the presented microphotographs (Fig. 2) the powder under study is prone to agglomeration. The microstructure of the particles is heterogeneous, the particle shape is irregular. The average particle size is 60-150 nm, the size of agglomerates is 200 nm–2.5 μm (Fig. 3).

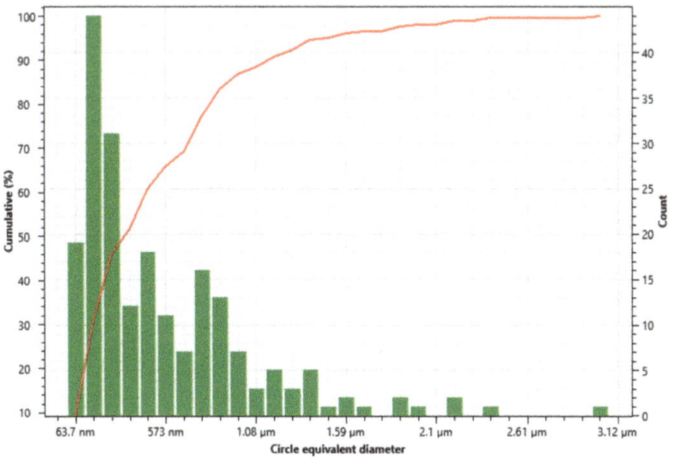

Fig. 3. *Titanium carbide powder particle size distribution.*

Elemental analysis of the powder (Fig. 4) showed the presence of impurities of molybdenum and tungsten, which is associated with the initial composition of raw materials used for the synthesis of titanium carbide powder.

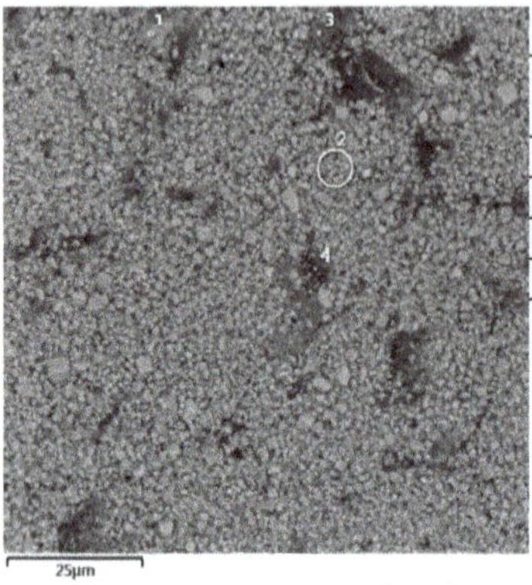

Spectrum name	Elemental composition, wt.%				
	C	O	Ti	Mo	W
Spectrum 1	21.63	4.53	60.03	6.15	7.66
Spectrum 2	18.10	7.07	64.92	6.33	3.58
Spectrum3	55.58	8.30	30.94	3.35	1.82
Spectrum 4	58.29	6.79	29.49	3.55	1.88

Fig. 4. *Results of spectral analysis of titanium carbide powder*

Differential thermal analysis (DTA) and thermogravimetric analysis (TGA) of the powder under study are shown in Figure 5.

Analysis of the thermogravimetric curve analysis (Fig. 5) showed that when increasing the temperature to 1000 °C the sample practically does not lose mass (-0.32%), indicating its relatively stable behavior in this temperature range. The endothermic process peaks at 133.1 °C, associated with the removal of adsorbed water. Differential thermal analysis (DTA) results indicate the presence of two exothermic effects with peaks at 441.9 °C and 476.4 °C. The first peak is associated with the oxidation of free carbon, with a loss of mass, the second peak indicates the oxidation of titanium carbide. The increase in mass during oxidation of the sample begins at 476.4 °C and is 1.26 % up to 900 °C, which indicates partial oxidation.

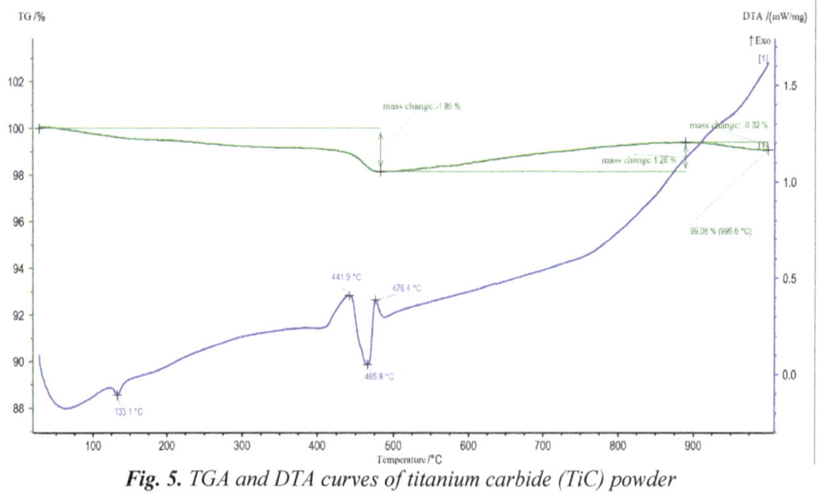

Fig. 5. *TGA and DTA curves of titanium carbide (TiC) powder*

Conclusions

The powder obtained from TN 20, TN-25, TN-30 carbide wastes contains single-phase titanium carbide and impurities of molybdenum and tungsten. The carbide particles have the size of 60...150 nm and are mainly agglomerated. The particle size of agglomerates is 200 nm...2.5 μm. It is relatively stable in the high temperature range.

Thus, titanium carbide powder obtained from carbide wastes can be used as an alternative to those available on the market as a component in the production of heat-resistant hard alloys, reinforced alloys, abrasive materials, coatings for cutting tools, etc., but this requires further research.

References

[1] A.N. Zhigach, I.O. Leipunsky, M.L. Kuskov, N.G. Berezkina, E.S. Afanasenkova, O.A. Safronova, B.V. Kudrov, G.W. Lopez, E.A. Skryleva, Synthesis of pure titanium carbide and titanium carbide/hydride core-shell nanoparticles via the flow-levitation method, and their characterization, J. Alloys and Compounds. 819 (2020) 153054. https://doi.org/10.1016/j.jallcom.2019.153054

[2] A.M. Elshazli, R.N. Elshaer, A.H.A. Hussein, S.R .Al-Sayed, Laser Surface Modification of TC21 (α/β) Titanium Alloy Using a Direct Energy Deposition (DED) Process, J. *Micromachines*. 12 (2021) 739. https://doi.org/10.3390/mi12070739

[3] S.N. Grigoriev, A.B. Nadykto, M.A.Volosova, A.A. Zelensky, P.M. Pivkin, WEDM as a Replacement for Grinding in Machining Ceramic Al_2O_3-TiC Cutting Inserts, J. *Metals*.2021, *11 (2021)* 882. https://doi.org/10.3390/met11060882

[4] S.C. Mojaki, Shivani B. Mishra, Ajay K. Mishra, Synthesis, characterization and material properties of titanium carbide nanocomposite derived from biochar, J. Materials Letters. 264 (2020) 127317. https://doi.org/10.1016/j.matlet.2020.127317

[5] Sreetama Ghosh, PremRanjan, AneeshKumaar, R. Sarathi, SundaraRamaprabhu, Synthesis of titanium carbide nanoparticles by wire explosion process and its application in carbon dioxide adsorption, J. Alloys and Compounds. 794 (2019) 645-653. https://doi.org/10.1016/j.jallcom.2019.04.299

[6] S. Nam, H.W. Lee, I.-H. Jung, Y.-M. Kim, Microstructural Characterization of TiC-Reinforced Metal Matrix Composites Fabricated by Laser Cladding Using FeCrCoNiAlTiC High Entropy Alloy Powder, J. *Appl. Sci. 11 (*2021) 6580. https://doi.org/10.3390/app11146580

[7] Y. Li, D. Zhang, H. Wang, W. Cong, Fabrication of a TiC-Ti Matrix Composite Coating Using Ultrasonic Vibration-Assisted Laser Directed Energy Deposition: The Effects of Ultrasonic Vibration and TiC Content. J. *Metals. 11 (*2021) 693. https://doi.org/10.3390/met11050693

[8] D. Kiesler, T. Bastuck, R. Theissmann, et al. Plasma synthesis of titanium nitride, carbide and carbonitride nanoparticles by means of reactive anodic arc evaporation from solid titanium, J. Nanopart Res. 17 (2015) 152. https://doi.org/10.1007/s11051-015-2967-8

[9] Hyunho Shin, Jun-Ho Eun, Titanium Carbide Nanocrystals Synthesized from a Metatitanic Acid-Sucrose Precursor via a Carbothermal Reduction, J. Nanomaterials. 2015 (2015) 469121. https://doi.org/10.1155/2015/469121

[10] P.N. Karakizis, D.I. Pantelis, G. Fourlaris, et al. The role of SiC and TiC nanoparticle reinforcement on AA5083-H111 friction stir welds studied by electron microscopy and mechanical testing. Int J AdvManuf Technol. 94 (2018) 4159–4176. https://doi.org/10.1007/s00170-017-1147-7

[11] J.M. Kim, J.S. Park, H.S. Yun, Microstructure and Mechanical Properties of TiC Nanoparticle-Reinforced Iron-Matrix Composites, J. Strength Mater. 46 (2014) 177-182. https://doi.org/10.1007/s11223-014-9533-y

[12] Sławomir Dyjak, Małgorzata Norek, Marek Polański, Stanisław Cudziło, Jerzy Bystrzycki, A simple method of synthesis and surface purification of titanium carbide powder, Int. J. Refractory Metals and Hard Materials. 38 (2013) 87-91. https://doi.org/10.1016/j.ijrmhm.2013.01.004

[13] Mohsen Mhadhbi, MiloudDriss, Titanium Carbide: Synthesis, Properties and Applications, J. Brilliant Engineering. 2 (2021) 1-11. https://doi.org/10.36937/ben.2021.002.001

[14] A. P. Amosov, A. R. Samboruk, A. A. Samboruk, A. A. Ermoshkin, D. V. Zakamov, and K. S. Krivolutskii, Self Propagating High Temperature Synthesis of Titanium Carbide Nanopowder from the Granulated Charge, J. Non Ferrous Metals. 56(1) (2015) 79–85. https://doi.org/10.3103/S1067821215010034

[15] L.Y. Fedorov, I.V. Karpov, A.V. Ushakovet al. Influence of pressure and hydrocarbons on carbide formation in the plasma synthesis of TiC nanoparticles. J. Inorg Mater. 51 (2015) 25–28. https://doi.org/10.1134/S0020168515010057

[16] N. RahnemaiHaghighi, R. Poursalehi, Effect of C/H and C/O ratios on the arc discharge synthesis of titanium carbide nanoparticles in organic liquids, J. ApplNanosci.9 (2019)411–421. https://doi.org/10.1007/s13204-018-00946-7

[17] EP 3138932 (A1) EC

Modern Trends in Manufacturing Technologies and Equipment Materials Research Forum LLC
Materials Research Proceedings **21** (2022) 220-223 https://doi.org/10.21741/9781644901755-39

Studies of the Influence of Temperature on the Photocatalytic Activity of Tungsten Oxide

N. Derbasova[1,a], V. Gavrish[1,b], T. Chayka[1,c*] and O. Gavrish[1,d]

[1]Sevastopol State University, 33 Universitetskaya Street, Sevastopol, 299053, Russia

[a]derbasovanm@mail.ru. [b]vmgavrish@sevsu.ru l, [c]KimChayka@yandex.ru, [d]olga_gavrish_53@mail.ru

Keywords: Tungsten Oxide, Organic Pollutants, Photocatalytic Activity, Methylene Blue

Abstract. The results of photocatalytic activity of tungsten oxide in visible light are analyzed. It was shown that tungsten oxide has photocatalytic activity under the action of visible light in the photodegradation of the dye methylene blue. It was determined that the concentration of 15 mg/L provides the highest rate of photodecomposition of methylene blue. The optimum temperature for the process was determined. The process of methylene blue decomposition is most effective at 25 °C.

Introduction

In recent years, considerable attention has been paid to obtaining materials with unique properties. Over the years, a large number of semiconductors have been developed and used as photocatalysts. Among various semiconductors, TiO_2 is considered the best photocatalyst and is widely used to detoxify water from a number of organic pollutants [1-3]. However, photocatalysis using titanium dioxide has a number of significant drawbacks. Thus, the energy-gap width of titanium dioxide is 3.0-3.2 eV; light absorption by titanium dioxide lies in the UV region of the spectrum, that is why the efficiency of photocatalysts under visible radiation is less than 10%. There is also insufficiently high quantum yield of phototransformation, which is connected with the high degree of recombination of charge carriers, low specific surface, as well as low adsorption capacity of TiO_2 [4,5]. Therefore, it became necessary to find an alternative to TiO_2. In turn, photocatalysts based on tungsten oxide WO_3 are among the best materials operating under visible light [6-8]. WO_3 has a energy-gap width of 2.6 eV, is stable under acidic and oxidizing conditions, has chemical stability in the pH range (pH <8), significant photocurrent conversion efficiency, is resistant to corrosion processes, stable in various electrolytes and, most importantly, is considered harmless.

The analysis of works of similar studies revealed that mainly the studies of photocatalytic activity of tungsten trioxide are aimed at studying and developing methods of chemical synthesis of nanodispersed tungsten oxide in forms suitable for its subsequent use [9-13]. Meanwhile, there are no studies on the photocatalytic activity of tungsten trioxide obtained by recycling tungsten carbide waste. Studying the properties of tungsten trioxide will enable the development of more effective photocatalysts for the oxidation of organic pollutants to CO_2 and H_2O.

Methodology for investigating the photocatalytic activity of tungsten oxide nanopowder

The decomposition process of methylene blue was studied by spectrophotometry and by the color change of the solution during the photocatalytic reaction. A sample of nanopowder weighing 0.2 g was placed in a round flat-bottomed flask and 200 ml of methylene blue solution (solution concentrations of 9, 13, 15, and 20 mg/L) was added.

The solution was irradiated with a standard white LED lamp with a temperature of 4000 K in the wavelength range of 380 ... 780 nm. Stirring was performed on a magnetic stirrer "Ritm-01" with a speed of rotation of the armature of 500 rpm. Centrifugation was performed on a "CentriFuga 80-1" centrifuge, at 3000 rpm. UV-vis spectra were recorded for aqueous solutions on spectrophotometer "Expert 003" in quartz cuvettes with a width of 1 cm.

Sampling was performed every 2 hours to control the course of the photochemical reaction and to observe changes in the concentration of the pollutant in the solution, and 10 ml of the sample was centrifuged for 15 min at 3000 rpm, separating the sediment. After that the solution was decanted and the dye concentration was evaluated on a photometer "Expert 003" by the degree of discoloration of the solution. After analysis, the sample was returned to the original solution.

From the works [14,15] it was revealed that the decomposition of solutions of methylene blue passes through the kinetics of the first order, which is expressed by the equation:

$$\ln(C_0/C) = k_{app} \cdot t$$

Where C_0 – is the initial concentration of the methylene blue solution;

C – concentration at a given time;

k_{app} – decomposition rate constant.

To determine the rate constant of decomposition in our case, we plotted $\ln(C_0/C)$ as a function of time shown in Figure 1.

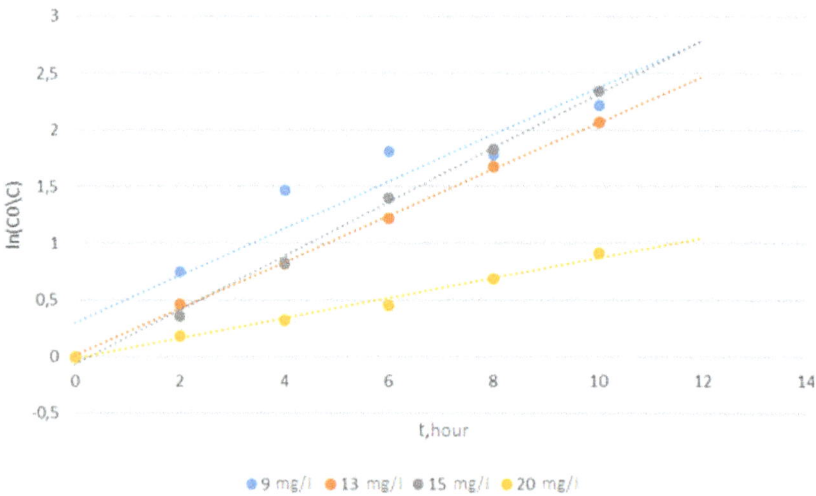

Fig. 1. *- Rate constant of decomposition at different concentrations of methylene blue*

The linear dependence of the data between $\ln(C_0/C)$ and irradiation time confirmed the applicability of this model with correlation coefficient values of 0.9489; 0.9995; 0.9984 and 0.9931. The decomposition rate constant (k_{app}) obtained from the slope of the plot was 3.7 × 103 min-1, 3.45 × 103 min-1, 3.9 × 103 min-1, and 1.53 × 103 min-1.for concentrations of 9, 13, 15, and 20 mg/L, respectively. Thus, the higher the k_{app}, the shorter the time required for complete decomposition of methylene blue.

The presented data show that the photocatalytic activity of tungsten oxide is most effective for the concentration of methylene blue 15 mg/l.

Study of photocatalytic activity at temperature exposure

We also studied the effect of temperature on the photocatalytic activity of tungsten oxide. For this purpose, aqueous solutions of methylene blue with a concentration of 15 mg/L were heated with a magnetic stirrer at constant stirring for 10 hours at temperatures 15, 25, and 30 ^{0}C. Methylene blue concentration drop was measured after 10 hours.

The results of the calculations are shown in figure 2.

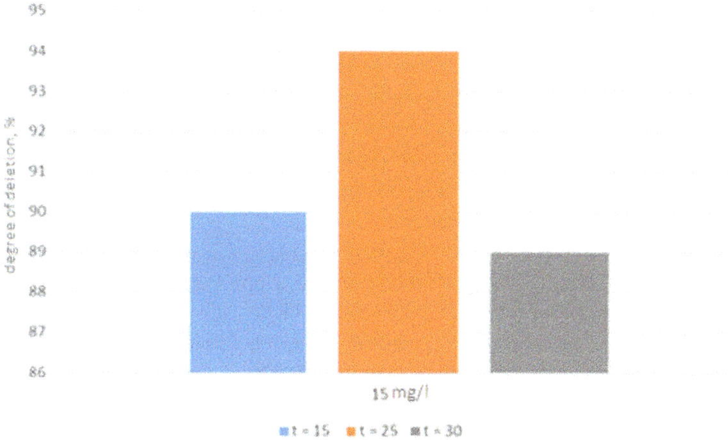

Fig. 2. Photocatalytic activity data for different temperatures

After 10 hours of continuous process, the dye spectrum value decreased by 90, 94 and 89%. From the analysis of the data obtained, we can conclude that the decomposition process of methylene blue is most effective at a temperature of 25 ^{0}C.

Conclusions

It was found that tungsten oxide has maximum photocatalytic activity under the action of visible light in the photodegradation of the dye methylene blue at a concentration of 15 mg/l. It was shown that the temperature optimum of the process is 25 ^{0}C.

References

[1] Carbuloni, C.F., Savoia, J.E., Santos, J.S.P., Pereira, C.A.A., Marques, R.G., Ribeiro, V.A.S.,Ferrari, A.M. Degradation of metformin in water by TiO_2-ZrO_2 photocatalysis. J. Environ. Manag. 262 (2020) 110347.https://doi.org/10.1016/j.jenvman.2020.110347

[2] Qiong Sun, Songhao Wu, Kaijing Li, Bing Han, Yingjie Chen, Beili Pang, Liyan Yu, L. Dong The favourable synergistic operation of photocatalysis and catalytic oxygen reduction reaction by a novel heterogeneous $CoFe_2O_4$-TiO_2 nanocomposite Appl. Surf. Sci., 516 (2020)146142. https://doi.org/10.1016/j.apsusc.2020.146142

[3] E. Mrotek, S. Dudziak, I. Malinowska, D. Pelczarski, Z. Ryzynska, A. Zielinska-Jurek Improved degradation of etodolac in the presence of core-shell $ZnFe_2O_4/SiO_2/TiO_2$ magnetic photocatalyst Sci. Total Environ., 724 (2020) 138167. https://doi.org/10.1016/j.scitotenv.2020.138167

[4] H. Moradi, A. Eshaghi, S.R. Hosseini, K. Ghani Fabrication of Fe-doped TiO2 nanoparticles and investigation of photocatalytic decolorization of reactive red 198 under visible light irradiation Ultrason. Sonochem., 32 (2016) 314-319. https://doi.org/10.1016/j.ultsonch.2016.03.025

[5] J.G.S.M.J. Valero-Romero, L. Oar-Arteta, L. van Koppen, D.Y. Osadchii, J. Gascon, F. Kapteijn Photocatalytic properties of TiO_2 and Fe-doped TiO_2 prepared by metal organic framework-mediated synthesis Chem. Eng. J., 360 (2019) 75-88. https://doi.org/10.1016/j.cej.2018.11.132

[6] C. Yu, F. Chen, Z. Liu, K. Yang, H. Ji, D. Li, W. Xie, S. Li Facile synthesis of a robust visible-light-driven $AgCl/WO_3$ composite microrod photocatalyst J. Alloys Compd., 809 (2019) 151844-151854. https://doi.org/10.1016/j.jallcom.2019.151844

[7] Cherkashina N.I., Pavlenko V.I., Yastrebinsky R.N. Phase transitions and change of electrophysical properties of WO3 in the temperature range 83-673 K // Proceedings of higher educational institutions. Physics. T. 62. № 5 (737) (2019)126-131. https://doi.org/10.1007/s11182-019-01789-7

[8] Pavlenko V.I., Bondarenko G.G., Cherkashina N.I. Physical and mechanical characteristics of the composite based on polyimide matrix filled with tungsten oxide, Perspective Materials № 7 (2019) 15-25. https://doi.org/10.30791/1028-978X-2019-7-15-25

[9] S. Adhikari, D. Sarkar High efficient electrochromic WO_3 nanofibers Electrochim. Acta., 138 (2014) 115-123. https://doi.org/10.1016/j.electacta.2014.06.062

[10] J.C. Murillo-Sierra, A. Hernández-Ramírez, L. Hinojosa-Reyes, J.L. Guzmán-Mar A review on the development of visible light-responsive WO_3-based photocatalysts for environmental applications Chem. Eng. J. Adv., 5 (2021) 100070-100091. https://doi.org/10.1016/j.ceja.2020.100070

[11] M.A.M. Khan, S. Kumar, T. Ahamad, A.N. Alhazaa Enhancement of photocatalytic and electrochemical properties of hydrothermally synthesized WO_3 nanoparticles via Ag loading J. Alloys Compd., 743 (2018) 485-493. https://doi.org/10.1016/j.jallcom.2018.01.343

[12] M. Ahmadi, R. Younese, M.-J.-F. Guinel Synthesis of tungsten oxide nanoparticles using hydrothermal method at ambient pressure J. Mater. Res., 29 (2014) 1424-1430. https://doi.org/10.1557/jmr.2014.155

[13] R. Ponnusamy, A. Gangan, B. Chakraborty, C.S. Rout Tuning the pure monoclinic phase of WO_3 and WO_3-Ag nanostructures for non-enzymatic glucose sensing application with theoretical insight from electronic structure simulations J. Appl. Phys., 123 (2018) 024701-024710. https://doi.org/10.1063/1.5010826

[14] R. Xu, M. Su, Y. Liu, Z. Chen, C. Ji, M. Yang, X. Chang, D. Chen, J. Clean. Prod. 242 (2020) 118366. https://doi.org/10.1016/j.jclepro.2019.118366

[15] Q. Wang, X. Lei, F. Pan, D. Xia, Y. Shang, W. Sun, W. Liu, Colloid. Surf. A. 555 (2018) 605–614. https://doi.org/10.1016/j.colsurfa.2018.07.016

On the Chemical Heterogeneity of Austenite in Maraging Steel

T.M. Makhneva[1,a*] and V.B. Dementyev[1,b]

[1]Udmurt Federal Research Center of the Ural Branch of the Russian Academy of Sciences", 34, T. Baramzinoy St., Izhevsk, 426067, Russia

[a]mah@udman.ru, [b]demen@udman.ru

Keywords: Maraging Steel, X-Ray Spectral Microanalysis, Phase Transformation, Liquation Austenite

Abstract. The change of Ni-, Cr-, Cu- contents in maraging steel composition occurring on heating in the subcritical and intercritical interval has been studied by the X-ray spectral microanalysis. Heating in the temperature range from 490° to 550°C has resulted in increasing of Ni- Cu- concentrations in the liquation austenite when the latter is present in the steel structure as a consequence of several reasons (the large ingot, low level of forging reduction ratio, etc.). The significant enrichment of surface layers of austenite inclusions may probably occur if there are great differences between interphase and intraphase diffusion rates. By varying the thermal treatment and thus the Ni-diffusion in austenite it is possible to create austenite layers with different Ni-contents within a grain or massive martensite and it is also possible to control the material properties.

Introduction

Fe-Ni binary system state diagram, the kinetics of phase transformations in binary alloys in the "iron corner" of the diagram have been thoroughly studied since 1940-50 by [1-5].

The available information about Fe-Ni-system is quite sufficient except the definition of conditions of $\gamma\to\alpha$ - transformation process where one can observe the occurrence of decomposition, i.e. $\gamma \to\alpha+\gamma_{ern}$ where γ_{ern} is nickel enriched austenite. All practically realized transformation models with γ_{ern}-occurrence include two stages: (1) $\gamma\to M$ (on cooling bellow M_s - temperature), and (2) $M\to\alpha+ \gamma_{ern}$ (sufficiently slow heating above A_H - temperature).

Data on the study of the kinetics *of* $\gamma\to\alpha\to\gamma$- transformations in iron-nickel alloys are given in [6-9]. Modern methods investigated the inhomogeneity of the γ-phase [6] as a result of the redistribution of system components between the phases [7]. In work [8] the assessment of the degree of heterogeneity in terms of nickel, depending on the mode of heat treatment of low-carbon chromium-nickel steel, was carried out.

Iinteresting data on $\gamma\to\alpha\to\gamma$ transformation kinetics study in Fe-Ni alloys have been presented in work [9]. The investigation of cosmic iron containing nickel has shown that realization of $\gamma\to\alpha\to\gamma_{ern}$ scheme is possible after million years under the condition of cooling rate in order of several degrees Celsius. In the work the new data on Fe-Ni diagram obtained by electron spectral analysis using Ø 0,03 μm (10^{-6} m) sonde are also presented. According to the refined data in the temperature range about 500°C and below this temperature the equilibrium line between α and $\alpha+\gamma$-regions becomes vertical increasing differences equilibrium nickel concentrations in α and γ -phases. The authors as it seems to us have unreasonably averaged the values of nickel concentrations at the interphase boundary and inside the phases having thus distorted the positions of equilibrium lines.

Experience in the manufacture of forgings with a low reduction ratio has shown that their structure often contains residual nickel-enriched liquation austenite (γ_{liq}). Despite multiple heatings to high enough temperatures for deformation (1200°C) and the deformation with forging reduction ratio of 3 the homogeneous austenite is very often unobtainable [8]. It was of interest to study the chemical composition of liquation austenite in one of the maraging chromium-nickel steel.

Liquation austenite (γ_{liq}) is a metastable phase. It is known that interphase equilibrium lines may be drawn further in the diagram area where the temperatures are lower and phase ratio is different. With these lines it is possible to identify metastable phase compositions.

Let us consider the alloy with C_o, concentration at the temperature *t*. For the bulk mass of the sample the heating temperature *t* lies below A_s temperature and the state is stable. When the liquation austenite is present in the steel, its structure presents some sort of conglomerate of two steel compositions, i.e. C_o, and C_1, or C_2 (fig.1). For the liquation austenite two positions (above and below A_H) and two ways of transformation are possible under these conditions. In the case of liquation austenite composition corresponding to the point *"1"* with C_1 concentration (i.e. it is to the left from equilibrium line and lower than A_H) it is possible to expect two processes to occur: either the dissolution of liquation austenite, or the appearance at the interface of the concentration *"C_3"*.

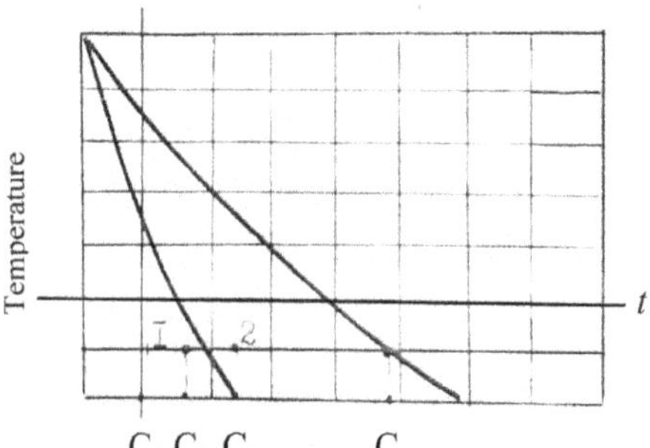

Fig. 1. *Nickel heterogeneity formation (scheme)*

Transformation ($\gamma\rightarrow\alpha$) at the temperatures above M_s practically cannot be realized. It would be therefore unreasonable to expect the disappearance of γ_{liq} which has not been dissolved before at the higher heating's. The viability of γ_{liq} with C_2-composition locating at point *"2"* depends on its sizes. If the size of the area is more than the critical value then *"C_3"* concentration must be expected to be at the interphase boundary. When the size of γ_{liq} is less than the critical value then one could expect the beginning of $\gamma\rightarrow\alpha$-transformation, and the probability of this process has already been discussed.

Thus, nickel concentration in the outer layers of the liquation austenite may be considerably increased if the duplex structure has artificially been formed. The actual degree of their enrichment must depend on Ni-diffusion rate from α-phase and the rate of Ni-atoms removal into the inner bulk of the area of γ_{liq}. Hence, it is more probable to get nickel encroached layer at the lower temperatures as the difference of nickel content in α- and γ-phases, being in equilibrium, is increasing with decrease in the temperature and the rate of Ni atoms removal from the phase boundary into the austenite is decreasing.

Even in work [2] A.A. Popov has shown that the phase equilibrium at the interphase boundary is rapidly reached. The diffusion rate inside a grain is determined by the temperature of heating, it is possible therefore to change concentration gradients in the grain considerably by varying thermotemporal parameters. In relatively short period of time the great difference in nickel content may occur at the interphase boundary. The removal of Ni-atoms from the boundary in the austenite would be defined by the rate of diffusion. The formation of the considerable layer of the austenite at the interphase boundary at the lower temperatures is expected to take a certain period of time as well. When the thickness of the enriched layer is several lattice parameters, such a layer cannot be observed using the available methods. Consequently, the thickness of the nickel-enriched (detectable) layer should be of the order of 10^{-1} m with a probe diameter of Ø 3 10-6 m and a nickel resolution of no worse than 0.1%.

Material and experimental procedure

In order to check the above mentioned statements we have examined 08X15H5D2T steel having the liquation austenite in its structure. Samples of steel in the hotforged state have been heated up to 490°C, 520°C and 550°C with holding of 20 hours, and to 650°C with holding of 6 hours. Heating below 450°C has not resulted in any appreciable redistribution of nickel in the given steel. Polished sections have been exposed to weak electrolytic etching in 10% oxalic acid. Chemical analysis of the structure components after regimes has been carried out with X-ray spectral microanalyses.

X ray spectrometer JEOL used in the present work has the sonde diameter of 3 10^{-6} m therefore much time is required to get the enriched layer of the indicated thickness at the temperatures about 500°C. The examinations with the help of the magnetometer have shown that at the temperatures around 550° the formation of the enriched austenite in the given steel starts after about 6 hours [10].

Results and discussion

X-ray spectral microanalysis of the areas of the polished microsections, containing the liquation austenite, has shown approximately 1% increase in nickel in them, i. e. about 5,5% for medium nickel-content in melting about 4,5%. This minor, at the first sight, change in Ni -content should not result in the qualitative changes in the structure. Our magnetometric investigations have shown that this is the nickel content interval for the given steel where M_s temperature transition into the range of the lower temperatures (below 0°C) has taken place. Indeed, the sub-zero treatment has resulted in the occurence of the martensite pattern on the surface of the liquation austenite areas.

In figures 2 A) B) C) the results on mesuring Ni- Cr- Cu- concentrations in α and γ-phases are summerized. In fact, at the temperature 490°C the conditions are such that due to differences in diffusion rates of nickel at the interphase boundary and in the austenite the accumulation of Ni in the surface layer of the austenite area occurs. With the temperature increase from 490°C to 650°C, with the acceleration in the intraphase diffusion and decrease in the magnitude of the equilibrium concentration of nickel in the austenite the nickel "peak" is smearing and gradually the change in nickel concentration is normalized and becomes characteristic of the monophase dissolusion. The exposed appreciable copper diffusion from α-into γ-phase is of interest. Copper concentration sometimes is twice as big as that in ferrite. Probably, this diffusion is made easier by isomorphism of copper and austenite. The change of chromium concentration occurs gradually. The inner layers of the liquation austenite contain approximately 1% less of chromium than the adjacent α-phase does.

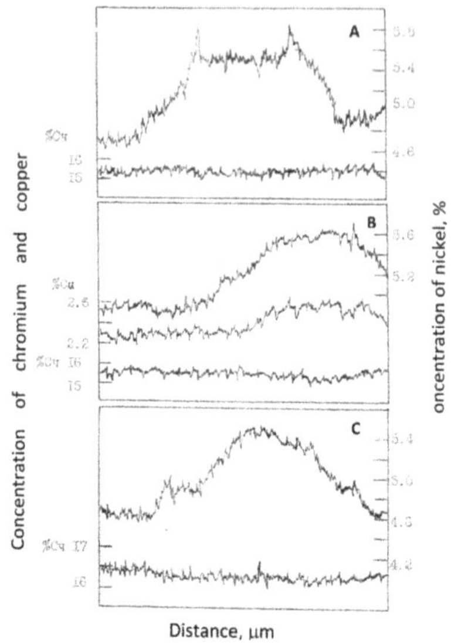

Fig. 2. *Redistribution of nickel, copper and chromium during heating of 08Kh15N5D2T steel into the two-phase α + γ-region.*
A - 490°C, 20h; B - 520°C, 20h; C - 550°C, 20h.

Enrichment of austenite with nickel in the two-phase region occurs in the steel under study, both in the presence of liquation [8] and without it, stabilizing the γ-phase [11]. This is also evidenced by the data of [7], in which the morphology study of α→γ-transformation in massive martensite carried out by V.I. Zeldovich has shown that the interphase boundary may lie inside the massive martensite itself [7]. Electron microscopy study of steel structure after heating to 650°C with 6 hours holding, which causes the formation of about 50% of austenite and after cooling 25-30% of austenite remains, has shown the formation of the duplex structure. Outer layers of the structure element have remained in the austenite state alter cooling, and inner layers have been in the martensite state

Thus the variation of the thermotemporal parameters under the thermal treatment 08X15H5D2T steel has made it possible to obtain different and among them anomalous distributions of nickel in the steel microstructure and to form the duplex structure. On the slow heatings the liquation austenite is "replenished" by nickel, starting from the subcritical temperatures, this may result in increasing of the austenite stability.

References

[1] N.P. Allen, C.J. Early, Iron Steel Inst., 28 (1950) 455.

[2] A.A. Popov, Diffusion movement of the border of the phase separation, Phase transformations in iron-carbon alloys. Mashgiz. (1950) 18-35.

[3] E. Scheil, Über die Irreversibilität der Eisen-Nickel-Legierungen, Archiv für das Eisenhüttenwesen 24. Jahrgang. Heft ¾- Märtz / Ahril (1953). https://doi.org/10.1002/srin.195301444

[4] Ya.M. Golovchiner, DAN SSSR. 107 (1953)109.

[5] Ya.M. Golovchiner, Problems of metal science and physics of metals, Metallurgizdat. 7. (1962) 281.

[6] V.A. Shabashov, V.V. Sagaradze, E.E. Yurikov, A.V. Savelyeva, Mesbauer and electron microscopic studies in iron-nickel alloys, FMM. 44. 5 (1977)1060-1070.

[7] V.I. Zeldovich, V.D. Sadovsky, Redistribution of components between α -and γ- phases at $\alpha \rightarrow \gamma$ - transformation in iron-nickel alloys, FMM. 28 4 (1969) 715-722.

[8] T.M. Makhneva, Investigation of chemical heterogeneity with respect to nickel and chromium in the structure of steel 08Kh15N5D2T, IM Ural Branch of the Russian Academy of Sciences. "Chemical Physics and Mesoscopy". 15-1 (2013) 91-100.

[9] A.D. Romig, J.I. Coldstein, Determination of the Fe-Ni and Fe-Ni-P Phase Diagrams at Low Temperatures (700 to 300°C). Metallurgical Transactions. 11. 7 (1980) 1151-1159. https://doi.org/10.1007/BF02668139

[10] E.S. Makhnev, T.M. Makhneva, Reverse martensitic transformation in steel VNS-2USH, Martensitic transformations in metals and alloys. Sat. report int. conf. Kiev: Naukova Dumka, 1979, pp. 180-184.

[11] T.M. Makhneva, A.A. Sukhikh, V.B. Dement'ev, Inverse martensitic $\alpha \rightarrow \gamma$- transformation in nanostructured maraging steels, Metal Science and Heat Treatment, Kluwer Academic/Plenum Publishers (United States) (2017) 1-6.

Analysis of the Corrosion Behavior of the TiNi Alloy in the Coarse-Grained State

Anna Churakova[1,2,a*], Elina Kayumova[3,b], Evgeni Vorobiev[2,c] and Nafisul Haque[4,d]

[1]Institute of Molecule and Crystal Physics - Subdivision of the Ufa Federal Research Center of the Russian Academy of Sciences, 151 pr. Oktyabrya, Ufa, 450075, Russia

[2]Ufa State Aviation Technical University, 12 K. Marx str., Ufa, 450008, Russia

[3] Ufa State Petroleum Technological University, 1 Kosmonavtov st., Ufa, 450064, Russia

[4]Department of Metallurgical Engineering, NEDUET, University Road, Karachi 75270, Pakistan

[a]churakovaa_a@mail.ru, [b]kayumkaaaa@gmail.com, [c]z.vorobyov@mail.ru
[d]engrnafis@gmail.com

Keywords: Shape Memory Alloys, Corrosive Behavior, Pitting Corrosion, Gravimetric Method, Passivating and Activating Solutions, Microstructure

Abstract. In this work was investigate the corrosion behavior of the TiNi alloy in a coarse-grained state in inorganic field with different concentration and holding time. An increase in the concentration of the solution leads to a significant acceleration of corrosion processes in the $Ti_{49.1}Ni_{50.9}$ alloy with a high Ni content, including until the samples are completely dissolved. It was revealed that solutions of 1 M sulfuric and hydrochloric acids after a month's exposure did not change in color and no precipitations were found, while solutions of 5 M hydrochloric and sulfuric acids acquired a violet and then green color, which is due to the predominant release of titanium ions (+4) and nickel (+2).

Introduction

In recent decades, materials with an ultrafine-grained structure have been created, having a grain size of 100 - 300 nm, which have a unique structure and properties, they change fundamental characteristics, such as the Debye and Curie temperatures, saturation magnetization, etc. Shape memory materials (SME) have already found wide application in medicine as long-term functioning materials implanted into the body. A special class of shape memory alloys are nickel and titanium alloys - alloys (NiTi). The range of their application depends on the temperature of martensitic transformation and mechanical properties [1-6]. They exhibit high elastic properties, were able to change their shape when the temperature changes and do not collapse under conditions of alternating load. Phase transitions in such alloys are characterized by a wide hysteresis and a long temperature range, in which the material exhibits shape memory and superelasticity effects. It is known that various metal alloys have the ability to change their shape as a result of temperature changes. This effect is called "austenitic transformation", and another transformation, from austenitic state to martensitic state, is called "martensitic transformation" [7-8]. The biochemical compatibility of physiological fluids and metal implants is largely determined by the electrochemical interaction between them, which usually leads to the transfer of metal ions into tissue fluids. In this case, the implant may also contain heavy elements that are toxic to the body. However, it is impossible to assess biochemical compatibility by the concentration of toxic elements, especially if their introduction into the composition of the implant leads to a significant increase in its corrosion resistance [9-14]. Since the corrosive properties of the implant are the

Materials Research Proceedings **21** (2022) 229-236	https://doi.org/10.21741/9781644901755-41

most important indicators of biochemical compatibility, it is necessary to take into account both the direct corrosion processes associated with the transition of ions through the interface, and the reactions leading to the formation of poorly conducting protective films. As a rule, such protective films delay the release of toxic ions into the tissue, as a result, the implant containing toxic elements weakly interacts with the surrounding tissues and becomes practically inert with respect to biological media [15-20].

Corrosion resistance is largely determined by the degree of defectiveness of the material and the peculiarities of its implantation into the human body as one of the most active carriers of aggressive media. Tissues are a complex biological system that reacts to the introduction of an implant by changing its own structure up to physical and mechanical destruction [21-23]. Therefore, it is important to know the features of corrosive behavior, affecting biochemical and biomechanical compatibility with body tissues.

Material and methods

The study material was a two-component alloy: $Ti_{49.1}Ni_{50.9}$ alloy, which has a B2 austenite structure at room temperature, with a bcc lattice of the CsCl type. To form a solid solution based on TiNi and to exclude the prehistory of obtaining the material, the alloy was quenched from the homogeneity region (from 800 °C) into water. The average grain size of the hardened alloy was 200 μm. To reveal the microstructure of the initial titanium nickelide, an etchant of the following composition was used: 60% H_2O + 35% HNO_3 + 5% HF. The study of the microstructure was carried out on an optical metallographic microscope OLYMPUS GX51, as well as using scanning electron microscopy (SEM) JEOL JSM-6490LV, an inverted microscope AXIO OBSERVER Z1M in the dark field mode (the studies were carried out in the laboratory of solid state physics of the Institute of Physics of Molecules and Center of the Russian Academy of Sciences). Before testing, pre-prepared samples were weighed on an analytical balance, the samples were placed in a desiccator, where the test sample was in contact with an aggressive medium at a temperature (25 °C). The samples were placed in the solution for a certain time until the appearance of noticeable traces of corrosion. After testing, the samples were washed with water, treated with alcohol, dried, and weighed on an analytical balance.

Results

Figure 1 shows optical microscopy of the surface of samples of the $Ti_{49.1}Ni_{50.9}$ alloy in a coarse-grained state.

Fig. 1. *Photographs of the surface of $Ti_{49.1}Ni_{50.9}$ alloy specimens before corrosion tests in coarse-grained state*

Figure 2 shows the surface structures of alloy samples in a coarse-grained state after corrosion tests in a 1 M H_2SO_4 solution with different holding times. At a given concentration of the solution, no noticeable corrosion damage is observed; only corrosion products are observed on the surface of the samples. At the same time, with an increase in the exposure time in the solution, the amount and volume fraction of corrosion products increase linearly.

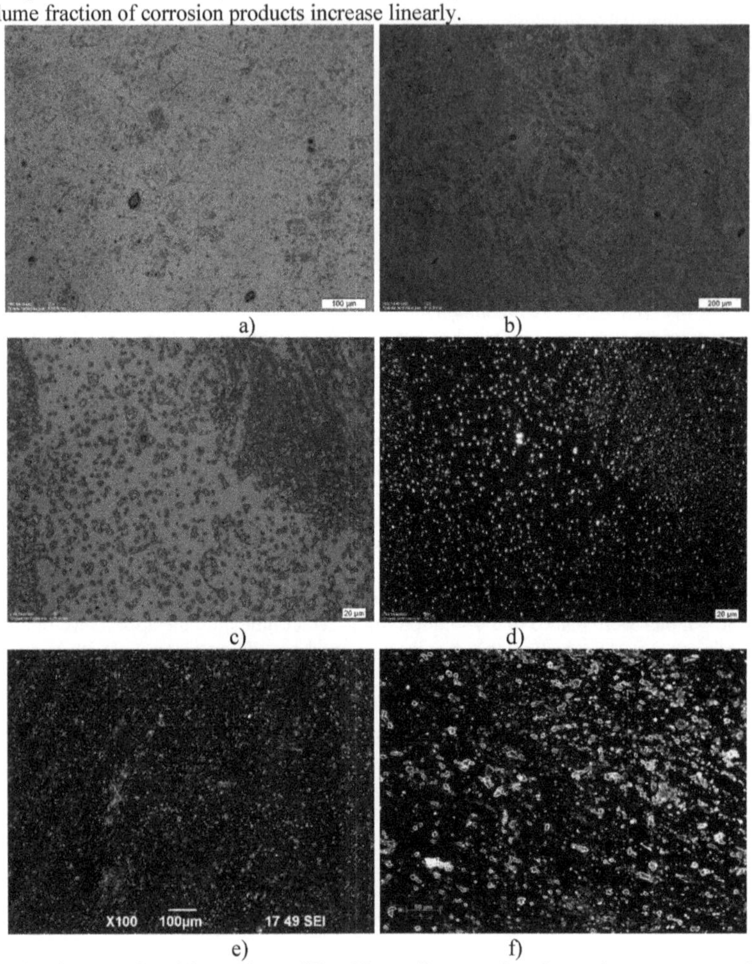

Fig. 2. *Photographs of the surface of $Ti_{49.1}Ni_{50.9}$ alloy samples obtained using an optical microscope (a-d), SEM (e) and an inverted microscope (dark field) (f) in a solution of 1 M H_2SO_4 in a coarse-grained state: a) 1 day; b) 3 days; c-d) 15 days; e-f) 30 days*

Corrosion tests in 1 M HCl also do not lead to damage to the samples, accompanied by an insignificant loss of sample weight, corrosion products are observed on the surface, the volume fraction is maximum at the maximum exposure time, small pitting is observed in some areas (Fig. 3, e).

Materials Research Proceedings 21 (2022) 229-236 https://doi.org/10.21741/9781644901755-41

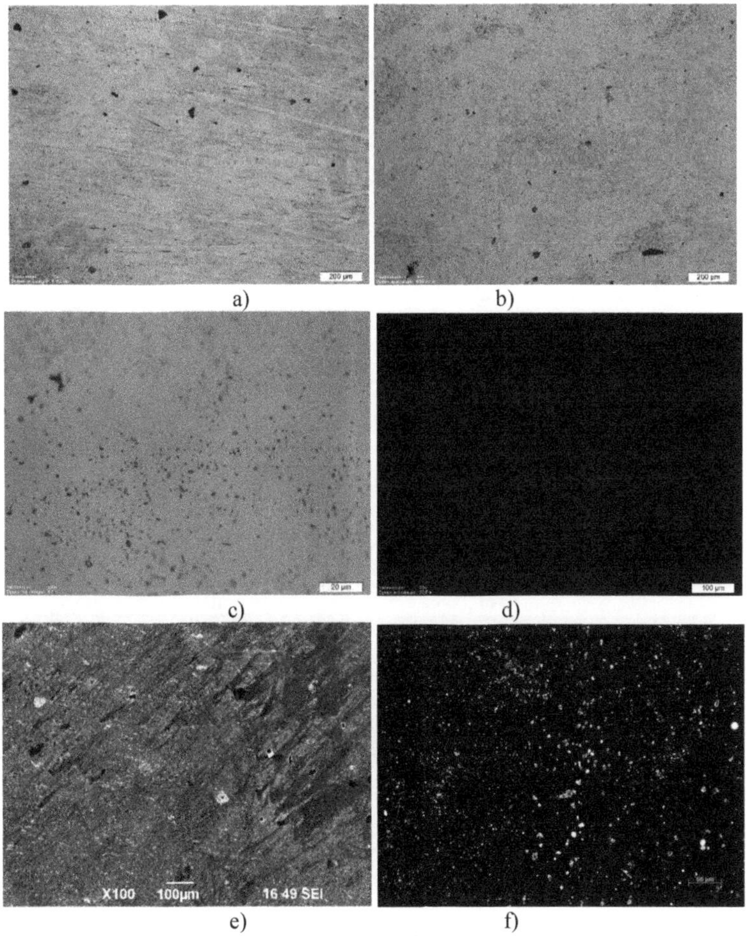

Fig. 3. *Photographs of the surface of $Ti_{49.1}Ni_{50.9}$ alloy samples obtained using an optical microscope (a-d), SEM (e) and an inverted microscope (dark field) (f) in a solution of 1 M HCl in a coarse-grained state: a) 1 day; b) 3 days; c-d) 15 days; e-f) 30 days*

Studies in 3% NaCl showed that, compared to tests in acid solutions, there is a smaller amount of corrosion products (Fig. 4). In this case, in the structure of the alloy, regions of different contrast with a size of about 400 ± 30 μm are distinguished at the maximum exposure time in the solution (Fig. 4, e). The maximum amount of corrosion products on the surface of the samples is observed after 15 days of exposure (Fig. 4, c-d).

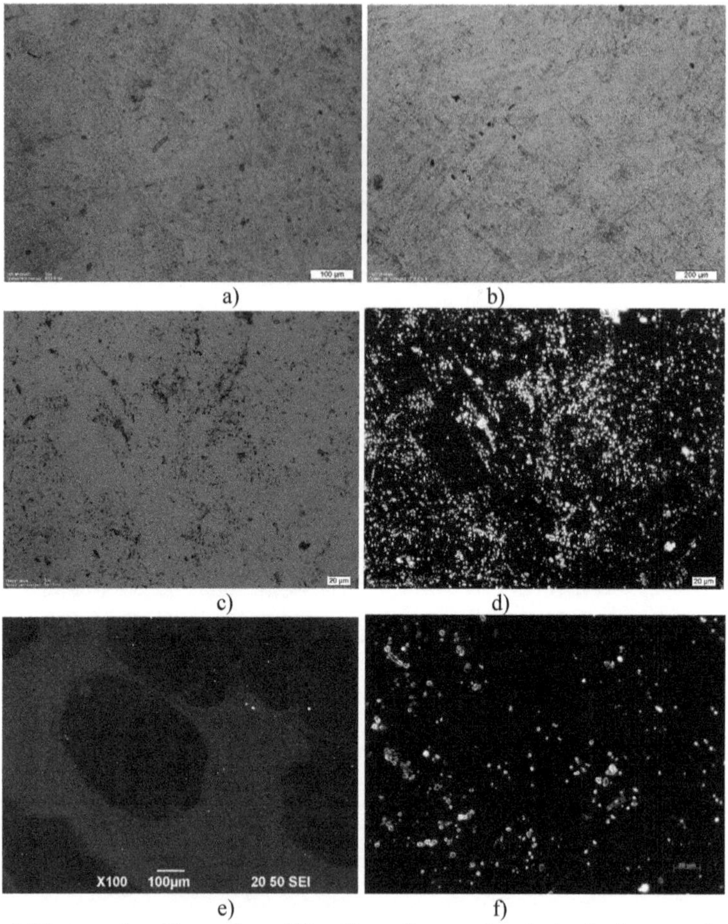

Fig. 4. *Photographs of the surface of $Ti_{49.1}Ni_{50.9}$ alloy samples obtained using an optical microscope (a-d), SEM (e) and an inverted microscope (dark field) (f) in a solution of 3% NaCl in a coarse-grained state: a) 1 day; b) 3 days; c-d) 15 days; e-f) 30 days*

Investigations of samples after corrosion tests in 5 M HCl on an inverted microscope (Fig. 5, f-g) made it possible to determine the nature of corrosion damage. In the case of a coarse-grained state, deep pits are observed, occupying more than 50% of the sample surface. Studies using scanning electron microscopy also made it possible to estimate the average size and depth of pitting corrosion pits in the coarse-grained state (Fig. 5, e). The average size of the holes is about 100 ± 10 μm. After 1 day of holding the samples in the solution, a large number of corrosion products are observed on the surface, significant damage with the formation of corrosion pits is observed on the 15th day of exposure (Fig. 5, c-d).

Materials Research Proceedings **21** (2022) 229-236 https://doi.org/10.21741/9781644901755-41

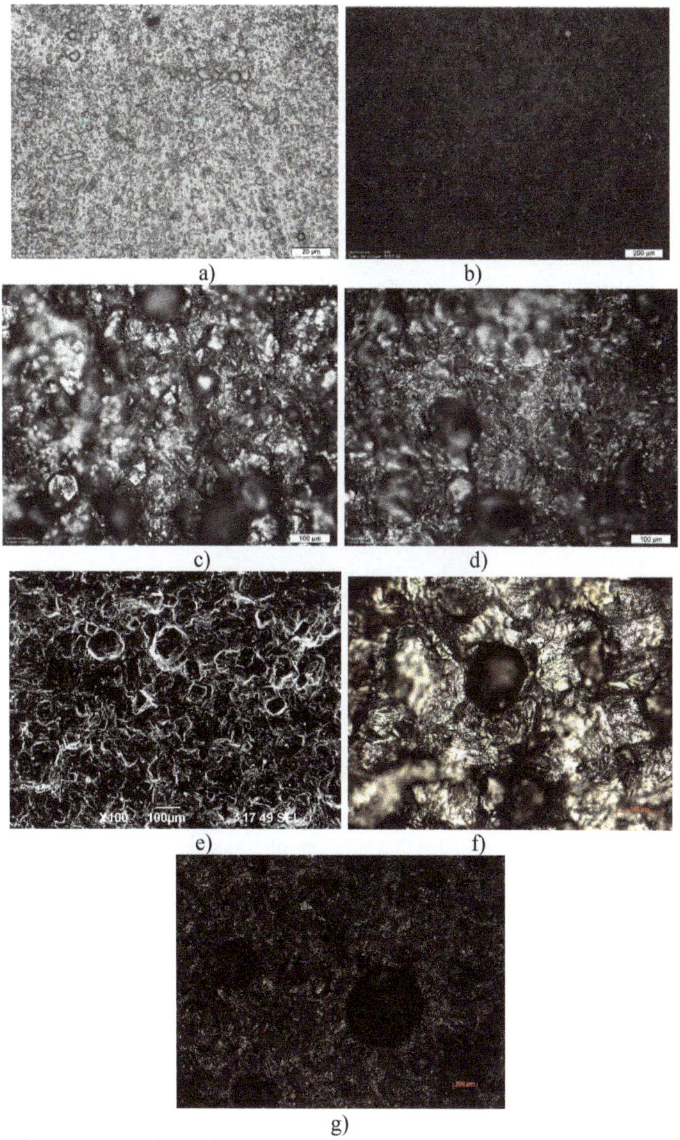

Fig. 5. *Photographs of the surface of $Ti_{49.1}Ni_{50.9}$ alloy samples obtained using an optical microscope (a-d), SEM (e) and an inverted microscope (dark field) (f-g) in a solution of 5 M HCl in a coarse-grained state: a) 1 day; b) 3 days; c-d) 15 days; e-g) 30 days*

Studies in solution showed that after 1 day, significant corrosion damage to the sample is observed (Fig. 6, a); after 3 days, the sample weight decreased by 50%, followed by complete dissolution of the alloy sample in a 5 M H_2SO_4 solution.

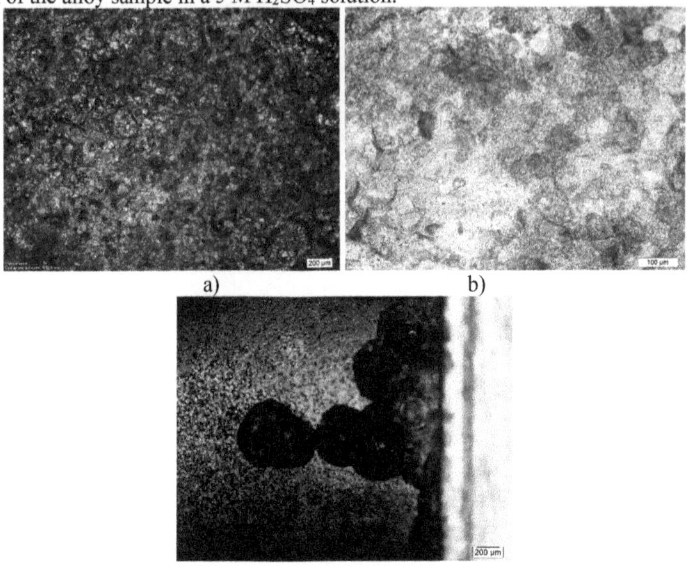

a) b)

c)

Fig. 6. *Photographs of the surface of $Ti_{49.1}Ni_{50.9}$ alloy samples obtained using an optical microscope in a solution of 5 M H_2SO_4 in a coarse-grained state: a, b) 1 day; c) 3 days*

Conclusions

An increase in the concentration of the solution leads to a significant acceleration of corrosion processes in the $Ti_{49.1}Ni_{50.9}$ alloy with a high Ni content, including until the samples are completely dissolved (5 M H_2SO_4 after 3 days of exposure; 5 M HCl is a significant proportion of corrosion damage after 30 days). It was revealed that solutions of 1 M sulfuric and hydrochloric acids after a month's exposure did not change in color and no precipitations were found, while solutions of 5 M hydrochloric and sulfuric acids acquired a violet and then green color, which is due to the predominant release of titanium ions (+4) and nickel (+2). The process of dissolution of the $Ti_{49.1}Ni_{50.9}$ alloy in acid solutions at high concentrations follows the mechanism of pitting, which is confirmed by the photographs of the microstructures presented.

Acknowledgment

This work was supported by the Council for Grants of the President of the Russian Federation for State Support of Young Russian Scientists - Candidates of Science (MK-6202.2021.1.2).

References

[1] K. Otsuka, X. Ren, Prog. Mater. Sci. 50 511–678 (2005). https://doi.org/10.1016/j.pmatsci.2004.10.001

[2] V. Brailovski, S. Prokoshkin, P. Terriault, F. Trochu, Shape Memory Alloys: Fundamental, Modeling and Applications (Ecole de Technologie Superieure, Quebec) 2003.

[3] G.V. Kurdyumov, L.G. Khandros, Sov. Phys. Dokl. 66 211–214 (1949).
[4] J.W. Christian, The Theory of Transformations in Metals and Alloys (Elsevier Science, Oxford) 2002. https://doi.org/10.1016/B978-008044019-4/50022-2
[5] Z.L. Xie, B. Sundqvist, H. Hanninen, J. Pietikainen, Acta Metall. Mater. 41 2283–2290 (1993). https://doi.org/10.1016/0956-7151(93)90310-O
[6] V.V. Rubanik, V.V. Klubovich, V.V. Rubanik Jr., J. Phys. IV 112 249–251 (2003). https://doi.org/10.1051/jp4:2003876
[7] S.P. Belyaev, R.F. Konopleva, I.V. Nazarkin, A.I. Razov, V.L. Solovei, V.A. Chekanov, Phys. Solid State 49 1969–1972 (2007). https://doi.org/10.1134/S1063783407100265
[8] K. Inoue, K. Enami, Y. Yamaguchi, K. Ohoyama, Y. Morii, Y. Matsuoka, K. Inoue, J. Phys. Soc. Jpn. 69 3485–3488 (2000). https://doi.org/10.1143/JPSJ.69.3485
[9] Meisner L.L.,2006 Corrosion properties of TiNi-TiAu quasi-binary cut alloys in biochemical solutions Physics and chemistry of material processing 1 78-84
[10] Ustinskaya T.N.,1987 Composition, electrochemical and protective properties of anode films on the TiNi intermetallide Electrochemistry 23 254-259
[11] Kossy G.G.,1978 Protection of metals 14(6) 662-666. https://doi.org/10.1109/TAES.1978.308691
[12] Stepanova T.P.,1978 Protection of metals 14(2) 169-171
[13] Deryagina O.G., 1980 Electrochemical behavior of anodically oxidized Ni-Ti alloys in sulfate solutions containing chlorine ions Electrochemistry 16(12) 1828-1833
[14] Tan L., 2003 Corrosion and wear – corrosion behavior of NiTi modified by plasma source ion implantation Biomaterials 24 3931-3939. https://doi.org/10.1016/S0142-9612(03)00271-0
[15] Okazaki Y., 1998 Corrosion resistance, mechanical properties, corrosion fatigue strength and cytocompatibility of new Ti alloys without Aland V Biomaterials 19 1197-1215. https://doi.org/10.1016/S0142-9612(97)00235-4
[16] Hofman A., 1996 Classes of materials used in medicine Biomaterials Science Academic Press 37-50. https://doi.org/10.1016/B978-0-08-050014-0.50007-9
[17] Shabalovskaya S.A., 2002 Surface, corrosion and biocompatibility aspects of nitinol as an implant material Bio-Medical materials and Engineering 12 69-109
[18] Liu Chenglong, 2006 In vitro electrochemical corrosion behaviour of functionally graded diamound-like carbon coatings on biomedical Nitinol alloy Thin Solid Films 496 457-462. https://doi.org/10.1016/j.tsf.2005.09.109
[19] Shevchenko N., 2004 Studies of surface modified NiTi alloy Applied Surface Science 235 126-131. https://doi.org/10.1016/j.apsusc.2004.05.273
[20] Vandenkerckhove R., 2004 Corrosion behaviour of a superelastic Ni-Ti alloys Materials Science and Engineering 378 532-536. https://doi.org/10.1016/j.msea.2003.11.072
[21] Denton M., 2005 Corrosion evaluation of wear tested nitinol wire Materials Science and Engineering 25 276-281. https://doi.org/10.1016/j.msec.2004.12.009
[22] Xiao Xu, 2004 Shock synthesis and characterization of nanostructured NITINOL alloy Materials Science and Engineering A 384 194-201. https://doi.org/10.1016/S0921-5093(04)00812-3
[23] Primak O., 2005 Morphological characterization and vitro biocompatibility of a porous nickel-titanium alloy Biomaterials 26 5801-580. https://doi.org/10.1016/j.biomaterials.2005.02.029

Why we Need a Heated Chamber for 3D Printing with 'High Performance' Polymers?

Evgeny Boytsov[1,a], Sergei Blaginin[1,b] and Alexey Sinkov[2,c*]

[1]LLC "VOLTEHNO", 404106, Volzhsky, Prospekt Lenina st., 308k, Volgograd region, Russia

[2]Volzhsky Polytechnic Institute (branch) of VolgSTU, 404121, Volzhsky, Engels st., 42a., Volgograd region, Russia

[a]diamagnetik@inbox.ru, [b]demilano@mail.ru, [c]sinkov73@mail.ru

Keywords: High Performance Polymers, High-Temperature Polymers, 3D Printer, Heated Chamber, Glass Transition Temperature, PEEK, FDM, FFF, Filament, Viscosity, Shrinkage Rate, Warpage, Annealing, Extrusion

Abstract. This paper is dedicated to the features of the processes of 3D printing of polymers with a high melting point such as PEEK, CarbonPEEK, ULTEM, PPSU using FDM (FFF) technology. The results of 3D printing with high-performance polymers using different conditions of the heated chamber are presented. Conclusions about the advantages of a heated chamber are made.

Introduction

Additive manufacturing technologies, in particular FDM or FFF 3D printing technologies, have become widespread and are of great interest in industry, medicine, science and other fields of activity due to their potential to simplify and reduce the cost of complex industries and the attractiveness of specialized research.

FDM technology implies the creation of 3D objects by applying successive layers of material that follow the contours of the digital model (Fig. 1). Typically, thermoplastic polymers supplied in the form of filaments or rods are used as printing materials.

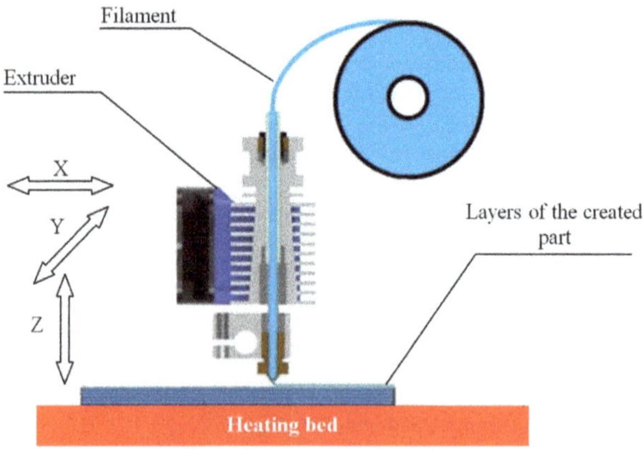

***Fig. 1.** FDM printing scheme. X, Y, Z - directions of movement of the extruder.*

Currently, the manufacture of parts and devices using FDM technologies has reached a new level in connection with the use of high-performance polymers such as PEEK [1], CarbonPEEK [2], ULTEM1010, ULTEM9085, PPSU, etc. as filaments (fig. 2). Products made from such materials have a thermal resistance above 100°C, have good chemical resistance to many aggressive environment, and in terms of strength characteristics are approach to the properties of some metals [3, 4].

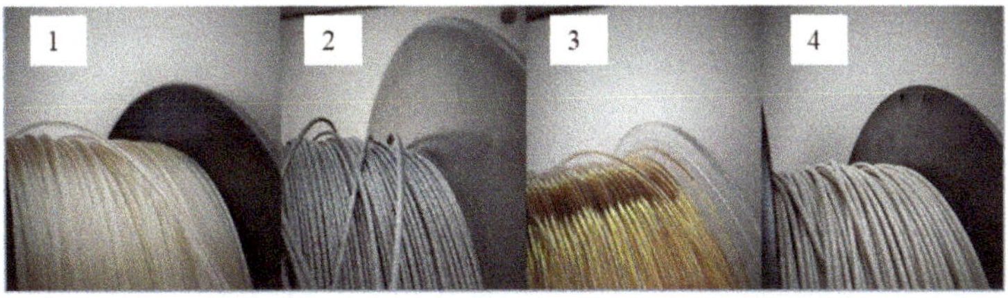

Fig. 2. *Filaments of high-performance polymers: 1 - PPSU; 2 - CarbonPEEK; 3 - ULTEM; 4 – PEEK*

For example, Airbus (a European manufacturer of aviation equipment), under the program for the development of plane of the A-350 type, has mastered the manufacture of more than 3,000 parts from high-performance polymers using FDM technologies. These are elements of the interior of the passenger compartment, fasteners, fittings, air ducts and supporting unloaded elements.

It is known that 3D printing with high-performance polymers is associated with some technological difficulties. To overcome such difficulties, a 3D printer must have certain characteristics. For a better understanding of what characteristics a 3D printer that prints with high-performance polymers should have, we will consider the main aspects of 3D printing technology with such materials.

When using the FDM technology, the filament is first heated above the melting temperature and the polymer passes from a solid to a liquid state. Liquid polymer is extruded through a thin channel - a nozzle. After extrusion, the polymer is cooled in the environment and a reverse transition from liquid to solid occurs. Thus, in the process of 3D printing, there are two phase transformations of the polymer: from solid to liquid and from liquid to solid.

High-performance polymers have a very high melting point, and the extrusion process is carried out at even higher temperatures. For example, PEEK polymer is extruded at temperatures around 450°C. When such a polymer is extruded into the environment at room temperature, a large temperature difference (temperature gradient) is obtained between the plastic and the environment. The high temperature gradient between the liquid polymer and the environment leads to intensive cooling of the surface layers of the product, during which there is a rapid transition from the liquid phase to the solid phase.

During rapid polymerization, the molecules of the material remain in a disordered position. This phenomenon causes increased stresses in the surface layers of the material. On finished products, this manifests itself in the form of delamination or cracks between the printed layers, a strong deviation from the initial dimensions and shape of the virtual model, high surface roughness and other negative effects are observed. As a result, such products have reduced thermal and mechanical properties.

Materials Research Proceedings **21** (2022) 237-246
https://doi.org/10.21741/9781644901755-42

High-performance polymers have increased viscosities resulting in high shrinkage rates on rapid cooling. It also makes it difficult to obtain the desired dimensional tolerances on the manufactured parts.

Thus, the presence of a high temperature difference between the extruded polymer and the environment leads to a number of negative effects.

To overcome or minimize these problems, several functions can be implemented to ensure that high-performance polymers are 3D printed with the best possible quality.

There are currently only a few companies capable of printing high-performance polymers with high precision and high quality. The reason for this lies in the knowledge of the chemical and physical features of these high-performance materials and the ability to translate these features into the designs and settings of 3D printers, allowing to provide the necessary conditions for the production of quality products.

Effect of polymer viscosity

When extruded, the filament melts and becomes a liquid, so viscosity will affect its behavior.

The viscosity of a liquid polymer is the amount of resistance to flow due to internal friction between adjacent layers of liquid sliding against each other.

In a traditional extruder design, the flow of liquid polymer exiting the extruder nozzle is limited by the friction generated on the inner surface of the channel. This leads to a decrease in the flow area of the nozzle channel and periodic clogging of the extruder. In addition, during the passage of the melt inside the nozzle channel, air enters together with the polymer, creating air cavities between the polymer layers. This causes a sharp deterioration in the mechanical properties of the part being created. As the diameter of the extruder nozzle decreases and the print speed increases, the negative effect of polymer viscosity affects print quality more intensively.

To solve the problems arising from the negative influence of polymer viscosity, a new extrusion system was needed. The research department of VOLTEKHNO LLC has developed its own design of the extruder (Fig. 3), which is completed with the 3DLIFE THERMO printer [5].

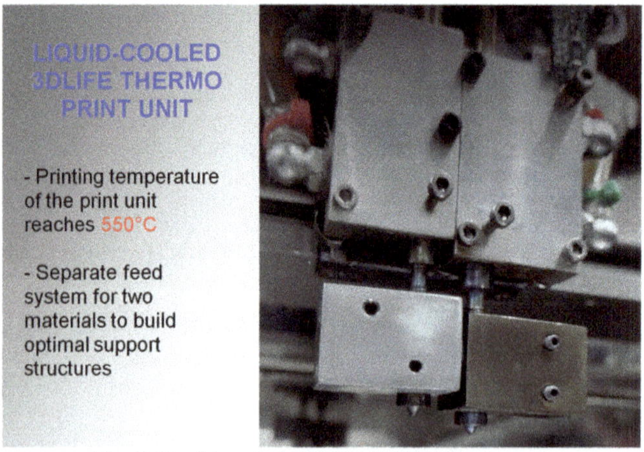

Fig. 3. *Liquid-cooled 3DLIFE THERMO print unit.*

Materials Research Proceedings **21** (2022) 237-246
https://doi.org/10.21741/9781644901755-42

The use of a friction-reducing ceramic inner bore facilitates the extrusion of highly viscous polymers. The special profiling of the inner channel allows for an acceleration of the polymer flow. The narrowing of the extrusion section leads to an increase in the shear stress acting on the polymer melt, which reduces its viscosity. Equipping the extruder with liquid cooling is very important for the thermoregulation of individual parts of the extruder, especially when the printing temperature reaches 450°C.

The technical solutions used in the design of the extruder allow creating an optimized polymer flow, reducing the negative phenomena of shrinkage, ensuring low dimensional tolerances for the finished product and maintaining the chemical-physical properties of the material [6].

Shrinkage rate and warpage of polymers

Shrinkage rate is a property of polymeric materials that determines the accuracy and size of parts, expressed as a percentage. This parameter is defined as the rate of volumetric contraction during the transition of the polymer from the liquid phase of the melt to the solid phase of the cooled state.

When the polymer is extruded into an environment with room temperature (20-30°C), there is a large temperature difference between the melt and the medium, which leads to a high cooling rate of the polymer. The cooling rate affects the processes occurring during the crystallization of the material, namely: the higher the printing temperature, the higher the cooling rate and the higher the shrinkage rate.

The shrinkage rate of polymers depends on the material and varies for most thermo polymers in the range from 0.2 to 1%. Shrinkage of high-temperature polymers can be up to 2%, which causes certain difficulties in printing.

Parts printed with highly shrinkable resins often have a warping effect. The warping effect is manifested in the detachment of plastic layers from the surface of the heated table on which the product is printed. The warping effect is especially pronounced at the edges of the first layer around the perimeter of the part being created. This is due to the fact that when the first layer is printed, the melt is extruded onto the table surface, and upon cooling, the polymer adheres to the table material. Thus, during the phase transition of a liquid into a solid state, adhesion of dissimilar bodies occurs: liquid and solid due to the forces of interfacial adhesion. Further cooling of the polymer is accompanied by shrinkage, which results in increased stresses in the surface layers of the material. When the second polymer layer is applied to the first, similar processes occur, which lead to an increase in surface stresses in the first layer. If the forces of surface stresses in the first layer exceed the forces of interfacial adhesion of the polymer to the table material, then the first layer will detach from the table surface. This leads to the curvature of the lower layers of the product, that is, to the appearance of the warping effect shown in Figure 4. With a further increase in the number of printed layers of the product, the stresses from the curvature of the first layer decrease, the warping effect decreases and the layers are aligned.

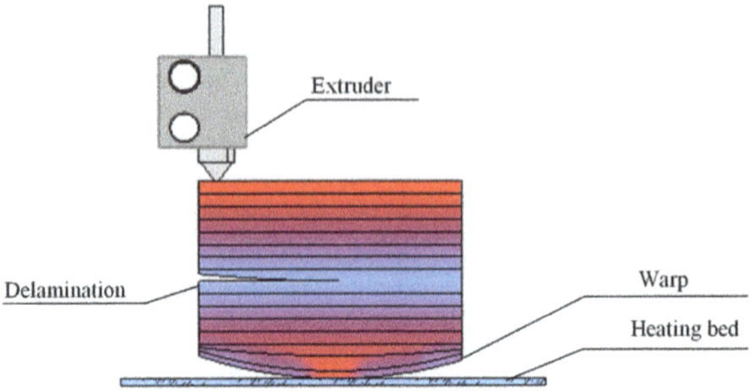

Fig. 4. *Effects of warping and delamination of the product.*

The high degree of polymer shrinkage rate, at which increased stresses arise between the layers of the product, affects not only the appearance of the warping effect. If surface stress forces appear in the product between the polymer layers, which exceed the interfacial adhesion forces of these layers, then the delamination effect will be observed, that is, the layers will detach (detach) from each other. Figure 4 schematically shows the effect of delamination on the surface of a product. Figure 5 shows the delamination effect of a PEEK polymer product.

Fig. 5. *The delamination effect of a PEEK polymer product*

Annealing of polymer products

When the polymer is extruded into an environment below the glass transition temperature (Tg), there is a large temperature difference between the melt and the environment, resulting in a high cooling rate of the polymer.

The high polymerization rate is accompanied by increased shrinkage, causing increased internal stresses between the layers. This leads to the appearance of the effects of warping and delamination and, as a consequence, to significant deviations of the geometric dimensions from the given shape of the product.

Figure 6 shows the main characteristics of the printing processes of high-performance polymers at temperatures below the glass transition temperature.

Materials Research Proceedings **21** (2022) 237-246
https://doi.org/10.21741/9781644901755-42

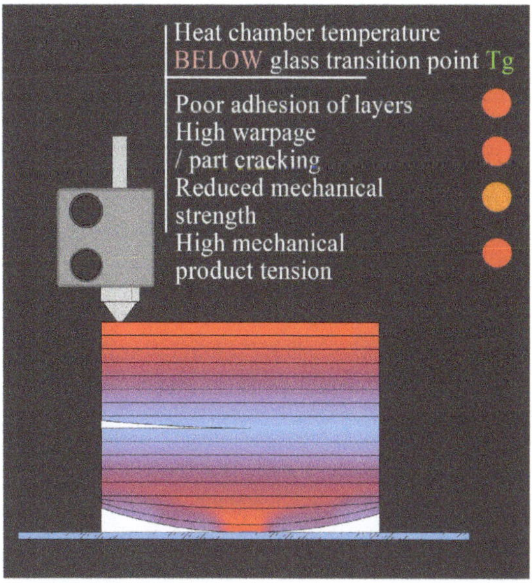

Fig. 6. *Characteristic of the printing process below the glass transition temperature.*

When printing with high-performance polymers at ambient temperatures below the glass transition temperature, parts are not completely crystallized. An example of such a PEEK polymer print is shown in Figure 7.

Fig. 7. *Sample of PEEK polymer 3D printing at ambient temperature below glass transition temperature.*

To improve the quality of products printed by the method described above, thermal annealing is used.

Annealing is a special heat treatment process that optimizes the molecular structure of the polymer and reduces the interlayer stresses that occur during rapid polymerization.

This process is carried out as follows. The product is placed in an oven, the air temperature is raised above the glass transition temperature of the polymer, and held for a certain period of time. The holding time is calculated depending on the minimum wall thickness of the product. This is followed by slow cooling at a rate of about 10°C per hour to ambient temperature.

Figure 8 shows the temperature changing of time during the thermal cycle of annealing PEEK polymer parts [7].

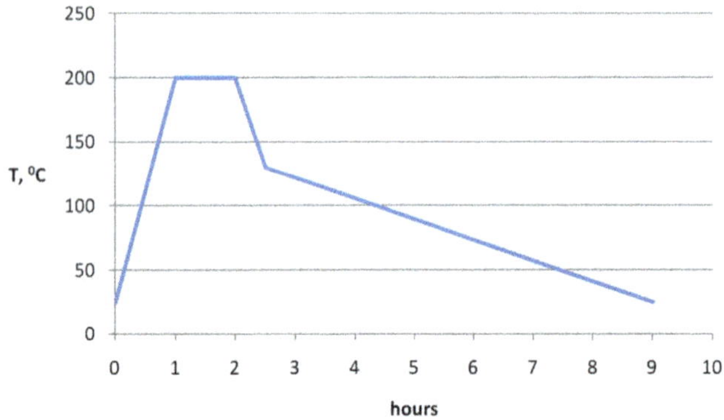

Fig. 8. *Annealing thermal cycle for PEEK parts.*

When using thermal annealing, some disadvantages of this process must be taken into account:
- annealing takes quite a long time and can last more than eight hours, which is not always economically feasible, especially under conditions of mass production;
- in the process of annealing, some change in the size of the product occurs, caused by the rearrangement of polymer grains at the microscopic level;
- heterogeneity in the crystal structure of the product in the presence of walls of different thickness. Thicker walls remain unannealed;
- using thermal annealing, it is impossible to eliminate the distortion of the product shape caused by the effects of warping and delamination.

Heated chamber and what is it for

The use of a heated chamber in the printing process allows you to change the ambient temperature in which the polymer is cooled. That is, in the heated chamber, it is possible to create favorable conditions for the desired cooling rate, depending on the properties of the polymer to be printed. Providing a controlled polymerization process, allows you to control the degree of polymer shrinkage, leads to a decrease in interlayer stresses and the elimination of the causes of the effects of warpage and delamination. In this case, the process of polymer crystallization occurs with the formation of regular crystal lattices, which provides a more uniform structure of the material and the product does not require further thermal annealing.

When comparing the results of printing products from high-performance polymers made without using a heated chamber and with a heated chamber, it is observed that the product made in a heated chamber minimizes deviations of geometric dimensions from a given shape, reduces surface roughness, and products have improved thermal and mechanical properties.

To solve the problems associated with the negative effects of a high degree of shrinkage of high-performance polymers, the research department of VOLTECHNO LLC has developed a heated chamber (Fig. 9), which is used in the design of 3DLIFE THERMO printers. The insulated heated chamber allows you to maintain a constant air temperature up to 250°C. This makes it possible to provide a controlled printing process with high-performance polymers with good repeatability of the geometric shapes of the finished products. The air is heated to the required temperature by heaters located inside the chamber. Uniform heating of the entire volume of the chamber is carried out using a fan built into the chamber, providing convective heat exchange between the heating elements and the air of the chamber.

Fig. 9. *Heated chamber of the 3DLIFE THERMO printer*

It has been found that the optimum ambient temperature maintained in a heated chamber is the glass transition temperature (Tg) of the polymer used.

The Tg parameter is one of the most important thermophysical properties of amorphous polymers. At this temperature, the polymer chains are randomly oriented and can move freely, while the polymer is in structural relaxation and cools smoothly and uniformly. Thus, the strength and dimensional accuracy of 3D printed high performance polymer parts is achieved [8].

Optimum chamber temperatures for some high performance polymers: PPSU - 225 ° C; PEI 1010 - 220 ° C; PEKK / PEEK - 160 ° C.

Figure 10 shows the main characteristics of the printing processes of high-performance polymers at temperatures above the glass transition temperature.

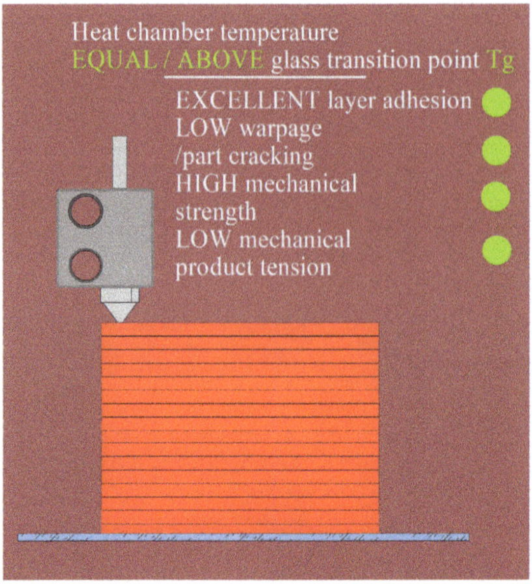

Fig. 10. *Characteristic of the 3D printing process above the glass transition temperature*

When printing with high-performance polymers at an ambient temperature above the glass transition temperature, the parts are completely crystallized. An example 3D print from PEEK is shown in Figure 11.

Fig. 11. *Sample of PEEK polymer 3D printing at ambient temperature above glass transition temperature*

Summary

The main advantages of the heated chamber:

1) a high degree of crystallinity of the product is achieved, which leads to improved chemical, mechanical and thermal characteristics;

2) the uniformity of the crystalline structure of the material is ensured, this leads to the same properties throughout the entire volume of the product and eliminates the negative effects of warping and delamination;

3) there is no need for additional heat treatment processes, such as annealing, which leads to a reduction in the production time of finished products;

4) there is no limitation on the minimum wall thickness due to the lack of heat treatment after printing the product;

5) no deformation due to heat treatment, resulting in better shape tolerances and dimensional accuracy.

References

[1] PEEK: Ensinger extends its hollow rod portfolio. https://www.ensingerplastics.com/en/press-and-news/press-releases/peek-ensinger-extends-its-hollow-rod-portfolio

[2] https://visionminer.com/blogs/articles/carbon-fiber-3d-printing

[3] K. Friedrich, Polymer composites for tribological applications. Institute for Composite Materials (IVW GmbH), Technical University Kaiserslautern, 67663 Kaiserslautern, Erwin Schroedinger Str. 58, Germany. / Advanced Industrial and Engineering Polymer Research 1 (2018), p. 3-39. https://www.sciencedirect.com/science/article/pii/S2542504818300083?via%3Dihub

[4] Polyketone: Compounds to meet mechanical, tribological and chemical demands.

https://www.ensingerplastics.com/en/press-and-news/press-releases/polyketone-compounds-to-meet-mechanical-tribological-and-chemical-demands

[5] E.P. Boytsov, S.I. Blaginin, A.V. Sarazov, A.V. Sinkov, R.N. Titov. Russian Federation patent for useful model No. 189218 "Printing unit for 3D printing with high-temperature polymers using FDM technology" dated 28.01.2019.

[6] A Sviridov, Iu Lopatina, Iu Kurganova. 3D-printed polyether ether ketone samples mechanical properties estimation. Modern Engineering, IOP Publishing, IOP Conf. Series: Materials Science and Engineering 589 (2019) 012021. https://doi.org/10.1088/1757-899X/589/1/012021

[7] https://www.roboze.com/en/resources/peek-3d-printing-what-does-a-3d-printer-need-to-print-peek.html

[8] https://polymerdatabase.com/polymer%20physics/GlassTransition.html

Contemporary Methods for Production of Ultramarine

A.V. Vyboishchik[1,a*] and M.Yu. Popov[1,b]

[1]South Urals State University, 76 Lenin Ave., Chelyabinsk 454080, Russia

[a]alex_vyb@list.ru, [b]pmik0@yandex.ru

Keywords: Ultramarine, Ultramarine Blue, Ultramarine Green, Kaoline, Diatomite, Malleablizing

Abstract. The article describes the features of ultramarine, its modifications, viz. ultramarine blue. The requirements for ultramarine blue grades are observed, the one-stage and two-stage technologies for the production of ultramarine blue are described, the advantages and disadvantages of both methods are listed, new recommendations for the production of ultramarine blue are offered.

Introduction

Ultramarine is a natural pigment applied since the middle ages [1, 2, 3]. Chemically, ultramarine is a sodium aluminosilicate of the chemical composition $(Na_2O \cdot Al_2O_3 \cdot mSiO_2) \cdot x \cdot Na_2S_n$ containing sodium sulphides and polysulphides. This inorganic pigment is found naturally as a lazurite mineral in various mountainous countries extracted from lapis lazuli rocks [4, 5] and, due to the time-consuming extracting process, is very expensive [6].

The applications of ultramarine are as follows: the production of plastics, paints, powder coatings, rubber and thermoplastic elastomers, detergents, cosmetics and soaps, art colours, leather finishes, synthetic fibers, etc.

Ultramarine, depending on customer needs, comes in the following colour modifications: ultramarine blue, ultramarine green, ultramarine purple and ultramarine red, of which ultramarine blue has found the widest application in various industries [7, 8, 9, 10, 11, 12].

Properties of Ultramarine Blue Grades

Blue ultramarine having the composition of m=2.5...3.0; n=2...5, x=2.5...3.0, is produced in the following grades [13, 14] as shown in Table 1.

Table 1. *Basic requirements for ultramarine grades*

Parameter	Ultramarine grade				
	UHK-A	UHK	US	UM-1	UM-2
Mass content of sulphur, no more than [%]	0.02	0.03	0.10	0.6	0.75
Intensity in relation to the standard, no less than [%]	240	200	180	140	120
Ratio of the remainder after wet sieving, no more than [%]	–				
0056K	0.75	0.1	0.1	0.7	10
0045K	–	–	–	–	–
016K		–	–	–	0.03

Materials Research Proceedings **21** (2022) 247-250 https://doi.org/10.21741/9781644901755-43

UHK-A ultramarine grade is applied in art watercolours, UHK grade is applied in all other art colours, the US grade is applied in sugar production, UM-1 and UM-2 grades are applied in paint-and-varnish and other industries. In oxidizing environment, ultramarine is stable up to 300°C, and it does decompose up to 500…600°C, whereas and in carbonic environment it is stable even at 700°C. The quality parameters of ultramarine are significantly affected by the chemical composition of the used raw materials and the production technology.

Methods for the Production of Ultramarine Blue

Ultramarine is yielded from kaoline having the chemical formula $Al_2O_3 \cdot 2SiO_2 \cdot 2H_2O$ [5, 6, 13], and also diatomite, sulphur and sodium carbonate as raw materials by means of heat treatment. All the above-listed raw materials must match the following requirements:

1. All the raw materials must be dry to avoid decomposition of the resulting sodium polysulphides.
2. The soda must not contain bicarbonates.
3. The kaoline and diatomite must not have iron impurities.
4. The sulphur must not contain arsenic.

The ratio of the raw materials used in ultramarine charge is maintained within the following requirements [13, 14] as given in table 2.

Table 2. *The formulation of raw materials in ultramarine charges [%]*

Raw material	Charge #1	Charge #2	Base material
Kaoline	30.3	25.6	96.9
Sulphur	29.4	34.6	99.5
Soda ash	28.9	28.8	97.7
Diatomite	5.5	7.5	86.9
Pitch	3.6	3.5	97.8

Ultramarine is yielded not only from kaodine, but also from amorphous sodium alumosilicate, the mixture of sodium aluminate and silica, the mixture of amorphous alumina and silicon dioxide, as well as from the mixture of aluminuim-containing and siliferous waste after special treatment [15, 16, 17, 18, 19, 20, 21]. For example, while yielding UM-1 and UM-2 ultramarine grades, silicon is substituted with ferrosilicon dust, technical sodium aluminate is substituted with alkaline-etched aluminium products, and while yielding UHK and UHK-A ultramarine grades, amorphous aluminium is substituted with dead aluminuim-containing catalyst.

The heat treatment of ultramarine charge is performed in either the one-stage or the two-stage technology.

According to the one-stage technology, the charge is fed from the fireclay into the porous crucibles, with each crucible containing a specified amount of charge. The crucibles are installed in the furnace in rows, one on top of the other. The roasting process consists of three steps.

The first step proceeds from the beginning of the roasting process during the furnace temperature increase up to 450°C, without air access in the reducing medium. The result of the first step is the obtaining of sodium sulphides and polysulphides.

The second step proceeds during the reaction of polysulphides (obtained at the first step) with kaoline, and during the furnace temperature increasing to 725…780°C, without air access into the reducing medium, and with the attained temperature maintained for several hours. The result of the second step is the obtaining of ultramarine green.

The third step, also known as malleablizing, proceeds during the oxidation of ultramarine green (obtained at the second step), and during the furnace cooling with atmospheric oxygen access, which is maintained for 8…15 days. The result of the third step is the obtaining of ultramarine blue.

The main disadvantage of the one-stage technology is the yielding of a heterogeneous product, which is due to the uneven furnace temperature. Another disadvantage of the one-stage technology is its long duration, which is very time-consuming. The two-stage technology, which is invented recently, obtains ultramarine green and ultramarine blue separately with the overall process duration decreasing to 40…50 hours, which is 2…8 times less time-consuming than the one-stage technology.

As a result of either of the two above-described technologies, only a semi-finished ultramarine product is yielded, which requires further washing from impurities, drying and, finally, grinding.

Besides, other disadvantages of the above-described technology of yielding ultramarine are, firstly, the process stoppability, or discontinuity; secondly, the application of ball mills for grinding raw materials and semi-finished products, thus providing insufficient dispersion of the products; and, thirdly, comparatively low drying temperatures of 150...200°C in drum dryers, thus reducing the process productivity.

In this regard, the introduction of a continuous process for yielding ultramarine with the use of advanced types of equipment, such as bead mills and combined spray dryers, will increase not only the process productivity, but also the final product quality.

Conclusions

The above-stated research allow us to make the following conclusions.

1. Ultramarine is a pigment used in various industries.
2. Ultramarine blue, as the most applicable ultramarine pigment, is traditionally yielded by the one-stage technology consisting of heating, reducing and malleablizing steps.
3. The disadvantage of the one-stage technology is the long process duration equal to 8…15 days due to the malleablizing step.
4. The invented two-stage technology separates the yielding processes of ultramarine green and ultramarine blue, and reduces the overall process duration to 40…50 hours.

References

[1] Gerhard Pfaff, Ultramarine pigments, Physical Sciences Reviews, 2021. https://doi.org/10.1515/psr-2020-0201

[2] Pawel Rejmak, Structural, Optical and Magnetic Properties of Ultramarine Pigments: A DFT Insight, The Journal of Physical Chemistry C. 122(51) (2018). https://doi.org/10.1021/acs.jpcc.8b09856

[3] Stanislaw Kowalak, Aldona Jankowska, Sabina Ł, Preparation of various color ultramarine from zeolite A under environment-friendly conditions, Catalysis Today. 90(1) (2004) 167-172. https://doi.org/10.1016/j.cattod.2004.04.023

[4] Robert Coenraads, Claudio Canut de Bon, Lapis Lazuli from the Coquimbo Region, Chile, Gems and Gemology. 36(1) (2000) 28-41. https://doi.org/10.5741/GEMS.36.1.28

[5] Jose Pedro Sancho, Oscar Restrepo, P. García, J. Ayala, B. Fernández, Luis Felipe Verdeja, Ultramarine blue from Asturian "hard" kaolins, Applied Clay Science. 41(3-4) (2008) 133. https://doi.org/10.1016/j.clay.2007.10.005

[6] Vahideh Tajer Kajinebaf, Fereshteh Rezaeian, Masoud Rajabi, Saeid Baghshahi, Replacing nano-clay for kaolin in ultramarine pigments, Pigment and Resin Technology. 43(1) (2013). https://doi.org/10.1108/PRT-03-2013-0020
[7] N. Gobeltz, A. Demortier, J.-P. Lelieur, A. Lorriaux, C. Duhayon, Presence of iron(III) impurities in ultramarine green pigments, New Journal of Chemistry. 20(1) (1996) 19-22.
[8] Eleanor Cato, Antonella Rossi, Nadim C Scherrer, Ester S.B Ferreira, An XPS study into sulphur speciation in blue and green ultramarine, Journal of Cultural Heritage. (2017). https://doi.org/10.1016/j.culher.2017.09.005
[9] Costanza Miliani, Alessia Daveri, Bruno Brunetti, Antonio Sgamellotti, CO2 entrapment in natural ultramarine blue, Chemical Physics Letters. 466(s 4–6) (2008) 148–15. https://doi.org/10.1016/j.cplett.2008.10.038
[10] Eleanor Cato, Nadim C Scherrer, Ester S.B Ferreira, Raman mapping of the S 3 − chromophore in degraded ultramarine blue paints, Journal of Raman Spectroscopy. 48(11) (2017). https://doi.org/10.1002/jrs.5256
[11] Lucy Goncalves, D. Moronta, Freddy Ocanto, Carlos Linares, Synthesis of ultramarine blue type pigments using several raw materials, Revista de la Facultad de Ingenieria. 25(1) (2010) 25-31.
[12] Dafna Arieli, David E. W. Vaughan, D Goldfarb, New Synthesis and Insight into the Structure of Blue Ultramarine Pigments, Journal of the American Chemical Society. 126(18) (2004) 5776-88. https://doi.org/10.1021/ja0320121
[13] A. I. Shereshevsky, T. P. Unanjanc, G. Ya. Baharovsky, *Himicheskie tovary* (*Chemical Products*) (Goshimizdat, Moscow, 1961)
[14] O. V. Orlova, T. N. Fomicheva, *Tehnologiya lakov i krasok* (*Technology of Varnishes and Paints*) (Himiya, Moscow, 1990)
[15] A.U. Rahman, F. Khan, M. Riaz, A. Latif, Synthesis of blue pigment from kaolin, Pakistan Journal of Scientific and Industrial Research. 52(1) (2009) 15-17.
[16] Ashok Borhade, G. Dholi Arun, G. Wackchaure Sanjay, R. Tope Dipak, Synthesis of intense blue ultramarine from coal fly ash as an aluminosilicate source under environment- Friendly conditions, Research Journal of Chemistry and Environment. 17(1) (2013) 14-20.
[17] Ashok Borhade, Tushar Kshirsagar, Arun G. Dholi, Novel synthesis of ultramarine blue from waste coal fly ash via thiocyanate aluminosilicate sodalite, Journal of Sulfur Chemistry. (2016). https://doi.org/10.1080/17415993.2016.1173215
[18] Yin-Hsiu Hsiao, yu-How Shen, Dah-Tong Ray, Synthesis of Ultramarine from Reservoir Silts, Minerals. 7(5) (2017). https://doi.org/10.3390/min7050069
[19] Haibo Wang, Shan Zhang, Hu Songyuan, Zhiwei Zhen, Mario A Gomez, Shuhua Yao, A systematic study of the synthesis conditions of blue and green ultramarine pigments via the reclamation of the industrial zeolite wastes and agricultural rice husks, Environmental Science and Pollution Research. 27(2–3) (2020). https://doi.org/10.1007/s11356-020-07624-8
[20] D. G. Booth, S. E. Dann, Mark T. Weller, The effect of the cation composition on the synthesis and properties of ultramarine blue, Dyes and Pigments. 58(1) (2003) 73-82. https://doi.org/10.1016/S0143-7208(03)00037-8
[21] Andreas A. Landman, Danita de Waal, Fly ash as a potential starting reagent for the synthesis of ultramarine blue, Materials Research Bulletin. 39(4-5) (2004) 655-667. https://doi.org/10.1016/j.materresbull.2003.12.002

Injection Molding Technology

Dmitry A. Kurasov

Kurgan State University, Kurgan region, Kurgan, 640020, Russia

naukka@mail.ru

Keywords: Injection Molding, Injection Molding Automation, Mold, Plastic, Metal

Abstract. The injection molding process is one of the most efficient and economical casting processes. The process is becoming increasingly common in various industries in large-scale and mass production of castings. It should be noted that by having great advantages over other methods of obtaining high-quality castings of higher accuracy, injection molding makes it possible to bring the dimensions of the castings as close as possible to the dimensions of the finished parts.

Introduction

The article discusses the injection molding process, the general presentation and its features. The main types of this technology and its classification are considered. Conclusions on positive and negative aspects of each technology are formed. Special attention is paid to the capabilities of mathematical and computer modeling, which allows to ensure control and optimization of the parameters of the injection molding process.

Principle of Injection Molding Process Implementation

The injection molding process is one of the most efficient and economical casting processes. The process is becoming increasingly common in various industries in large-scale and mass production of castings [1-3]. It should be noted that by having great advantages over other methods of obtaining high-quality castings of higher accuracy, injection molding makes it possible to bring the dimensions of the castings as close as possible to the dimensions of the finished parts.

Injection molding is a very difficult process to control, since a number of manufacturing factors constantly change in the process, on which the quality of the resulting product depends, such as the temperature ratio or injection rate of molten metal, the accuracy of the manufacture of the matrix and its lubrication, and many other dynamically changing factors. In conventional manufacturing, success depends on the experience and qualifications of specialists, sometimes parts require secondary processing to eliminate defects and ensure their compliance with the highest quality standards.

The classic injection molding machine is shown in Fig. 1.

Injection molding is a relatively old technology, it has been used since the late 1800s. Injection molding was first used by G. Brouss in 1838 in the manufacture of a letter depicting letters for newspaper printing machines. In 1839, the first patent was taken for a piston machine for pouring metal under pressure. In mechanical engineering, injection molding began to be used in 1849 for the production of small parts from tin-lead alloys. The V Sturgis machine used for these purposes had a manual piston drive, with the help of which a pressure of 100-150 Pa was created in the pressing chamber located inside the crucible with molten metal.

In injection molding machines, a huge screw (screw) is installed, which directs molten plastic into the mold under high pressure. This screw transmission method was invented in 1946, and is still used. Injection molding machines are, of course, not the same as modern high-tech 3D printing machines [4]. There is nothing innovative in them, but the use of injection molding is a prerequisite for creating most new innovative products. The mold consists of two halves (die and punch), which, when closed, form a cavity in the form of a desired part. Hot liquid plastic is poured into it under high pressure. High pressure is required for the molten plastic to fill each portion in the mold cavity. When the plastic cools, the two mold halves are moved apart, and the finished part is removed from them.

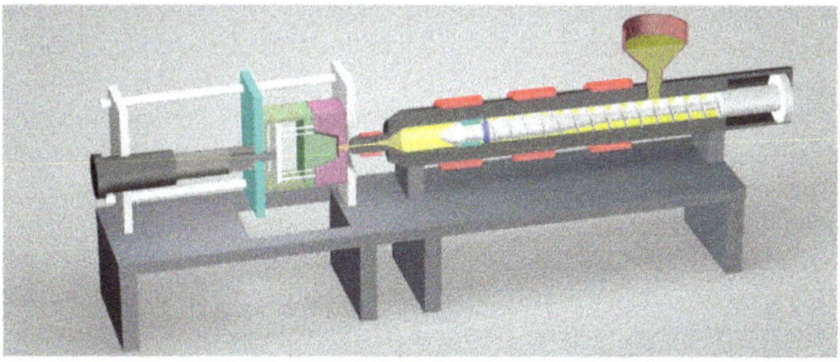

Fig. 1. *Machine of Injection Molding*

Developing the design and design of the case for mass production is a rather difficult task, and the cost of the molds themselves is in the tens of thousands of dollars. At the same time, injection molding remains one of the most popular technologies, because only it allows you to produce millions of identical parts at an incredibly low price per piece.

The technology of injection molding and high pressure casting is described in the following works [5-7].

Advantages and disadvantages of injection molding technology

The advantages of the injection molding method include the possibility of obtaining complex thin-walled castings, which are quite difficult to obtain by other casting methods; high physical and mechanical properties of castings due to specific conditions of casting formation; strength characteristics of parts as compared to casting into sand molds increase by 15-20%; obtaining high-quality castings with low surface roughness and high dimensional accuracy; high productivity of the process due to minimum casting cycle duration, complex mechanization and automation of the production process; low cost of casting due to reduced labour input for manufacturing parts; ensuring favorable sanitary and hygienic working conditions of workers.

The disadvantages of the injection molding method include: the high cost of molds, the complexity and duration of their manufacture; low resistance of molds, especially at casting of alloys with high melting point (steel, cast iron, copper alloys); difficulty of casting with complex cavities and undercuts; unsuitability of the metal mold, which contributes to the appearance of stresses in castings during cooling; therefore, injection molding alloys must have a narrow crystallization interval, high fluid flow, and sufficient strength and ductility.

As a rule, in Russia, the raw material for the production of cast parts is scrap metal of non-ferrous metals or "churks" of aluminum, or other non-ferrous metal, which are bought by the relevant smelters of this metal (more often scrap metal). The raw material is processed into a powder - a finished raw material for the casting process.

The injection molding process includes:

1) mixing metal powders with polymers such as wax and a polypropylene binder to form a mixture;

2) injecting the liquid material into the mold by casting machines;

3) the molded part or "raw part" is cooled and pushed out of the mold;

4) removal of the binder by means of solvents, catalytic method or combination of methods.

The resulting part is brittle and porous (40 percent of the volume of "air"). The finishing processes of parts after casting vary greatly depending on what the final product is required. The most common option: the part is transferred to the agglomerate furnace in a protective atmosphere at a temperature close to the melting point of the material. To seal the particles using capillary forces. The high diffusion rate leads to shrinkage and thickening of the material.

In addition, the difference in injection molding processes is due to the variety of alloys from which castings are made. Along with zinc, magnesium and aluminum alloys, widely used in production, alloys of copper, iron, titanium, heat-resistant alloys are increasingly being used. Special conditions for filling the mold with melt, which include evacuation and oxygen blowing, are also introduced into the technological processes.

There are several different injection molding technologies. The variety of injection molding processes is primarily associated with different types of pressing chambers and the arrangement of the pressing mechanism relative to the mold in the machine. The most common technological processes are carried out on machines with hot and cold horizontal and vertical pressing chambers. Machines with a hot horizontal pressing chamber and machines of vertical layout are not widely used.

Injection molding is divided into high pressure casting (HPDC), injection molding (LPDC), injection molding, gravitational casting (GDC).

High-pressure casting (HPDC) is divided into two types: cold-cast or hot-cast. Almost most often, the third version of the heated form is also used: (warm). In this method, the desired temperature is maintained by means of thermoregulators which are connected directly to the mold.

HPDC is an important process in the production of large volumes and inexpensive components for automotive, household appliances and electronics.

Advantages and disadvantages of injection molding technology

Digital manufacturing is the concept of technological preparation of production in a single digital environment using the tools of planning, verification and modeling of production processes, the so-called Industry 4.0 [8-9]. The CPU involves a single information space and end-to-end automation of all production processes, which is possible due to the translation of all information about the product, production processes and operation into digital form, that is, a parallel product creation chain is created, in other words, a digital twin of the enterprise and the product is created.

For example, using specialized CAE systems (AFS, SIMTEC, ProCAST, NovaFlow, PowerCAST, CastCAE) [10] allows you to visualize the injection molding process. Due to this, it is possible to identify various defects and deformations, residual stresses, determine volumetric and linear shrinkage, optimize casting and solidification modes, and, based on the results of the visualization, make adjustments at the stage of designing the casting design and process, which will significantly reduce the percentage of scrap, and therefore, there will be significant savings in metal, working time and electricity.

Automation of injection molding [11] provides increased resistance of mold parts and pressing chamber due to stable thermal conditions and regular lubrication of working and friction surfaces, improving surface quality and density of castings by regulating and stabilizing the thermal condition and lubrication of the mold and the pressing chamber; reducing losses due to metal burning, increasing productivity due to combining several operations in a casting cycle; reduced costs of muscular energy of the foundry and increased safety of its operation.

Automation of injection molding is a multifaceted and difficult task. Depending on the specific production conditions, the quality of castings is managed at three different levels. In order to ideally follow the given process and the corresponding quality of the part, it is necessary to simultaneously monitor all three levels of control of the injection molding process with an instant reaction when detecting deviations taking into account the influence of the parameter of one level on another.

Conclusions

The most accurate machine-building workpieces are obtained by the injection molding method, its main advantages are the possibility of obtaining workpieces with minimum allowances for mechanical treatment or without it and minimum roughness of unprocessed surfaces, ensuring high productivity and low labor input of making parts, mainly from alloys based on zinc and aluminum. The main parameters affecting the filling and casting process are: the pressure on the metal during filling and underpressure, the pressing speed, the design of the runner-ventilation system, the temperature of the poured alloy, lubrication and vacuumizing modes.

The presence of the digital twin of the casting section will allow real-time monitoring of readings of sensors whose data are in a single information space from the casting machine installed in the pressing chamber, the mold, on manipulators designed to pour metal and extract the casting. This will allow an analysis of the injection molding process and quality control of castings, which will provide an opportunity to carry out an automated correction of the casting process modes without stopping production.

References

[1] T. Kovalyova, E. Eremin, S. Arinova, I. Medvedeva, A. Dostayeva, Enhancing surface roughness of castings when sand-resign mold casting, Metallurgy. 1-2 (2017) 135-138.

[2] N.V. Nemchinova, A.A. Tyutrin, Metallographic study of samples of aluminum rhondoles, Basic research. 3 (2015) 124-128.

[3] A.N. Romashkin, V.S. Oak, I.A. Ivanov, S.I. Markov, A.N. Malginov, D.S. Tolstoy, Development of a through technological process for the production of blanks for mechanical engineering based on compute-tert modeling, Metallurg. 9 (2014) 109-117.

[4] Additive technologies: 3D printing capabilities and perspectives, Pro-work. 2 (2017) 38-44.

[5] G. Singh, A. Verma, A. Verma, A Brief Review on injection moulding manufacturing process, Materials Today: Proceedings. 4 (2017) 1423-1433. https://doi.org/10.1016/j.matpr.2017.01.164

[6] B. Pratap, R.K. Gupta, A. Yadav, M. Nag, Plastic injection molding and its process parameters, AIP Conference Proceedings. 2273 (2020) 050053. https://doi.org/10.1063/5.0024291

[7] S. Tripathia, C. Mittermayrb, D. Muhra, H. Jodlbauera, Large scale predictability analysis of process variables from injection molding machines, Procedia Computer Science. 180 (2021) 545-560. https://doi.org/10.1016/j.procs.2021.01.274

[8] D.A. Kurasov, Computer-aided manufacturing: Industry 4.0, IOP Conference Series: Materials Science and Engineering. 1047 (2021) 012153. https://doi.org/10.1088/1757-899X/1047/1/012153

[9] D.A. Kurasov, Digital technologies Industry 4.0, CEUR Workshop Proceedings. 2843 (2021).

[10] CAD & Graphics log [Electronic Resource]. Access mode: http://sapr.ru/, swb.

[11] N.A. Spirin, V.V. Lavrov, Information technologies in metallurgy. (2004) 495.

Modern Trends in Manufacturing Technologies and Equipment Materials Research Forum LLC
Materials Research Proceedings **21** (2022) 255-258 https://doi.org/10.21741/9781644901755-45

Elastomeric Composites Surface Structure Study by Scanning Electron and Atomic Force Microscopy

Yu. V. Kornev[1,a], H.H. Valiev[1,b*], A.N. Vlasov[1,c], Yu. N. Karnet[1,d] and N.A. Semenov[1,e]

[1]Institute applied mechanics of the Russian Academy Sciences, Leningradskiy prospect 7, Moscow, Russia

[a]yurikornev@mail.ru, [b]hhvlv@mail.ru, [c]iam@iam.ras.ru, [d]iam@iam.ras.ru, [e]semenov.n@iam-ras.ru

Keywords: Elastomeric Composites, Rice Husk Fillers, Scanning Electron and Atomic Force Microscopy, Surface Structure, Elastic Strength Properties

Abstract. New elastomeric composites were synthesized with filler particles from rice husk products both in micro and sub-micron sizes. The obtained composites surface structure was investigated by scanning electron and atomic force microscopy. The application as a prospective strengthening sub-micron sizes filler for the composite complexes creation with high elastic and strength properties have been shown.

Introduction

When creating elastomeric composites, structures and products made from such materials, considerable attention is paid to environmental issues, as well as to improving economic efficiency and energy efficiency in its production. The current interest in elastomeric composites is the problem to find effective fillers [1-6]. In this work we have investigated the surface structure of elastomeric composites based on industrial synthetic butadiene styrene rubber SBR 1500 with filler particles from rice husk products [7-8], both in micro and sub-micron sizes. These experiments are important to understand the structure-property relationships for polymer composites and are taken into account in the mathematical models creation for predicting these materials mechanical properties [9-10].

Materials and methods

Two type composites were prepared: samples 1 with micro dispersed fillers and samples 2 with sub-microdispersed fillers. High-dispersed filler particles were obtained from the micro dispersed feed using a PM100 planetary ball mill (Retsch, Germany). The composites main components were: rubber SBR 1500 -100 wt %, fillers-65 wt % (filling 30% by volume). The obtained composites surface structure was investigated by JSM-6700F-field emission scanning electron microscope with JED-2300F energy dispersion microanalyzer (Jeol, Japan). The tests were carried out in high vacuum conditions. Platinum sputtering 50 nm thickness was preliminarily applied on the samples surface. The sample surfaces were examined using an atomic force microscope (AFM) easyScan (Nanosurf, Switzerland) operating in half-contact mode in air at room temperature. The phase-contrast mode was additionally used in the AFM. The AFM was protected from external excitations using a TS-150 dynamic anti-vibration stage (Fabrik am Weiher, Switzerland).The vulcanizes elastic-strength properties were tested on a UTS-10 (Zwick Roell, Germany) tensile machine.

Results and discussions

The elastomeric composites surface structure SEM image for samples with micro- or sub-microdispersed fillers is shown in Fig. 1 a, b. The filler particles (light areas) in the elastomeric matrix (dark areas) are visualized in these images. The determined sizes filler agglomerates in the rubber matrix with SPIP program (Image Metrology, Denmark) for samples 1 are in the range from 200 nm to 10 mkm, distances between fillers about 500 nm; for samples 2 – sizes from 30 nm to 1 mkm, distances - about 100 nm. The surface structure AFM images of these industrial synthetic butadiene styrene rubber SBR 1500 composites with a micro- and nanostructured amorphous silicon dioxide filler are shown in Fig. 2 a, b (left - topography, right - phase contrast). In these images also the filler particles are visualized as a light area in the elastomeric matrix dark areas. The AFM image data analysis with respect to the filler sizes are in agreement with the SEM data.

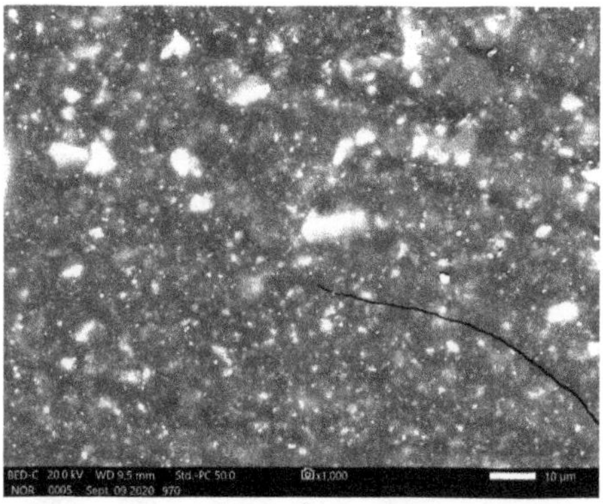

a)

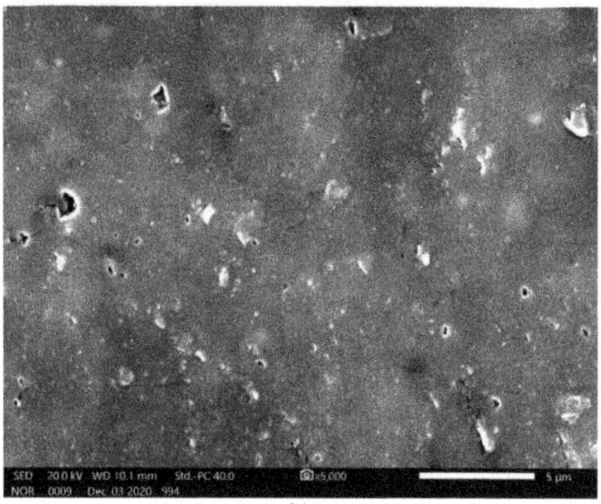

b)

***Fig. 1**. SEM images of elastomeric composites with different fillers: a) microdispersed amorphous silicon dioxide, b) sub-microdispersed amorphous silicon dioxide.*

Materials Research Proceedings **21** (2022) 255-258 https://doi.org/10.21741/9781644901755-45

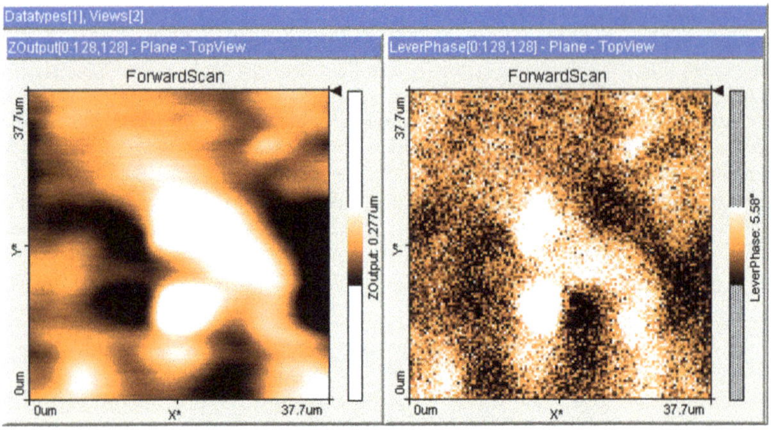

a)

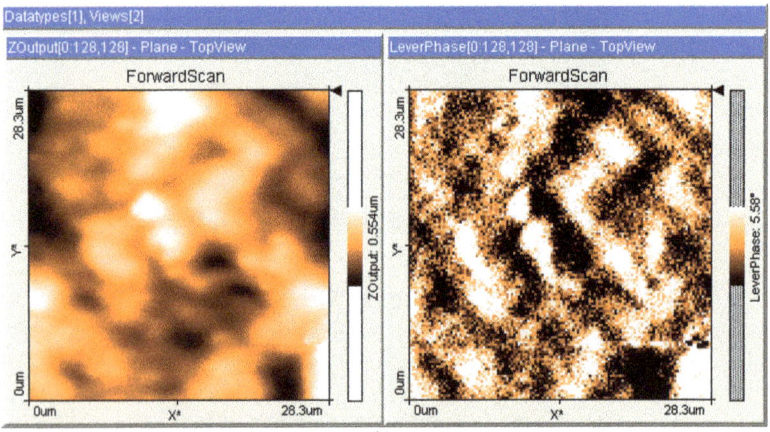

b)

Fig. 2. *AFM images of the surface structure on industrial synthetic butadiene styrene rubber SBR 1500 filled with:*
a) microdispersed amorphous silicon dioxide. Scans 37.7 x37.7 μm;
b) sub-microdispersed amorphous silicon dioxide. Scans are 28.3 x28.3 μm.
(In each image on the left is topography, on the right-phase contrast).

The data analysis images presented in Figs. 1-2 show that when the filler particles size decreases, there is a noticeable decrease in the inclusions size (particle agglomerates), which indicates a noticeably better filler particles distribution in the elastomeric matrix. For example, sample 2 filled with sub-micron particles shows a decrease in the agglomerate average size to about 30 nm from the initial 200 nm obtained for sample 1 filled with micro filler particles. As follow from our strain – stress measurements on these composites performed on UTS-10 tensile machine this leads to a significant improvement in the elastic strength properties in sample 2 compared to sample 1. Thus, sample 2 shows a 5-fold increase in strength compared to sample 1, and 1.5-fold increase in tension at 300% elongation.

Conclusion

The synthesized elastomeric composites study by scanning electron and atomic force electron microscopy shows a noticeable influence the surface structure parameters of the fillers from rice husk on the complex mechanical properties. The data obtained on the used fillers distribution features in the elastomeric matrices correlate with the measured physical and mechanical properties of the synthesized composites. The efficiency of directional control over the initial particles size in amorphous silicon dioxide derived from renewable raw materials (rice husks) and their use as promising strengthening filler for creating composites with reinforced mechanical properties have been shown.

Acknowledgements

This work has been carried out on the basis of the State Assignment for the period of 2019-2021 in the IPRIM RAS. State Registration Number: AAA-A19-119012290177-0.

References

[1] Matthews F.L., Rawlings R.D. Composite Materials: Engineering and Science. N. Y., Woodhead Publishing, 1999.-470 p.

[2] Kornev A.E., Bukanov A.M., Sheverdyayev O.N. Technology of elastomeric materials. Moscow: NPPA "Istek", 2009.-504 p.

[3] Farzana Hussain, Mehdi Hojjati, Masami Okamoto, Russell E. Gorga Review article: Polymer matrix Nanocomposites, Processing, Manufacturing, and Application // Journal of Composite Materials. 2006. V. 40. P. 1511. https://doi.org/10.1177/0021998306067321

[4] Yanovsky Yu.G., Grigoryev F.V., Nikitina E.A., Vlasov A.N., Karnet Yu.N. Nanomechanical properties of polymer composite nanoclusters // Physical Mesomechanics. 2008. V. 11. № 5-6. P. 247-259. https://doi.org/10.1016/j.physme.2008.11.005

[5] Yanovsky Yu.G. Nanomechanics and strength of composite materials, Moscow: IPRIM RAS, 2008, -179 p.

[6] Vlasov A.N., Volkov-Bogorodsky D.B., Yanovsky Y.G., Kornev Y.V., Karnet Y.N. Mechanical properties estimation in hyperelastic composites with small disperse fillers additives // Mechanics and nanomechanics of structurally complex heterogeneous media. Successes, problems and prospects. 2015. P.219-226.

[7] Bakar R.A., Yahya R., Gan S.N. Production of high purity amorphous silica from rice husk //Procedia Chem.-2016.-V.19.-P.189-195. https://doi.org/10.1016/j.proche.2016.03.092

[8] Liu X., Chen X., Yang L.,Chen H., Tian Y., Wangn Z. A review on recent advances in the comprehensive application of rice husk ash //Res. Chem. Intermed.-2016.-V.42.-P.893-913. https://doi.org/10.1007/s11164-015-2061-y

[9] Kornev Y.V., Semenov P.E., Semenov N.A., Solyaev Y.O., Kh. Valiev Kh. Mechanical properties of elastomeric composites reinforced by particles of mineral fillers based on silicon oxide and carbon // Collection: XII All-Russian Congress on Fundamental Problems of Theoretical and Applied Mechanics - Proceedings. In 4 volumes. 2019. P. 1433-1435.

[10] Valiev H.H., Vlasov A.N., Kornev Yu.V. et al. Scanning probe microscopy of elastomers with mineral fillers //Composites from Renewable and Sustainable Materials.-London: Intechopen, 2019.-P. 39-50.

Oscillatory Changes in Structural Characteristics in Multicomponent Ceramics, YBCO

Boris Kodess[1,2,a*] and Pavel Kodess[2,b]

[1]VNIIMS, 46 Ozernaya Str, 119361, Moscow, Russia

[2]ICS&E, 919 Telluride St., 80011 Aurora, CO, USA

[a]kodess@mail.ru, [b]icse96co@gmail.com

Keywords: Multicomponent Ceramics, Yttrium Barium Cuprate, YBCO, X-Ray Irradiation, Long-Time Series, Active Oxygen, Breathing Modes of Behavior, Compression and Expansion Waves, Macrostresses, Complexity, Nonlinear Dynamics, Self-Organization Process

Abstract. The structural characteristics of samples of a four-component superconducting material (YBCO) after exposure to X-ray irradiation during a long time are investigated. The effect of X-ray beam processing on angular positions (corresponding parameters of the crystal lattice) and the width of Bragg reflections is established. The phenomenon of oscillatory behavior in the unit cell dimension with long-time irradiation is found. The analysis of the profiles of reflection also demonstrates the presence of reversible changes phase composition with the exposure time. The observed phenomena reflect the presence of a nontrivial and specific process of compression and expansion of the unit cell due to the accumulation and then disengagement outside of ionized oxygen, which is formed under such irradiation exposure on the surface of the samples.

Introduction

Irradiation processing is widely used for alloying. This type of impact on materials needs to be considered in some applications. For example, after the exposure to neutron irradiation of substances in which there are covalent chemical bonds, there happen either a change in the unit cell dimension [1-2], or even transformation of part of the volume into another element.

There is a class of substances in which, along with rigid covalent and ionic types of chemical bonds, there also exist weaker chemical bonds. We observed the impact of soft X-ray irradiation on the structural characteristics of the known material DKDP (potassium dihydrogen phosphate) for the formation of biocrystals and the used in laser devices [3]. We also [4-5] studied the effect of maturing in the long-term X-ray exposure on the well-known superconducting four-component ceramic material YBCO (YBa2Cu3O7-x, yttrium barium cuprate).

The distinctive feature of this substance is that oxygen, when chemically bound with these metals - Y, Ba, Cu - exist in three different charge states. These include oxygen bound with copper atoms in unit cell and which is in a nearly neutral state. Also, in [4] the changes were observed after a long maturing of the samples. Note that breathing modes of behavior of atoms are inherent, and most likely observed, for large macromolecules in bioactive substances. This phenomenon is unique for inorganic materials.

In this article, we present the results from a new time-series experiment for YBCO samples, when measurements were carried out continuously, and discuss the possible mechanism of the observed oscillatory processes in it.

Samples and procedures

YBCO samples were taken from a batch prepared as certified standard materials of composition and diffraction properties (GSO PRF-23) and had the composition YBa2Cu3O6.93. The superconducting transition temperature, Tc = 91,5 K, has been maintained for over 30 years since the creation of this batch. Also during this period, the characteristics of the full diffraction pattern were repeatedly measured. The calculations of the structural characteristics, carried out using the Rietveld method, confirmed the invariability of the composition of these samples.

More detailed continuous measurements of the characteristics of the diffraction pattern during the X-ray exposure on the surface of the samples were carried out on the DRON-4 diffractometer (CuKα-radiation) and followed by additional measurements having much softer VKα-radiation and CrKα-radiation and then much harder MoKα-radiation. The oscillating changes in the angular positions and the shape of the profiles were revealed; however, after some time in the course of the measurements, the angular position of the reflections, corresponding to the change in the crystal lattice parameter, began to return to the initial state. A typical picture is shown in Fig. 1.

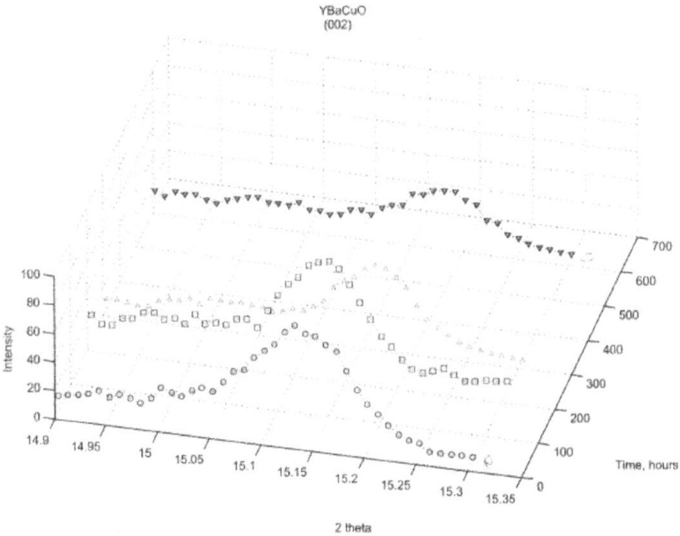

Fig. 1. An example of changes in a diffraction reflection with Miller indices (002) at different time intervals of observation.

To analyze the profiles the «Profan» program was used. The results showed that after the exposure to the X-ray beam, they can be described as a set of several profiles and in terms of the appearance of reflections corresponding to compositions with higher oxygen content. For softer VKα- and CrKα-radiations, the changes started faster and were more pronounced, and the opposite picture was observed under MoKα-radiation.

Discussion

The structural characteristics and properties of ceramic and single-crystal YBCO samples with different oxygen contents were studied in great detail by various methods [6]. When studying the

Materials Research Proceedings 21 (2022) 259-263 https://doi.org/10.21741/9781644901755-46

effect of laser irradiation of YBCO films, a change in the relief was found due to growth defects which are the results of relaxations of the accumulating strains [7].

The investigation of the photoresponse of YBCO films [8] showed that the relaxation of excited quasiparticles can occur owing to localized states near the Fermi level. The mechanism of similar photoinduced reactions based on the results of X-ray crystallographic and spectroscopic data was considered for proteins [9] and other substances [15].

The X-rays with different wavelengths, applied by us, gave the penetration into the analyzed sample volume of different depths. The obtained results indicate that the observed phenomenon occurs mostly in the near-surface regions of the samples. The appearance of additional reflections, reflecting higher oxygen content, means the oxygen penetration into the layers of surface. This process is possible because when exposed to X-rays, the air is noticeably ionized, and there is a part of more active oxygen (ozone) in it. Such oxygen saturates vacancies in the basal plane of the unit cell. In that part of the vacancies some also contained oxygen with the charge of such atom being close to neutral. An anomalously high diffusion rate was then observed several times for this compound of this atomic plane.

An increase in the oxygen content in a unit cell leads to a noticeable compression waves in this cell and the appearance of micro- and macrostresses in the sample. Indeed, we observed spontaneous bending of plates made of this material after a long maturing, which may be associated with relaxation of macrostresses due to the arising of compression waves. The presence of the compressed powder in quartz cuvettes for X-ray diffraction also led to cracking of the cuvettes holding this powder. This observation reflects a very high level of such macrostresses, capable of destroying a sufficiently strong material.

The YBCO unit cell appeared because of the replacement of lanthanum and strontium in the LaSrCuO4-x compound for yttrium and barium, which led to a strong chemical contraction of this cell, along with an increase in Tc from 39 to 92K. We considered a similar chemical contraction effect for another superconducting compound V3Si [10]. In this case, due to the chemical contraction of V atoms by Si atoms with covalent bonds, the value of Tc also increased almost three-fold.

Moreover, we observed the phenomenon of strong swelling of V3Si, milled to a fraction of 20 microns and pressed in cuvettes when the maturing process took place. In some cases, when the milled particles decreased to 3-5 microns, the cracking of quartz cuvettes was also observed, similar to that observed for YBCO.

The nonlinearity of the response of a multicomponent system to external influences at the electronic and atomic level of structure was inherited during the transition to a larger scale of the structure [11-12] and vice versa. For several samples this manifested itself in the appearance of chaotic phenomena and avalanches when exposed to a strong magnetic field [13-14] of near the superconducting state.

It should be noted that the process of self-organization in highly nonlinear systems can be associated with bifurcations and different trajectories of the dynamic process. Therefore, the longer observation we made, we were able to see different sequences of changes in the diffraction pattern and the contribution of the emerging regions with different oxygen contents. However, despite the different levels of oscillations of the structural characteristics, they all returned to their initial state.

The oscillatory phenomena are well studied in the reaction-diffusion systems [15]. In [16], on the basis of synergetic approaches, as a theory of self-organizing systems, it is shown that the controlling of the structure and properties is related to establishing conditions of spontaneously released elastic energy as the system loses its space-time symmetry being an independent source of energy.

At the atomic level, however, this observation is considered to be rather unique. Therefore, the YBCO samples were also provided to the Division of Material Reliability, NIST, USA, where they were analyzed on their X-ray diffractometer. The presence of oscillatory changes in the characteristics of the diffraction pattern was confirmed, which shows the reproducibility of the observed phenomenon.

Summary

The emerging of spatio-temporal patterns demonstrates the process of self-organization in this ceramics at the atomic and microscopic levels as well as the presence of peculiar 'breathing' modes of behavior of this multicomponent material. At the same time, the added complexity of the structure of the material by several components with variable concentration leads to the dynamics of processes, which has a more pronounced nonlinear character, and a variety of the observed spatio-temporal patterns.

Acknowledgments

The authors are grateful to Prof. D. Balzar (NIST, Denver University, CO, USA) for reproducing part of a time-series of YBCO samples, and to I.L. Kommel and A.A. Ignatov for their help in carrying out very long-time series of measurements, and G.V. Bitukova, I. P, Jouravlev and G.V. Guschin for their help in processing a large amount of accumulated data.

References

[1] L.A. Vasilyeva, M.I. Boychuk, S.A. Mikaeva, Radiation resistance of quartz generators. Book of reference. Engineering Journal with supplement, 8 (2020) 23-29. https://doi.org/10.14489/hb.2020.08.pp.023-029

[2] B.N. Kodess, V.A. Sarin. A Neutron diffractometer for determining the structural characteristics of single crystals, Measurement Techniques, 57(11), (2015) 1299-1303. https://doi.org/10.1007/s11018-015-0624-3

[3] N.S. Kolesova, G.P. Revkevich, B.N. Kodess, A.A. Katznelson, Effect of soft X-ray radiation on structural characteristics of KD_2P0_4. Moscow University, Physics Bulletin, Sect. 3, Physics and Astronomy 50(4), (1995) 107

[4] B.N. Kodess, M.T. Medetbekov, The Processes Self-Organization of Structural Parameters Investigation in HTSP materials. VNIIMS, 25 December 1995, Moscow, Depozit VINITI, 04G113Dep., N 2075-B96 (rus); (1996), 48p.: B.N. Kodess, Self-Oscillations in a crystalline substance. Nature, 04/09/1995, K096, (1995) 1-6

[5] V.M. Avduchina, N.S. Kolesova, B.N. Kodess, A.A. Katznelson, G.P., Revkevich, S.M. Zivotov, Structure Changes in Crystals with Various Nonstoichiometries under Long Aging and Exciting Influence. Surface Investigation (Physics, Chemistry, Mechanics), 12 (1997) 695-709

[6] B.N. Kodess, L.A. Butman, M.A. Poray-Koshits, On Yttrium Metalloxide Stability. (Engl. Transl.) Sov. Physics DOKLADY, 35(1), (1990) 60-62

[7] V.D. Okunev, Z.A. Samoilenko, S.J. Lewandowski, T.A. D'yachenko, V.A. Isaev, E.I. Pushenko, V.N. Varyukhin, P. Gierlowski, S. Barbanera, Effect of laser irradiation on the structure and properties of dielectric column clusters in the YBaCuO superconducting films. Technical Physics, 58(2), (2013) 231-238. https://doi.org/10.1134/S1063784213020163

[8] J. Demsar, M. Zavrtanik, B. Podobnik, V. I. Dediu & D. Mihailovic, Photoresponse of Y-Ba-Cu-O thin rims on the Ms time scale-evidence for localized states near E_F. Journal of Superconductivity, 10(4), (1997) 455-458. https://doi.org/10.1007/BF02765737

[9] M.G. Khrenova· A.V. Nemukhin, B.L. Grigorenko, A.I. Krylov, T.M. Domratcheva, Quantum chemistry calculations provide support to the mechanism of the light-induced structural changes in the flavin-binding photoreceptor proteins. J. Chem. Theory Comput. 6(8), (2010) 293–2302. https://doi.org/10.1021/ct100179p

[10] B.N. Kodess, F.A. Sidorenko, Anisotropy of Magnetic Susceptibility of the V3Si Silicide, Physics of Metals and Metallography, 122(4), (2021) 345-350. https://doi.org/10.1134/S0031918X21040037

[11] B.N. Kodess, Multi-factor analysis in metrology of materials and its correlation with the System of Units of measurements, Legal and applied metrology, (1), (2021) 46-51

[12] B.N. Kodess, P.B. Kodess, Reference data and multi-factorial methods for development of modern materials, Material Science Forum, K0646, (2022) 1-10 (to published)

[13] V.V. Chabanenko, B.N. Kodess, N.V. Kuzovoi, H. Szymczak, Fine Structure of Thermal Runaway Process in the V3Si Single–crystal Superconductor as a Result of Pinning Center Response, Physics Procedia, 36 (2012) 634-637. https://doi.org/10.1016/j.phpro.2012.06.181

[14] A. Nabialek, V. Chabanenko, S. Vasiliev, G. Shushmakova, H. Szymczak, B. Kodess, Two components of the magnetostriction of the crystalline metallic V3Si superconductor, Journal of Applied Physics, 105(6), (2009) 063918-063918. https://doi.org/10.1063/1.3093696

[15] V.K. Vanag, Waves and patterns in reaction–diffusion systems. Belousov–Zhabotinsky reaction in water-in-oil microemulsions, Physics-Uspekhi 47(9), (2004) 923-941. https://doi.org/10.1070/PU2004v047n09ABEH001742

[16] V.S. Ivanova, A.A. Shanyavsky, Nanoscience as a theory of control over the structure and properties of nanosystems, Metallurgia mashinostroyeniya 6, (2008) 31-35 (rus)

Impact of Machining and Phase Composition Change of Titanium Alloyed with Molybdenum

Boris Kodess[1,2,a*], German Teterin[2,b], Pavel Kodess[2,c] and Nikolay Dormidontov[3,d]

[1]VNIIMS, 46 Ozernaya str., 119361, Moscow, Russia

[2]ICS&E, 919 Telluride St., 80011 Aurora, CO, USA

[3]IMET RAS, 49 Leninskii Pr., 119334, Moscow, Russia

[a]kodess@mail.ru, [b]gteterin@hotmail.com, [c]icse96co@gmail.com, [d]ontip@mail.ru

Keywords: Titanium–Molybdenum, Single-Crystal, Beta-Phase, Lattice Constant, Orthorhombic Phase, Strength

Abstract. The titanium samples alloyed with molybdenum and aluminum are used to conduct full high-precision X-ray experiments enabling to determine the characteristic of the atomic interrelations - the unit cell dimension, and to establish the phase components of the doped single crystal. An orthorhombic phase is found, the volume of which increases after the deformation impact.

Introduction

The refractory elements added to the alloys of titanium with aluminum change their strength and other functional characteristics. The alloying significantly extends the application of these materials in biomedicine and other applications. Currently, in order to create new compositions based on titanium, a comprehensive study of these alloys by various methods and modeling of their characteristics based on quantum-mechanical calculations is used [1-2].

The concentration dependence of microhardness, corrosion as well as other characteristics of the atomic and electronic structure are studied in [2] for alloys of the Ti-Mo binary system. Both the calculations and the experimental data allowed to establish the factors of the relationship between the energy chains of composition-structure properties, ranging from the atomic and electronic level to the meso-, micro- and macrolevels of the real macroscopic properties of products made of these alloys. The influence of molybdenum additions on the phase composition, microstructures, and mechanical properties of initial composition of the Ti6Al4V with a gradual increase in-the Mo content up to 6 wt.% is studied in [3]. It is shown that the element hardening the solid solution such as Mo improves mechanical properties like yield strength, ultimate tensile strength, ductility, and hardness, compared with the pre-alloyed Ti6Al4V alloy.

The impact of mechanical treatment on the strength properties of samples of titanium with a higher molybdenum content is studied in this work. For the beta-phase of the samples, high-precision experiments were carried out for the first time, using single crystal X-ray diffractometry, and the comparison was made with the undoped composition. These experiments enabled to determine the basic thermodynamic characteristic of atomic interrelation after crystallization of alloys – the unit cell dimension (the parameter of the formed crystal lattice) [4].

Samples and procedure

To investigate microhardness by the Vickers (HV50) method the plates cut from ingots were used. The electric spark cutting was followed by electropolishing with the surface layer of 20 μm and

60 μm being removed. Then the surface was treated with an emery cloth coated with corundum of an average size of 40 μm, after which the surface was polished using diamond paste (with the particle size of 9 μm) and colloidal silicon. 49 times the measurements (matrix 7*7 on each plate) were carried out using a PMT3 microhardness-meter with the applied load of 50 grams. The deformed layer having been measured, the electropolishing was again carried out bringing the surface to the initial state, and the measurements of the microhardness of the same samples were repeated.

For the measurements performed on a four-circle single-crystal diffractometer ("Xcalibur" Rigaku-Oxford Diffraction) cubes with the face size of 1 mm were cut from large grains of a composition containing 15% Mo and 6% Al. To improve the measurement accuracy, spherical samples with a diameter of no more than 0.3-0.4 mm were prepared from them. After rolling, the surface layer was removed by etching. The preliminary diffraction patterns exhibited high quality of the analyzed samples and suggested a reliable and detailed analysis of the intensity distribution in each recorded Bragg reflection.

In the measurements a short-wavelength MoKα1 radiation with a wavelength of 0.070932 nm was used, which enabled the recording of the diffraction pattern of total volume analyzed samples. The samples were sequentially rotated making possible to record the intensity of the reflected beams on the CCD-detector with a full display of the Ewald sphere, during 80-180 hours (two control reflections were recorded every 50 frames to take into account possible drifts; however, voltage fluctuations did not exceed 0.1 V at a voltage of 50 kV).

From 8 to 24 sets of the independent system of Bragg reflections were obtained (depending on the Miller indices of these reflections), which ensured additionally an increase in the accuracy and reliability of the measurement results. The lattice parameters of the beta-phase were calculated in accordance with the Bragg-Wolf law and cubic symmetry of the unit cell. Unlike powder diffractometry, when from 6 to 40-60 averaged reflections are recorded, the use of the geometry of 4-circle diffractometer and a single-crystal form of the samples made possible to record more than 2600 Bragg reflections in the diffraction pattern of each sample at room temperature (297-300) K.

Results and discussion

The study of alloys based on titanium, including bimetallic products, showed the presence of synergy of effects at the micro- and macro-level of the material structure during various kinds of machining [5-8]. For the investigated composition, also considered sequentially were the data at the microstructural and then at the atomic level of the structure.

The microhardness of the samples remained invariable after a layer of thickness of 20 μm and 60 μm was removed. The use of diamond pastes for surface treatment increased microhardness and the use of larger corundum particles of an average size of 40 μm, which are commonly used in surface treatment technology, led to an additional increase in microhardness more 11%.

It was shown in [3] that phase shifts of diffraction reflexes occurred when the stabilizer Mo from 2 to 6 wt.% was added, exhibiting the occurrence of the lattice distortion. It was found that the lattice parameter changed from 0.3228 nm for Ti6Al4 to 0.3232 nm, 0.3243 nm, and 0.3251 nm for Ti6Al4V-2Mo, -4Mo, and -6Mo, which the authors considered was due to the increase of stabilizer Mo content from 2 to 6 wt.%, which acting as hardening agent in the solid-solution in the pre-alloyed Ti6Al4V alloy, and which is seen in the improved mechanical properties of the alloys - microhardness. Ti6Al4V-6Mo had the highest value among these compositions.

The lattice parameters for our alloy containing 15% molybdenum and 6% aluminum, demonstrated a further increase in the crystal lattice parameters of the beta-phase - 0.330595 (2) nm and 0.330628 (2) nm for two studied samples.

An analysis of references was also carried out, for related to the crystal lattice parameters of the alpha-phase of pure hexagonal titanium [9] and the beta-phase [10] of titanium. Two methods were used to estimate the values of the lattice parameter of the beta-phase at room temperature. For the undoped beta phase, in one method the lattice parameter at room temperature was determined by extrapolating the values measured for the beta-phase at high temperature, which gave a value of a_{extr}= 0.3276 nm; in the other method it was near-this values of a_{extr}= 0.3279-0.3282 nm. They were obtained by extrapolating data from the dependences of the lattice parameters on the concentration of substituent varied elements (Mo, Nb, V, Zr, Cr, Ta, Fe, Al).

The experimental data in [11] indicate that as the Mo content increases, on parts of the volume of sample a martensitic transformation can occur, including under the action of the applied stress. The martensite has a hexagonal structure denoted as α`-martensite, in dilute compositions with a higher solute content as ortorhombic α``-martensite in alloys.

It should be noted that in this study the full diffraction pattern from the analyzed sample is recorded. The use of single-crystal samples increases the intensity of the observed reflections by 3-4 orders of magnitude. The software (CrysAlis) allows for such samples to detect the presence of a second phase, though it may be in small quantities. In fact, along with the predominant beta-phase, a family of reflections was found in the samples, with a unit cell belonging to the ortorhombic phase. The amount of this phase may differ in different parts of the sample and probably be the cause of small scattering of values of the parameters of the two studied samples. Preliminary estimates show that the number of reflections of this second phase can be more than 25% of the number of reflections of the beta phase.

It is well known that the presence of a metastable anisotropic phase in samples leads to the appearance of noticeable distortions in the matrix and to an increase in the density of dislocations in the matrix. This, in turn, can be one of the reasons for the observed increase in strength characteristics during deformation action on the surface of the samples.

Summary

The use of single crystal samples enables to increase the accuracy and reproducibility of the results of diffraction experiments. The addition of molybdenum results in an increase in the crystal lattice parameter of the beta-phase of titanium and in the appearance of a second metastable orthorhombic phase in this alloy. The mechanical surface treatment, inducing more micro-distortions and providing the formation of a second phase, leads to an increase in the strength characteristics of the surface of products made from this alloy.

References

[1] T.E. Jones, M.E. Eberhart, S. Imlay, C. Mackey, G.B. Olson, Better alloys with quantum design, Physical Review Letters, 109(12) (2012) 125506. https://doi.org/10.1103/PhysRevLett.109.125506

[2] L.N. Grishchishyna, The admixture engineering: the energy factors influence from atomic and electronic systems to grain hardness of the titanium alloys, Electron microscopy and strength of materials. Ser. Physical Materials Science, Structure and Properties of Materials, 18 (2012) 170-183.

[3] M. Rajadurai, A. Muthuchamy, A.R. Annamalai, D.K. Agrawal, C.P. Jen, Effect of Molybdenum (Mo) Addition on Phase Composition, Microstructure, and Mechanical Properties

of Pre-Alloyed Ti6Al4V Using Spark Plasma Sintering Technique, Molecules **26**(10), 28946 (2021) 1-13. https://doi.org/10.3390/molecules26102894

[4] B.N. Kodess, O.P. Lazukina, E.N. Volkova, K.K. Malyshev, Impurity Composition and Lattice Parameters of High-Purity α-Mn, Inorganic Materials, 56 2020) 512-517. https://doi.org/10.1134/S0020168520050076

[5] L.A. Kommel, G.P. Teterin, B.N. Kodess, Influence of Thermophysical Properties of Titanium Alloys on Electric Upset Forging Parameters, Forging and Stamping Production, 7 (1998) 29-34.

[6] B.N. Kodess, L.A. Kommel, G.P. Teterin, V.K. Ovcharov, Microstructure Evolution in TI-Alloys During Severe Deformation by Electric Upsetting and Im-pact Fused-Forging Modeling, NATO Science. Series 3. High Technology. 80. "Investigations and Applications of Severe Plastic Deformation". Ed. T.C. Lowe and R.Z. Valiev. Kluwer Academic Publishers. Dordrecht/Boston/London, (2000) 211-218. https://doi.org/10.1007/978-94-011-4062-1_28

[7] P. Kodess, G. Teterin, B. Kodess, Evolution of the microstructure of bimetallic valves, Acta Cryst. Ser. A: Foundation and Advanced, A67 (2011) 439. https://doi.org/10.1107/S0108767311088945

[8] B.N. Kodess, M.T. Medetbekov, The distribution of phases, composition and properties of Ti-Al alloys after various perturbation action, VNIIMS, GosStandart of Russia, **14**(96) (1996) 1-45 (rus).

[9] B.N. Kodess, Standard Reference Data. Titanium. Lattice constants for the temperature range from 5 K to 300 K. Linear temperature expansion coefficient for the temperature range from 5 K to 1200 K. VNIIMS, M. (2017) 33 p.; Nat. Standard-R 8.935 (2017) 12 p.; allgosts.ru›17/020/gost_r_8.935-2017.

[10] B.N. Kodess, Standard Reference Data. Titanium with additives that stabilize the beta-phase. The crystal lattice parameters of the phase with a molybdenum concentration of up to 15 at. %, aluminum up to 6 at. %. Linear Thermal Expansion Coefficient Ti beta-phase (VT1), VNIIMS, M. (2019) 41p.

[11] T. Zhou, M. Aindow, S.P. Alpay, M.J. Blackburn, M.H. Wu, Pseudo-elastic deformation behavior in a Ti/Mo-based alloy, Scripta Materialia, 50(3) (2004) 343-348. https://doi.org/10.1016/j.scriptamat.2003.10.012

Materials Research Proceedings **21** (2022) 268-273 https://doi.org/10.21741/9781644901755-48

Testing and Diagnostics of Carbon Polymer Composite Materials by Radio Wave Method Used in Aircraft Construction

German V. Dmitrienko[1,a*], Dmitry V. Mukhin[1], Aleksandr A. Fedorov[1]

[1]Ulyanovsk Technical University. Institute of Aviation Technologies and Management, 13A, Sozidateley Avenue, Ulyanovsk, Russia, 432072

[a]dmitrienko.german@yandex.ru

Keywords: Polymer Composite Materials, Carbon Plastics, Dielectric Constant

Abstract. The article is devoted to the applicability of radio wave methods of testing and diagnostics of aviation composites for newly launched products. The description of the measurement process is also considered by the method of integral equations.

Introduction

Recently, fundamentally new types of radio-absorbing structural composite materials (CMs) for aviation purposes have appeared, such as: conductive polymers, carbon nanotubes, nano-structured CMs. In electrodynamics, such materials are classified as low impedance –polymer composite materials (PCMs), which are characterized by large values of the relative permittivity ε and the tangent of the dielectric loss angle $tg\delta$ simultaneously. The carbon-fibre-reinforced plastics (CFRPs) are widely used in aviation technology as structural materials for creating load-bearing structures, and as radio engineering materials: to reduce the effective reflective surface of aircraft, to protect radio engineering systems from powerful electromagnetic radiation.

In order to increase the accuracy of processing the measurement results and control the characteristics of the product by its dielectric parameters, a more rigorous from the electrodynamic point of view, a mathematical method, the method of integral equations (MIU), is used [1-5]. The procedure for drawing up integral equations is carried out through surface currents induced on the surface of the PCM sample and the metal structure of the waveguide sensor by a probing electromagnetic wave. The dielectric parameters of the PCMs and the characteristics of the reflecting surface are set through the surface impedance. The information parameter is the reflected wave, and the measuring problem is reduced to the problem of exciting a circular waveguide by currents induced in the PCMs by the field of the incident wave. The measurement results are processed by numerical methods. The methodology of drawing up integral equations for bodies of revolution is given in [1-5]. On its basis, a multilayer model of a PCMs sample is compiled.

$$E(p)=\int_L \xi^E(p,q,\varepsilon_a,\mu_a)J^E(q)+\xi^M(p,q,\varepsilon_a,\mu_a)J^M(q)dl, \qquad (1)$$

$$H(p)=\int_L h^E(p,q,\varepsilon_a,\mu_a)J^E(q)+h^M(p,q,\varepsilon_a,\mu_a)J^M(q)dl, \qquad (2)$$

A model of a PCM sample with a rough surface (RS) is represented as a multilayer structure consisting of homogeneous layers, the number of layers is determined by the discrete partition in ε. Integral equations for a multilayer structure are based on the equations obtained earlier (1) and

(2) for each layer, using both of its boundaries. The method of drawing up integral equations for a multilayer medium is carried out as for a single layer, taking into account the distribution of currents at the interfaces. The compound is carried out according to the boundary conditions of Shchukin-Leontovich since PCMs haveε,μ - electrical and magnetic permeability and σ - conductivity. The sample is divided into *N-th* number of layers; their boundaries have no common points. Each layer can contain third-party sources; the layers are numbered in the direction from the observation source into the interior of the PCM sample. Equations are drawn up, for layers K and K+1 having a common boundary Σ_K using (1), (2), the equivalence theorem is applied to the K-layer. The normal at the interfaces between the layers is directed towards the primary sources.

For the K-layer, the equations are:

$$\frac{1}{2}E_K(p)=\int_{\Sigma_K}\left[\xi^E J_K^E(q)+\xi^M \tilde{J}_K^M(q)\right]d\sigma_q+\int_{\Sigma_{K-1}}\left[\xi^E J_{K-1}^E(q)+\xi^M \tilde{J}_{K-1}^M(q)\right]d\sigma_q+E_K^{\Pi}(p)\ ,p\in\Sigma_K$$

$$\frac{1}{2}H_K(p)=\int_{\Sigma_K}\left[h^E J_K^E(q)+h^M \tilde{J}_K^M(q)\right]d\sigma_q+\int_{\Sigma_{K-1}}\left[h^E J_{K-1}^E(q)+h^M \tilde{J}_{K-1}^M(q)\right]d\sigma_q+H_K^{\Pi}(p)\ ,p\in\Sigma_K \quad (3)$$

Integral equations are written in matrix form for the K-layer, similarly written for the K+1-layer, the integrals are taken over the surfaces Σ_K and Σ_{K+1} as (3).

$$\frac{1}{2}E_{K+1}(p)=\int_{\Sigma_{K+1}}\left[\xi^E J_{K+1}^E(q)+\xi^M \tilde{J}_{K+1}^M(q)\right]d\sigma_q+\int_{\Sigma_K}\left[\xi^E J_K^E(q)+\xi^M \tilde{J}_K^M(q)\right]d\sigma_q+E_{K+1}^{\Pi}(p),\ p\in\Sigma_{K+1}$$

$$\frac{1}{2}H_{K+1}(p)=\int_{\Sigma_{K+1}}\left[h^E J_{K+1}^E(q)+h^M \tilde{J}_{K+1}^M(q)\right]d\sigma_q+\int_{\Sigma_K}\left[h^E J_K^E(q)+h^M \tilde{J}_K^M(q)\right]d\sigma_q+H_{K+1}^{\Pi}(p),\ p\in\Sigma_{K+1} \quad (4)$$

In order to obtain excitable currents on the surface, equation (3) is conjectured vectorially on the normal ε_{K+1}, but (4) on the normal ε_K, then equations (3) and (4) are added.

In the process of describing the RS ε will change from layer to layer, but ε will be constant inside the PCM sample. Integral equations inside the layer will be simplified. In the description of the reflection process, one can replace the internal structure with one layer, and in the description of the process, the transmission can be replaced by one layer of finite thickness. An integral equation is drawn up for each layer, for the boundaries Σ_1 и Σ_{N+1} there are no integral equations, as well as the boundaries themselves. At the outer boundaries of the PCM sample, the Shchukin-Leontovich boundary conditions are applied. Equations are stitched inside the polymer composite materials. The multilayer model described by the International Innovative University is convenient for describing, predicting and modeling the temperature characteristics of PCM.

Discussion of the results

Integral equations are solved by numerical methods. The most common is the projection numerical method (Galerkin method) [1].

According to the given theory, a control was carried out and the value of the relative dielectric constant of the PCM sample was determined, material U-PU ε=950±30% and the tangent of dielectric loss angle tgδ=1,3±40%, having a reflection coefficient modulus of 0.94 [6].

The main problem of constructing simple design engineering formulas. Difficulties are caused by the surface roughness (RS) of the measured material. For conducting polymers, it can be of 4-

Materials Research Proceedings 21 (2022) 268-273 https://doi.org/10.21741/9781644901755-48

8 classes, which significantly introduces uncertainty in the final result. The final result is the complex dielectric constant $\underline{\varepsilon}$ of the PCM sample. We measured PCM samples with a rough surface, the level of which was 5-7 microns, which corresponds to an additional phase incursion of the order of 0,1° in the three-centimeter wavelength range. The RS level of the reference standard, relative to which the control is carried out, is 1 micron.

The assessment of the level of surface roughness of the measured sample must be carried out in the instrumental part - the introduction of additional intermediate exemplary measures, as well as in the methodical part - the introduction of corrective additives, taking into account the surface roughness of the measured PCM sample. Other types of uncertainties are present in traditional methods and ways to measure and can be considered known.

Analysis of the influence of the RS of PCM on the control results

To control the dielectric characteristics of PCMs, first of all, the quality of their technological preparation, that is, the roughness of the measured surface, will be of interest. Methods for measuring RS of products regulated by GOST 2789-73 and GOST 27964-88 [7], turned out to be unsuitable for controlling the surface roughness of PCM products, because measurement of RS is carried out at some points and in small areas. In addition, the surface condition of PCM products is characterized by the following features: characteristic color, porosity, cracks, texture of the reinforcing filler, cavities, chips, sooty formations, and roughness [9-11]. Some of these features relate to the structural features characteristic of a given composite material, the other is due to the modes of mechanical processing of the product, the rest of the features refer to the technology of manufacturing the product from this composite. The RS is an initial requirement for the technological process of surface formation, and it is directly or indirectly related to various indicators of its quality. The control of the RS of PCM products is the most difficult due to the peculiarities of the PCM structure. So, for example, the structural features of the surface of carbon composite materials are: pores, shrinkage cavities, pile of fibrous fillers, coke grains up to 3 mm in size, etc. The RS, in the form of inhomogeneities, with sizes from 0.5 mm to 3 mm. For example, in the technical documentation for a product made of carbon-containing PCM, the RS parameter is usually set by the numerical characteristic R_z, which is determined on the basis of a profilogram. The RS of PKM can vary from 1 micron to 0.5 mm, so that the uncertainty of measurement of the RS will be determined by the uncertainty of the measured device. PCM samples for measurement are checked for conformity to the plane, according to the RS, in terms of geometric dimensions, in homogeneity, in the absence of mechanical damage, etc. The uncertainty in the manufacture of samples is ±0,005 mm.

In the process of preparing PCM samples for measurements, they have an unremovable RS, which leads to additional instrumental and methodological uncertainties. As noted earlier, the uncertainties caused by the RS of the measured PCM sample are predominant. The uncertainty of the phase of the reflected signal caused by the displacement of the reflection plane is large, comparable in magnitude with the value of the instrumental uncertainty in measuring the phase. Since in dielectric measurement it is considered that the plane of reflection from the dielectric material coincides with the plane of contact with the measuring sensor, the RS of traditional materials (ε<100) is not taken into account. In reality, when measuring PCM with the RS, the question arises about the location of the reflection plane. Depending on the nature of the irregularities of their height on the surface, the geometry, the reflection plane will be above or below the median surface of the PCM irregularities. The RS model, presented in the form of a multilayer structure, makes it possible to verify this numerically. Therefore, it is necessary to carry out an analysis to take it into account in the process of calculating of ε.

There is a known method of representing the RS through an intermediate layer, in which ε changes, along a gradient deep into the material. This method is complicated in the process of numerical implementation. A more accurate representation of this rough layer is represented as a multi-layer structure in which each layer has its own average value of ε.

This method of representing RS gives the best accuracy, the disadvantage of this method is the need for accurate knowledge of the geometry of the relief of a rough surface in order to determine the average value of ε in each layer.

To improve this technique, it is proposed to represent the WB in the form of a multilayer structure. From a mathematical point of view, a WB consisting of sublayers of thickness d_i - can be considered with an average value of ε calculated by the Lichtenecker equation [12] for complex dielectrics

$$\varepsilon_\Sigma=\varepsilon_0 V_1+\varepsilon_1 V_2, \qquad V_1+V_2=1,$$

where ε_0 is the dielectric constant of the consolidated space, ε_1 is the dielectric constant of the PCM, V_1 is the filling factor of the sublayer with the consolidated space from the entire volume, V_2 is the filling factor of the sublayer of the PCM from the entire volume of the sublayer. As a result, we get an intermediate layer representing the RS between free space and PCM, consisting of several thin sublayers in which ε changes according to a certain mathematical law.

The distribution law ε over the layer is mathematically represented as a sum of sublayers $\sum_{i=1}^{N}\left[\varepsilon_1 V_{1i}+\varepsilon_2 V_{2i}\right]$, where V_{1i}, V_{2i} are the filling factors *i-sublayer* from the entire volume. The RS of PCM is presented in the form of a multilayer structure, in which each of the layers has a different filling of the PCM. A mathematical model of a complex RS is presented, which is divided into several sublayers, representing it in the form of a plane-parallel structure. Each sublayer of such a structure is characterized by a different coefficient of filling the layer with a dielectric, and its ε is considered averaged over the layer, while the sublayer is considered homogeneous and isotropic. The reflection of a plane wave from such a structure is calculated through a transmission matrix describing each sublayer. The number of layers is selected depending on the level of the RS and the distribution of irregularities was uniform over the selected discrete $\Delta\varepsilon$.

Each sublayer is described through a characteristic matrix [8], which is a transmission matrix. The characteristic of each layer is represented by the transmission matrix of the sublayer depending on the filling factor of the PCM sublayer. So for the *i-sublayer* of the transfer matrix M_i is equal to

$$M_i=\begin{bmatrix} ch\left(jkd_i\sqrt{\varepsilon_i}\right) & \frac{1}{\sqrt{\varepsilon_i}}sh\left(jkd_i\sqrt{\varepsilon_i}\right) \\ \sqrt{\varepsilon_i}\,sh\left(jkd_i\sqrt{\varepsilon_i}\right) & ch\left(jkd_i\sqrt{\varepsilon_i}\right) \end{bmatrix}, \qquad (5)$$

where k is the wave number, d_i is the thickness of *the i-sublayer*, ε_i is the dielectric constant of the *i-sublayer*. The generalized matrix of the entire rough layer is represented as a product of intermediate matrices (7).

$$M=\prod_{i=1}^{N}M_i=\begin{bmatrix} M_{11} & M_{12} \\ M_{21} & M_{22} \end{bmatrix}. \qquad (6)$$

Materials Research Proceedings **21** (2022) 268-273
https://doi.org/10.21741/9781644901755-48

From the general matrix, there is the reflection coefficient of the electromagnetic wave from the RS of PCM according to the formula (7) and the transmission coefficient according to the formula (8)

$$\underline{R}=\frac{(M_{11}+M_{22})-(M_{12}+M_{21})}{M_{11}+M_{22}+M_{12}+M_{21}}, \tag{7}$$

$$\underline{T}=\frac{2}{M_{11}+M_{22}+M_{12}+M_{21}}. \tag{8}$$

Therefore, in order to accurately determine the plane of reflection from a rough surface, it is necessary to introduce a definition of ε_Σ to represent the phase characteristics of the RS through a multilayer structure. The introduction of the location of the plane of reflection from the RS makes it possible to compose a simpler mathematical model for describing the formation of the front of the reflected wave reflected from a smooth surface. That is, a smooth surface moves away or approaches the source of the sounding wave by a certain amountΔ. Then the rough layer is expressed as a layer having a certain average value ε over the layer. The RS of PCM varies from 1 micron to 0.5 mm, the geometry depends on the manufacturing technology of the PCM (internal structure) and the method of machining. According to the analysis of PCM samples, it was revealed that: the range of values of $\underline{\varepsilon}$ ranges from 100 to 1500; The RS varies from 5 to 500 microns. On the basis of the multilayer structure of the RS, mathematical modeling was carried out to find the complex reflection coefficient $\underline{R}$ from different levels of the RS.

The obtained calculated dependencies make it possible to correct the measured value of the complex reflection coefficient. The procedure for using phase correction is carried out during the control of the measured sample of PCM with the RS using a traditional exemplary standard. The next step is to measure the profile of the RS by measuring the profilogram of the PCM sample, and using the profilometer, the level of the RS is determined. According to the measured data, the distribution and the level of the RS, the corresponding value of the phase shift is found, according to the value of which the result of the complex reflection coefficient from the PCM sample with the RS is refined. As a result, the value of the complex reflection coefficient from a PCM with a smooth surface is obtained. Then ε is calculated. The result will be corrected and it will have reduced methodological uncertainty.

References

[1] Vasil'yev, E.N. Excitation of bodies of revolution., Radio i svyaz', Moscow, 1988.

[2] Kislyuk, M.ZH. The method of secondary waves in problems of electrodynamics Leningrad, 1970.

[3] Malushkov, G.D. Methods for solving problems of electromagnetic excitation of bodies of revolution., Izv. VUZov. Ser. Radiofizika. T.18(1975) 1563-1587. https://doi.org/10.1007/BF01037102

[4] Nikol'skiy, V.V. Projection methods in electrodynamics., Vysshaya shkola, Moscow, 1977.

[5] Khizhnyak, N.A. Integral equations of macroscopic electrodynamics, Naukova Dumka, Kyiv, 1986.

[6] Dmitrienko, G.V., Anisimov, V.G. Measurement of the permittivity of low-impedance composite materials at microwave frequencies., Measurement Techniquesthis, 52(2)(2009) 177–183. https://doi.org/10.1007/s11018-009-9236-0

[7] GOST 27964-88. Measurement of roughness parameters., Izd-vo standartov, Moscow, 1989.

[8] Kornblit, S. Microwave optics. Optical principles in application to the design of microwave antennas. Svyaz', Moscow, 1980.

[9] Kuchin, A.A. Obradovich K.A. Optical devices for measuring surface roughness. Mashinostroyeniye, Leningrad, 1981.

[10] Baryshnikov I.V., Datskovskiy V.A., Upolovnev A.V. Vliyaniye sherokhovatoy poverkhnosti volnovedushchey sistemyna VCH-poteri. [Influence of the rough surface of the waveguide system on RF losses] / // Radio engineering and electronics. T.33(1988) 2029-2034.

[11] Danilov, YU.I. On the problem of microroughness of current-carrying surfaces in the microwave range of radio waves., Trudy LIAP. V.59 (1968), 92-96.

[12] Bogoroditskiy, N.P. Materials in radioelectronics., Gosenergoizdat, Moscow, 1975.

Materials Research Proceedings **21** (2022) 274-279 https://doi.org/10.21741/9781644901755-49

The Pusher as an Object of the Simulation Model in the Tasks of Studying Cyclic Mechanisms Dynamics

Viktor V. Telegin[1,a] and Igor V. Telegin[1,b*]

[1]Lipetsk State Technical University, Lipetsk, Russia

[a]vv.telegin@yandex.ru, [b]igor.v.telegin@gmail.com

Keywords: Pusher, Cyclic Mechanism, Dynamic Processes, Object-Oriented Technologies

Abstract. The cyclic mechanism is one of the basic automatic machines components. It is the cyclic mechanisms that define the performance and reliability limits of these machines. The methods for assessing the automatic machines reliability are based on data of dynamic forces acting on the mechanism's links, leading to their deformation and possible destruction under certain operating conditions. Simulation modeling of dynamic processes in cyclic mechanisms is based on its representation in the form of interconnected typical objects, the allowable properties and limits of which are known. The article presents the dynamics study results one of the basic high-speed automatic machines mechanism in the object-oriented representation. Mechanism performs translational motion and allows the possibility of one-way kinematic connection breaking.

Introduction

Any mechanism can be considered as a collection of some elements, interconnected by various kinds of connections. Through these links, the elements of the mechanism interact with each other, thus ensuring its functioning [1, 2]. The term element will denote a single mechanism detail, its fragment, or, conversely, several details at once. This definition of an element corresponds to the concept of an object, one of the basic in the component modeling theory of complex dynamic systems and object-oriented technologies [3, 4].

For a visual representation of mechanisms, schemes (principle, kinematic, hydraulic, pneumatic and others), flat and three-dimensional drawings, as well as dynamic models are used. Let us juxtapose each element of one mechanism with an object, according to the term given above. Obviously, this object can contain as one single detail or even a fragment of it, or several details. Any processes in such an object can be modeled in any way and, in accordance with the principles of encapsulation and polymorphism, at this stage, their nature and description, mathematical, software or physical, is not considered.

This approach is of interest both when solving problems associated with the development of dynamic and mathematical mechanisms models, and, which is especially important, their software implementation for further computer processing [5, 6].

Mathematical model of the pusher object

A pusher is a translationally moving link, for example, the cam mechanism pusher of an ejector from the dies (Fig. 1) of a cold stamping machine (CSM) for nuts manufacture [7, 8]. The pusher object provides the possibility of one-way kinematic connection breaking with the object preceding it (Fig. 1, positions 1 and 2).

Materials Research Proceedings **21** (2022) 274-279 https://doi.org/10.21741/9781644901755-49

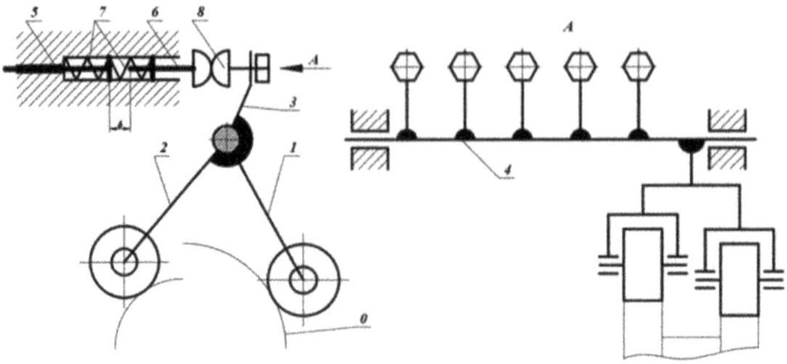

Fig. 1. *Kinematic diagram of the CSM ejection mechanism:*
0 *– block of cams,* ***1, 2*** *– forward and reverse levers,* ***4*** *– axle,* ***5, 6*** *– parts of a composite ejector rod,* ***7*** *– springs,* ***8*** *– adjusting screw.*

A conventional image, the object itself and its model are shown in Fig. 2 and 3.

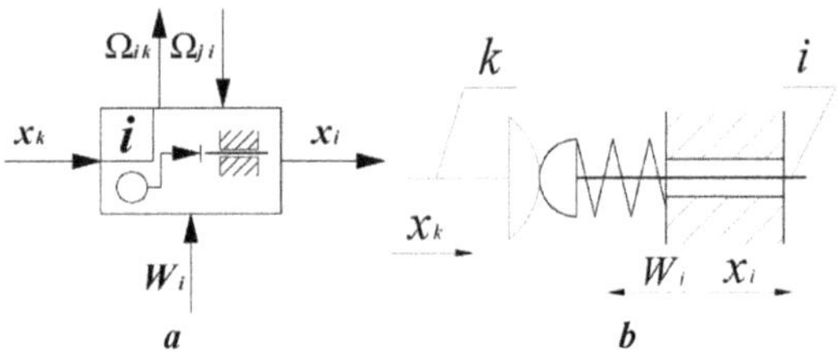

Fig. 2. *Pusher object: a – object model (i-th object), b – kinematic diagram.*

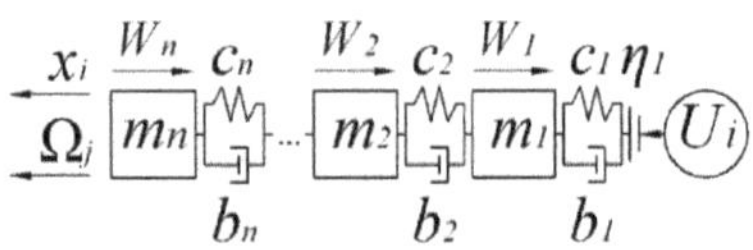

Fig. 3. *The dynamic model of the pusher object.*

Masses motion equations of the object dynamic model [3]:

$$\begin{cases} m_1\ddot{x}_1 = -c_1(x_1 - U_i(x_k)) - b_1(\dot{x}_1 - U_i'(x_k)\dot{x}_k) + \\ \qquad + c_2(x_2 - x_1) + b_2(\dot{x}_2 - \dot{x}_1) + W_1 \\ m_2\ddot{x}_2 = -c_2(x_2 - x_1) - b_2(\dot{x}_2 - \dot{x}_1) + \\ \qquad + c_3(x_3 - x_2) + b_3(\dot{x}_3 - \dot{x}_2) + W_2 \\ \dots \quad \dots \quad \dots \\ m_n\ddot{x}_n = -c_n(x_n - x_{n-1}) - b_n(\dot{x}_n - \dot{x}_{n-1}) + W_n + \Omega j \end{cases} \tag{1}$$

In Eq. 1, as in Fig. 3, the parameter x_k is not necessarily linear. For example, if the object corresponds to a cam disc mechanism with a linear follower, x_k is the cam angle. Rigidity c_1 is a piecewise linear function, the value depends on the size of the gap η_1.

Stability of the pusher object model

The main parameters that determine the object behavior under the dynamic force influence are its mass and stiffness. The currently used 3D modeling methods in the design of mechanical engineering products [9, 10, 11] allow to be calculated these parameters with a high degree of accuracy. The object under study is adequate to the real link of the mechanism if, within the specified limits of its parameters variation, the values of the criteria for its stability are close when using models of varying complexity in calculations [12, 13].

In Fig. 4 shows the details of the CSM ejection mechanism (see Fig. 1, positions 5, 6) and the tested object, ***20*** *mm* in diameter and ***200*** *mm* length. The object mass is $m \approx 0.49$ *kg*, the rigidity is $c \approx 3.3 \cdot 10^8$ *N/m* (density of the steel: $\rho \approx 7800$ *kg/m*^{3}, Young's modulus $E \approx 2.1 \cdot 10^{11}$ *N/m*^{2}).

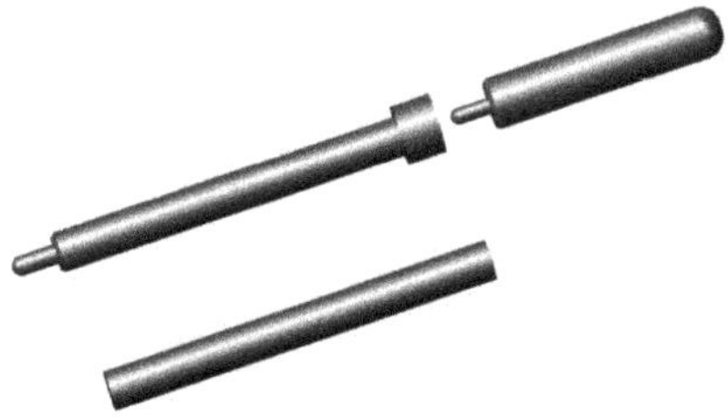

Fig. 4. *Test object and details of the CSM ejection mechanism.*

The maximum speed of the model output link and the accuracy of its positioning δ_{max} are taken as criteria for the object stability. The parameters of the model varied within the following limits: weight – ***0.05...1.5***, rigidity – $10^8 ... 4 \cdot 10^9$.

We will assume that the movement of the link is carried out according to a harmonic law with amplitude equal to ***0.01*** *m*, which, in particular, corresponds to the stroke of the CSM ejector. The pusher movement speed (ω) is assumed equal to ***10*** strokes per second (***600*** strokes per minute), closing load ***1000*** *N*, dissipation coefficient $\psi = 0.4$.

$$\begin{aligned} U &= 0.01 \cdot \sin 2\pi\omega t, \\ U' &= 0.01 \cdot \cos 2\pi\omega t. \end{aligned} \tag{2}$$

The calculation data carried out on models of varying complexity (single-mass, dual-mass, and ten-mass) in the Maple software [14, 15] turned out to be practically the same (coincidence in the

Materials Research Proceedings **21** (2022) 274-279 https://doi.org/10.21741/9781644901755-49

sixth decimal place). Fig. 5-6 show the calculated dependences of the objects kinematic parameters and positioning accuracy.

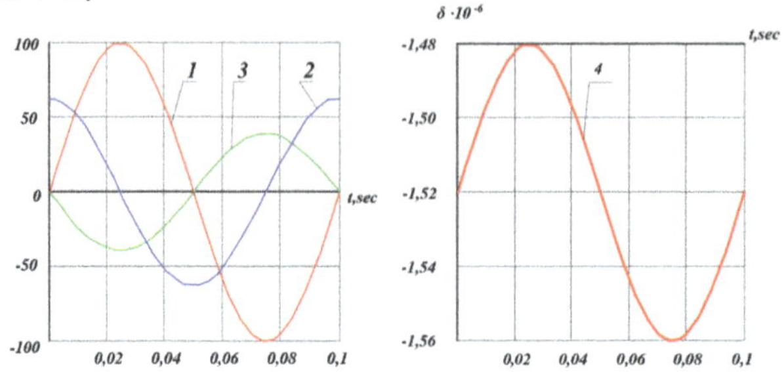

Fig. 5. *Kinematic characteristics of the object:* ***1*** *– displacement (**10^{-4}** m),* ***2*** *– speed (**10^{-2}** m/sec),* ***3*** *– acceleration (1 m/sec^2),* ***4*** *– positioning accuracy (1 m).*

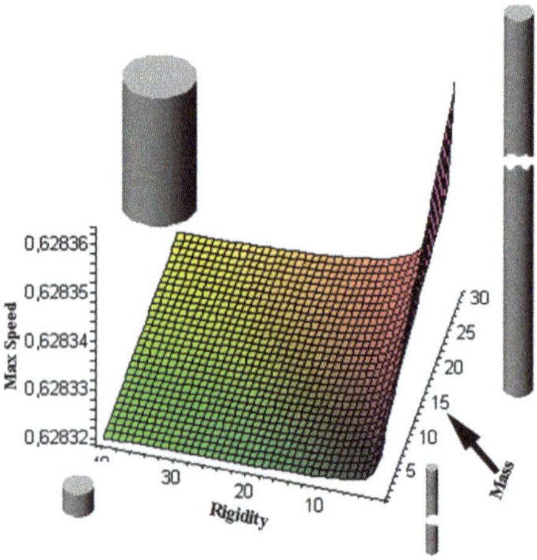

Fig. 6. *To calculate the stability of an object:* ***ω = 10, ψ = 0.4, P = 1000*** *N.*

It should be noted that the influence of the dissipation coefficient on the behavior of this object is insignificant. At the same time, the elastic-inertial parameters of the object noticeably affect the dynamic processes occurring in the object in a mode when a break in the kinematic chain of the object with the previous link is possible. This situation is shown in Fig. 7. The closing force in this case is ***20*** *N*. At the same time, there is a significant difference in the calculation results obtained for the single-mass and dual-mass models. The calculation results obtained on the basis of the two-mass and ten-mass models are close (the difference is about 10%).

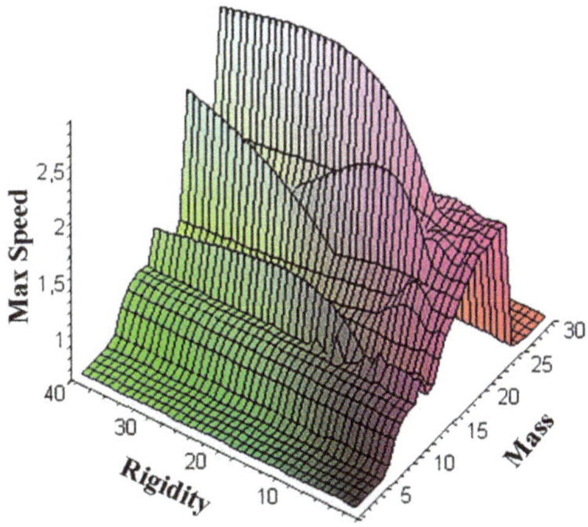

Fig. *7. To the calculation of the object stability. Ten-mass model:* $\omega = 10$, $\psi = 0.4$, $P = 20$ *N.*

Conclusion

A technique is proposed for assessing the limits of objects applicability, from which a simulation model of a mechanical system (mechanism) is formed.

The software implementation of this technique can be used as an integral part of software systems for the dynamic analysis of mechanical systems, which are based on the component modeling methods of object-oriented technologies.

References

[1] I.I. Wolfson, Oscillations of Machines with Cyclic Action Mechanisms, Mashinostroenie, Leningrad, 1990.

[2] 2. V.L. Weits, I. A. Gidaspov, G.V. Tsaryov, Dynamics of Drives with Closed Kinematics, Publishing House of Mordovia University, Saransk, 1991.

[3] I. Telegin, A. Kozlov, A. Zhirkov, Component simulation in problems of calculated model formation of automatic machine mechanisms, MATEC Web Conf. International Conference on Modern Trends in Manufacturing Technologies and Equipment (ICMTMTE 2017). 129 (2017) 03016. https://doi.org/10.1051/matecconf/201712903016

[4] V. Telegin, A. Kozlov, T. Shumilova, Development and research of the rotating lever object as a dynamic model of a cycle mechanism, International Conference on Modern Trends in Manufacturing Technologies and Equipment (ICMTMTE 2018). 224 (2018) 02078. https://doi.org/10.1051/matecconf/201822402078

[5] V.V. Telegin, Performance Assessment in Design of High-Speed Automatic Machines Based on Analysis of Their Dynamics, Proceedings of the 6th International Conference on Industrial Engineering (ICIE 2020), Lecture Notes in Mechanical Engineering. (2021) 1000–1007. https://doi.org/10.1007/978-3-030-54814-8_115

[6] I.V. Telegin, Dynamics of the crank press, RU Patent 2013618511. (2013).

[7] Yu.A. Miropolsky, Cold Die Forging on Automatic Machines, Mechanical Engineering, Moscow, 2001.

[8] G.A. Navrotsky, Forging and stamping machines, Mechanical engineering, Moscow, 1965.

[9] V.G. Kontsevich, Solid State Modeling of Engineering Products in Autodesk Inventor, DMK Press, Moscow, 2007.

[10] A.A. Alyamovsky, Engineering Computations in SolidWorks Simulation, DMK Press, Moscow, 2010.

[11] V.V. Telegin, A.M. Kozlov, A.V. Kirichek, Solid Modeling in Autodesk Inventor at Initial Stage of Training of Specialists in Field Mechanical Engineering, Proceedings of the 4th International Conference on Industrial Engineering (ICIE 2018), Lecture Notes in Mechanical Engineering. (2019) 1241–1247. https://doi.org/10.1007/978-3-319-95630-5_130

[12] A.V. Ryabova, V.Yu. Tertychny-Dauri, Elements of the theory of stability, Textbook, ITMO University, St. Petersburg, 2015.

[13] A.S. Volmir, Stability of deformable systems, Part 1, textbook for universities, Yurayt Publishing House, Moscow, 2021.

[14] V.Z. Aladiev, V.K. Boyko, E.A. Rovba, Programming in Maple and Mathematica Packages: Comparative Aspect, Monograph, Grodno State University, Grodno, 2011.

[15] Ch.G. Edwards, D.E. Penny, Differential Equations and Boundary Value Problems, Modeling and Computation with Mathematica, Maple and MATLAB, Dialectics-Williams, Kiev, 2007.

Determining the Diameter of Reinforcing Bars Inside Concrete Structures Regardless of Magnetic Properties

Alexey V. Mikhailov[1,a*], Yuri L. Gobov[1,b] and Leonid V. Mikhailov[1,c]

[1]M.N. Miheev Institute of Metal Physics of Ural Branch of Russian Academy of Sciences, Russia

[a]mikhaylov@imp.uran.ru, [b]go@imp.uran.ru, [c]mikhaylov_lv@imp.uran.ru

Keywords: Non-Destructive Testing, Testing of Reinforcement in Concrete, Dipole Model

Abstract. In this work, using a series of numerical experiments, the dependence of the magnetic field strength of the response of the reinforcing bar to an external magnetizing field on the magnetic properties of the reinforcing bar was investigated. The possibility of determining with high accuracy the diameter of reinforcing bars, regardless of the magnetic properties of the material from which they are made, has been proven.

Introduction

An urgent task in the field of materials processing and materials science is to assess the current state of reinforced concrete structures reinforced with steel reinforcement. The solution to this problem consists in finding the location of the reinforcing mesh, determining the strength characteristics of concrete, detecting delamination of reinforcing bars from concrete, as well as determining such parameters as the depth (thickness of the concrete cover) and the diameter of the bars. It is believed that the strength of concrete after pouring only increases over time, so the current state of the concrete itself can be estimated based on the initial mechanical properties of concrete (concrete grade). For this, a destructive method is used, based on casting samples in the form of cubes, followed by placing them under a press. The value of the critical load at which concrete destruction occurs is an indicator of its grade.

During the operation of reinforced concrete structures, it is necessary to monitor the separation of steel reinforcement from concrete. The integrity of the reinforced concrete structure must be preserved, therefore, methods and means of non-destructive testing play an important role in the operation of reinforced concrete products. If the method of impedance spectroscopy is often used to determine delamination [1], then to determine such parameters as the diameter of reinforcing bars and the distance from the reinforcing mesh to the surface of a reinforced concrete product, the greatest success has been achieved using the magnetic method of non-destructive testing. The essence of the method is reduced to the formation of a primary magnetic field of a certain configuration (or a set of configurations), registration of the spatial distribution of the measured parameter and, on this basis, reconstruction of the initial three-dimensional distribution of ferromagnetic objects in the controlled area. The thickness of the concrete cover and the location of the steel reinforcement in the structure are usually determined based on the experimentally established relationship between the parameters of the stray field and the specified controlled parameters of the structure [2]. To determine the distance, you need to know the value of the stray field, and the diameter of the reinforcement can be calculated from the magnitude of the stray field only with a known distance to the measuring point. The magnitude of the measured stray field will be influenced by both the magnetic properties of the armature and the magnitude of the magnetizing field.

Today, there are a number of ready-made devices that implement the magnetic method of non-destructive testing of reinforcement in concrete (for example, IPA-MG4.1 (SKB Stroypribor), Profoscope (Proceq, Switzerland), Ferroscan PS 250 (Hilti, Liechtenstein)). Despite the great variety, all existing products and technologies solve the problem with approximately the same accuracy. In [3], an experimental study of the magnitude of the error that accompanies the magnetic control method in determining the diameter of reinforcement and a protective layer of concrete using IPA-MG4.1 and Profoscope was carried out. As a result, it was concluded that the investigated devices determine the depth of the reinforcement with an error of 5-10%, and the diameter of the reinforcing bars with an error of 15-30%. It should be noted that the experimental results were obtained under conditions close to ideal: the depth of occurrence was determined with a known diameter of the reinforcement; the diameter was determined at a known burial depth; there are no adjacent parallel and perpendicular reinforcing bars. In real studies, it is almost impossible to achieve these conditions, therefore, the error in determining the depth of occurrence can be 50-60%, and the diameter of the rods - up to 100%.

Earlier in [4], a dipole model of local magnetization of reinforcement was proposed, which makes it possible to significantly increase the reliability of the application of the magnetic method for determining independently of each other the distance to the reinforcement and the diameter of the reinforcement. To determine the distance, the amplitude of the stray magnetic field from the locally magnetized section of the armature and the gradient of the magnetic field were measured. A permanent magnet acted as a magnetizing system; magnetic field sensors were located on the side of the magnet. It was shown that the magnitude of the magnetizing field and the diameter of the reinforcement do not affect the determination of the distance. Also, a scientific position was formulated that the diameter of the reinforcement can be determined regardless of the magnetic properties of the reinforcement, provided that such fields are created in the material, close to the coercive force, at which the magnetic permeability of the reinforcing bar tends to its maximum value ($\mu(H) \rightarrow \mu_{max}$). The purpose of this work is to confirm the advanced scientific position, as well as to study the influence on the measured field of the geometric parameters of the magnetized section of the reinforcement and its magnetic properties.

Main part

The magnetic induction in the magnetized section of the reinforcement, taking into account the demagnetization coefficient, can be calculated by the formula:

$$B = \mu_0 \mu_{body} H_e = \frac{\mu_0 H_e}{\frac{1}{\mu(H)} + N\left[\frac{\mu(H)-1)}{\mu(H)}\right]}, \quad (1)$$

where H_e – field created by a permanent magnet dipole:

$$H_e = \frac{2 \cdot q_M \cdot \frac{1}{2} L_M}{\left(z^2 + \frac{1}{4} L_M^2\right)^{\frac{3}{2}}} \quad (2)$$

Provided that $\mu(H) \rightarrow \mu_{max}$, that is $\mu_{max} \gg 1$, then:

$$B \approx \frac{\mu_0 H_e}{\frac{1}{\mu(H)} + N}. \quad (3)$$

If we assume that the magnetized section of the reinforcement is an ellipsoid, then the demagnetizing coefficient can be easily estimated as the ratio of the diameter of the reinforcement to the length of the magnetized section.

$$N \approx \frac{D}{L} \cdot \alpha. \quad (4)$$

Then, if the length of the magnetized section of the reinforcement does not exceed ten of its diameters, then $N \approx 0.25...0.33$ and $N \gg \frac{1}{\mu(H)}$, those:

$$B \approx \frac{\mu_0 H_e}{N}. \quad (5)$$

In reality, the shape of the locally magnetized section of the reinforcement can differ significantly from the ellipsoid, therefore, N and $\frac{1}{\mu(H)}$ may be comparable in magnitude and it is necessary to be guided by formula (3). In addition, the length of the magnetized section varies depending on the length of the magnet and the distance from the magnet to the reinforcing bar, therefore, using formula (4), only a qualitative estimate can be obtained. More precisely, the demagnetizing coefficient should be determined based on the results of numerical simulation by the finite element method.

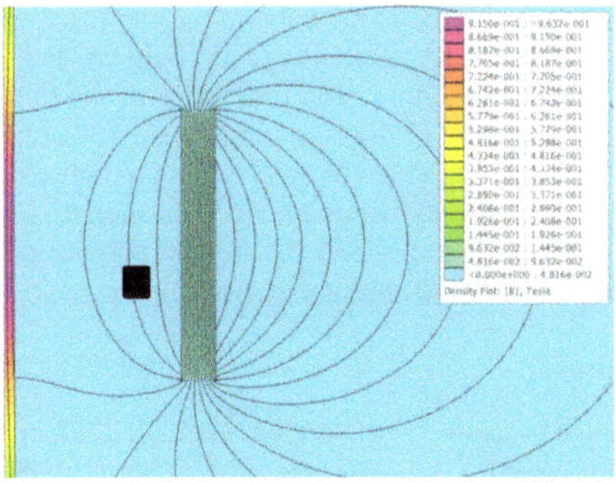

Fig. 1. *Distribution of magnetic induction obtained in the FEMM software package 4.2*

This work contains the results of a series of calculations in the FEMM 4.2 software package. The response of a reinforcing bar to an external magnetic field created by a permanent magnet is considered for various reinforcement diameters of 8, 12 and 16 mm and a range of μ = 100 ... 5000. The magnetic field strength was determined at a point near the bottom surface of a permanent magnet (black square in Fig. 1): at this point, the location of the magnetic field sensor in a real device that implements this method is assumed. For example, in this Fig. 1 shows the distribution of magnetic induction when solving an axisymmetric problem for a reinforcing bar with a diameter of 8 mm and an annular magnet with a section of 100x12.5 mm and μ =1000. The calculation results fully correspond to the planar problem with the magnet dimensions 100x100x12.5 mm, since the characteristics of the permanent magnet were selected based on the correspondence of the volumes of the sector of the ring magnet and the magnet in the form of a parallelepiped. The volume of the latter is 0.125 dm^3 (mass 1 kg), while the volume of the ring magnet is 10 times larger, therefore, in the calculation, the coercive force of the magnet was reduced 10 times compared to the nominal value. In this case, the distributions of the magnetic field from the sector of the ring magnet, which occupies 1/10 of its total volume, and the magnet in the form of a parallelepiped, will be similar.

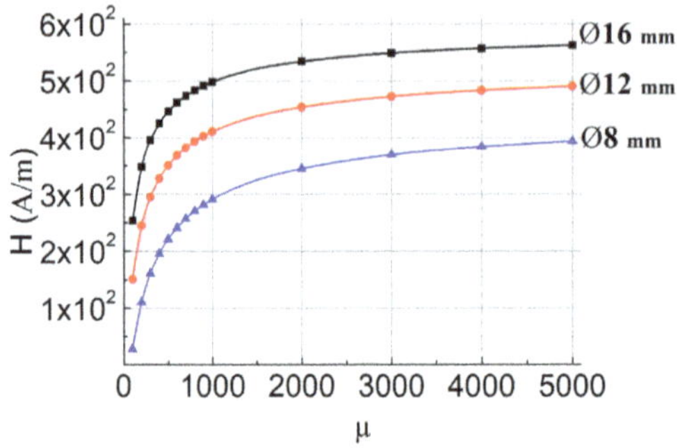

Fig. 2. *Magnetic field strength of the response of the reinforcing bar to an external magnetizing field, depending on the magnetic properties of the reinforcing bar. Sectional dimensions of the ring magnet 100x12.5 mm*

The results of calculating the strength of the magnetic field created by the magnetized section of the reinforcement, depending on the magnetic properties of the reinforcing bar, are shown in Fig. 2. The cross-sectional dimensions of the ring magnet are 100x12.5mm. It can be seen from the figure that there is an interval of μ values, at which it is possible to unambiguously determine the diameter of the reinforcement at a known magnetic field strength - from 2000 to 5000. In this interval, the graphs have a linear character and the unique correspondence of the *H* value to a single μ value on one of the graphs for different *d*.

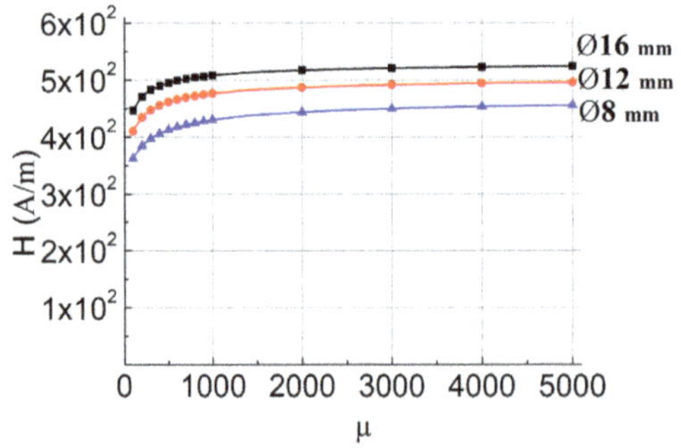

Fig. 3. *Magnetic field strength of the response of the reinforcing bar to an external magnetizing field, depending on the magnetic properties of the reinforcing bar. Sectional dimensions of ring magnet 50x25 mm*

Structural steels from which reinforcing bars are made, according to national standards, can have μ_{max} values from 1500 to 3500, therefore, it is necessary to expand the range of unique definitions of reinforcement. This can be done, according to expressions (4, 5), by changing (downward) the length of the magnetized section of the reinforcement. Such a case was also considered, the results are presented in Fig. 3. The cross section of the ring magnet did not change, and the dimensions were 50x25 mm. Analyzing the graphs, we can say with confidence that the minimum of the interval for the unambiguous determination of d has expanded to $\mu = 1000$, therefore the proposed method can be applied to all structural steels with the properties stated above

Conclusion

1) In the work, using a series of numerical experiments, it was proved that due to only one measurement of the absolute value of the magnetic field strength, it is possible to determine the diameter of the reinforcing bars with great accuracy, regardless of the magnetic properties of the material from which they are made.

2) Expansion of the range in which the diameter of the reinforcement is determined unambiguously is possible by reducing the length of the magnetized section of the reinforcing bar. This is achieved by changing the geometric parameters of the magnetizing system.

Acknowledgments

The work was performed in accordance with the theme “Diagnostics” No. AAAA-A18-118020690196-3.

References

[1] S. Hong, R. Harichandran, Sensors to Monitor CFRP/Concrete Bond in Beams Using Electrochemical Impedance Spectroscopy, J COMPOS CONSTR. 9 (2005) 515-523. https://doi.org/10.1061/(ASCE)1090-0268(2005)9:6(515)

[2] GOST 22904-93. Konstrukcii zhelezobetonnye. Magnitnyj metod opredelenija tolshhiny zashhitnogo sloja betona i raspolozhenija armatury.

[3] Vasil'ev A.I., Ocenka korrozionnogo iznosa rabochej armatury v balkah proletnyh stroenij avtodorozhnyh mostov, Beton i zhelezobeton. 2 (2000) 20-23.

[4] Gobov Y.L., Mikhailov A.V., Smorodinskii Y.G., Magnetic method for nondestructive testing of rebar in concrete, Russian journal of nondestructive testing. 54 (2018) 871-876. https://doi.org/10.1134/S1061830918120057

Influence of Acid-Activated Micro-Dispersed Additive on the Properties of Cement Sulphate-Resistant Compositions

Victoria Petropavlovskaya[1,a], Tatiana Novichenkova[1,b*], Kirill Petropavlovskii[1,c] and Maria Zavadko[1,d]

[1]Tver State Technical University, 22, Af. Nikitin, Tver, 170026, Russia

[a]victoriapetrop@gmail.com, [b]tanovi.69@mail.ru, [c]raikiri@inbox.ru, [d]79043517876@yandex.ru

Keywords: Ash, Waste, Activation, Microdispersed Additive, Microspheres, Cement

Abstract. Research is devoted to the possibility of obtaining a composition based on a microdispersed ash additive and a plasticizer. Waste from sewage sludge ash used. They are formed at thermal power plants. Such ashes are distinguished by the presence of a large amount of organic matter and a low calcium content. Therefore, they are rarely used in construction. The activation of the waste fuel and the use of a plasticizer can improve the performance of cement compositions with the inclusion of sewage sludge ash.

Introduction

Modern approaches to construction production are based on the formation of comfortable conditions for the existence of humans, animals and plants, provided that natural resources and energy are saved. Therefore, the attention of researchers and manufacturers is directed to the involvement of waste in the resource-intensive production of building materials and products. One of such wastes is ash waste from the fuel industry. It is formed when coal is burned. More than 1 billion tons of ash have been accumulated in Russia. It occupies large areas of the urban area and is the cause of environmental pollution [1-9]. Ash can be a source of air and groundwater pollution with heavy metals, which pose a potential threat to the environment even in small quantities [2, 3, 4].

At the same time, in the construction industry, fuel ash is used as a raw material for cements [8], clinker-free binders, concretes [5-9], the production of porous aggregates, silicate materials, ceramic products, as well as for thermal insulation and other materials [1, 10, 11, 12, 13]. Ash is also used in compositions for roadway construction, strengthening of soils and foundations. However, not all ashes are used in production, in most cases due to their composition.

Waste fuel can have a different chemical and grain composition. Fly ash consists mainly of highly basic aluminosilicate glass, of several crystalline phases, including quartz, hematite, magnetite, mullite, anhydrite [5], and may contain varying amounts of unburned coal in the form of a coarse fraction (depending on the conditions combustion) and other components [1, 6]. Unburned coal - organic residue can reduce the corrosion resistance of concrete, can swell when interacting with water, have low adhesion with a binder, and can form hazardous compounds. These compounds can destroy the cement stone. The efficient use of fly ash requires the removal of unburned coal and the separation of an aluminosilicate microsphere [6]. Aluminum-containing microstructures - glass-crystalline aluminosilicate balls (microspheres) are the most valuable component in ash [7]. According to the chemical composition of the shell, the microspheres are oxides of silicon and aluminum in combination with small amounts of oxides of iron, calcium, magnesium, sodium, etc. Aluminosilicate granules can participate in the formation of the structure of an artificial stone and contribute to an increase in strength without disturbing the technology of its cementation.

Dispersion and granulometry of fuel ash are also the most important characteristics [11].

Fly ash as a raw material component of geopolymer binders is responsible for the reactivity in the medium of the alkaline activator and the formation of the optimal pore structure of the composite [11].

The value of fly ash from the standpoint of obtaining aerated concrete composites based on geopolymer binder is in obtaining cheap raw materials. A raw material source in the synthesis of energy-efficient cementless cellular materials can be acidic fly ash [11]. The amount of acidic fly ash in aerated concrete can vary from 15 % to 80 % by weight of the cement. Replacing part of the cement with ash can reduce production costs and improve the environmental situation.

The use of high-calcium ash with a significant content of free CaO as a component in various cements is hampered by their low durability. This is due to an increase in the volume of lime during its delayed hydration, which leads to the formation of cracks and a decrease in strength [9, 13].

The formation of cracks in artificial cement stone using high-calcium ash can be avoided. For this, ash grinding to a higher specific surface area can be used [13]. Additionally, it is proposed to grind ash together with a superplasticizer. In this case, a nanomodified organomineral additive is formed, which has a high homogeneity and chemical activity [14].

Ash, when combined with a superplasticizer, is capable of at the same time reducing the water demand of the concrete mixture [14]. According to K.S. Stenechkina the use of ash-based modifiers reduces delamination, water separation, and also reduces the formation of fine-grained concrete efflorescence [14].

According to S.V. Samchenko [15, 16], the use of a superplasticizer can lead to the formation of a large number of crystallization centers and the growth of fine-fibrous or fine-acicular crystals of the ettringite phase. This will contribute to the formation of a dense and durable structure of the future stone [15, 16, 17]. Conditions can be created in the system under which the rate of nucleation exceeds the rate of crystal growth [15, 16].

Partial replacement of cement with fly ash reduces heat generation during the hydration of the binder [2], therefore, the risk of cracking of concrete at an early stage is reduced, following this one can consider their use in massive structures, for example, in foundations, platinum used up to 50 % of ash by weight of the binder [4, 9].

The durability of concrete is additionally favorably influenced by calcium hydrosilicates. When a nano-modifier based on mechanically activated fly ash is introduced into a multicomponent binder, a greater number of crystals are formed. In a concrete system, an increase in strength and density occurs [14, 18]. In addition, according to the authors of [1] fly ash has the properties of an independent hydration-hardening binder, which are enhanced by mechano-chemical activation of ash by additional grinding with cement and a setting regulator.

The introduction of up to 50 % fly ash contributes to the preservation of the strength of fine-grained concrete [1, 14]. The introduction of up to 40 % of activated fly ash contributes to an increase in strength up to 50 %. The strength of concrete is increased by 30-50 % if a nanomodifier is used. It is an additive of mechanically activated fly ash in the amount of 0.5-1.0 % in combination with a superplasticizer [14]. It is noted that the superplasticizer is able to improve the technological properties of the concrete mixture and the physical and mechanical characteristics of the concrete stone at the same time [10].

However, it is necessary to establish limits for the use of ash in cements and concrete. An increase in the proportion of ash in the binder to 50 % reduces the strength of concrete after heat treatment by 6-7 MPa [18]. But it is noted that the proportion of replacement of Portland cement with ash does not affect the strength of concrete after 6 months of hardening. Therefore, the crushed ash is considered as a slowly hardening binder component. On the other hand, the progress of the reaction with ash depends on the presence of capillary pores where hydrates are precipitated, i.e. the degree of ash reaction is higher in a system with a large amount of water-filled space [19].

Another type of waste is sewage sludge ash.

Such ashes are composed of inhomogeneous particles with a rough surface texture and a porous microstructure. This can lead to high water uptake and increased water demand. For example, in the composition of cement stone and concrete. This was established in [20]. To achieve the desired workability of the mixture with hydro removal ash, a higher water content was required compared to the control mixtures with fly ash. This was expected given the benefits of fly ash on the rheological properties of the mixtures.

In addition to the high organic content of sewage sludge ash, the main oxides are SiO_2, Al_2O_3 and CaO. Oxides Fe_2O_3, Na2O, MgO, P_2O_5, SO_3 and others are present in smaller quantities. The average aluminum content is approximately 14 %. This indicates the suitability of such ash for the production of aerated concrete. The high aluminum content of acidic sewage sludge ash can also incr21ease the resistance of concrete to chloride attack due to the ability of amorphous alumina to bind chlorides.

With a higher content of sewage sludge ash - up to 11 % - the content of heavy metals, sulphates and in particular phosphorus becomes excessively high. This leads to an increase in the setting time and suppression of the development of strength.

The characteristics of such ash, in particular its fineness, suggest that this material may be suitable for use in concrete. Ash is introduced as filler and fine aggregate. Thus, the use of sewage sludge ash led to an increase in strength (up to 70 %) due to the filling effect from a higher content of fine particles. Despite the increase in the permeability of the resulting mixtures. There is a decrease in the thermal conductivity of concrete due to the porous nature of the organic part of sewage sludge ash [20].

With the dispersion of the hydro removal ash, the area of intergrowth contacts between the particles increases and the transfer of electrons in the surface layer becomes much easier [2, 14]. This is reflected in the change in the performance characteristics of the created material in comparison with the original. Fibrous crystals are observed in the microstructure of ash-cement stone [20]. They "pierce" the pore space of the artificial cement stone, contributing to its strengthening.

When ash is used as an ultradispersed filler, the physical and mechanical characteristics of ash-cement stone, ash-cement-sand concrete are improved. At the same time, for ash-cement-sand concrete, the compressive strength increases to 35 %, and in bending up to 32.4 %. In the case of ash-cement stone, the compressive strength increases up to 30 %, and the thermal conductivity decreases by 6.5 %. The optimal ratio of ash to cement in this study was 30:70 [21].

Fly ash, hydraulic removal ash - each of them has its own advantages and disadvantages. Therefore, when designing compositions, it is necessary to provide for measures to improve the quality indicators of composites, concretes and solutions based on them.

In this work, we investigated the possibility of obtaining an ash-cement composition based on activated acidic sewage sludge ash to obtain a sulfate-resistant binder.

Materials and methods

In the studies, the following raw materials were used to obtain ash-cement compositions: sewage sludge ash (fig. 1) – waste from the Kashirskaya SDPP, Moscow Region. Ash dispersion varied from 0 μm to 100 μm; cement – manufacturer “LafargeHolcim” company, Kolomna, Moscow region, standard compressive strength - at least 50 MPa; quartz sand was used as a fine aggregate (fig. 2) (Tver region). The work also used an organic plasticizing additive - hyperplasticizer (produced in St. Petersburg). In the composition of raw mixtures, the ratio of cement and sand was taken constant – Cement : Sand = 1: 3. The conditions for varying the components in the composition of the adopted mixtures are shown in Table 1.

Table 1. *Composition of mixtures*

№ composition	Composition components			
	Cement	Sand	Ash,%	Plasticizer,%
1	+	+	5	0
2	+	+	15	0
3	+	+	5	0,2
4	+	+	15	0,2
5	+	+	5	0,1
6	+	+	15	0,1
7	+	+	10	0
8	+	+	10	0,2
9	+	+	10	0,1

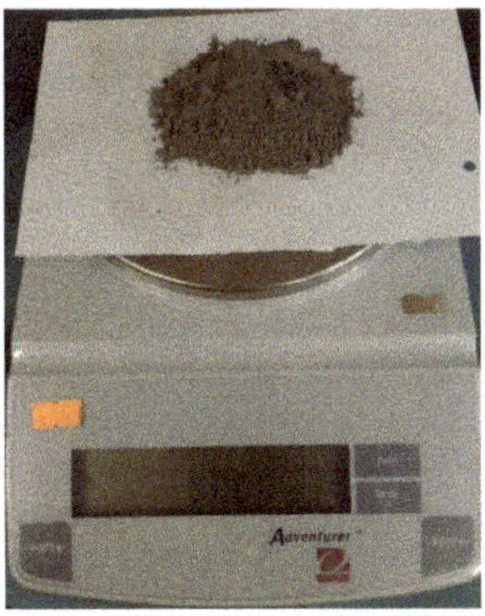

Fig. 1. *Sewage sludge ash*

Fig. 2*. Quartz sand*

Studies of the physical and mechanical characteristics of cement compositions were carried out at the age of 7 days (fig. 3). Compressive strength tests of specimens with natural moisture were carried out on a hydraulic press (fig. 4).

Materials Research Proceedings **21** (2022) 285-292 https://doi.org/10.21741/9781644901755-51

Fig. 3. *Testing a sample on a hydraulic press*

Fig. 4. *Fracture of sample*

The amount of ash and organic plasticizing additive in table 1 is given as a percentage of the weight of the cement.

To determine the ultimate compressive strength and average density of the cement composition containing a complex additive, standard cubes (fig. 5) were made from mixtures with a normal consistency. All investigated compositions of cement compositions before testing were stored in a chamber with a humidity of at least 90 % at a temperature of (20 ± 2) °C (fig. 6).

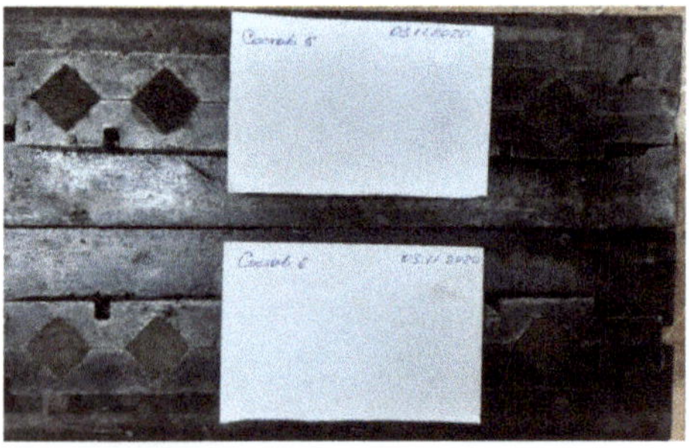

Fig. 5. *Sample molding*

Fig. 6. *Samples in the hardening chamber*

Results and Discussion

In order to improve the properties of the ash component, its mechanical activation was carried out. Based on its results, a microdispersed additive was obtained with a wide particle size distribution and a decrease in the average particle diameter in the mixture to 13 μm.

The results of studies of physical and mechanical characteristics are shown in table 2.

Table 2. *The results of testing the compositions of cement compositions with additives*

№ composition	1	2	3	4	5	6	7	8	9
Compressive strength, MPa	15.4	17.9	22.1	14.1	22.4	24.9	22.5	21.4	20.3
Density, kg / m^3	1851	1946	1963	2092	2168	2186	2184	2213	2184

The effect of the material composition of cement compositions on the compressive strength is shown in fig. 7. The change in the average density of artificial stone depending on the material composition of cement compositions is shown in (fig. 8).

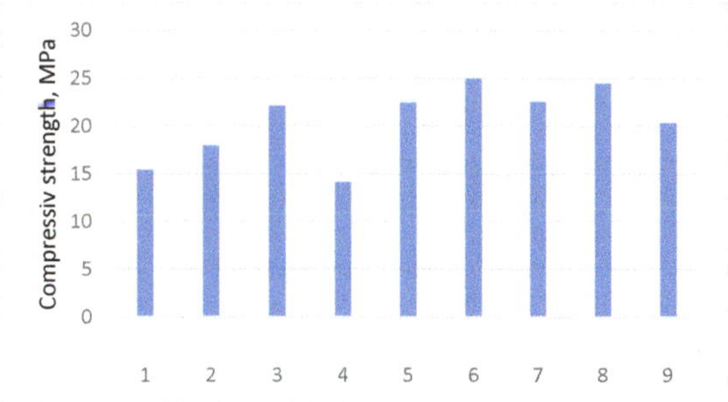

Fig. 7. *Compressive strength of cement compositions.*

According to the results of density studies, it was found that the average density of the compositions increases with an increase in the content of the ash component. The maximum density value is achieved by adding 10 % ash (composition No. 8). The content of the hyperplasticizer in it was maximum – 0.2 %. With the introduction of ash over 10 %, the density of the composition decreases monotonically. The minimum density value has been reached for composition No. 1. The ash content in it was 5 %, organic additive – 0 %.

The maximum value of the ultimate compressive strength of the samples was obtained for composition No. 6 with an ash content of 15 % and a plasticizer - 0.1 % by weight of cement. The average density of this composition has reached a value of 2186 kg / m^3. The decrease in density relative to the maximum achieved result was less than 1 %.

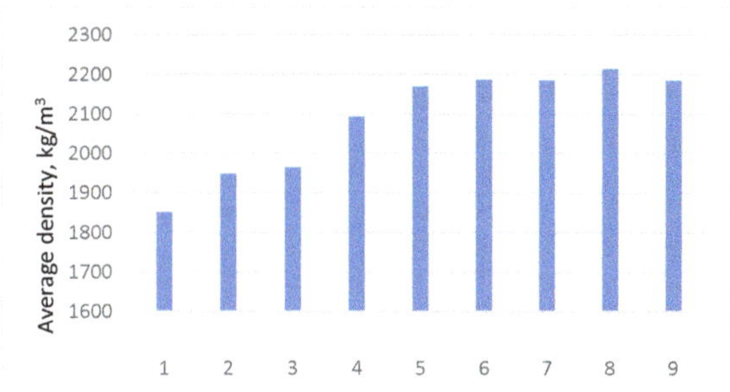

Fig. 8. *Average density of cement compositions*

Analysis of the conducted studies on the strength and density of cement compositions with the addition of acid ash and a plasticizer showed that the best results were achieved when testing composition No. 8. An increase in strength correlates with an increase in the density of artificial cement stone. This is due to the formation of a stone structure with a dense packing of particles due to the presence of ash particles as a mechanical filler at different structural levels. In addition to compaction, activated ash acts as a physicochemical agent. She participates in physical and chemical transformations in the hardening cement system. This is confirmed by the test results.

The process of structure formation is activated. The introduction of an active mineral additive in the form of ash accelerates mineral formation. The results of tests of this composition for sulfate resistance for 7 months confirm the high resistance of the cement composition with an activated microdispersed ash additive. Cement stone shows no signs of sulfate corrosion. Research is ongoing. The compressive strength of composition No. 8 on the 7th day of hardening was 24.4 MPa, the density of the stone was 2213 kg / m^3.

Conclusion

The analysis of the combined effect of the activated microdispersed additive and the plasticizer on the properties of cement sulfate-resistant compositions showed that the addition of ash and an organic additive can both worsen and improve the performance of cement stone and concrete. The optimum amount of activated ash is 10% with a plasticizer content of 0.2 %.

The results obtained are in good agreement with the results of studies of the effect of mechanical activation of fly ash with a superplasticizer on the properties of the material obtained in [6]. The combined plasticizing effect of ash and organic polycarboxylate component favorably affects the properties of the resulting cement stone. All of its main characteristics are increased. The activation of the ash component allows to improve the structure of the stone, to increase its strength due to the formation of additional structural bonds. The combination of high density and strength relative to other series of samples also predetermines the best operational characteristics of this composition - water absorption, water resistance, frost resistance, etc.

Acknowledgements

This work was supported by the Russian Science Foundation (project No 21-79-30004).

References

[1] E.A. Belyakova, R.N. Moskvin, V.S. Belyakova, Ash and slag wastes of CHP and prospects for their disposal, Education and science in the modern world. Innovations 5 (2016) 151-157.

[2] Z.T. Yao, X.S. Ji, P.K. Sarker, J.H. Tang, L.Q. Ge, M.S. Xia, Y.Q. Xi, A comprehensive review on the applications of coal fly ash, Earth-Science Reviews 141(2015) 105-121. https://doi.org/10.1016/j.earscirev.2014.11.016

[3] Kemal Celik, Cagla Meral, A. Petek Gursel, P. Kumar Mehta, Arpad Horvath, Paulo J.M. Monteiro. Mechanical properties, durability, and life-cycle assessment of self-consolidating concrete mixtures made with blended portland cements containing fly ash and limestone powder, Cement & Concrete Composites 56 (2015) 59-72. https://doi.org/10.1016/j.cemconcomp.2014.11.003

[4] N.I. Kozhuhova, D.N. Danakin, I.V. ZHernovskij, Features of geopolymer aerated concrete production based on fly ash of Novotroitskaya CHPP, Construction materials 1-2 (2017) 113-117.

[5] M.O. Korovkin, V.I. Kalashnikov, N.A. Eroshkina, Effect of high calcium fly ash on the properties of self-compacting concrete, Regional architecture and construction 1(22) (2015) 49-53.

[6] K.S. Stenechkina, Organomineral additive based on CHP fly ash, Velez 4-1 (46) (2017) 102-106.

[7] Ciarán J. Lynn, Ravindra K. Dhir, Gurmel S. Ghataora, Roger P. West, Sewage sludge ash characteristics and potential for use in concrete, Construction and Building Materials 98 (2015) 767–779. https://doi.org/10.1016/j.conbuildmat.2015.08.122

[8] A.F. Kosach, M.A. Rashchupkina, I.N. Kuznecova, M.A. Darulis, Influence of ultradisperse filler based on water removal ash on cement stone properties, Bulletin of Tomsk State University of Architecture and Construction 1(21) (2019) 150–158. https://doi.org/10.31675/1607-1859-2019-21-1-150-158

[9] I.YU. Petrik, V.N. Gubar', S.V. Kornienko, Corrosion resistance of concrete with high content of TPP fly ash, Bulletin of the Donbass National Academy of Construction and Architecture 4 (126) (2017) 103-107.

[10] S.V. Samchenko, E.M. Makarov, Formation and growth of ettringite crystals in the presence of polymeric functional additives, Successes of modern science and education 5(12) (2016) 118-122.

[11] V.P. Kuz'mina, Fillers for dry building mixtures, Dry building mixtures 3 (2017) 8-15.

[12] M.K. Kuderin, K.D. Babiev, Aluminium silicate microsphere in energy saving and energy efficiency of buildings and structures, Science and technology of Kazakhstan 1 (2019) 94-101.

[13] A.Yu. Bashkurov, Selection of optimal composition of lightweight grout for well cementing, Exploration and protection of subsoil 5 (2019) 53-56.

[14] Van Lam Tang, Zoan Nguen, Lam Tung, S.V. Samchenko, Effect of ash and slag waste additive on properties of sulphoaluminate Portland cement, Bulletin Moscow State University of Civil Engineering 8(14) (2019) 991-1003. https://doi.org/10.22227/1997-0935.2019.8.991-1003

[15] Yu.M. Bazhenov, V.V. Voronin, L.A. Alimov, A.M. Bahrah, O.A. Larsen, V.N. Solov'ev, Dyk Vin' Kuang Nguen, High-quality self-compacting concretes using coal combustion waste, Bulletin Moscow State University of Civil Engineering 12 (111) (2017) 1385-1391. https://doi.org/10.22227/1997-0935.2017.12.1385-1391

[16] E. Berodier, K. Scrivener, Evolution of pore structure in blended systems, Cement and Concrete Research 73 (2015) 25-35. https://doi.org/10.1016/j.cemconres.2015.02.025

[17] V.B. Petropavlovskaya, T.B. Novichenkova, A.F. Buryanov, K.S. Petropavlovskii, Self-hardening of a gypsum, Key Engineering Materials 729 (2017) 517. https://doi.org/10.4028/www.scientific.net/KEM.737.517

[18] Xu Gang, Shi Xianming, Characteristics and applications of fly ash as a sustainable construction material: A state-of-the-art review, Resources, Conservation & Recycling 136 (2018) 95-109. https://doi.org/10.1016/j.resconrec.2018.04.010

[19] R.S. Fedyuk Design of cement composites of increased impermeability, Bulletin Moscow State University of Civil Engineering 5 (2016) 72-81. https://doi.org/10.22227/1997-0935.2016.5.72-81

[20] Yu.A. Belencov, O.M. Smirnova, D.D. Shamanina, Self-compacting concrete using ground limestone filler, Construction – 2016: Materials of the II Bryansk International Innovation Forum (2016) 22-25.

[21] M.S. Ageeva, S.M. Shapovalov, A.N. Bocman, A.V. Ishchenko, On the use of industrial waste in the production of cementitious substances, Bulletin of Belgorod State Technological University 9 (2016) 58-62.

Thermal Treatment of Carbidostals TiC – I2XI8H10T

Leonid M. Savinykh[1,a*], Stanislav Y. Pomyalov[1,b], Tatyana A. Dudorova[1,b] and Tamara A. Verzhbalovich[1,b]

[1]Kurgan State University, Kurgan region, Kurgan, Russia, 640020

[a]savinih_LM.polytech@kgsu.ru, [b]pomialov_45@mail.ru, [c]dudorovatan@yandex.ru, [d]Zhuksks@list.ru

Keywords: Carbidostal, Heat Treatment, Impregnation, Annealing, Quenching, Tempering, Sintering, Strength, Austenite, Ferrite, Hardness

Abstract. An important task of thermal treatment of carbidostals having different composition of titanium carbide and steel I2XI8H10T is the possibility of improving their strength properties during operation under abrasive wear conditions. This question is new and little studied.

Introduction

Carbidostals obtained by impregnating the carbide carcass with a steel melt generally have a higher density and toughness than those obtained by conventional pressing and sintering methods. This is achieved by obtaining a structure with isolated fine inclusions in the steel matrix, a significant technological advantage of the process of producing carbidostal by impregnation is the fact that when forming a carbide frame, small pressures are used to compress titanium carbide powders, this makes it possible to manufacture articles of complex shape.

Carbidostals, like alloyed steels, are subjected to all types of heat treatment (annealing, quenching and tempering), the structure of the carbidostals depends significantly on their heat treatment, which can also cause diffusion redistribution of components between the refractory phase and the steel matrix. The physical and mechanical properties may depend on the method of producing the carbidostals.

Therefore, the study of heat treatment modes for this carbidostal chemical composition was a pressing issue. The ability of carbidostals to anneal allows them to reduce their hardness so much that they can be machined and the possibility of using one or another machining after annealing is determined by the ratio of carbide and steel components, the greater the content of refractory phase, the higher the hardness of the carbidostal and the more difficult its treatment in annealed state. The annealing modes of the carbidostal are determined by the steel class used as a binder and generally correspond to the annealing modes of these steels.

Theoretical provisions

All types of heat treatment (annealing, quenching and tempering) were carried out in a resistance furnace in an argon atmosphere. The prototypes were 16 x 16 x 10 mm in size.

The annealing operation was carried out to homogenize, relieve internal stresses, and obtain a more uniform structure. Annealing mode is selected according to literature data [1, 2]. Heating was carried out to 1140 K, holding – 1 hour, cooling – with the furnace to 990 K, holding at this temperature b hour, subsequent cooling with the furnace. Microstructures of carbidostal TiC – 45% I2XI8H10T after impregnation and annealing are given in Figure 1. As can be seen from the above micrographs, after annealing there were no noticeable changes in the structure of the carbidostal on the non-etched grinding. Quantitative analysis of structural constituents (Table 1) showed a tendency to increase carbide grains after annealing. The ratio of carbide to metal phases after annealing also remained substantially unchanged (Table 2). However, the composition of the metal matrix appears to have changed as the austenite became ferrite after annealing. If before annealing, according to the X-ray structural analysis, the austenite lattice parameter a = 3,602 A, then after annealing, doped ferrite was fixed with the lattice parameter a = 2,864 A. Magnetic weighing of the samples confirmed the data of X-ray structural analysis: after impregnation, the carbidostal was not attracted by the magnet, after annealing, it was attracted well. Results of magnetic weighing before annealing – 4,4 g, after – 404,5 g.

The conversion of γ-α during annealing can be explained by the fact that during annealing, part of the titanium and chromium carbides leave the solid solution, carrying carbon with them. attracted, after annealing - attracted well. Results of magnetic weighing before annealing – 4,4 g, after – 404,5 g.

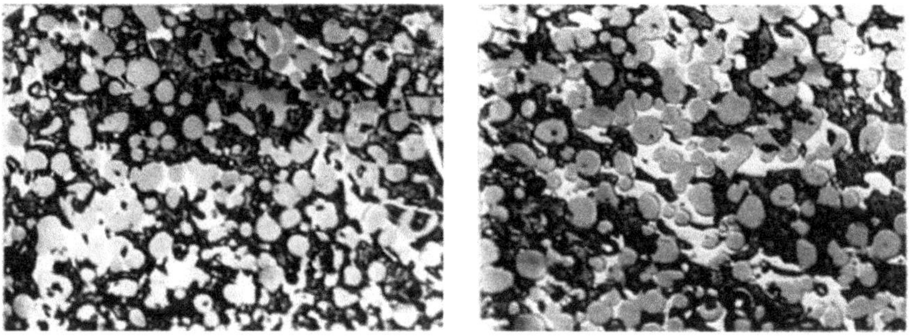

a b

Fig. 1. *The microstructure of the carbidostat TiC - 45% X18H10T after impregnation (a) and annealing (b), x 800*

A decrease in the carbon content of stainless steel and possibly an increase in the titanium content which is transferred from the carbide of the non-stoichiometric composition leads to a narrowing of the γ-region and an expansion of the α-iron region. This is the reason for the conversion of austenite to ferrite during annealing. The separation of titanium carbides and possibly chromium from the solid solution during annealing is indirectly confirmed by some decrease in the microhardness of the metal bond (Table 1).

It should be noted that with sufficiently slow cooling of the samples after impregnation, a ferrite structure of the steel matrix is also obtained. Since the ferrite structure reduces the wear resistance of the carbidostal, it is necessary to apply heat treatment after impregnation to form the austenitic structure of the steel matrix. As experiments have shown (Table 2), such heat treatment is quenching followed by tempering. On the other hand, the presence of ferrite, that is, the possibility of carbidostal being magnetic, is a positive point. In some cases, the article must be subjected to grinding treatment. Consequently, the article will be held firmly on the magnetic plate of the grinding machine.

The hardening modes of the carbidostals are also determined by the composition of the steel bond. Carbidostals are characterized by no changes in linear dimensions after quenching, in order to carry out the following task, wrapping the hardened parts in stainless steel foil or heating neutral.

The most complete dissolution of carbides in austenite is observed when heating the carbidostal under quenching to a temperature of 1270 K and holding at this temperature for 1 hour [3].

Magnetic weighing after quenching in this mode detected the minimum weight of the a-phase, and the lattice parameters of the alloyed austenite after quenching - a = 3,656 A.

Table 1. *The dimensions of carbide grains in the carbidostal are TiC-45% I2XI8HI0T*

Alloy State	Sample area	Grain distribution TiC by size %					Average grain size, μm
		Under 2 μm	2-3 μm	3-4 μm	4-5 μm	5-7,5 μm	
After impregnation	Top	63,0	27,3	7,6	2,0	no	3,10
	Middle	71,5	25,6	2,8	1	no	2,95
	Bottom	78,5	18,7	2,0	1	no	2,82
After annealing	Top	46,9	31,2	17,4	0,7	0,34	3,28
	Middle	57,5	26,0	11,8	4,0	no	3,25
	Bottom	72,8	22,3	3,2	0,7	0,20	2,87
After quenching	Top	46,6	45,1	7,8	0,4	no	3,27
	Middle	51,2	39,7	8,4	0,7	no	3,23
	Bottom	49,8	47,4	8,6	0,2	no	3,46

Thus, the hourly holding at the quenching temperature allows the required amount of titanium carbides to be dissolved in the austenite, and subsequent cooling at a high rate helps to fix a sufficiently large amount of the carbide phase in the solid solution, which provides the necessary austenitic structure to the steel matrix.

The samples were quenched from the heating temperature into oil. No leash, warps and changes in sample sizes were detected during the hardening process.

Table 2. *The characteristic of the steel binder is carbidostal TiC - 45% I2XI8HI0T*

Alloy State	Sample area	Content, vol.%		Average size steel interlayer, μm	Length of contact line TiC - bundle, mm^{-1}	Steel matrix microhardness, kgf/mm2
		TiC	Steel bundle			
After impregnation	Top	32,5	63,5	2,80	30,9	400-550
	Middle	31,7	68,3	2,50	29,6	400-550
	Bottom	40,4	59,6	2,20	31,5	400-550
After annealing	Top	36,6	61,2	3,05	24,4	360-380
	Middle	39,6	60,4	3,00	26,4	360-380
	Bottom	38,9	61,2	2,89	28,2	360-380
After hardening and vacation	Top	37,9	63,1	2,36	25,9	650-750
	Middle	40,6	59,4	2,46	26,8	650-750
	Bottom	40,4	59,6	2,41	30,9	650-750

The conversion of $\alpha \rightarrow \gamma$ during hardening of the carbidostal TiC - I2XI8HI0T is explained by the dissolution of titanium carbides in austenite during the heating and holding of the alloy. As a result of rapid cooling, carbon and titanium remain in the solid solution. This is indirectly confirmed by the increase in the microhardness of the metal matrix (Table 2) as well as the total hardness of the carbidostal. As shown in Tables 1, 2 and Figure 2, the dimensions of carbide grains and steel bundle change little after quenching.

The microstructure of the carbidostal after quenching on untwisted samples also differs little from the starting ones. After etching the slips, you can see the areas sharply delineated by concentration lines (Figure 2). In addition, significant amounts of fine carbides are observed to precipitate.

To relieve stress, the samples after quenching were tempered at 800 K, held at this temperature for 1 - 2 hours and cooled in oil.

Table 3 shows the hardness of the carbidostal after sintering and heat treatment.

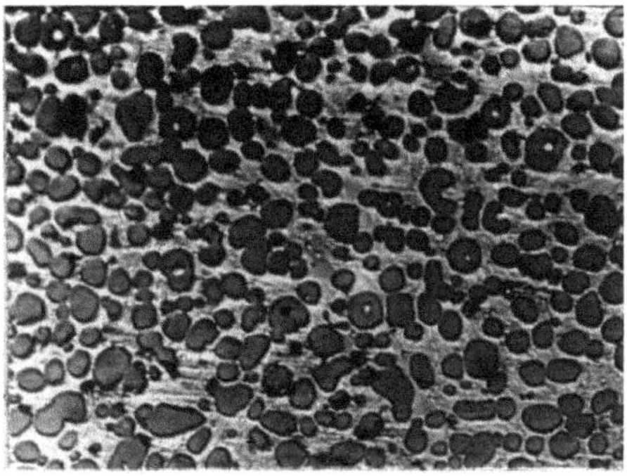

a

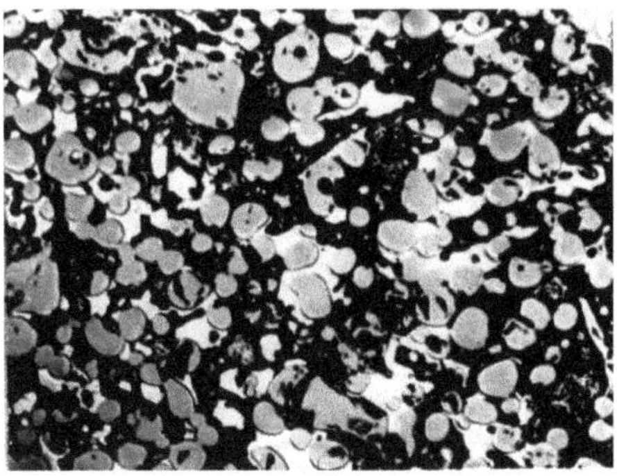

b

Fig. 2. *The microstructure of the carbidostal TiC - I2XI8H10T after quenching: a - untwisted samples, x 800; b – samples etched, x 800*

Based on the results of the study, a technology for producing carbidostal TiC - I2XI8HI0T by impregnation of a green carbide framework [3, 4] was developed.

Table 3. *Carbidostal hardness I2XI8HI0T*

Composition, vol.%		HRC hardness after			
TiC	Steel	Impregnations	Annealing	Trainings	Holidays
55	45	83,7±0,8	82,4±0,8	85,28±0,16	85,02±0,18
45	55	81,71±0,72	80,7±0,7	84,28±0,5	83,82±0,38

Conclusions

Thus, quenching and low tempering of the carbidostal TiC - I2XI8HI0T provide an austenitic structure to the metal matrix and increase the hardness of the carbidostal.

References

[1] S.S. Kiparisov, V.K. Narva, L.I. Dalyaeva, News universities. Non-ferrous metallurgy **2**, 136-140 (1976)

[2] S.S. Kiparisov, I.V. Paisov, V.K. Narva, Powder metallurgy **9**, 10-13 (1969)

[3] Yu.G. Gurevich, V.B. Akimenko, N.R. Frage, L.M. Savinykh, Steel **10**, 77-78 (1981)

[4] Yu.G. Gurevich, L.M. Savinykh, Powder Metallurgy, 13-14 (1984)

Modern Trends in Manufacturing Technologies and Equipment | Materials Research Forum LLC
Materials Research Proceedings **21** (2022) 298-303 | https://doi.org/10.21741/9781644901755-53

Algorithm Synthesis of Controlling a Transport Unit of a Coextrusion Flexible Manufacturing Section for Processing Multicomponent Materials

Valery V. Dyadichev[1,a*], Sergey S. Stoyanchenko[2,b] , Roman L. Plomodyalo[3,c] and Aleksandr V. Dyadichev[1, d]

[1]V.I. Vernadsky Crimean Federal University, 2 Pavlenko Street, Simferopol, Russia

[2]Lugansk State University named after Vladimir Dal, 20-A Molodezhnyi Sq., Lugansk, Ukraine

[3]Kuban State Technological University (KubSTU), Moskovskaya str. 2, Krasnodar, Russia

[a]mr.dyadichev@mail.ru, [b] santono@mail.ru, [c] ndt-2018@mail.ru, [d] adyadichev@mail.ru

Keywords: Algorithm, Mixture of Materials, Dosing Device, Multicomponent Materials, Synthesis, Coextrusion Manufacturing Section

Abstract. The paper offers a mathematical model of the process of coextrusion flexible manufacturing section functioning. On the basis of dynamic programming method an algorithm of controlling transport unit of coextrusion flexible manufacturing system is developed. The objective of the current research is the development of the mathematical model of the flexible manufacturing complex for processing multicomponent materials functioning process, and construction of the algorithm of controlling a transport unit of a coextrusion flexible manufacturing complex for processing multicomponent materials on this basis. The paper offers the criterion of evaluating the quality of controlling the transport unit. The chosen variant of controlling the transport unit has to meet many requirements. This paper con-siders the issues connected with the search for an optimum algorithm of control-ling a transport unit according to the set criterion. Most works offer analytical models which are based on the assumption of the current processes' stability. Real manufacturing conditions are characterized by the effect of numerous perturbing factors. Under such conditions the assumption of the current processes' stability makes the obtained models almost untrue.

Introduction

Using the automation of transport work in the GPS allows eliminating the hard work of auxiliary workers and freeing them to participate in the main process, saves time for the main workers, who spend up to 20% of the total fund of time on the receipt and delivery of blanks, parts, equipment.

Upon the arrival of goods to the GPS, after work on a batch of parts completion, the vehicles are the main executive devices that implement control algorithms for the transfer of goods in the system. Cargo flows in the GPS are extremely diverse - these are billets, semi-finished products, finished parts, equipment, production waste. Universal vehicles can be used for these material flows, but separate transport subsystems are often used [1-6].

The interconnection of transport in the GPS with the control system, storage devices for the elements of the material flow, various design solutions are significantly complicate the tasks of developing a transportation system of GPS. For type site as GPS we can distinguish non-systemic, interoperational and operational transport. Non-systemic transport provides the connection of the GPS with other production units, and above all with a higher-level system, which is responsible for providing the GPS with blanks and equipment, determines the further movement of finished parts to the consumer, ensures the delivery of cutting tools and other equipment. The role of off-system transport varies depending on the degree of independence of the GPS. The organization of retrieval system work determines greatly the work efficiency of the flexible manufacturing section. The efficiency of separate units of the techno-logical equipment very often differs considerably. The only transport unit quite often operates several units of the technological equipment. Under such conditions the solution to the problem of controlling the transport unit of the production complex is of a peculiar difficulty. The chosen variant of control-ling the transport unit has to meet many requirements. This paper considers the issues connected with the search for an optimum algorithm of controlling a transport unit according to the set criterion.

Analysis of Scientific Publications

Many scientific researches [7-17] are dedicated to scrutinizing the issues of mathematic modelling of multifunctional manufacturing systems. The scientific work studies the issues of analyzing the structures of the traffic in flexible manufacturing systems. Most works offer analytical models which are based on the assumption of the current processes' stability. Real manufacturing conditions are characterized by the effect of numerous perturbing factors. Under such conditions the assumption of the current processes' stability makes the obtained models almost untrue.

The Objective of the Work

The objective of the current research is the development of the mathematical model of the flexible manufacturing complex functioning process, and construction of the algorithm of controlling a transport unit of a flexible manufacturing complex on this basis. The paper offers the criterion of evaluating the quality of controlling the transport unit.

Research

While developing technological structures of manufacturing systems, they often use the variant when one transport unit serves several technological modules (see Fig. 1). Under wrong management of the work by the transport unit downtimes of technological modules may happen due to the absence of workpieces and too often rearrangements of the transport unit. In such systems it is necessary to manage the work of the transport unit in order to prevent or minimize the downtimes of technological modules and minimize the amount of rearrangements of the transport units. The analysis of the task of search for a rational strategy of controlling the transport unit shows that this task belongs to the class of dynamic tasks of discreet optimization [18-20].

To formulate the task of search for a strategy of controlling the transport unit in terms of dynamic programming we will introduce a range of hypotheses. Let the average time of transporting be equal to all the technological sections and be t time units. The duration of the transport unit arrangement for serving i technological module is Tai time units. We will designate by tai the number showing how much Tai is bigger than t. Let also ti be the number showing how many times an average duration of the cycle i of the technological module is bigger than t. Let us designate by Ei the piece capacity of the input for i module of the storage (I=1,M, M is the amount of the technological modules, served by the transport unit).

Let us consider the functioning process of the flexible manufacturing section at the time interval from 0 to T and divide it into a range of pieces Δt. The duration of each piece equals to the duration of the transport operation.

The control over the transport unit will be set in a form of a vector X, the dimension of which equals to the amount K of pieces, into which the considered interval is divided $0..T$. The elements X_j j=1, K of this vector reflect taking a definite decision at a corresponding piece Δt, the decision concerns the control over the transport unit. Let us assume that X_j equals to l, then it means that at j-piece one of the possible decisions concerning the control over the transport unit was taken:

$1 \le l \le M$ - is the transportation at j-piece of one manufactured item for l-section;

$(M+1) \le l \le 2 \cdot M$ - is the downtime of the transport unit, arranged for transportation of items for (l-M) technological module;

$(2 \cdot M + 1 \le l \le 3 \cdot M$ - is the rearrangement of the transport unit for transporting items for (l-$2M$) technological module.

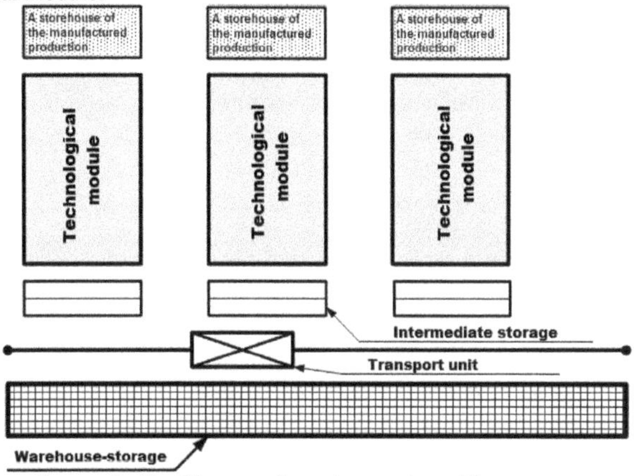

Fig. 1. *The scheme of a flexible manufacturing section with a common transport unit*

The space of states, in which the studied system may be at each step, is described by the group of three <A,B,C>. Here A is the number of the technological module, at the items' transportation of which the transport unit is arranged ($1 \le A \le M$). B and C are the vectors of the dimension M. Each component bi (I=1,M) of the vector B numerically equals to the amount of the manufactured items, contained in the input for i-section storage ($0 \le b_i \le E_i$). The component ci (i=1,M) of the vector C shows how many steps (pieces) are left to the delivery of the signal by i-technological module for the supply of the next item ($0 \le c_i \le t_i$) to this module. The general number of states in which the system may be at each step is determined by the next expression:

$$M \cdot \prod_{i=1}^{M}((E_i+1)\cdot(t_i+1)) \quad (1)$$

The common costs F, connected with the operation of the flexible manufacturing complex, can be presented as a sum:

$$F=F1+F2 \quad (2)$$

Here F1 is the losses determined by the downtimes of the technological modules:

$$F_1 = \sum_{i=1}^{M} e_i \sum_{j=1}^{k} q_{ij}, \quad (3)$$

where:
q_{ij}=1, if $b_{ij} + c_{ij} = 0$,

Materials Research Proceedings **21** (2022) 298-303 https://doi.org/10.21741/9781644901755-53

0, if not,

e_i – is the cost connected with the downtime of i-technological module during one piece Δt.

F_2 is the losses connected with the rearrangement of the transport unit

$$F_2 = e_p \sum_{j=1}^{k} r_j \tag{4}$$

where r_j – 1, if $x_j \leq 2M$ and xj+1>2M,

0 - if not.

In multistep processes with a subsequent decision taking the transition of the system from one step to another and from one state to another is described by a functional equation. For the considered task the functional equation of the dynamic programming is presented as follows:

$$F_k(s) = \min_{\substack{s \in S \\ s' \in S}} [g_k(s) + F_{k-l}(s')], \tag{5}$$

where:

F_k(s) is the value of the target function (1), obtained at the step k of the optimization to the state S,

g_k(s) is the increment of the target function under the transition of the manufacturing system form the state s' at (*k-l*) –step in to the state s at k-step,

S' is the multitude of states at (*k-l*) –step, from which the transition into the state S at k-step is possible,

S is the multitude of possible states of the manufacturing system at each step.

l is tai, if in the state s of the manufacturing system the transition is performed by the rearrangement of the transport unit for transporting the manufactured items for i technological module,

1 if not.

Fig. 2 shows the work scheme of the obtained method of algorithm synthesis of controlling the transport unit of the flexible manufacturing complex.

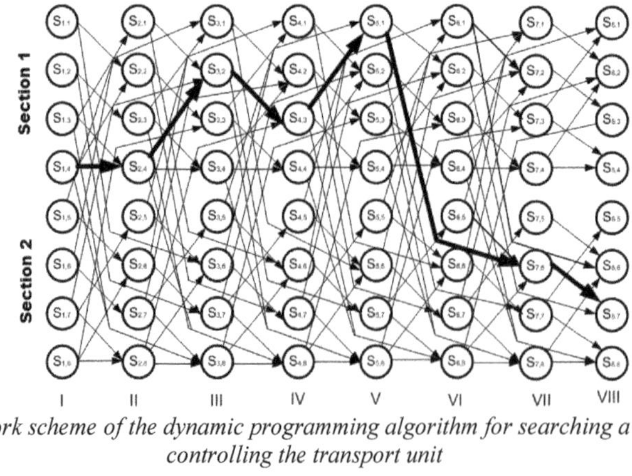

Fig. 2. *The work scheme of the dynamic programming algorithm for searching a strategy for controlling the transport unit*

Materials Research Proceedings 21 (2022) 298-303 https://doi.org/10.21741/9781644901755-53

The calculating experiments with the computer implementation of the obtained algorithm of controlling the transport unit showed that its use reduces the losses connected with the operation of the flexible manufacturing complex by 25-20%. It is necessary to mention specially that the developed algorithm has a high performance. This fact allows using it in such manufacturing systems where the manufacturing tasks are often changed.

Displayed equations are centered and set on a separate line.

$$x + y = z \quad (6)$$

Acknowledgement

The study was carried out with the financial support of the Ministry of Education and Science of the Russian Federation within the framework of the scientific project FZEG-2020-0030.

References

[1] B. Bauer, R.F. Bullnheimer, C. Hartl, Strauss. Minimizing total tardiness on a single machine using ant colony optimization, Proc. of the 1999 Congress on Evolutionary Computation (CEC99), Washington D.C., USA (1999) 1445–1450

[2] J. Blazewicz, et all, Handbook on Scheduling. From Theory to Practice, Springer Verlag (2019). https://doi.org/10.1007/978-3-319-99849-7

[3] O. Brucker, Scheduling Algorithms. Springer Verlag (2007)

[4] P. Brucker, S. Kravchenko, Scheduling Jobs With Release times on Parallel Machines to Minimize Total Tardiness. OSM Reihe P, Heft 258, Universit¨at Osnabr¨uck, Fachbereich Mathematik/ Informatik (2005)

[5] S. Chang, Q. Lu, G. Tang, W. Yu, On decomposition of total tardiness problem, Oper. Res. Lett. 17 (1995) 221-229. https://doi.org/10.1016/0167-6377(95)00027-H

[6] F. Della, A. Croce, V. Grosso, Paschos, lower bounds on the approximation ratios of leadingheuristics for the single-machine total tardiness problem, J. of Scheduling 7 (2004) 85-91. https://doi.org/10.1023/B:JOSH.0000013056.09936.fd

[7] A.P. Gavrish, B.M. Voronets, Robotized Machine Building Construction Complexes for Mechanical Processing (in Russian), Technika, Kyev (1984)

[8] E.N. Khobotov, Some models and solution methods for scheduling tasks in discrete manufacturing systems, A&T 12 (2007)85-100

[9] I.A. Klausov, I.S. Blyakherov, G.M Varyash, A.A. Ivanov, Automatic Loading of Technological Machines (in Russian), Reference book, Mashinostroyeniye, Moscow (1990)

[10] M.S. Kunyaev, A.M. Sidorenko, A.S. Firsov, E.N. Khobotov, Methods of schedule building in manufacturing systems, MSTU named after N.E. Bauman (2009)

[11] A. Lazarev, A. Kvaratskhelia, Tchernykh, Solution algorithms for the total tardiness scheduling problem on a single machine, Workshop Proc. of the ENC'04 Int. Conf. (2004) 474–480

[12] D. Merkle, M. Middendorf, An Ant Algorithm with a New Pheromone Evaluation Rule for Total Tardiness Problem, Evo Work Shops, LNCS 1803, Springer, Verlag (2000) 287–296. https://doi.org/10.1007/3-540-45561-2_28

[13] M.L. Pinedo, Planning and Scheduling in Manufacturing and Services, Springer Dordrecht Heidelberg ,London, New York (2009). https://doi.org/10.1007/978-1-4419-0910-7

[14] M.L. Pinedo, Scheduling. Theory, Algorithms, and Systems, Springer, Dordrecht Heidelberg London, New York (2008)

[15] K. Rikhter, Dynamic problems of discrete optimization (in Russian), Radio and communication, Moscow (1985)

[16] C. Schwindt, et al., Handbook on Project Management and Scheduling 1, Springer International Publishing, Berlin, Heidelberg, New York (2015). https://doi.org/10.1007/978-3-319-05443-8

[17] A.M. Sidorenko, E.N. Khobotob, An aggregation in scheduling on machine building enterprises, Control theory and systems 5 (2013) 132-144. https://doi.org/10.1134/S1064230713050110

[18] Z. Tian, C.T. Ng, T.C.E. Cheng, An O(n2) algorithm for scheduling equal-length preemptive jobs on a single machine to minimze total tardiness, J. of Scheduling 9(4) (2006) 343–364. https://doi.org/10.1007/s10951-006-7039-6

[19] W. Szwarc, A. Grosso, F. Della Croce, Algorithmic paradoxes of the single machine totaltardiness problem, J. of Scheduling 4 (2001) 93-104. https://doi.org/10.1002/jos.69

[20] V.A. Yegorov, V.D. Lusanov, S.M. Scherbakov, Transporting Storing Systems. FMS (in Russian), Mashinostroyeniye, Leningrad (1989)

Materials Research Proceedings 21 (2022) 304-309 https://doi.org/10.21741/9781644901755-54

Development of a Method for Obtaining a Wear-Resistant Coating for a Cutting Tool

A. E. Litvinov[1,a*], V. U. Buzko[2,b], E. Yu. O. Balaev[1,c] and A. I. Goryachko[2,d]

[1]Institute of Mechanical Engineering and Car Service, Kuban State Technological University, Krasnodar, Russia

[2]Kuban State University, Krasnodar, Russia

[a]artstyleone@mail.ru, [b]buzkonmr@mail.ru, [c]balaev1122@mail.ru, [d]a_i_goryachko@mail.ru

Keywords: High-Entropy Alloys, Mechanical Activation, Mechanical Alloying, Coatings, High-Velocity Oxygen-Fuel Spraying, Structure, Phase State

Abstract. The article presents a method for producing a nanostructured wear-resistant high-hard coating with high physicomechanical and strength characteristics, resistance to shock and vibration loads. The result is an increase in adhesion between the substrate and the coating, as well as an increase in microhardness. One of the common methods of metal cutting is band-cutting machines that use closed band saws as cutting tools. Since materials with high physicomechanical characteristics (hardness, strength, etc.) are increasingly being used in modern production, which significantly complicates the cutting process and makes increased demands on the cutting tool. To expand the range of processed materials for which the productive use of band-cutting machines is possible, it became necessary to create a band saw with higher cutting characteristics. At the same time, the specificity of the working conditions of the band saw shows that the blade should have such characteristics as increased vibration resistance, resistance to alternating and dynamic loads, and the cutting part of the saw should have increased resistance to shock, dynamic, alternating loads, have high hardness, as well as increased wear resistance.

Introduction

One of the common methods of metal cutting is band-cutting machines that use closed band saws as cutting tools. Since materials with high physicomechanical characteristics (hardness, strength, etc.) are increasingly being used in modern production, which significantly complicates the cutting process and makes increased demands on the cutting tool. To expand the range of processed materials for which the productive use of band-cutting machines is possible, it became necessary to create a band saw with higher cutting characteristics. At the same time, the specificity of the working conditions of the band saw shows that the blade should have such characteristics as increased vibration resistance, resistance to alternating and dynamic loads, and the cutting part of the saw should have increased resistance to shock, dynamic, alternating loads, have high hardness, as well as increased wear resistance. [1,2]

The idea of creating a layered composite is one of the right directions for solving this problem. So, the layered composite material proved to be workable under difficult loading conditions on other engineering products and firmly strengthened its position in the production, since with the correct analysis of the working conditions of the product, whether it be a part or a cutting tool, in particular a band saw blade, it can be found that this or that zone or layer can work on a specific type of loading. Therefore, it can be made of a material with characteristics that satisfy these load conditions. Thus, on the basis of the analysis of loading conditions during the operation of the band saw blade, the idea of bimetallic saw was proposed, which is the right direction to solve the problem, but the main disadvantage is the presence of a welded joint sensitive to vibration, dynamic and shock loads, and the difficulty of obtaining a defect-free welded joint seam. An alternative could be surface plasma alloying of the teeth of the band saw blade, which allows significantly changing the physicomechanical characteristics of the material of the cutting edge of the teeth (which will be the material of the whole band saw blade - spring-spring steel). At the same time, the output is a layered composite with high adhesion, comparable in value to metallurgical, since the band saw blade material is alloyed, which is also the base material, and with correctly selected modes it is possible to reduce the values of possible residual deformation as a result of heat treatment to zero.[1-4]

Research materials and methods

For the research, the following equipment was used: a JET band saw (testing the obtained saws under conditions close to factory conditions), a Falcon 500 micro hardness meter (micro hardness tests for preliminary analysis of wear resistance), an Instron 8801 testing machine (fatigue test), installation of a magnetron radio frequency spraying Q150T ES (coating process), scanning electron microscope "JEOL JSM - 7500F" (studying the microstructure of coatings), Bruker Vertex 70 FTIR IR spectrometer (for analyzing the composition of coatings).

Discussion of the results

To apply nanostructured and wear-resistant high-hard coatings by magnetron sputtering on the surface of cutting tools, a sequential technological cycle is used in the work, which includes: 1) Preliminary etching of the product surface with low-temperature argon plasma to improve the adhesion of the applied protective coating; 2) Application by the method of magnetron-plasma spraying of a nanostructured film of metal or alloy on the surface of the product; 3) Thermal oxidative phase-forming annealing for the appearance of nanoparticles of high-hard oxides of a cubic system (aluminum or chromium oxides) and nanoparticles of medium-hard oxides (oxides of titanium, niobium, vanadium), which also leads to an increase in the adhesion of the protective coating, an increase in hardness, wear resistance and fatigue strength due to the recrystallization process applied coating.

A method of obtaining a wear-resistant coating for a cutting tool, comprising applying a coating on the surface of a cutting tool based on an alloy of titanium, aluminum and niobium, while obtaining a coating on the surface of the cutting tool is carried out by magnetron-plasma spraying including preliminary plasma etching of the surface of the cutting tool in a vacuum chamber with accelerated ions at a pressure of 1-3 Pa, after which there is an ion-plasma spraying of the coating carried out at a pressure of 1-3 Pa, current 100-150 mA to obtain a coating of the composition TiAlVNbTiB2 in the following ratio, weight%: Al 5.5-6.5; V 7-8; Nb 2-4 TiB2 1-2; Ti - the rest, and subsequent phase-forming thermal oxidative annealing at a temperature of 550-650 ° C for 1-2 hours. [5]

Preliminary etching of the surface of the tool allows you to increase the adhesion strength between the coating and the cutting tool due to the microstructure of the surface of the cutting tool and allows you to easily clean the surface of the product from contaminants and grease residues. At the same time, ion-plasma etching (surface ablation of the material) occurs, which allows you to change the structural and mechanical properties of the product, increase the roughness, which will improve the adhesion between the surface of the metal of the cutting tool and the applied material. Plasma treatment can be applied to a wide range of types of cutting tools of any composition and complex geometric shapes. Ion-plasma etching can improve the physico-mechanical properties of the resulting cutting tool.

Materials Research Proceedings **21** (2022) 304-309
https://doi.org/10.21741/9781644901755-54

Using magnetron-plasma spraying, nanostructured metal thin-layer films of the desired chemical composition and thickness are applied with high adhesion, which is close in value to metallurgical. In the method for producing a coating, thermal cutting of the cutting tool does not occur, so that no residual stresses occur on the surface of the tool and along the product-coating interfaces. In turn, the resistance of the coated cutting tool to fatigue cracking increases. Also, ion-plasma deposition occurring at room temperature provides a coating having a nanoscale structure in the size range of 5-15 nm. Thin nanostructured coatings, with a certain shape (cubic and tetragonal syngonies) and the size of nanoparticles (belonging to the region of maximum realization of the Hall-Petch effect), show ultra-high hardness, high fatigue strength and increased wear resistance.

Obtaining on the surface of a cutting tool a coating based on titanium aluminum alloy can increase the heat resistance and heat resistance of the cutting edge, which allows for a long time to save the geometry of the cutting edge of the tool and, as a result, the cutting characteristics of the tool.

Alloying titanium-aluminum alloy with vanadium to 8% allows the crystal lattice of the TiAl alloy to be transferred from a tetragonal to cubic form, which in turn improves the plasticity of the resulting coating, thereby increasing the resistance of the coating to fatigue cracking. Also, the transfer from a tetragonal to a cubic lattice allows TiAl alloys to increase the microhardness of the coating, which also increases the wear resistance of the tool[6-8].

Alloying a titanium-aluminum alloy with niobium leads to an increase in the thermodynamic activity of Al compared to Ti, thereby contributing to the formation of a stable protective Al2O3 layer on the coating surface, while the Al2O3 oxide film has a high microhardness and a dense structure that blocks further oxidation of the coating, including titanium, contained in the coating, also the Al2O3 oxide film increases the corrosion resistance and wear resistance of the coating.

The addition of a small amount of titanium diboride TiB2 to the coating composition, which is a strengthening dispersion phase that is in equilibrium with an alloy of titanium and aluminum (titanium aluminum) with good physicochemical and mechanical compatibility, can increase the strength characteristics of the coating and microhardness and maintain ductility and viscosity, increase heat resistance and heat resistance, due to the blocking of grain growth of the structure with increasing temperature, which always accompanies the process of metal cutting.

Thermal oxidative phase-forming annealing makes it possible to create a surface layer of high-hardness, highly heat-resistant nanoparticles of cubic phase oxides on the surface of a metal adhesive film of a tool, which results in high-hardness coatings with high wear resistance and considerable fatigue resistance.

Obtaining a wear-resistant coating for a cutting tool, as follows, at the first stage, the product surface is pre-etched with low-temperature argon plasma to improve the adhesion of the protective coating in the vacuum chamber by accelerated ions at a pressure of 1-3 Pa, after which a nanostructured film is applied by magnetron-plasma spraying from TiAlVNbTiB2 alloy in the following ratio, wt%: Al 5.5-6.5; V 7-8; Nb 2-4; TiB2 1-2; Ti - the rest, by transferring from the target surface a similar composition to the coating onto the surface of the cutting tool, after which thermal oxidative phase-forming annealing is carried out at a temperature of 550-650 ° C for 1-2 hours for the appearance of nanoparticles of high-hardness cubic oxide oxides (aluminum oxides), also leading to an increase in the adhesion of the protective coating, an increase in hardness, wear resistance, corrosion resistance and fatigue strength due to the process of recrystallization of the applied coating.

Example 1. Obtaining a wear-resistant coating for a cutting tool, as follows, at the first stage, the surface of the product is pre-etched with low-temperature argon plasma to improve the adhesion of the applied protective coating in the vacuum chamber by accelerated ions at a pressure of 3 Pa, after which a nanostructured TiAlVNbTiB2 alloy film is applied by magnetron-plasma spraying in the following ratio, weight%: Al 5.5; V 7; Nb 2; TiB2 1; Ti - the rest, by transferring from the target surface a similar composition to the coating onto the surface of the cutting tool, after which thermal oxidative phase-forming annealing is carried out at a temperature of 550 ° C for 1 hour for the appearance of nanoparticles of high-hardness cubic oxide oxides (aluminum oxides), which also leads to an increase adhesion of the protective coating, an increase in hardness, wear resistance, corrosion resistance and fatigue strength due to the process of recrystallization of the applied coating

Example 2. Obtaining a wear-resistant coating for a cutting tool, as follows, at the first stage, the surface of the product is pre-etched with low-temperature argon plasma to improve the adhesion of the protective coating in the vacuum chamber by accelerated ions at a pressure of 2 Pa, after which a nanostructured TiAlVNbTiB2 alloy film is applied by magnetron-plasma spraying in the following ratio, weight%: Al 6; V 7.5; Nb 3; TiB2 1.5; Ti - the rest, by transferring from the target surface a similar composition to the coating onto the surface of the cutting tool, after which thermal oxidative phase-forming annealing is carried out at a temperature of 600 ° C for 1.5 hours for the appearance of nanoparticles of high-hardness cubic oxide oxides (aluminum oxides), which also to increase the adhesion of the protective coating, increase hardness, wear resistance, corrosion resistance and fatigue strength due to the process of recrystallization of the applied coating

Example 3. Obtaining a wear-resistant coating for a cutting tool, as follows, at the first stage, the product surface is pre-etched with low-temperature argon plasma to improve the adhesion of the protective coating in the vacuum chamber by accelerated ions at a pressure of 1 Pa, after which a nanostructured TiAlVNbAl2O3 alloy film is applied by magnetron-plasma spraying in the following ratio, weight%: Al 6.5; V 8; Nb 4; TiB2 2; Ti - the rest, by transferring from the target surface a similar composition to the coating onto the surface of the cutting tool, then conduct thermal oxidative phase-forming annealing at 650 ° C for 2 hours for the appearance of nanoparticles of high-hardness cubic oxide oxides (aluminum oxides), which also leads to an increase adhesion of the protective coating, an increase in hardness, wear resistance, corrosion resistance and fatigue strength due to the process of recrystallization of the applied coating (table 1)

Table 1. *Test results.*

Material		The processing resistance of 30HGSA, min.	Microhardness, GPa
TiAlVNbTiB2	Example 1	82-84	47,3
	Example 2	85-87	47,8
	Example 3	88-90	49,2
Prototype		81	46,7

The microstructure of the obtained coatings (Figure 1) was studied using a JEOL JSM7500F scanning electron microscope. The microstructure of the obtained coatings on the saw teeth is shown in Figure 1.

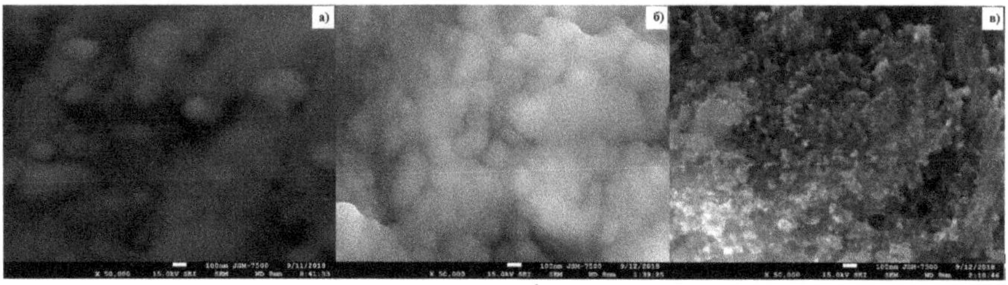

a b c

Fig. 1. *Microstructure of the obtained coatings a - nichrome Kh15N60; b - Ti6Al4V alloy; c - thermally oxidized aluminum (Al_2O_3).*

Analysis of the data presented in the table allows us to conclude that the cutting tool with a wear-resistant coating obtained by the present method are characterized by higher physicomechanical characteristics compared to inserts made by known methods. Thus, the combination of the claimed features allows to achieve the technical result.[9-12]

Conclusions

As a result of the research, a method for producing a wear-resistant coating for a cutting tool was developed, including applying a coating on the surface of a cutting tool based on an alloy of titanium, aluminum and niobium, characterized in that the coating on the surface of the cutting tool is carried out by magnetron-plasma spraying including preliminary plasma etching of the surface cutting tool in a vacuum chamber with accelerated ions at a pressure of 1-3 Pa, after which ion-plasma variable coating spraying carried out at a pressure of 1-3 Pa, a current of 100-150 mA to obtain a coating of the composition TiAlVNbTiB2 in the following ratio, weight%: Al 5.5-6.5; V 7-8; Nb 2-4 TiB2 1-2; Ti - the rest, and subsequent phase-forming thermal oxidative annealing at a temperature of 550-650 ° C for 1-2 hours.[13-15]

Acknowledgment

This work was supported by a grant from the President of the Russian Federation (contract No.SP-399.2019.1)

References

[1] S.V. Kiselev, A.V. Blakhin, A.F. Dulevich, "Band saws with high durability". Bulletin of BSTU. Vol 4, pp. 24-27, April 2007.

[2] A.E. Litvinov, "Improving tool life and machining precision in band saws". Russian Engineering Research.. Vol 36, pp. 760-761, 2016. https://doi.org/10.3103/S1068798X16090124

[3] Z.M. Blednova, P.O. Rusinov, E.U. Balaev, "Quantification of hereditary regularities of surface layer formation and transformation made of multicomponent shape memory materials in a high-energy impact" Materials Today: Proceedings . p. 4652-4657. 2017. https://doi.org/10.1016/j.matpr.2017.04.046

Materials Research Proceedings **21** (2022) 304-309 https://doi.org/10.21741/9781644901755-54

[4] L.X. Baldaev, V. Borisov, Gas thermal spraying. Textbook. 2007. p. 344.

[5] A. Hasui, O. Morigaki, Surfacing and spraying. Mechanical. Moskow, 1985 p. 240

[6] G.P. Laskin, A.P. Kuznechov, "Simulation of ion-plasma doping process". Bulletin of TumSU Vol 6, p 64-67, 2010

[7] V.A. Putilin, A.V. Kamashev, A.V. Putilin, "Alloyng of surface of metals under application of short palse laser" Mechanical engineering 2016. p. 344.

[8] M.K. Gusejnov, M.K. Kurbanov, B.A. Bilalov, G.K. Safaraliev, "Magnetron deposition of thin films of solid solution (SiC)1-x(AlN)x." Physics and technology of semiconductors. Vol 44, p. 841-844, 2010. https://doi.org/10.1134/S1063782610060217

[9] E.V. Rublevskaja, P.A. Komarova., A.V. Shcherbakova, T.D. Husnutdinov, E.V. Ramenskaja, "Analysis of parameters of cutting process by band saws" Actual problems of aviation and cosmonautics Vol 1, p.35-36, 2017

[10] V.P. Astahkov, Tribology of metal cutting p. 419, 2016

[11]L. Gan, R.D. Gomez , A. Castillo , P.J. Chen , C.J. Powell , W.F. Egelhoff Jr. Ultra-thin aluminum oxide as a thermal oxidation barrier on metal films. Thin Solid Films 415 (2002), pp. 219–223. https://doi.org/10.1016/S0040-6090(02)00622-3

[12] Pouilleau J, Devilliers D, Garrido F, et al. Structure and composition of passive titanium oxide films, Mater Sci Eng B. 1997; 47, pp. 235–243. https://doi.org/10.1016/S0921-5107(97)00043-3

[13] Dorian A. H. Hanaor & Charles C. Sorrell. Review of the anatase to rutile phase transformation, Journal of Materials Science volume 46, pp. 855–874 (2011). https://doi.org/10.1007/s10853-010-5113-0

[14] C. A. Vasantha Kumar, J. S. Rajadurai. Influence of rutile (TiO2) content on wear and microhardness characteristics of aluminium-based hybrid composites synthesized by powder metallurgy, Transactions of Nonferrous Metals Society of China, 2016, Volume 26, Issue 1, pp. 63-73. https://doi.org/10.1016/S1003-6326(16)64089-X

[15] JiaqinLiao, Rou Tan, Zhixiang Kuang, Chunyu Cui, Zengxi Wei, Xiaolan Deng, Zhanheng Yan, Yuezhan Feng, Fang Li, Caiyun Wang, Jianmin Ma. Controlling the morphology, size and phase of Nb2O5 crystals for high electrochemical performance, Chinese Chemical Letters, 2018, Volume 29, Issue 12, pp. 1785-1790. https://doi.org/10.1016/j.cclet.2018.11.018

Modern Trends in Manufacturing Technologies and Equipment | Materials Research Forum LLC
Materials Research Proceedings **21** (2022) 310-316 | https://doi.org/10.21741/9781644901755-55

Automatic Control System for Bodies of Revolution Processing

Oleg I. Drachev[1,a*], Boris M. Gorshkov[2,b] and Natalia S. Samokhina[2,c]

[1]Togliatti State University, Togliatti, Russia

[2]Volga Region State University of Service, Togliatti, Russia

[a]doi05@mail.ru, [b]kaf_ekis@tolgas.ru, [c]skipert-99@ya.ru

Keywords: Metalworking Processing of Bodies, Self-Centering Steady Rests, Automatic Control System, Manufacturing Non-Rigid Shaft-Type Parts, Level of Residual Stresses

Abstract. This research is related to metalworking processing of bodies of revolution with the help of universal lathe machines. The technology includes the application of two types of vibrations to the working tool and the processed surface error measurement. To increase the manufacturing accuracy, the workpiece processed surface error is measured while a workpiece is being rotated; this rotation is performed with the workpiece being rigidly fixed in end supports and at the same time being damped in the sections between these supports. Furthermore, the parameters of vibrations applied to the tool working travel are defined by the workpiece form error and the nature of distribution of stresses that appear when the workpiece is fixed; the nature of the workpiece processed surface form error is extrapolated from the data obtained in the workpiece sections between the supports. Before manufacturing, the workpiece is corrected while being fixed in rigid supports, and the correction itself is performed as the function of magnitude and vector of the workpiece maximum deflection plane. The workpiece may be fixed in rigid supports; steady rests with double rollers may be used as such supports. The workpiece dampening in its sections between end supports may be performed using self-centering steady rests.

Introduction

Non-rigid shaft-type vital parts are used in modern-day mechanical engineering, machine tool engineering and instrument engineering. The continuous increase of their production volume is induced by the improvement of strength calculations and designs, by the constant decrease of the products' metal intensity and by the increase of high-speed precision machines production.

Imbalances in the design parameters of non-rigid bodies of revolution (NRBR) create serious technological difficulties in the production process; the main reasons for these difficulties are the following:

a) significant elastic and plastic deformations at all stages of the parts' processing, assembly and operation;

b) low vibration resistance and various types of pliancy of the technological system elements;

c) significant influence of technological heredity on the performance reliability;

d) workpieces distortion caused by uneven residual stresses added at all the stages of processing;

e) the parts' low thermal resistance.

The negative effect of the above-mentioned factors during the production of non-rigid bodies of revolution leads to the technological bases dysfunctions, the parts' form and size errors, surface defects, cutting modes limitations and, ultimately, to the decrease of operation accuracy and reliability [1-15].

Relevance and goal of this research

The known methods of manufacturing axisymmetric bodies of revolution according to the traditional technology using standard steady rests are not efficient for manufacturing non-rigid shaft-type parts. At the present moment the matter of producing these types of parts is solved through the selection of processing modes and the application of manual correction operations, which do not ensure the required product quality. The improvement of technological processes of the NRBR mechanical processing is an important goal that increases the production efficiency and the mechanical engineering products quality. The goal of our research is to increase the accuracy of processing, to maintain it after processing and to obtain the following technical results: to increase the accuracy of the part form in cross-section by stabilizing the cutting force during the workpiece rotation, to increase the stability of the part geometry over time by forming a more uniform pattern of stress distribution throughout the workpiece volume. The use of an automatic control system simplifies the error measurement due to the application of the method of extrapolating the measured signals in the sections under the steady rests. The controlled steady rests decrease the vibrations level in the cutting zone.

Automatic control system

This article focuses on the automatic control system (ACS) of the technological processing system in lathes for bodies of revolution of non-rigid parts. The ACS operates as follows. A shaft workpiece (Fig.1) is fixed in supports 1; depending on the process pressure in drive 2 (Fig.2) of the clamp, these supports can operate in two modes: damper mode or rigid supports mode.

Intermediate supports act as dampers 3. Two end supports act as rigid supports 4. The shaft is rotated the required number of times at low speed. The sensor system 5 (Fig.2), which is installed on the body 6 of supports 1, measures the displacement of levers 7 (changes of gap clearances S2, S3) and of rod 8 (change of gap clearance S1) of supports 1 under the influence of profile errors (workpiece surface wobbling against the bases – two end steady rests acting as rigid supports; and deviation from the out-of-roundness in the cross-section); the obtained data is processed in control unit 9.

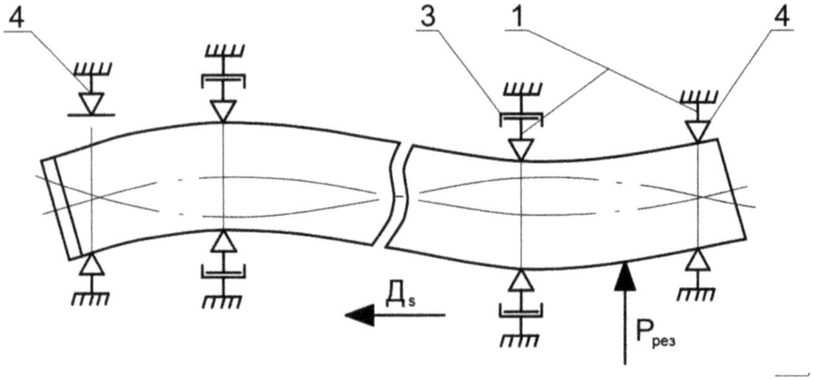

Fig. 1. *Shaft installation diagram for error measurement.*

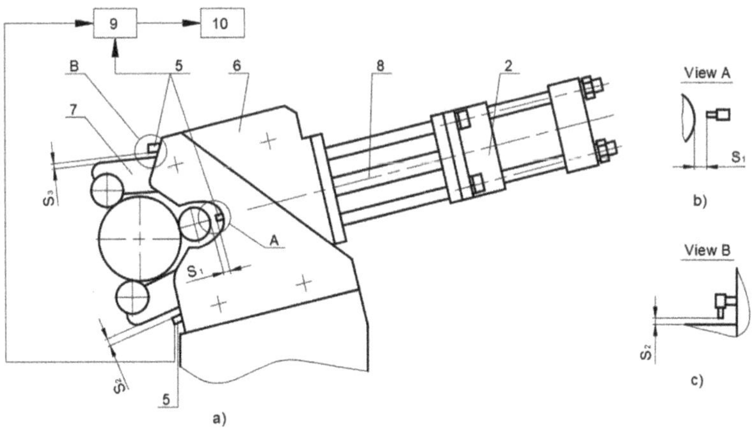

Fig. 2. *a – General view of a steady rest with a measuring system, b and c – location of sensors.*

A harmonic analysis of the obtained information is carried out to identify the first and second harmonics of vibrations of the workpiece profile errors. A general picture of the spatial attitude of the processed surface in relation to the base necks under the steady rests (working in the rigid supports mode) and its profile is formed in control unit 9. The nature of the error location in the workpiece sections between the steady rests is determined by extrapolating the data obtained in the sections of the steady rests installation, because the cutting pattern is strictly regular in the longitudinal direction (along the axis along the helix line with a certain pitch). The data is transmitted from control unit 9 to feed drives input 10, the change of feed Ds (Fig.1) is set during the turning process in order to control the cutting force (to stabilize it) as a function of the predicted allowance change. To stabilize the workpiece axis during the manufacturing process, supports 1 are sequentially switched into the rigid supports mode along the workpiece length as the cutting zone moves, so that it is always located between rigid supports 4. The tool is assigned with vibrations with two components. One of them is determined by spatial deviations, the second one is determined by the nature of the stress state in the workpiece volume. Taking into account the spatial errors (Fig.3), the first component of the tool vibrations is set (its form is Δf and its location is e).

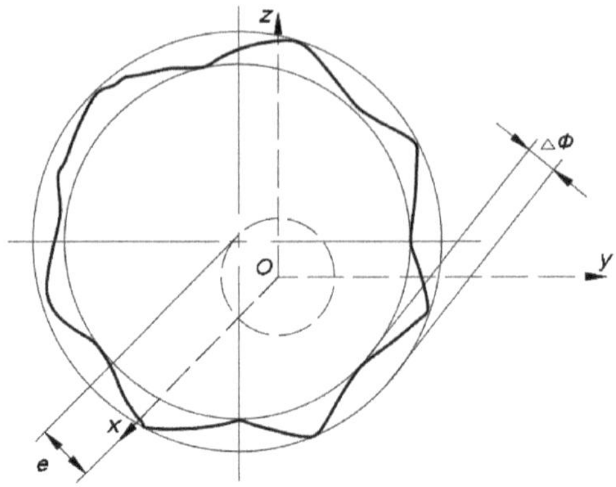

Fig. 3. *View of the errors of bodies of revolution.*

Furthermore, the feed control is carried out as a function of the cutting force stabilization (when the allowance is increased, the feed is decreased and vice versa). When a workpiece is clamped, the deflection of its axis leads to the formation of stresses in the workpiece volume (compression stress on the convex side and tension stress on the concave side). Once the workpiece is released, these tensions lead to the workpiece distortion. The second component is applied during the manufacturing process in order to equalize these stresses. To balance the residual stresses, the feed value is reduced in the workpiece part with the residual tension stresses and is increased in the part with the compression stresses, thus equalizing the stress state along the workpiece cross-section.

The tool vibrations may be reduced if a workpiece is processed preliminarily. Before a shaft is processed with the dressing roll installed into the support, it is processed in the midspans to equalize the stress state in the workpiece volume and to decrease the spatial deviations by bending the shaft in the direction that is opposite to the maximum deflection and that lies in its plane. This way we form the stresses opposite to the stresses appearing when a workpiece is clamped. To compensate the plastic deformations, a shaft workpiece is deformed during the manufacturing process with these deformations taken into account: it is bended by a certain value.

The number of supports 1 (Fig.1) and their location along the workpiece length depend on the function of its length and diameter, taking into consideration that the ratio of the span length to the workpiece diameter is no more than 5. The end steady rests have double rollers (not shown) in order to increase the reactive torque when the workpiece is rotated and to stabilize its axis.

Research

The development of adaptive control systems, including in particular the ACS for the workpiece-support subsystem, is a radical method of increasing vibration resistance and accuracy.

To assess the effectiveness of the ACS, we carried out the experiments to determine the value of the nonuniformity of the residual stresses that exist in a workpiece and appear in it after the turning process.

The goal of our experiment was to ascertain the reason why the shaft bending changes during the machining process. Our task was to determine what influences the shaft form the most: the residual stresses in the workpiece or the stresses that appear during the turning process.

For this experiment we used the samples made of steel of austenite class 12X18H10T. The 1500 mm long samples were cut from a supplied bar with the diameter of 30 mm. These samples were divided into two batches, 12 pieces in a batch. Each batch was processed in its own mode:

The 1st batch: supplied material. The lathe with the ACS was used for turning: cutting depth t = 0.5 – 0.75 mm, the part's rotation speed π = 800 rpm, tool advance S = 0.063 mm per rotation. We used carbide tools with carbide material VK13, with the cutting edge angle α = 450. The sample warping was measured after each pass.

The 2nd batch: full annealing of all 12 samples was performed (heating up to the temperature of 1060° C, holding for 2-3 hours, cooling in the furnace, the samples were hung vertically); therefore, the residual stresses in the workpieces were eliminated or nominal. The parts' warping was measured before and after the samples' annealing. Then the samples were processed like those from the 1st batch.

The value of the deflection of a part's axis was taken as a measure of a part's warping (axial mode changes); this value was measured in a vertical position in five sections with the help of dial gauges with the graduation mark of 0.01 mm, the results were entered into a table. The deflections were measured immediately after the treatment process and after a period of time up to 240 hours in order to determine if there was the relaxation of stresses after the treatment.

The experiment's results were used to create the graphs of changes of the samples' deflections as the samples were turned; Fig.4a shows the results of the 1st batch processing, the supplied material samples; Fig.4b shows the results of the 2nd batch processing, the annealed samples. Lines of different configuration represent the samples with various serial numbers (1...6).

Both graphs show that the shaft form during the treatment process is not stable; the deflection tends to appear during the following treatment. This phenomenon is more noticeable when one looks at the unannealed samples. The shaft deflection decreases down to 0.4...0.6 mm (for the unannealed samples) and down to 0.2...0.4 mm (for the annealed samples) during the turning process (in the beginning, with removal t = 1...2 mm).

Therefore, we can make a conclusion that there are residual stresses present in a workpiece. These stresses amount to 40% of the stresses in the unannealed samples.

During further turning the shaft form becomes even less stable and the deflection value increases.

Beside that, the graphs show that the samples, both annealed and unannealed, that have a small deflection in their initial state have a lesser deflection during the machining process. The initially straight samples may provide a deflection close to the allowable one: 0.01...0.04 mm per one meter of the shaft length.

We should also note that there exists the residual stresses relaxation: over time the deflection change ranges widely, from 0 to 100%, but in average it is 13 ... 24%.

Let us define the value of the nonuniformity of the residual stresses in the test sample using the above-mentioned theoretical calculations. The shaft deflection with the ratio l/d – 50 is 0.5 mm in average in our experiments. This corresponds to the residual stresses nonuniformity of 20% with their minimal value σres = 100 N/mm2. The nonuniformity value should be even less with the residual stresses minimal value σres = 20. 30, 100 N/mm2.

Additionally, we carried out the experiments with and without the use of the ACS with the following cutting modes: π = 1440 rpm, S = 0.11 - 0.22 mm per rotation; without the ACS: two passes with cutting depth t1= 1.5 mm, t2 = 2.8 mm; with the ACS: one pass with cutting depth t3 = 3 mm. Cutter material is T15K6. This experiment allowed us to obtain roughness Ra 0.8, to increase the longitudinal accuracy up to 10 μm/m with the diameters range of 30 - 60 mm and the lengths range of 1.5 – 4 m and to eliminate the finished product residual deformations during operation.

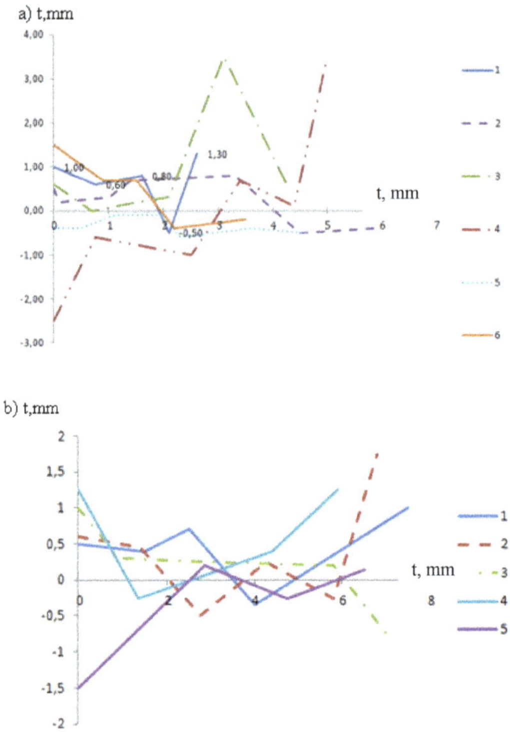

Fig. 4. *Measurements of the samples' deflections during the turning process: a – supplied material samples, b – annealed samples.*

Conclusion

The experiments we carried out prove beyond all doubt that it is necessary to control the course of the technical process, ranging from the correct operation to the thermal power processing and the finishing. Only by decreasing the level of the residual stresses and equalizing them along the entire workpiece length we can keep the processes of relaxation and creeping to a minimum, i.e. we can stabilize the finished part form for the entire period of operation.

References

[1] Solomentsev Y.M., Mitrofanov V.G., Protopopov S.P. and others. Adaptive process control. Moscow: Mashinostroenie, 1980.

[2] Drachev O.I. Manufacturing techniques for low-stiff axisymmetric parts. Saint Petersburg: Politekhnika, 2005.

[3] Drachev O.I. &Kravtsov A.N. Automatic control of the process of low-stiff parts turning. Togliatti: VSTU, ZAO «ONIKS», 2012.

[4] Artamonov E.V., Tveryakov A.M., Shtin A.S. Definition of the temperature of the ultimate operational capability of instrumental hard alloys. Obrabotkametallov (tekhnologia, oborudovanie, instrumenty) - Metals processing (Technology, Equipment, Tools), vol. 20 (2018) 47-57.

[5] Bobrowski A.V., Drachev O.I., Gorshkov B.M. Improving the operational accuracy of shafts by volumetric plastic deformation. Available online, 2020. https://doi.org/10.1016/j.matpr.2020.01.458

[6] Zakovorotniy V.L., Voloshin D.A., Rodriges S.B., Flek M.B. Method of modeling elastic deformations of cylindrical end mills in the process of milling. Investigation of machines and tools for machining complex and precise surfaces. Saratov: SSTU, 2003.

[7] Vladimirov A.A. &Sergiev A.P. Influence of vibrational amplitude on component functions of the complex vibrational effect of low-frequency vibrations. IzvestiyaVolgGTU – The Bulletin of VolgGTU, 7 (2018) 13-17.

[8] Zakovorotniy V.L., Ladnik I.V., Dhande S.G. A method for characterization of machine-tools dynamic parameters for diagnostic purposes. Journal of Materials Processing Technology, vol. 53, 1995. https://doi.org/10.1016/0924-0136(94)01745-M

[9] Zakovorotniy V.L. &Bordatchev E.V. Computer aided system for analysis of the cutting motion nonunformity during machine tool approval tests. Modelling, Measurement and Control B, AMSE Press, vol. 56 (3), 1994.

[10] Bobrovsky A. V., Drachev O.I., Kravtsov A.P. Improving the accuracy and quality of machining by controlling the cutting process, IOP Conference Series: Materials Science and Engineering 709 (2020) 044088. https://doi.org/10.1088/1757-899X/709/4/044088

[11] Palagnyuk G.G. Theoretical substantiation of the method of reducing the energy intensity of the cutting process and its multiparameter active control. Diagnostics and control in technical systems (1995) 74 – 82.

[12] Starkov V.K. Machining, stability and quality management in automated production. Moscow: Mashinostroenie, 1989

[13] Chemavskiy F.G. Modern trends in the construction of precast cutters with non-reversible plates, Review. Moscow: NIIMash, 1983

[14] Savilov A.V., Svinin V.M., Timofeev S.A. Investigation of output parameters of titanium reverse turning. Journal of Physics: Conference Series. International Conference "Information Technologies in Business and Industry" 2018 - Enterprise Information Systems. https://doi.org/10.1088/1742-6596/1015/4/042055

[15] Svinin V.M., Samsonov A., Savilov A.V., Pyatykh A. (2018). Self-oscillation suppression when turning non-rigid shafts using spring tools and the spring headstock center. MATEC Web of Conferences, vol. 224 (2018) 01059. https://doi.org/10.1051/matecconf/201822401059

Model of Filling the Internal Structure of Workpiece with Curved Layers for 3D Printing

A.N. Grechukhin[1,a*] and V.V. Kuts[1,b]

[1] Southwest State University, 50 Years of October St., 94, Kursk, Russia

[a] agrechunin@mail.ru, [b]kuc-vadim@yandex.ru

Keywords: Additive Technologies, Layer-By-Layer Synthesis, Curved Layer

Abstract. This paper presents a solution to the problem of filling the internal structure of the workpiece with curve layer in 3D printing. A generalized model of filling the internal structure of workpiece with curve layer is designed. The results are presented for solving the problem on the example of curved layers of a conical shape with filling along a helical line. The research results can be in demand in the development of algorithms and software for technological equipment. They allow to ensure the formation of the internal structure of products in curved layers during 3D printing.

Introduction

The development of additive technologies is aimed at expanding the area of their application for solving various problems, including for obtaining structures with increased physical and mechanical properties. FDM printing technology is one of the ways to create such structures by filling the internal structure of products with curved layers formed by sections of various geometric shapes [1-9].

Relevance of the research topic

Research carried out by research teams [11-15] shows that the formation of the internal structure of products with curvilinear layers makes it possible to obtain structures with higher strength characteristics compared to filling with flat layers. The development of models, algorithms and software is an urgent task. It allows curve layers to form the internal structure of the product during 3D printing.

The preparation includes the following stages in printing using FDM technology in general: the separation occurs of a three-dimensional digital model, as a rule, in STL format, which is a list of coordinates of the vertices of triangular faces (facets) forming a surface into many curvilinear layers; the task is to select the trajectory of the extruder relative to the part; finding is carried out by areas of filling with material in accordance with the shape of the part [16-22]. The layer is applied one by one until the object is completely built.

An urgent task is to determine the trajectories of movement of the extruder, as well as areas of filling with material during the formation of the internal structure of products with curved layers.

Materials and methods for solving the issue

Work [10] proposes a technique for dividing a three-dimensional model into curved layers for 3D printing, for example, a conical shape (fig. 1).

a) b) c)

Fig. 1. *The section of the original part is represented by a cone: a) the position of the original part and the cutting cone; b) facets forming the intersection of the secant cone with the original part; c) a lot of sections of the original part with a cone*

The task is to divide the model into curved conical layers for 3D printing by forming an array of facet numbers (*M*) that have intersection points with a secant conical surface. It is described by the equation:

$$r_0(u,v)=\begin{bmatrix}\sin u\cdot\sin(2\varphi)v;\\ -\cos u\cdot\sin(2\varphi)v;\\ \cos(2\varphi_1)v+v+Z_0\end{bmatrix}, \tag{1}$$

where *u*, *v* – cone parameters; φ – half the cone angle; Z_0 – cone vertex coordinate.

In order to prepare for the printing of FDM technology, the following tasks are required. It consists in setting the trajectories of the extrude. Such trajectories can be described by different spatial curves, and their points are located on the cutting surface of the cone.

In particular, when dividing the model into conical layers, the trajectory can be set by the movement of the extruder by a conical helical line. It is described by the equation:

$$r(v)=\begin{bmatrix}\cos\left(\dfrac{k\cdot(v-Z_0)\cdot 2\pi}{P}+U_n\right)\cdot(v-Z_0)\cdot tg\varphi\\ \sin\left(\dfrac{k\cdot(v-Z_0)\cdot 2\pi}{P}+U_n\right)\cdot(v-Z_0)\cdot tg\varphi\\ v\\ 1\end{bmatrix}, \tag{2}$$

where *v* – screw parameter; Z_0 – cone vertex coordinate; φ – half the cone angle; *P* – screw pitch; U_n – Initial twist angle of helix; *k* – screw line torsion direction: *k*=-1 – clockwise, *k*=1 – counterclockwise.

It is necessary to create an array of points to determine the areas of filling of the spatial curve (an array of parameters), which determines the trajectories of the extruder, which are the intersection points of the specified spatial curve with the plane of facets, the numbers of which belong to the array *M*.

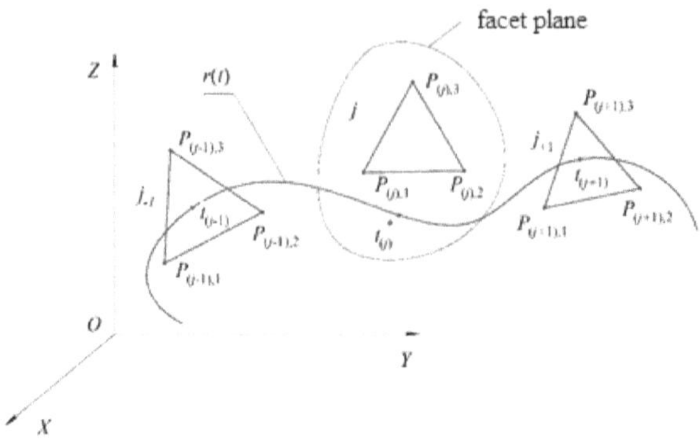

Fig. 2. Defines an intersection that represents a space curve and a face plane

Determination of the necessary intersection points of a given spatial curve with the facet planes is necessary for each facet included in the array M (fig. 2) by solving the equation:

$$A_j \cdot r(t) \cdot k_j = 0 \tag{3}$$

regarding parameter t, where A_j – matrix of transition to the coordinate system, built on the points of the j-th facet, whose coordinate center is located at the point P_{j1}, and the Z_j axis is perpendicular to the facet plane. Fig. 3 shows a scheme for determining the point of intersection of a spatial curve with a facet plane

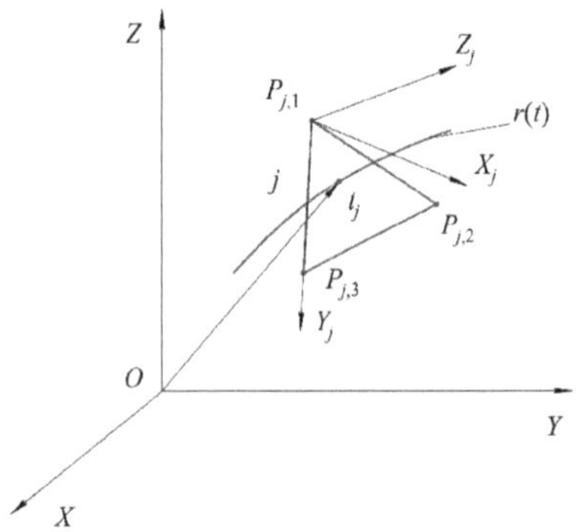

Fig. 3. The diagram shows the definition of the intersection point of the spatial curve with the facet plane

The set T of values of the parameter t_k can be found by solving the equation. There is a need to select from the set T subset of parameters for which the curve $r(t)$ as intersection points located inside the facet. Known methods [23] can be used to solve this problem.

We define the areas of filling with material by grouping the parameters of the curve $r(t)$, corresponding to the points of intersection of the curve with all facets. To do this, it is necessary to construct a variational series of parameters t_k, with properties $t_1 \le t_2 \le ... \le t_k \le ... \le t_{k-1} \le t_k$ and form an array of pairs of parameters $[t_1, t_2]$, $[t_3, t_4]$,... $[t_{k-1}, t_k]$. The obtained pairs of parameters t_k will correspond to the sections of the extruder movement trajectories, on which filling with the material takes place.

For example, a diagram is presented for determining the points of intersection of a helical line with a certain body in fig. 4.

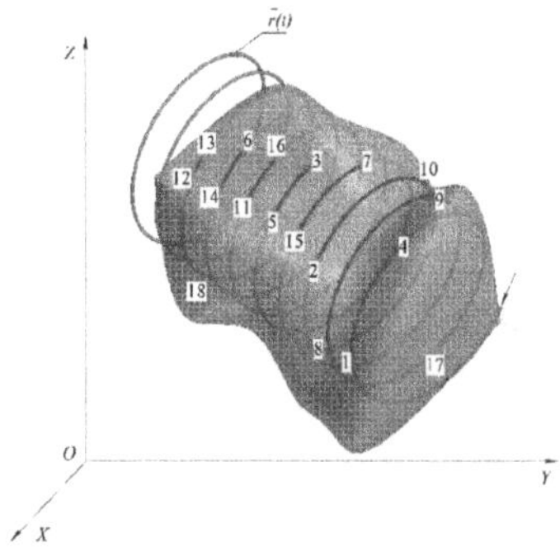

Fig. 4. *The diagram shows the intersection of a helix with a certain body*

18 points are determined as a result of intersection of a helix with a certain body. The variation series will take the following form: t_{17}, t_1, t_4, t_8, t_9, t_2, t_{10}, t_{15}, t_7, t_5, t_3, t_{11}, t_{16}, t_{14}, t_6, t_{12}, t_{13}, t_{18}. 8 pairs of points are defined by grouping intersection points: $[t_{17}, t_1]$, $[t_4, t_8]$, $[t_9, t_2]$, $[t_{10}, t_{15}]$, $[t_7, t_5]$, $[t_3, t_{11}]$, $[t_{16}, t_{14}]$, $[t_6, t_{12}]$, $[t_{12}, t_{18}]$.

Therefore, the filler is completed by the material on the spiral section corresponding to the obtained pair.

Conclusion

The model was developed to fill the internal structure of products with curved layers for 3D printing as a result of the research. The results can be applied to research in the development of algorithms and software for technological equipment, which ensures the formation of the internal structure of products by curved layers during 3D printing.

References

[1] Burns M. Automated Fabrication: Improving Productivity in Manufacturing. -Englewood Cliffs, N.J., USA: PTR Prentice Hall, 1993. - 369 pp.

[2] Improving the quality of additive methods for forming the surfaces of odd-shaped parts with the application of parallel kinematics mechanisms Kuts V.V., Razumov M.S., Grechukhin A.N., Bychkova N.A. International Journal of Applied Engineering Research. 2016. T. 11. № 24. C. 11832-11835.

[3] Dobroskok, V.L Rational orientation of products during their layer-by-layer shaping based on the initial triangulation 3d model [text] / V.L. Dobroskok, L.N. Abduraimov, S.I. Chernyshov // Scientific notes of the Crimean Engineering Pedagogical University. 2010. № 24. pp. 13-21.

[4] Algorithm for dividing a three-dimensional model into layers for 3D printing I. A. Gushchin, A. R. Avdeev, A. A. Shvets, A.V. Drobotov//News of TulSU. Technical sciences. 2016. Issue 11. Part 2. pp 99-105.

[5] Optimum part deposition orientation in stereolithography / S. K. Singhal, A. P. Pandey, P. M. Pandey, A. K. Nagpal // Computer-Aided Design & Applications. – 2005. – Vol. 2. – Nos. 1–4. – P. 319–328. https://doi.org/10.1080/16864360.2005.10738380

[6] Hong S. Byun. Determination of optimal build direction in rapid prototyping with variable slicing / Hong S. Byun, Kwan H. Lee // Int. J. Adv. Manuf. Technol. – 2006. – №. 28. – P. 307–313. https://doi.org/10.1007/s00170-004-2355-5

[7] Lan P.T. Determining fabrication orientations for rapid prototyping with stereolithography apparatus / S. Y. Chou, L. L. Chent, D. Gemmill // Computer- Aided Design. – 1997. – Vol. 29. – № 1. – P. 53– 62. https://doi.org/10.1016/S0010-4485(96)00049-8

[8] Massod S. H. A generic algorithm for part orientation system for complex parts in rapid prototyping / S. H. Massod, W. Rattanawong, P. Iovenitti // J. Mater. Process. Technol. – 2003. – Vol. 139. – № 1–3. – P. 110–116. https://doi.org/10.1016/S0924-0136(03)00190-0

[9] Masood S. H. A generic part orientation system based on volumetric error in rapid prototyping / S. H. Masood, W. Rattanawong // Int. J. Adv. Manuf. Technol. – 2002. – № 19. – P. 209–216. https://doi.org/10.1007/s001700200015

[10] Kuts V.V, Grechukhin A.N, Olesitsky A.V and others Algorithm for dividing a volume model into curvilinear layers for 3D printing, Fundamental and applied problems of engineering and technology 2021, pp 22-27.

[11] Grechukhin A.N, Kuts V.V, Razumov M.S and others Improving the accuracy of additive forming methods Innovation, quality and service in engineering and technology, 2018, pp 128-131.

[12] Grechukhin A.N, Kuts V.V, Razumov M.S Control of spatial orientation of robot units in the process of additive forming of products Bulletin of Voronezh state technical University, 2018, 4 pp 122-129.

[13] Grechukhin A.N, Kuts V.V, Razumov M.S Experimental determination of the cross-section parameters of a single layer in the additive forming products News of Tula state University. Technical science, 2019, 10 pp 264-270.

[14] Grechukhin A.N, Anikutin I.S, Byshkin A.S Management of space orientation of the end effector of generation of geometry system fiveaxis manufacturing machinery for additive generation of geometry MATEC Web of Conferences, 2018 Vol 7 pp 128-136 DOI 10.1051/matecconf/201822601004.

[15] Grechukhin A.N, Kuts V.V, Razumov M.S Ways to reduce the error of additive methods of forming – 2018 MATEC Web of Conferences Vol 7 pp142-150. https://doi.org/10.1051/matecconf/201822601002

[16] Grechukhin, A.N, Kuts, V.V, Oleshitskiy, A.V Development and Research of Technological Equipment that Implements Dynamic Control of Process of Additive Fabrication of Parts of Complex Spatial Shapes Based on Mechanisms with a Hybrid Layout. IOP

Conference Series: Materials Science and Engineering, 2020, 709(3), 033112. https://doi.org/10.1088/1757-899X/709/3/033112

[17] Kuts, V.V., Merkulov, V.S., Grechukhin, A.N., Privalov, A.S Investigation of the process of additive formation of fusible materials using a low-power solid-state laser. IOP Conference Series: Materials Science and Engineering, 2021, 1029(1), 012010. https://doi.org/10.1088/1757-899X/1029/1/012010

[18] Grechukhin, A.N., Kuts, V.V., Razumov, M.S Solving Problem of Curved Surface Approximation by Layers with Constant and Variable Sections During Forming by Additive Methods. Lecture Notes in Mechanical Engineering, 2020, pp. 239–248. https://doi.org/10.1007/978-3-030-22041-9_28

[19] Grechukhin, A.N., Kuts, V.V., Oleshitsky, A.V Control additive error of morphogenesis with the use of hybrid layout. AIP Conference Proceedings, 2019, 2188, 020003. https://doi.org/10.1063/1.5138377

[20] Grechukhin, A.N., Privalov, A.S., Garkavtseva, P.A. Experimental studies of the process of additive electric arc forming in the environment of protective gases AIP Conference Proceedings, 2019, 2188, 020004. https://doi.org/10.1063/1.5138378

[21] Grechukhin, A.N., Kuts, V.V., Oleshitsky, A.V. Control of geometrical parameters of a single layer at additive forming of products by FDM technology. IOP Conference Series: Materials Science and Engineering, 2019, 680(1), 012004. https://doi.org/10.1088/1757-899X/680/1/012004

[22] Grechukhin, A.N., Kuts, V.V., Razumov, M.S Calculation of the controlled parameters of the 6-coordinate robot in the process of additive forming of products. Journal of Physics: Conference Series, 2019, 1210(1), 012053. https://doi.org/10.1088/1742-6596/1210/1/012053

[23] Information on http://cyber-code.ru/tochka_v_treugolnike/

Investigation of Electromagnetic Properties of Tool Hard Alloys under the Influence of High Temperatures

E.V. Artamonov[1,a*], A.M. Tveryakov[1,b] and A. S. Shtin[1,c]

[1]Industrial University of Tyumen, Volodarskogo 38, Tyumen, Russia

[a]evgart2014@mail.ru, [b]tveryakov@mail.ru, [c]shtin_as@mail.ru

Keywords: Hard Alloys, Metalworking, Maximum Working Capacity Temperature, Electromagnetic Properties, Cutting

Abstract. The article shows the developed installation for determination of temperature of maximum operability of replaceable cutting hard-alloy plates on the basis of study of change of electromagnetic properties. The method of research is given. Tests of images were carried out to time of heating of the replaceable cutting plates from solid B35 alloy. The heating temperature interval was selected according to the temperature mode of the process of cutting difficult materials. Heating was carried out to 1000° C. The results of the study were obtained to determine the temperature of maximum operability of replaceable cutting hard alloy plates based on the study of the change in electromagnetic properties for alloy B35 amounted to 460-730 ° C.

Introduction

Today, under the conditions of sanctions, the Russian Federation needs more than ever the development of energy-saving technologies in various industries.

Metallurgy does not stand still, recently new brands of materials have appeared that have characteristics in terms of strength, hardness, ductility, heat resistance many times superior to their analogues, for example, Russian grades of alloys such as GDP, American "martensite."

Studies have shown that the selection of tool material, as well as the assignment of processing (cutting) modes, are the main problems in the processing of steels and alloys included in the group of difficult-to-process materials [1].

When assigning cutting modes for new materials, tool companies conduct tests for destruction of replaceable cutting inserts during cutting, and the obtained numerical values are published in catalogs [2]. But as practice shows, these numerical values do not always correspond to optimal cutting modes. In practice, the mode conditions for the tool are divided into machinability groups [3].

In their works, the problem of ensuring the fullest life of a cutting tool from various positions was studied: Vasin S.A. [4], Grigoriev C.N. [5], Coarse S.V. [6], Carceanu, I., Cosmeleat.0., G. [7], Ferri, C. [8], Kümmel, J. [9], Munish Kumar Gupta [

A literature analysis of the work of world-renowned scientists in the field of mechanical blade processing of metals by cutting showed that the temperature factor has a resulting effect on the cutting process of difficult-to-process materials. [4-18].

Tool materials (IM), like other materials, have physical and mechanical properties that can change under the influence of external factors. These properties have a decisive effect on the health and resource of the tool made of these materials. Studies have shown that the physical and mechanical properties of tungsten hard alloys in the process of operation, namely in the process of cutting hard-to-process materials under the influence of high temperatures, vary symmetrically [19]. The study of the processes inside the tool materials, including tungsten hard alloys, will allow, on the basis of a scientifically sound method, to determine and assign the operating modes

Materials Research Proceedings 21 (2022) 323-328
https://doi.org/10.21741/9781644901755-57

of this material taking into account its internal changes. This will increase the life of the cutting tool and create conditions under which the cutting tool will have maximum cutting properties when processing difficult materials [20].

The purpose of the work: to conduct research on the change in the electromagnetic field of eddy currents arising in replaceable cutting plates made of tungsten hard alloy B35 in the temperature range characteristic of the cutting process.

To achieve the goal, the following tasks were set and solved.

1. Carry out a literary review, as well as an analysis of existing work on the problem of rational selection of a tool material, as well as the purpose of cutting modes.

2. Conduct experimental studies of the change in the electromagnetic field of eddy currents occurring in solid alloys B35 under the influence of high temperatures characteristic of the cutting process.

Research methodology

To solve the second problem, it is necessary to conduct laboratory studies of changes in the magnetic field of eddy currents arising in replaceable cutting plates made of tungsten hard alloy B35. An analysis of existing solutions showed that such devices that meet our requirements do not exist. One of the requirements for the device is the possibility of using a replaceable multifaceted hard alloy plate as a test sample in the delivery state, to maximize the approximation of the results to real conditions.

During the development of the laboratory plant, a literary review was carried out, which showed that at the moment non-destructive testing methods of technological facilities have been widely used.

The "eddy current" [21] method of nondestructive testing is suitable for studies of internal changes occurring inside tool materials), with the possibility of testing samples in the state of supply (replaceable cutting hard alloy plates (SMTP). This type of nondestructive testing will allow conducting studies on samples in the form of replaceable cutting inserts in the state of supply, which allows reducing the time for the production of special samples. Also, this advantage allows us to bring the conditions in which laboratory studies are carried out closer to real conditions.

As the test images, SMTP from instrumental hard alloy B35 in the delivery state were chosen.

The research was carried out at a special laboratory installation, which was based on the eddy current method of non-destructive testing. The diagram of the proposed installation is shown in Figure 1. The laboratory plant is a system of instruments for measuring changes in the magnetic field of eddy currents arising in ferromagnets.

Results and discussion

It is recommended to use SMTP as the test objects in the delivery state, since the dimensions of the used coil with heat-insulating coating are limiting. When changing the coil, it is allowed to examine other forms of factors on samples.

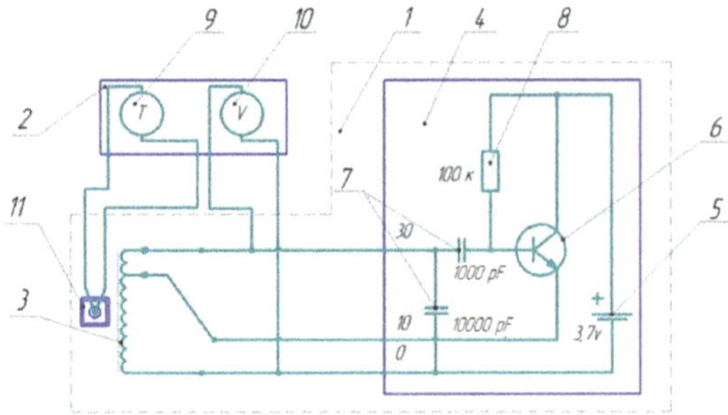

Fig. 1. *Scheme of the plant developed by the authors to determine the temperature of maximum operability of replaceable cutting hard alloy plates by electromagnetic properties: 1- Self-oscillating circuit; 2-Housing; 3-Coil; 4- PCB; 5- Power supply; 6- Transistor; 7- Capacitor; 8- Resistance; 9 - Thermometer; 10 - Voltmeter; 11- Replaceable cutting insert.*

The unit includes self-oscillating circuit (AC) and instrument board. To measure the quantitative complex characteristic of the change in the magnetic field of BT (eddy currents), a multi-voltmeter is used, and a digital thermometer was used to measure the state of temperature throughout the experiment. Electronic part of device represents electronic circuit consisting of two series-connected capacitors, resistance, transistor and power supply unit. AK is made in the form of autotransformer, like eddy current converter excitation winding. The winding is coated to prevent damaging effect of object heating and destruction of the winding itself.

Tests of the samples were carried out during heating of replaceable cutting inserts made of tungsten hard alloy B35.

The heating temperatures were selected in the range up to 1000° C. The studies were carried out as follows. The sample was heated, in our case the sample is a replaceable cutting hard alloy plate in the supply state. Then magnetic field is applied to analysed sample by non-contact method.

Changes in values indicating internal changes of the hard alloy are recorded on the voltmeter. The results are displayed on the PC monitor screen.

Based on the results of experimental studies, tables of the results were compiled. In them, with an interval of 10°S, the corresponding values of the magnetic field of eddy currents induced in replaceable cutting plates made of tungsten hard alloy B35 are indicated.

Using the capabilities of the mathematical apparatus of the MS Excel software product, the polynomial dependencies of the obtained data on the heating temperature were built, shown in Figure 2. The temperature range (Q) (Figure 3) in which the experimental studies were performed corresponds to the temperature range characteristic of the cutting process.

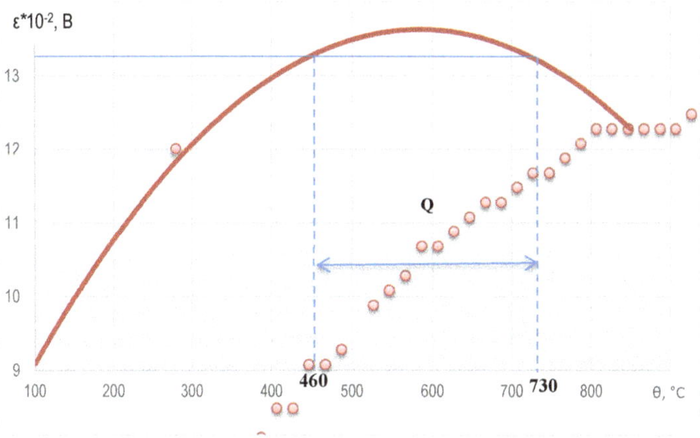

Fig. 2. *Obtained one-parameter dependence of change of AK EMF values during the alloy experiment B35 under the influence of temperatures*

The obtained graphical dependence, where a polynomial is selected as the approximating curve, is processed according to the rules of engineering studies, in accordance with 5% of the error allowed for such calculations.

To determine the boundary of the temperature range, a special technique is used. From the value of the maximum magnetic field, a 5 percent interval is deposited. A line parallel to the temperature axis is drawn through this point through the entire graph field. The points where the line intersects the line (which) of the graph are taken to be outside the range you want. From the obtained points, we draw projection lines on the temperature axis. The obtained temperature interval is taken as the interval of maximum life of replaceable cutting inserts made of hard alloy B35.

Conclusions

It is allowed to use the tool in the entire temperature range, but from the technological point of view, it is recommended to adhere to the maximum temperature, since it corresponds to the maximum permissible cutting speed.

During the work, experimental studies were carried out on changing the electromagnetic field of eddy currents arising in solid alloys of grade B35, under the influence of high temperatures characteristic of the cutting process, which made it possible to obtain a temperature interval for changing the electromagnetic properties of 460 ÷ 730 ° C.

References

[1] E.V. Artamonov, Strength and operability of replaceable hard alloy plates of prefabricated cutting tools, Tyumen, Vector Buk, 190. 2003 p. 135.

[2] N.N. Zorev Untersuchung der Kintakt-spannunger auf den Arbeits-flachen des Werkzeugs miteiner Schneidenabrundung, in: N.N. Zorev, M.H. Uteshev – Berichte der Internationalen Forschungesgemein-schaft fur mechanische produktionstechniktionstechnik Vol. 20-1 Schweiz, 1971, pp. 31-32.

[3] Information on http://www.kzts.ru/core/user_files/psm-2014.pdf

[4] S.A. Vasin Cutting materials: Thermomechanical approach to the system of relationships in cutting: a textbook for technical universities, in: S.A. Vasin, A.S. Vereshchak, V.S. Kushner - M.: Publishing House MSTU named after N.E. Bauman, 2001, p. 448

[5] V.A. Grechishnikov Formation processes and instrumental technique, in: V.A. Grechishnikov, N.A. Chemborisov, V.B. Stupko, D.T. Safarov, O.B. Kuchina, S.N. Grigoriev, A.G. Schirtladze. - Old Oskol: TNT, 2012. p. 328.

[6] V.V. Lapshin, Machining of linear negative matrices for fresnel lenses and prisms, in: V.V. Lapshin, E.M. Zakharevich, S.V. Grubyi - Russian engineering research, Allerton Press, Inc, New York, Vol. 36.10, 2016 pp. 826-830. https://doi.org/10.3103/S1068798X16100105

[7] I. Carceanu, G. Cosmeleată, B. Ghiban, M. Balanescu, I. Nedelcu, High performance composite materials created through advanced techniques, Materiale Plastice Vol. 44, Issue 4, 2007, pp. 321-325.

[8] C. Ferri, T. Minton, S.B. Ghani, K. Cheng, Internally cooled tools and cutting temperature in contamination-free machining. Proceedings of the Institution of Mechanical Engineers, Part C: Journal of Mechanical Engineering Science Vol. 228, Issue 1, 2014, pp. 135-145. https://doi.org/10.1177/0954406213480312

[9] J. Kümmel, D. Braun, J. Gibmeier, J. Schneider, C. Greiner, V. Schulze, A. Wanner, Study on micro texturing of uncoated cemented carbide cutting tools for wear improvement and built-up edge stabilization. Journal of Materials Processing Technology Vol. 215, 2015, pp. 62-70. https://doi.org/10.1016/j.jmatprotec.2014.07.032

[10] Munish Kumar Gupta, Machining of Aerospace Material - Ti (grade-2) alloy: Modelling and Optimization/ Pardeep Kumar Sood, Gurraj Singh, Vishal S. Sharma Sustainable// Journal of Cleaner Production. https://doi.org/10.1016/j.jclepro.2017.01.133

[11] K.S. Murthy, I. Rajendran, Optimization of end milling parameters under minimum quantity lubrication using principal component analysis and grey relational analysis. Journal of the Brazilian Society of Mechanical Sciences and Engineering Vol. 34, Issue 3, 2012, pp. 253-261. https://doi.org/10.1590/S1678-58782012000300005

[12] R. Neugebauer, C. Hochmuth, G. Schmidt, M. Dix, Energy efficient process planning based on numerical simulations. Advanced Materials Research 2011, Vol. 223, pp. 212-221, 17th CIRP Conference on Modelling of Machining Operations; Sintra; Portugal, 2011. https://doi.org/10.4028/www.scientific.net/AMR.223.212

[13] A. Nurulamin, I. Jaafar, W. Zubaire Role of discrete nature of chip formation and natural vibrations of system components in chatter formation during metal cutting, IIUM Engineering Journal. 2010, Vol. 11, pp. 124-126. https://doi.org/10.31436/iiumej.v11i1.53

[14] S. Tangjitsitcharoen, T. Moriwaki Intelligent monitoring and identification of cutting states of chips and chatter on CNC turning machine, Journal of Manufacturing Processes. 2008, Vol.10, pp. 40–46. https://doi.org/10.1016/j.jmatprotec.2007.04.043

[15] A. Patwari, Investigations of formation of chatter in a non-wavy surface during thread cutting and turning operations, in: A. Patwari, N. Amin, W. Faris – Text : electronic, Advanced

Materials Research. 2010, Vol.83, pp. 637–645. https://doi.org/10.4028/www.scientific.net/AMR.83-86.637

[16] W. Tillmann, C. Schaak, D. Biermann, R. Abmuth, S. Goeke, Robot based deposition of WC-Co HVOF coatings on HSS cutting tools as a substitution for solid cemented carbide cutting tools. IOP Conference Series: Materials Science and Engineering,Volume 181, Issue 1, 17 March 2017, Vol. 2011 19th Chemnitz Seminar on Materials Engineering; Chemnitz; Germany; 16 March 2017- 17 March 2017. https://doi.org/10.1088/1757-899X/181/1/012011

[17] H. Zhang, Z.Z. Fang, Q. Lu, Characterization of a bilayer WC-Co hardmetal using Hertzian indentation technique. International Journal of Refractory Metals and Hard Materials, Vol. 27, Issue 2, March 2009, pp. 317-322. https://doi.org/10.1016/j.ijrmhm.2008.07.014

[18] C. L. Zhang, A hybrid approach of ANN and HMM for cutting chatter monitoring, in: C. L. Zhang, X. Yue, Y. T. Jiang, – Text : electronic, Advanced Materials Research. 2010, Vol.97, pp. 3225–3232. https://doi.org/10.4028/www.scientific.net/AMR.97-101.3225

[19] E.V. Artamonov, Installation for diagnostics of temperature of maximum operability of replaceable cutting inserts based on change of electromagnetic properties of hard alloys, in: E.V. Artamonov, A.M. Tveryakov, A.S. Shtin Control. Diagnostics. 2018. Vol.12. pp. 54-57. https://doi.org/10.14489/td.2018.12.pp.054-057

[20] E.V. Artamonov Determination of Temperature of Maximum Working Capacity of Tool Hard Alloys, in: E.V. Artamonov, A.M. Tveryakov, A.S. Shtin – Text : electronic, Obrabotka metallov-metal working and material science, – 2018. – Vol. 20.3, pp. 47-57. https://doi.org/10.17212/1994-6309-2018-20.3-47-57

[21] E.V. Artamonov Development of installation elements for determining the temperature of maximum operability of cutting elements from hard alloys, in: E.V. Artamonov, A.M. Tveryakov, A.S. Shtin .Transport and mechanical engineering of Western Siberia. 2018. Vol. 1. – pp. 31-38.

Materials Research Proceedings **21** (2022) 329-334 https://doi.org/10.21741/9781644901755-58

Control of Deformation of Elastic Polymer-Abrasive Circles at their Wear

Dmitriy Podashev

Irkutsk National Research Technical University, 83 Lermontov street, 664074, Irkutsk, Russia

dbp90@mail.ru

Keywords: Elastic Polymer-Abrasive Circle, Wear of Instrument, Deformation Of Instrument, Deformation Regulation

Abstract. The article is devoted to the study of the wear intensity of elastic polymer-abrasive circles when processing the surfaces of parts made of high-strength aluminum alloys. Empirical dependencies of wear on cutting speed and tool deformation are obtained, on the basis of which method of tool deformation correction with long-term, continuous surface treatment is proposed. In practice, such a procedure is necessary due to the loss of process performance that occurs due to the reduction of tool deformation due to wear. The proposed technique allows to effectively control the finishing process as the tool is worn out.

Introduction

The problem of mechanization and automation of manual labor during finishing cleaning and finishing operations is relevant for modern mechanical engineering. For example, in the aviation industry, the design of any aircraft includes more than 50 large and long parts on which these operations are required. These parts are manufactured, usually by milling. Many works are devoted to the research of the milling process and the assurance of the quality of the surface layer of parts in this type of machining, for example [1, 2, 3]. However, even with a well-organized technology for manufacturing parts by milling, including on CNC machines, the surface roughness required in the drawing is not provided at the transition points when changing the direction of feeding, when processing curved surfaces, random defects associated with the actual state of the cutter, etc., also regularly occur.

Today, most finishing cleaning and finishing operations are performed manually, while the proportion of labor intensity of manual work on the finishing process of such parts can reach 65% of the total labor intensity of manufacturing parts after milling.

Machining methods, such as grinding, polishing, volumetric vibration machining, etc., are very effective ways to solve the problem of surface grinding. The works of many scientific teams are devoted to the study of these abrasive processing methods, for example [4, 5, 6, 7].

Each of these abrasive methods can be characterized by a different level of efficiency. Each method had certain property and disadvantages, as well as areas of effective application.

It should be noted that when machining with rigid tools, it is difficult to clear a thin surface layer (especially aluminum alloy parts widely used in the aircraft industry) due to the possibility of removing a certain layer of material and violating the required accuracy in size.

Volumetric vibration machining methods were very effective and were used for finishing metal parts with dimensions of up to 300 mm. However, for the processing of large and long parts, examples of which are given in Figure 1, the use of these methods is not economically feasible, since large and expensive equipment, as well as a large amount of preparatory and final work, are necessary for their implementation.

In connection with the above, the most promising direction capable of effectively solving these problems in ensuring the quality of finishing of large-sized, complex-shaped and long-sized parts taking into account the peculiarities of their sizes and structures is the treatment with polymer-abrasive circles with a bond of nonwoven materials that have high elasticity.

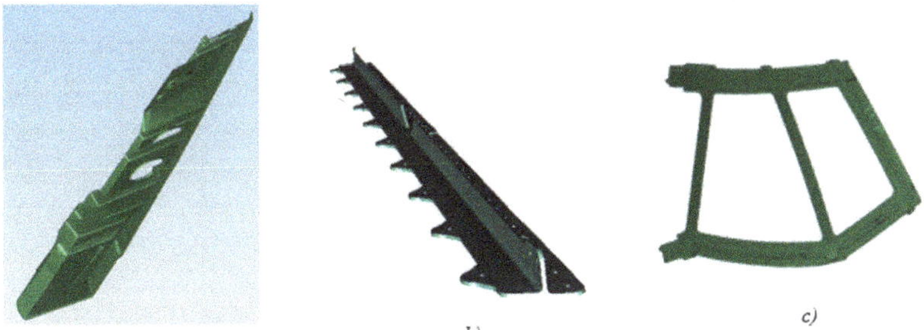

a) b) c)

Fig. 1. *Examples of complex profiled, large parts of the airframe :*
a) spar; b) cross section; c) light frame.

It should be noted that any polymer-abrasive tool (petal circles, radial and end brushes) during processing is worn out and during continuous operation, adjustment of such a treatment mode as tool deformation is required. In practice, when machining the surfaces of parts (see Figure 1), which can be as heavy as 10 m or more, such a procedure is necessary due to a decrease in the productivity of the machining process, which is due to a reduction in tool deformation. Consider the nature of wear of the tool depending on the processing modes and the method of correcting its deformation ΔY when treating the tops with elastic polymer-abrasive circles.

Correction of deformation of elastic polymer-abrasive circles due to their wear when machining surfaces

Research was conducted on the Deckel Maho DMC 635V universal milling machine on the simple samples representing plates with sizes 3 x 20 x 100 mm from V95pchT2 aluminum alloy.

a) b) c)

Fig. 2. *Elastic polymer-abrasive circles of company 3M:*
a) pressed circles of brand FS-WL; b) pressed circle of brand DB-WL; c) flap circle of brand CF-FB

The treatment was carried out with elastic polymer-abrasive circles of types FS-WL 8A MED, FS-WL 6S FIN, FS-WL 2S CRS, DB-WL 8S MED and CF-FB 0,5A FIN of company 3M (Minnesota Mining and Manufacturing Company) [8] (Figure 2), made of abrasive material Scotch-Brite™. This material consists of synthetic fibers forming a three-dimensional flexible nonwoven canvas throughout which abrasive grains are uniformly distributed. The main physical and mechanical data of the tools are given in Table 1. The work [9] is devoted to detailed research of physical and mechanical properties.

Table 1. *Parameters of elastic polymer-abrasive circles*

Circle	D_k, mm	B_k, mm	r_k, mm	d_k, mm	M_K, kg	γ_K, kg/m^3	Abrasive	Granularity Z, mkm
FS-WL-8AMED	140,5	26	17,5	25,4	0,278	712,77	Al_2O_3,	50-60
FS-WL-6SFIN	129,5	25,5	17,5	25,4	0,162	501,63	SiC	45-50
FS-WL-2SCRS	147,2	26	17,5	25,4	0,162	377,37	SiC	~100
DB-WL-8SMED	147,8	25,6	17,5	25,4	0,284	666,29	SiC	50-60
CF-FB-0,5AFIN	193	50	45	76,5	0,418	339,03	Al_2O_3,	45-50

D_k – circle diameter, mm; B_k – circle width, mm; r_k – radius of circle bushing, mm; d_k – diameter of hole, mm; M_K – circle weight, kg; γ_κ – density of circle's material, kg/m^3.

The circle wear was determined by weighing it before and after processing on the analytical scales of the Ohaus series Discovery (DV) model DV214C. Wear value I (mm/min) per unit time is used as circle wear index.

Value *I* was calculated by the formula:

$$I = \frac{G_1 - G_2}{\gamma_k \cdot \pi \cdot D_k \cdot B \cdot T}, \quad (1)$$

where γ_k – density of circle's material in g/mm^3 (see Table 1) ; *G* and G_2 – circle weight before and after treatment, respectively, in g; *B* – width of the sample to be treated, in mm; D_K – circle diameter, in mm (see Table 1); *T* – processing time of sample length l, in min.

The conclusion on the significance or insignificance of the influence of deformation of the circle of ΔY, velocity V and the supply of S on the wear of tool I was obtained using the Fisher criterion according to [10]. $F_{0,95}$ is Fisher's criterion at a fiducial probability of 0.95. This criterion at degrees of freedom $f_1 = k - 1 = 3$ and $f_2 = k(n - 1) = 8$ according to reference data [10] is equal to $F_{0,95} = 4{,}1$. It is established that the circle wear does not depend on the feed S.

Figure 3, *a, b* shows the dependencies of circle wear for 1 minute of operation on the deformation of circles.

It is stated that the wear of the tool has been found to increase with increasing deformation. This is due to the fact that the vertical component of the cutting force [11] increases with increasing deformation, and therefore the friction force also increases.

Figure 4, *a, b* shows the dependencies of circle wear in 1 minute of operation on the cutting speed. It has been established that tool wear is increasing with increasing speed.

This is due to the fact that with an increase in speed, the centrifugal and din components of the force of interaction of the abrasive grain along the treated surface grow. At the moment of meeting of abrasive grains of circle with treated surface impact occurs, as a result of which force becomes significantly more static [11]. At the same time, the centrifugal force and the impact pulse directly depend on the speed.

The experimental dependencies obtained were approximated and the following regression equation (wear *I* in mm/min) was obtained:

$$I = a_1 \cdot V^2 + a_2 \cdot \Delta Y^2 + a_3 \cdot V + a_4 \cdot \Delta Y + a_5 \cdot V \cdot \Delta Y + a_6, \quad (2)$$

where V – cutting velocity, m/min; ΔY – circle deformation, mm.

The values of the coefficients $a_{1\text{-}5}$ and the absolute terms a_6 of this equation for each of the instruments studied are shown in Table 2.

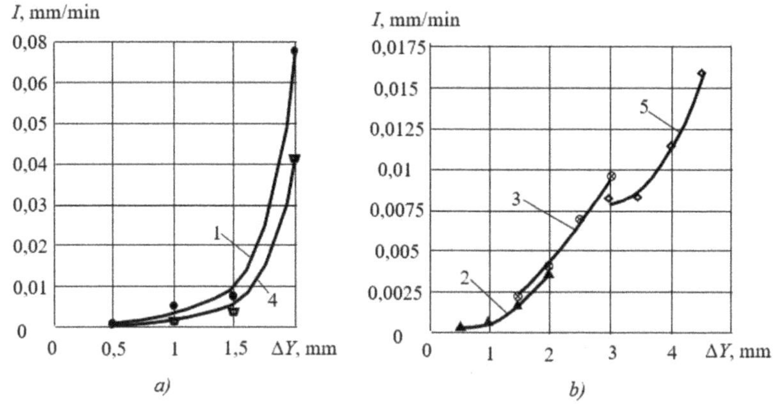

Fig. 3. Dependency of wear for 1 min of operation from deformation ΔY for circles: 1 – FS-WL 8A MED at V = 441, 4 m/min; 2 – FS-WL 6S FIN at V = 406, 8 m/min; 3 – FS-WL 2S CRS at V = 462, 4 m/min; 4 – DB-WL 8S MED at V = 464, 3 m/min; 5 – CF-FB 0,5A FIN at V = 606, 3 m/min.

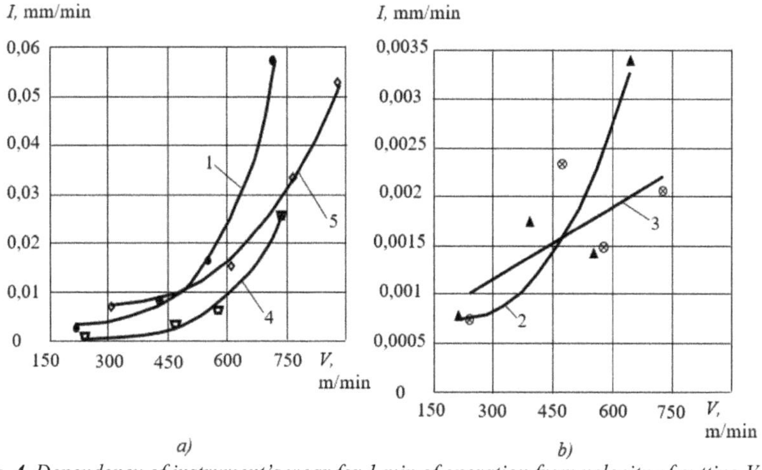

Fig. 4. Dependency of instrument's wear for 1 min of operation from velocity of cutting V for circles: 1 – FS-WL 8A MED; 2 – FS-WL 6S FIN; 3 – FS-WL 2S CRS; 4 – DB-WL 8S MED at deformation ΔY = 1,5 mm; 5 – CF-FB 0,5A FIN at deformation ΔY = 4,5 mm.

Table 2. *Values of coefficients and absolute terms in the equation (2)*

Coefficient	FS-WL 8A MED	FS-WL 6S FIN	FS-WL 2S CRS	DB-WL 8S MED	CF-FB 0,5A FIN
a_1	$6{,}3929\cdot10^{-8}$	$-8{,}34\cdot10^{-10}$	$-5{,}894\cdot10^{-9}$	$8{,}869\cdot10^{-9}$	$2{,}6883\cdot10^{-8}$
a_2	$-4{,}16136\cdot10^{-4}$	$4{,}2595\cdot10^{-4}$	$1{,}3418328\cdot10^{-3}$	$-5{,}78041\cdot10^{-7}$	$1{,}089303\cdot10^{-3}$
a_3	$-6{,}2725\cdot10^{-6}$	$2{,}62655\cdot10^{-6}$	$-1{,}65935\cdot10^{-6}$	$-9{,}2\cdot10^{-10}$	$-3\cdot10^{-5}$
a_4	$9{,}07143\cdot10^{-3}$	$-1{,}133675\cdot10^{-4}$	$-4{,}264372\cdot10^{-3}$	$8{,}843099\cdot10^{-3}$	$-6{,}745733\cdot10^{-3}$
a_5	$6{,}95575\cdot10^{-6}$	$9{,}94751\cdot10^{-7}$	$6{,}987568\cdot10^{-6}$	$1{,}183662\cdot10^{-6}$	$7{,}964633\cdot10^{-6}$
a_6	$-6{,}91871\cdot10^{-3}$	$-6{,}408575\cdot10^{-4}$	$3{,}15588\cdot10^{-3}$	$-1{,}668018\cdot10^{-3}$	0,01203567

It should be noted that due to wear of the tool, the deformation of the circle ΔY is gradually reduced, as a result of which the intensity of removal of the material decreases. Therefore, with a long processing process, it is necessary to periodically correct the deformation of the tool by the value of the worn out part of the l_{izn}.

$$l_{izn} = I \cdot T_{izn}, \tag{3}$$

where T_{izn} – time of treatment during which the wear l_{izn} occurs, min;
I – wear intensity which is calculated by formula (2), mm/min.

For the studied circles, the T_{izn} time (at the given allowable l_{izn} values) is given in Table 3 (as an example) for the smallest I_{min} (mm/min) and the highest I_{max} (mm/min) of the possible wear intensity.

Table 3. *Estimated time of the circle operation before correction of instrument tensioning ΔY*

Circle Type	Wear I by (2)		Wear l_{izn}, mm				
	I_{min}, mm/min	I_{max}, mm/min	0,20	0,35	0,5	0,75	1,0
			Time of treatment T_{izn}, min				
FS-WL 8A MED	0,002		100	175	250	375	500
		0,078	2,6	4,5	6,4	9,6	12,8
FS-WL 6S FIN	0,0005		400	700	1000	1500	2000
		0,0037	54,1	94,6	135,1	202,7	270,3
DB-WL 8S MED	0,002		100	175	250	375	500
		0,041	4,9	8,5	12,2	18	24,4
FS-WL 2S CRS	0,002		100	175	250	375	500
		0,01	20	35	50	75	100
CF-FB 0,5A FIN	0,0076		26,3	46,1	65,8	98,7	131,6
		0,016	12,5	21,9	31,2	46,9	62,5

Conclusion

The proposed method of correction of the ΔY treatment mode during long-term surface treatment with elastic polymer-abrasive circles ensures effective control of the finishing process as the tool is worn.

References

[1] Fomin A.A., Gusev V.G., Sadrtdinov A.R., Assurance of accuracy of longitudinal section of profile surfaces milled at high feeds, Lecture Notes in Mechanical Engineering. Part F4 (2019) 527-536. https://doi.org/10.1007/978-3-319-95630-5_55

[2] Fomin A.A., Gusev V.G., Timerbaev N.F., The processing of the profile surface of the work-pieces, characterized by low rigidity, Solid State Phenomena. 299 SSP (2020) 852-860. https://doi.org/10.4028/www.scientific.net/SSP.299.852

[3] Batuev V.A., Guzeev V.I., Batuev V.V., Determining the cut thickness in milling complex surfaces, Russian Engineering Research. 4 (2020) 343-346. https://doi.org/10.3103/S1068798X20040061

[4] Tamarkin M., Tishchenko E., Chukarina I., Sosnitskaya T., Parts processing technology for transport engineering, Advanced in intelligent systems and computing. 1115 AISC (2020) 913-922. https://doi.org/10.1007/978-3-030-37916-2_90

[5] Roshchupkin S.I., Sidorov D.E., Kolesov A.G., Assessment of the actual depth of cutting with cylindrical external grinding, IOP Conference Series: Materials Science and Engineering. International Conference on Modern Trends in Manufacturing Technologies and Equipment 2019, ICMTME 2019. Art. No. 022010 (2020). https://doi.org/10.1088/1757-899X/709/2/022010

[6] Makarov V.F., Muratov K.R., Gashev E.A., Optimal cutting time and speed in abrasive lapping of ceramic, Russian Engineering Research. 10 (2017) 916-918. https://doi.org/10.3103/S1068798X17100148

[7] Starodubtseva D.A., Koltsov V.P., Vinh L.T., Grinding of aluminum alloy panels after shot peen forming on contact type installations, IOP Conference Series: Materials Science and Engineering. 2019 International Conference on Innovations in Automotive and Aerospace Engineering, ICI2AE 2019. Art. No. 012109 (2019). https://doi.org/10.1088/1757-899X/632/1/012109

[8] Information on http://www.3mabrazives.ru/

[9] Dimov Yu. V., Podashev D.B., Properties of elastic abrasive wheels, Russian Engineering Research. 7 (2017) 631-637. https://doi.org/10.3103/S1068798X17070085

[10] Pystulnik E.I., Statistical methods of analysis and processing of observations, Nauka, Moscow, 1968.

[11] Dimov Yu. V., Podashev D.B., Cutting forces when machined with elastic abrasive circles, The messenger of Irkutsk National Research Technical University. 7 (2015) 47-55.

Wearproof Composite Coatings on Ti

Dmitry V. Mashtalyar[1,a*], Konstantine V. Nadaraia[1,b], Evgeny A. Belov[1,c], Sergey N. Suchkov[1,2,d], Arina I. Pleshkova[2,e], Sergey L. Sinebrukhov[1,f], and Sergey V. Gnedenkov[1,g]

[1] Institute of Chemistry FEB RAS, 159 pr. 100-letiya Vladivostoka, Vladivostok 690022, Russia

[2] Far Eastern Federal University, 10 Ajax Bay, Russky Island, Vladivostok 690922, Russia

[a]madiva@inbox.ru; [b]nadaraiakv@mail.ru; [c]belov_eal@mail.ru; [d]snsuchkov@yandex.ru; [e]pleshkova.ai@students.dvfu.ru; [f]sls@ich.dvo.ru; [g]svg21@hotmail.com

Keywords: Coating Materials, Oxide Materials, Polymers, Composite Materials, Titanium

Abstract. In this work the formation of protective coatings on VT1-0 commercially pure titanium by plasma electrolytic oxidation (PEO) and subsequent fluoropolymer treatment is presented. The structure, morphology, corrosion, and mechanical properties of the formed composite coatings were studied. It was established that PEO coatings are an excellent basis for the formation of a solid composite layer with high adhesion to its surface. The presence of polytetrafluoroethylene in the composition of the coating reduces the corrosion current density by 4 orders of magnitude and increases the wear resistance by 2 orders of magnitude in comparison with the base PEO coating.

Introduction

Currently, titanium and it alloys are widely used in advanced industries from missilery to implant surgery. The use of titanium and its alloys is due to its several unique properties, such as high corrosion resistance and strength, low density, high heat capacity, etc. [1–5]. However, titanium and its alloys require additional surface treatment for improving the tribological and corrosion properties [4,6–8].

The formation of protective coatings is widely used to modify the surface properties of titanium alloys. In industries is more common thermal oxidation, the summary of which lies in the high thermal oxidation of metal [4]. As a result, a durable oxide coating is formed on the titanium surface. However, this method is labor- and energy-intensive, and has several technological limits.

Currently, many researchers have developed a method for the formation of protective coatings on titanium alloys by plasma electrolytic oxidation (PEO) [1,4–6,8,9]. After the analysis of previous studies, we have found that such surface layers are effective for protecting titanium from wear, prevent corrosion of metals in galvanic couple with titanium, and have a suitable surface structure for the formation of composite coatings (CC) by incorporating substances into the PEO layer for additional improving of the mechanical and electrochemical characteristics of the surface of treated material [5,10].

Previously, we have developed a method to forming CC by immersion samples with base PEO coating in a suspension of superdispersed polytetrafluoroethylene (PTFE). The possibility of using some commercial fluoroplastic suspensions is interest too. The use of such suspensions can significantly expand the range of materials used to create polymer-containing layers.

Materials and methods

As a material for the studied samples the VT1-0 commercially pure titanium (wt. %: 0.25 Fe; 0.12 Si; 0.07 C; 0.12 O; 0.04 N; 0.01 H; balance: Ti) was used. As a sample for the formation of coatings the rectangular plates ($50\times50\times1$ mm^3) were used.

The formation of PEO coatings was carried out in a phosphate electrolyte, containing of 25 g L^{-1} trisodium phosphate (Na_3PO_4) in a monopolar potentiodynamic mode. In the first stage, the voltage was increased from 50 to 300 V, the duration of process was 120 s. At the second stage, the voltage dropped to 240 V, the duration of process was 580 s.

For the formation of composite coatings, a commercial aqueous suspension of PTFE F4-D (Halopolymer, Russia) was used. The coatings were formed by immersion in a suspension (exposure time was 1 min). The dipping speed was 100 mm min^{-1}, the extraction speed was 10 mm min^{-1}. Then, the samples were dried in air and thermally treated in an L3/11 muffle furnace (Nabertherm, Germany) at 365°C for 20 min.

A scanning electron microscope (SEM) Merlin (Carl Zeiss, Germany) was used to study surface morphology of the coatings. SEM image was processed by the ImageJ software (National Institutes of Health, USA).

The electrochemical parameters of the coatings formed on VT1-0 commercially pure titanium under the study were obtained by potentiodynamic polarization method. The coatings were studied using a VersaSTAT MC electrochemical system (Princeton Applied Research, USA). The tests were carried out in the three-electrode cell at room temperature in an aqueous solution of 3.5 wt. % NaCl. A platinized niobium mesh was used as a counter electrode. The saturated calomel electrode (SCE) was used as a reference electrode. The samples area exposed to electrolyte was equal to 1 cm^2.

Potentiodynamic polarization test was carried out at a scan rate of 0.167 mV/s from E_C –0.15 V to E_C +0.50 V vs. OCP. The Levenberg-Marquardt [11] method was used to calculate the corrosion parameters of studied samples by fitting the experimental data (i.e. values of potential E, and current density I) to the following Equation (1):

$$I = I_C\,(10^{\frac{E-E_C}{\beta_a}} + 10^{-\frac{E-E_C}{\beta_c}}). \tag{1}$$

This method makes it possible to obtain the best fit values of corrosion potential E_C, corrosion current density I_C, the slope of the cathodic polarization curve β_c, and the slope of the anodic polarization curve β_a. The application of the Levenberg-Marquardt method ensured a high accuracy of the conducted calculations of the corrosion parameters. The polarization resistance R_P was determined in a separate experiment by linear polarization test at sweep rate of 0.167 mV s^{-1} in the potential region $\Delta E = E_C \pm 20$ mV, in which the linear dependence $I = f(E)$ is observed [12,13]. The test was carried out after stabilization of the open circuit potential. Calculation of R_P values is carried out according to the Equation (2):

$$R_p = \frac{\Delta E}{\Delta I}, \tag{2}$$

The wear resistance of coatings was measured on a TRB-S-DE device (CSM Instruments, Switzerland). The tests were carried out in dry friction mode at a sliding speed of 50 mm s^{-1} and a load of 25 N. The corundum ball (α-Al_2O_3) was used as a counter-body. The diameter of the ball was 3 mm, and the diameter of the wear track was 5 mm. To estimate the wear, the track profiles were recorded after the wear tests using a Surtronic 25 profilometer (Taylor Hobson Ltd, England). The wear was estimated using the profiles obtained from the wear track cross-section, and thus the wear rate, P (mm^3 $(N{\times}m)^{-1}$), was calculated using Equation (3):

$$P = \frac{\Delta V}{N\,F}, \tag{3}$$

where ΔV is the worn volume (mm^3), N is the distance moved (mm), and F is the normal load (N). The wear volume was calculated using 2D profiles. In all the experiments carried out, the wear of the counter-body was not detected and was not considered at the calculation.

An adhesion characteristic of coatings was evaluated by Revetest Scratch Tester (CSM Instruments, Switzerland). The scratch length was 5 mm, the maximum applied load was 30 N. The rate of increase of the load was 30 N min^{-1}.

Results and discussion

Based on the analysis of SEM images (Fig. 1), it can be concluded that the fluoropolymer penetrates to the pores of the PEO coating and forms smooth thin film on the samples' surface (Fig. 1b). A significant number of pores of all sizes are equally sealed, and the surface becomes smooth and close to homogeneous. The polymer film in different areas has a different thickness, but whole thickness of the polymer film does not exceed 2 μm. The porosity of the coatings over the surface (Fig. 1) with and without the fluoropolymer is 14.45% and 0%, respectively.

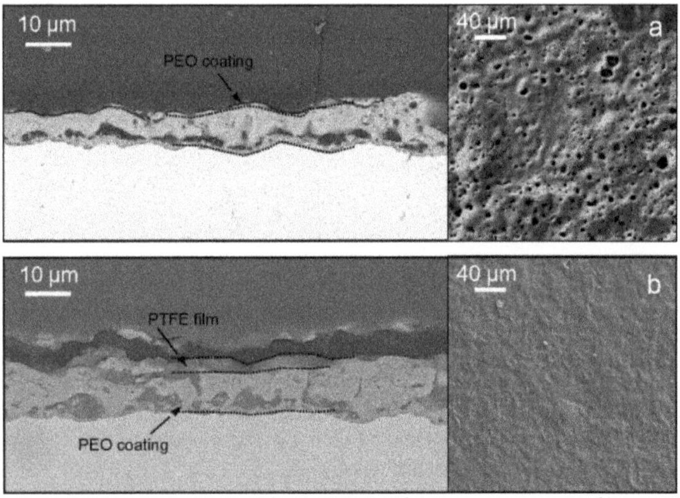

***Fig. 1.** SEM image of the PEO (a) and composite coatings (b)*

Potentiodynamic polarization data indicate an improvement in the corrosion properties of the samples after formation of CC on their surface. Evaluation of the polarization curves (Fig. 2) presents an increase of the corrosion potential E_C for PEO coating in comparison with the bare material; however, due to the high porosity and the presence of microdefects in the coating, the corrosion current density decreases insignificantly. Treatment of a PEO coating with a fluoropolymer leads to a strong decrease of the corrosion current density (by four orders of magnitude in comparison with that for a PEO coating).

This effect is associated with the incorporation of polytetrafluoroethylene to the coating. Due to the high strength and uniformity of the polymer distribution over the surface of the CC, as well as the absence of pores and defects (in comparison with the PEO coating), the value of the corrosion current density decreases (Table 1).

Table 1. *Electrochemical and mechanical properties for samples with different types of surface treatment.*

Sample	E_C (V)	I_C (A cm^{-2})	R_p (Om cm^2)	L_{C3} (N)	P (mm^3 (N m)$^{-1}$)
Ti	-0.38	$1.08 \cdot 10^{-7}$	$3.52 \cdot 10^5$	-	-
PEO	0.11	$3.06 \cdot 10^{-8}$	$9.15 \cdot 10^5$	14	$3.97 \cdot 10^{-4}$
CC	0.11	$1.02 \cdot 10^{-10}$	$7.16 \cdot 10^8$	23	$2.98 \cdot 10^{-6}$

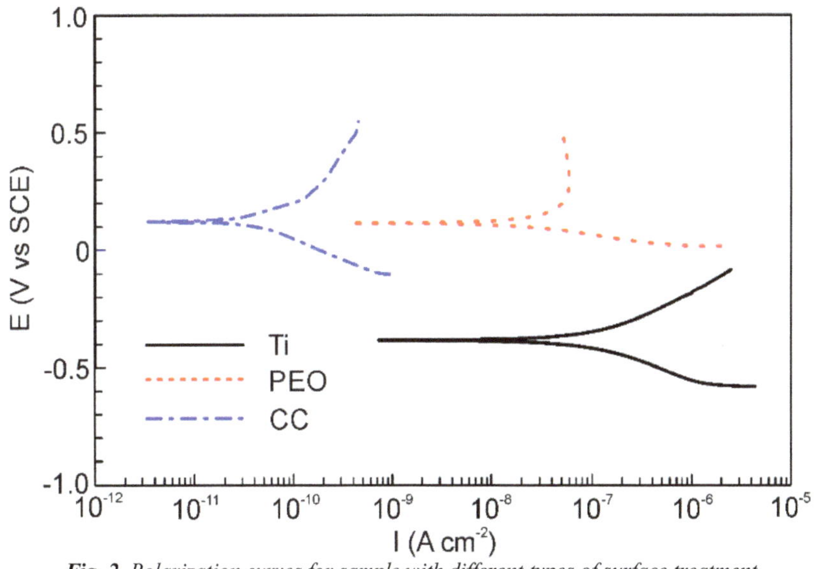

Fig. 2. *Polarization curves for sample with different types of surface treatment*

Figure 3 shows optical images of the scratch on the PEO coating (Fig. 3a) and CC (Fig. 3b). Analysis of these images clearly shows that the value of the load L_{C3}, at which the coating peels off, is higher for the CC (Table 1), and is about 23 N (for PEO coating is about 14 N). These values indicate high adhesion of both the polymer film to the base PEO coating and the whole composite coating to metal.

Analysis of tribological test (Fig. 4) made it possible to establish that the wear resistance of composite coatings is more than 2 orders of magnitude higher in comparison with PEO coating

(Table 1). It is associated with the formation of a dense polymer film that forms a uniform smooth surface with a low coefficient of friction (about 0.01).

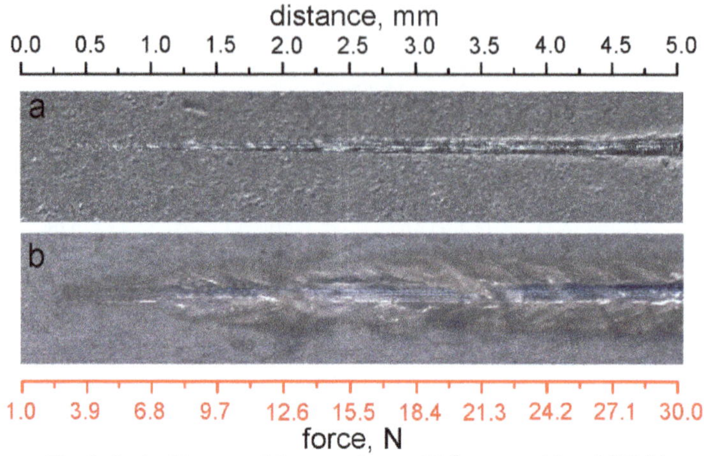

Fig. 3. *Optical images of the scratch on a PEO coating (a) and CC (b)*

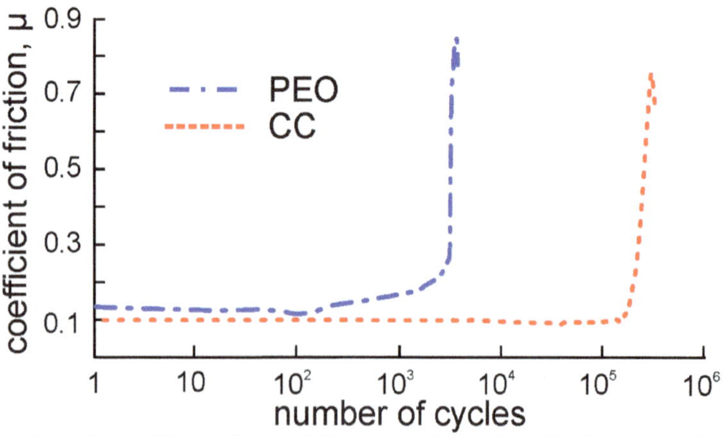

Fig. 4. *Dependence of the coefficient of friction μ on the number of cycles for samples with different types of surface treatment*

Summary

Composite coatings were obtained on VT1-0 commercially pure titanium by plasma electrolytic oxidation with following processing in suspension of PTFE (F4-D, Halopolymer, Russia). This method of treatment significantly affects the morphology, structure, and properties of the surface of the titanium, allowing its use in friction units and in the contact with another metals in corrosion environment.

Summarizing all the above, this enables one to conclude that processing of PEO coatings on VT1-0 titanium with PTFE suspension makes it possible to reduce the corrosion current density by 4 orders of magnitude, increase the adhesion of the coating to the metal surface and reduce surface wear by 2 orders of magnitude in comparison with PEO coatings.

Acknowledgements
This work was supported within the frames of the Grant of the Russian Science Foundation, project No. 20-13-00130. The electrochemical tests were carried out within the framework of the government assignments from Ministry of Science and Higher Education of the Russian Federation, project No. 0205-2021-0003.

References

[1] D. V Mashtalyar, K. V Nadaraia, A.S. Gnedenkov, I.M. Imshinetskiy, M.A. Piatkova, A.I. Pleshkova, E.A. Belov, V.S. Filonina, S.N. Suchkov, S.L. Sinebryukhov, others, Bioactive Coatings Formed on Titanium by Plasma Electrolytic Oxidation: Composition and Properties, Materials (Basel). 13 (2020) 4121.

[2] A. Fattah-alhosseini, M. Molaei, K. Babaei, The effects of nano- and micro-particles on properties of plasma electrolytic oxidation (PEO) coatings applied on titanium substrates: A review, Surfaces and Interfaces. 21 (2020) 100659. https://doi.org/https://doi.org/10.1016/j.surfin.2020.100659

[3] M. Mohedano, B. Mingo, H. Mora-Sánchez, E. Matykina, R. Arrabal, Effects of pre-anodizing and phosphates on energy consumption and corrosion performance of PEO coatings on AA6082, Surf. Coatings Technol. 409 (2021) 126892. https://doi.org/https://doi.org/10.1016/j.surfcoat.2021.126892

[4] D. V Mashtalyar, K. V Nadaraia, I.M. Imshinetskiy, E.A. Belov, V.S. Filonina, S.N. Suchkov, S.L. Sinebryukhov, S. V Gnedenkov, Composite coatings formed on Ti by PEO and fluoropolymer treatment, Appl. Surf. Sci. 536 (2021) 147976. https://doi.org/https://doi.org/10.1016/j.apsusc.2020.147976

[5] S. Aliasghari, A. Němcová, P. Skeldon, G.E. Thompson, Influcncc of coating morphology on adhesive bonding of titanium pre-treated by plasma electrolytic oxidation, Surf. Coatings Technol. 289 (2016) 101–109. https://doi.org/https://doi.org/10.1016/j.surfcoat.2016.01.042

[6] D. Sreekanth, N. Rameshbabu, K. Venkateswarlu, Effect of various additives on morphology and corrosion behavior of ceramic coatings developed on AZ31 magnesium alloy by plasma electrolytic oxidation, Ceram. Int. 38 (2012) 4607–4615. https://doi.org/https://doi.org/10.1016/j.ceramint.2012.02.040

[7] S. V Gnedenkov, S.L. Sinebryukhov, D. V Mashtalyar, K. V Nadaraia, D.P. Kiryukhin, G.A. Kichigina, P.P. Kushch, V.M. Buznik, Composite coatings formed on the PEO-layers with the use of solutions of tetrafluoroethylene telomers, Surf. Coatings Technol. 346 (2018) 53–62. https://doi.org/https://doi.org/10.1016/j.surfcoat.2018.04.036

[8] H. Sharifi, M. Aliofkhazraei, G.B. Darband, A.S. Rouhaghdam, Tribological properties of PEO nanocomposite coatings on titanium formed in electrolyte containing ketoconazole, Tribol. Int. 102 (2016) 463–471. https://doi.org/https://doi.org/10.1016/j.triboint.2016.06.013

[9] A. Fattah-Alhosseini, M.K. Keshavarz, M. Molaei, S.O. Gashti, Plasma electrolytic oxidation (PEO) process on commercially pure Ti surface: effects of electrolyte on the microstructure and corrosion behavior of coatings, Metall. Mater. Trans. A. 49 (2018) 4966–4979.

[10] M. Molaei, M. Nouri, K. Babaei, A. Fattah-Alhosseini, Improving surface features of PEO coatings on titanium and titanium alloys with zirconia particles: A review, Surfaces and Interfaces. 22 (2021) 100888. https://doi.org/https://doi.org/10.1016/j.surfin.2020.100888

[11] J.L. Aljohani, E.S. Alaidarous, M.A. Zahoor Raja, M.S. Alhothuali, M. Shoaib, Backpropagation of Levenberg Marquardt artificial neural networks for wire coating analysis in the bath of Sisko fluid, Ain Shams Eng. J. (2021). https://doi.org/https://doi.org/10.1016/j.asej.2021.03.007

[12] X. Zhang, G. Cai, Y. Lv, Y. Wu, Z. Dong, Growth mechanism of titania on titanium substrate during the early stage of plasma electrolytic oxidation, Surf. Coatings Technol. 400 (2020) 126202. https://doi.org/https://doi.org/10.1016/j.surfcoat.2020.126202

[13] C.E. Tanase, M. Golozar, S.M. Best, R.A. Brooks, Cell response to plasma electrolytic oxidation surface-modified low-modulus β-type titanium alloys, Colloids Surfaces B Biointerfaces. 176 (2019) 176–184. https://doi.org/https://doi.org/10.1016/j.colsurfb.2018.12.064

Additive Technologies in the Production of Vehicle Rubber with Sensory Properties

Evgeniy V. Stepanov[1,a*], Marek Schmidt[2,b], Dennis Bäcker[2] and Anton V. Tumasov[1]

[1]Nizhny Novgorod State Technical University n.a. R.E. Alekseev, st. Minina, 24, 603950 Nizhny Novgorod, Russia

[2]Chemnitz University of Technology, Straße der Nationen 62, 09111 Chemnitz, Germany

[a]evgeniystepanov1991@nntu.ru, [b] marek.schmidt@mb.tu-chemnitz.de

Keywords: Sensor Integration, Rubber with Sensory Properties, Piezoceramics, Vehicle, Additive Technologies

Abstract. Annotation. Rubber products are widely used in the construction of vehicles, for example, as sealing and protective devices, suspension joints and are the basis of automobile tires. Modern trends related to increasing the level of vehicle safety require the use of innovative approaches in the design and use of new materials with unique properties. This article proposes an approach to create a rubber with sensory properties that can be used in various automotive products and prevent situations that can harm both human health and lead to serious damage to the structure of the vehicle itself. We have developed an intelligent vehicle door seal to prevent injury to a person when the door is closed carelessly. The sealant, which reacts to deformation when a foreign body enters the seal site, consists of rubber with the addition of piezoceramic powder and two electrode layers. Each electrode layer has several parallel strip-like electrodes positioned along the perimeter of the seal. This document describes possible applications for rubber products with sensory properties and an additive method for making such rubber with the addition of piezoceramic powder.

Introduction

The creation of an intelligent and safe vehicle is an urgent task. A person getting injured when using a vehicle can be not only in a road accident, but also in normal situations related to the embarkation / disembarkation of passengers or loading / unloading a vehicle. Children are often injured when they close the vehicle door on their own themselves. To prevent such situations, this article proposes the use of an intelligent rubber door seal with sensory properties.

Another possible application of smart rubber to improve safety is to use this approach in the manufacture of automobile tires. A large number of accidents are associated with the destruction of tires when the car is moving, the consequences of such situations, as a rule, become critical for maintaining safe movement. The creation of an intelligent bus will allow monitoring the state of its structure and the loads on it during operation. If a defect in the structure or tire overload, which can lead to a critical situation, is detected, it becomes possible to inform the driver about the danger or prevent movement by turning the vehicle into autonomous mode (for example, for self-driving vehicles or vehicles equipped with ADAS).

Currently, there are various additive methods for the manufacture of thermoplastic elastomers [1]. However, for the additive manufacturing of vulcanizable rubber, no commercially viable solution is available on the market. Only a few research approaches can be found in the literature. This is caused by the rheological material behavior of vulcanizable rubber as well as the following

Materials Research Proceedings **21** (2022) 342-348 https://doi.org/10.21741/9781644901755-60

vulcanization process. This leads to specific challenges for applications related to additive manufacturing processes [2]. In particular, the viscosity of rubber compounds leads to the risk that components built up in layers will deform unintentionally above a certain layer height or collapse completely under their own weight. One possible solution is the parallel layer-by-layer build-up of an additional material that completely encloses and thus supports the rubber shape. The enclosing material also serves to stabilize the shape during the vulcanization process. As additional material for dimensional stabilization of the rubber shape there are investigations with thermoplastic material [3, 4] as well as molding sand, silicone rubber and plaster [5]. However, the above approaches are still subject of research and are not mature processes. An alternative variant is to design the rubber compounds so that self-vulcanization takes place when they are combined. During the additive manufacturing process, the materials are mixed together in a screw extruder and discharged in layers via a nozzle. Subsequently, the vulcanization begins on the printing platform, which keeps the geometrically desired shape stable [6]. However, the possible properties of the finished part remain very limited due to the predominantly predetermined chemical composition of the rubber compounds.

This article proposes to use methods of additive production of vulcanizable rubbers to create a flat composite intelligent material by adding a piezoceramic component to the composition. In 3D printing, rubber and piezoceramic powder is fed into the machine's extruder, which mixes the structures before the resulting material leaves the nozzle. Then, vulcanization of the composite rubber can be done in an oven. At the intermediate stage of 3D printing, it is possible to integrate electrode layers into the sensor rubber to obtain a signal upon deformation of the product, which is difficult to implement with other manufacturing methods. The rubber composite additive machine uses a screw extruder, which gives more flexibility in the shape of the feed material. The material can be used, for example, in the form of powder, granules, liquid or strips. In particular, extrusion-based 3D printing is already being used for non-thermoplastic materials such as ceramics [7] or composites [8].

The aim of this study is to study the technological process of 3D printing rubber with the addition of ceramic powder based on extrusion and testing a prototype rubber with sensory properties using the example of a vehicle door seal.

The structure of piezoceramic rubber and the principle of its manufacture

Figure 1 shows the schematic structure of a piezoceramic rubber. The rubber material contains piezoceramic particles inside. On the outside, the rubber is covered with a flexible electrode layer. When the rubber is deformed, pressure is applied to the piezoceramic particles and the direct piezoelectric effect takes place, resulting in a measurable electrical signal. The sensitivity of the signal depends on the amount and distribution of particles in the rubber. Due to the resilience of the rubber, the sensitivity generally does not come close to existing stiff sensors. However, stiff sensors are not designed for integration into rubber parts.

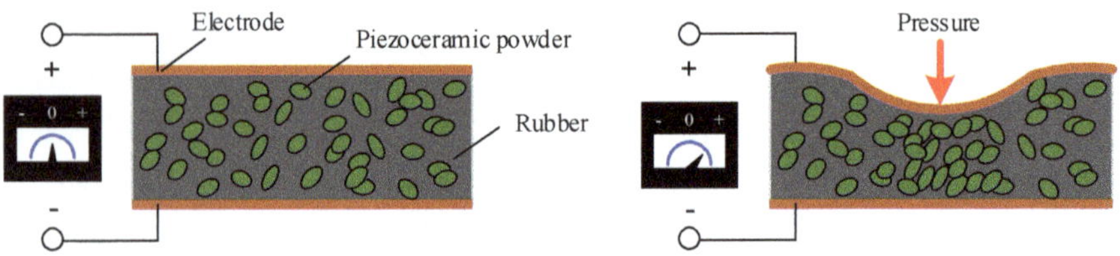

Fig. 1. Operating principle of a piezoceramic rubber

Materials Research Proceedings **21** (2022) 342-348 https://doi.org/10.21741/9781644901755-60

As described above, the biggest challenge for the additive manufacturing of vulcanizable rubbers is the maintenance of the shape during printing and vulcanization. Nevertheless, flat two-dimensional geometries can already be produced without support material [2]. The basis of the technological equipment for the production of composite rubber compounds is the FDM printer (see Figure 2). The printer had to be modified with a screw extruder, as is usually the case when processing rubbers by injection molding. The viscosity of the printing materials plays a particularly important role for the print quality. Nitrile butadiene rubber (NBR) with its viscosity at a printing temperature of 70°C has proven to be well suited [2] and was used for the printing tests. Piezoceramics in the form of powder were mixed into the rubber compound to create the sensory composite material.

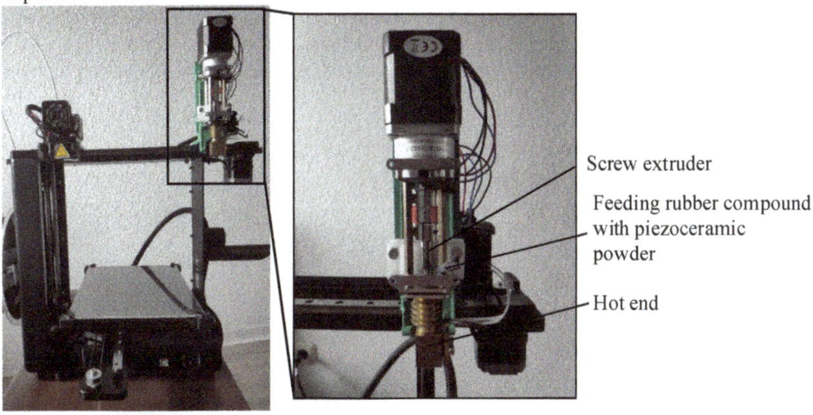

Fig. 2. The MAKERGEAR M3-ID desktop 3D printer modified with a screw extruder

A prototype sealant was printed on the FDM printer while feeding material into the screw extruder. Figure 3a presents first trial of printing rubber with piezoceramic powder. It can be seen that the piezoelectric component is not homogeneously distributed throughout the compound during the printing process. This process requires a significant improvement. Furthermore, it could be shown in further printing tests, that the printing on a metallic material (brass foil) is possible with an adhesive. This offers the possibility of attaching one of the two necessary electrodes in this way.

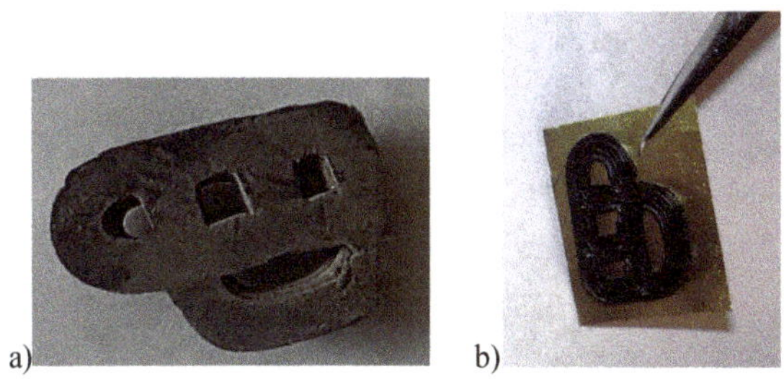

Fig. 3. a) Printed prototype sealant with rubber/piezoceramic powder mixture, b) Printed prototype sealant on a brass foil with an adhesive

Application of piezoceramic rubber in vehicle

A general view of the intelligent door seal system is shown in Figure 4. The microcontroller (item 1) is highly resistant to electromagnetic interference and vibration to be installed in a door or on a car body. This microcontroller (printed circuit board) uses a simple method to measure the current [9] flowing through the piezoceramic seal (item 2) when it is deformed. The sensory rubber used for an intelligent door seal must not only have the required elasticity to provide a sealing function, but also be highly sensitive to local deformation to detect foreign bodies around the perimeter of the car door. To prevent the door from closing completely (item 4) when a foreign body is detected at the junction with the body, an electromechanical actuator (item 3) is used, which is embedded in the swing mechanism and is kinematically connected to the body and the door. When a control signal is supplied from microcontroller 1, the actuator switches from passive mode to active mode and instantly creates a braking force that prevents further movement of the car door. This ensures the process of safely closing the door without causing injury to a person and no damage to the car body.

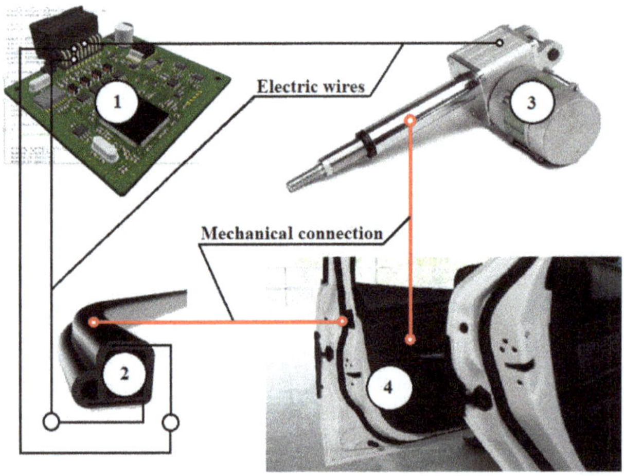

Fig. 4. Schematic diagram and composition of the intelligent door seal system

The system for monitoring the condition of automobile tires is shown in Figure 5. This system also has a microcontroller (item 1) that uses the method of measuring the current flowing through the piezoceramic layer of the tire (item 2) when it is critically deformed or damaged. The rubber with sensory properties used for a smart tire must have the required elasticity, high durability and low sensitivity to small deformations of the tire, allowing it to filter minor loads that cannot lead to damage to the tire structure.

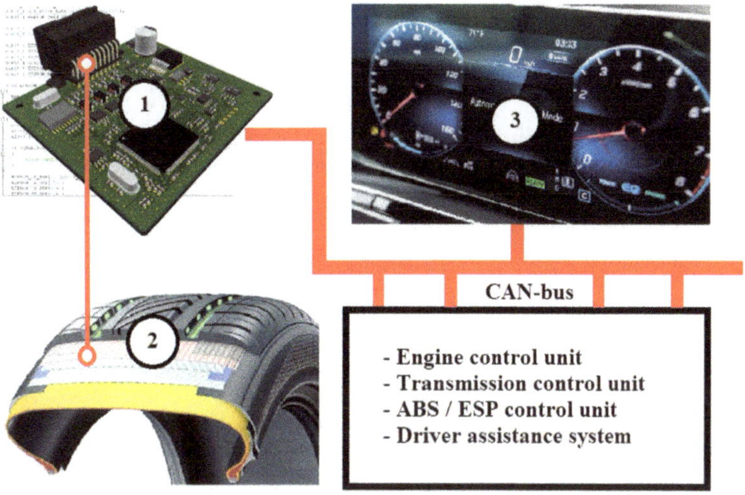

Fig. 5. Schematic diagram and composition of the intelligent vehicle tire system

Signal testing of piezoceramic rubber

The measurements are associated with many technical difficulties. The additive manufacturing has not yet been able to achieve uniform distribution of the piezoceramic powder (see Figure 3a). Therefore, many small samples ca. 10 mm x 7 mm (Figure 6) with different proportion of piezoelectric material were prepared by hand. If the ceramic content is very high, the rubber in case of strong deformation tends to easy break at the bending point. Unfortunately, at this stage, a reasonable electrical signal can only be detected with such samples. In the case of samples with a significantly lower proportion of ceramic, the measured signal is not discern at all or hardly distinguishable from the noise.

Fig. 6. Two test samples on the left side with piezoelectric material content without electrodes in comparison to the original rubber (right sample).

The Figure 7 shows the measured electrical signal during bending tests. It is obvious that the signal has a strong noise component. Nevertheless, two bending tests with approx. 15 mV and 20 mV can be identified.

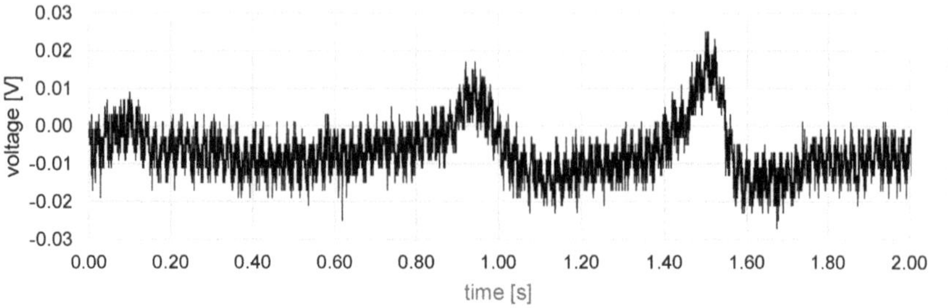

Fig. 7. Electrical signal measured during strong deformation of piezoelectric rubber.

Another difficulty lies in sufficient contacting of the surface. A simple variant of the electrode is the electrically conductive silver layer. The very thin layer due to insufficient adhesion flakes off the surface after a few bending attempts and contributes to the strong noise during the measurement. Therefore, contacting is still an important subject of investigation.

Conclusions

The results of this study open up a new perspective for the application of smart rubbers in automotive designs. The use of such a composite material is possible not only to improve safety for new cars, but can also serve as an indicator at various stages of the manufacturing process of rubber products. For example, in the manufacture of automobile tires, it is possible to trace the presence of defects in the finished product when compared to a reference tire using non-destructive testing methods by analyzing signals from the smart bus.

Given these circumstances, the present study points to the feasibility of realizing the use of the intelligent automotive door seal and intelligent tire as a commercial product for automotive rubber structures. In addition, the rubber functions implemented in this article can also be useful outside of the automotive industry.

References

[1] Cohen, D.; Sargeant, M.; Somers, K. 3-D printing takes shape. McKinsey Q. 2014, 1, 1–6.

[2] Drossel, G.-W.; Ihlemann, J.; Landgraf, R.; Oelsch, E.; Schmidt, M. Basic Research for Additive Manufacturing of Rubber.Polymers2020,12, 2266. https://doi.org/10.3390/polym12102266

[3] Wittek, H.; Klie, B.; Giese, U.; Kleinert, S.; Bindszus, L.; Overmeyer, L. Approach for additive Manufacturing of high-viscosity, curable Rubbers by AME Processing (Additive Manufacturing of Elastomers) — Rubber 3D.KGK Rubberpoint Mach. Euipments, 2019,10, 30–35

[4] Sundermann, L.; Klie, B.; Giese, Z.; Leineweber, S.; Overmeyer, L. Development, Construction and Testing of a 3D-Printing-Systemfor Additive Manufacturing of Carbon Black filled Rubber Compounds.KGK Rubberpoint Mach. Euipments2020,6, 53–56.

[5] Drossel, G.-W.; Ihlemann, J.; Landgraf, R.; Oelsch, E.; Schmidt, M. Media for Dimensional Stabilization of Rubber Compounds during Additive Manufacturing and Vulcanization.Polymers2020,12, 2266. https://doi.org/10.3390/polym12102266

[6] Scheungraber, P. Apparatus and Method for 3D Printing a Workpiece of a Rubbery Non-Thermoplastic. Material. Patent No. DE102017207737A1, 8 November 2018.

[7] Faes, M.; Valkenaers, H.; Vogeler, F.; Vleugels, J.; Ferraris, E. Extrusion-Based 3D printing of ceramic components. Procedia CIRP 2015, 28, 76–81. https://doi.org/10.3390/polym12102266

[8] Buswell, R.A.; Leal de Silva, W.R.; Jones, S.Z.; Dirrenberger, J. 3D printing using concrete extrusion: A roadmap for research. Cem. Concr. Res. 2018, 112, 37–49. https://doi.org/10.1016/j.cemconres.2018.05.006

[9] Reif, K., Ed. Automobilelektronik. Eine Einführung für Ingenieure; Vieweg und Teubner: Wiesbaden, Germany, 2007.

Development of an Experimental Robotic Complex for Direct Metal Deposition and Testing of Deposition Modes for Heat-Resistant Powder Material

Maxim Oleynik[1,a*], Alexander Khaimovich[2,b] and Andrey Balyakin[3,c]

[1]Department of Engine Production Technologies, Samara National Research University, Russia

[a]oleynik1997@mail.ru, [b]berill_samara@bk.ru, [c]a_balik@mail.ru

Keywords: Industrial Robot, Tilt-And-Turn Device, Direct Metal Deposition, Optimization

Abstract. The paper considers the issue of optimizing the movement of an industrial robot used in additive manufacturing in the technology of direct metal deposition of parts. The developed mathematical model that takes into account the joint work of a six-axis robot manipulator and a two-axis positioner is described. The algorithm for calculating the motion based on the relative position of two adjacent points of the working tool trajectory relative to the rotary axis of the positioner with a given accuracy is described. The simulation of processing is carried out both when working only with the manipulator, and when working together with a two-axis positioner, and control programs with recalculated coordinates and rotation angles of the positioner are obtained.

Introduction

The first industrial robots (IR) appeared in 1956 and were intended for the simplest manipulations of moving parts [1]. Due to the low rigidity and high inertia, as well as the low level of software development, it was not possible to use IR in more complex operations [2]. However, the further development of the industry allowed them to be used for welding [3, 4], coating [5], maintenance of machine tools and machines [6]. Currently, industrial robots are increasingly being used in medicine [7], as well as in the processing of materials. Such types of mechanical processing as milling [8, 9], polishing [10], as well as the use of robots in additive manufacturing, for example, in the technology of direct metal deposition of powder materials, are actively studied [11]. The use of direct metal deposition technology, in which controlled continuous multi-axis deposition of metal powder particles in the laser radiation field is carried out, provided by combining a gas-powder jet with a laser beam, is one of the progressive directions for the creation of aircraft engine parts [12-15].

Among the advantages of using modern industrial robots, we can highlight their high flexibility, which allows to quickly reconfigure production and ability to work out movements that are not available for traditional CNC machines due to of more degrees of freedom [2].

Despite the continuous improvement of the IR, the problems of rigidity and inertia are still inherent in robotic manipulators. In addition, there may be situations in which a robot singularity is observed - a phenomenon accompanied by the fact that the mathematical function used in the calculations of the IR's movement tends to infinity, which causes incorrect movements of the IR's links [16]. Therefore, most studies are aimed at optimizing the movements and improving the accuracy of the positioning of tool. Methods of programming sequential linearization were proposed, in which an approach was developed to ensure the smoothness of the trajectory over the entire path section [17] and a method for predicting the dynamics of the robot depending on its position [18].The creation of NC-programs can be performed both by means of offline

Materials Research Proceedings 21 (2022) 349-357 https://doi.org/10.21741/9781644901755-61

programming, implying the use of various CAM systems, and through online programming, that is, programming the robot directly at the installation site via control panel[19].

In this paper to improve the accuracy of the positioning of tool the optimization of movement when processing axisymmetric and / or close-shaped parts is considered.

Description of the Method

The object of the study is a robotic complex of direct metal deposition (RCDMD) developed by the Institute of Laser and Welding Technologies (ILWT, St. Petersburg) [11]. RCDMD consists of a six-axis robot manipulator Fanuc Robot M-20iA_20M with mounted a additive head on it and a table for workpieces mounted on a tilt-and-turn device (a two-axis positioner Fanuc 2-axis Arc Positioner).

A IR can be considered as a set of rigid links connected in series, each of which is equipped with a servo drive. The robot has six rotational kinematic pairs. The two-axis positioner has two rotary joints J_7 and J_8, which rotate around the *B* and *C* axes, respectively.

The task of kinematics is to describe the spatial position of the links of the mechanism. There are direct and inverse kinematics problems. The direct kinematics problems gives a decision about the position and orientation of the tool (in this work of the additive head), based on the given geometric parameters of the links and the vector of the attached angles.

The inverse problem gives a solution to the possible values of the attached angles that provide a given position of the tool and its orientation relative to the absolute coordinate system. In general, for a 6-axis robot equipped with a positioner, the mathematical model of kinematics for the inverse problem can be represented as:

$$\begin{cases} J_1 = J_1(x,y,z,i,j,k,\alpha_1,\theta_1,a_1,d_1,t); \\ J_2 = J_2(x,y,z,i,j,k,\alpha_2,\theta_2,a_2,d_2,t); \\ J_3 = J_3(x,y,z,i,j,k,\alpha_3,\theta_3,a_3,d_3,t); \\ J_4 = J_4(x,y,z,i,j,k,\alpha_4,\theta_4,a_4,d_4,t); \\ J_5 = J_5(x,y,z,i,j,k,\alpha_5,\theta_5,a_5,d_5,t); \\ J_6 = J_6(x,y,z,i,j,k,\alpha_6,\theta_6,a_6,d_6,t); \\ J_7 = J_7(x,y,z,i,j,k,\alpha_7,\theta_7,a_7,d_7,t); \\ J_8 = J_8(x,y,z,i,j,k,\alpha_8,\theta_8,a_8,d_8,t), \end{cases}$$

where x, y, z are Cartesian coordinates; i, j, k are the guiding orts of the tool axis; α_i, θ_i, a_i, d_i are the parameters of the links; t is the time; α_i is the angular displacement – the angle by which the z_{i-1} axis should be rotated around the x_i axis so that it becomes co-directional with the z_i axis; θ_i is the connecting angle by which the x_{i-1} axis should be rotated around the z_{i-1} axis so that it becomes co-directional with the x_i axis; a_i is the linear displacement – the distance between the intersection the x_{i-1} axis with the x_i axis and the beginning of the i-th coordinate system, counted along the x_i axis; d_i is the distance between the intersection of the z_{i-1} axis with the x_i axis and the beginning of the $(i-1)$-th coordinate system, calculated along the z_{i-1} axis. For rotational joints, the parameters α_i, a_i, d_i are constant values for each specific robot model that characterize the design of the links.

There are many methods for solving the inverse problem of kinematics: the method of inverse transformations, direct geometric methods, a method based on nonlinear mathematical programming. The last one has gained the greatest popularity, since with a large number of possible solutions, it is possible to impose restrictions that give an optimal solution [19].

First, the positioner rotation angles $J_7 = B$ and $J_8 = C$ are calculated. In this case, the optimization task is set, when the robot's movements are minimized and tool approach to the processing area is provided by the positioner. This method of setting the movement is explained by the presence of industrial robots with low (compared to conventional NC-machines) rigidity and high inertia of the joints. The essence of the method is that when processing axisymmetric (or close to such a shape) parts, the movement along a trajectory that has an axis of rotation that coincides or is close to the axis of rotation of the table is provided by the movement of the positioner. In this case, the robot will only make corrective movements. Figure 1 shows a diagram of the combined movement.

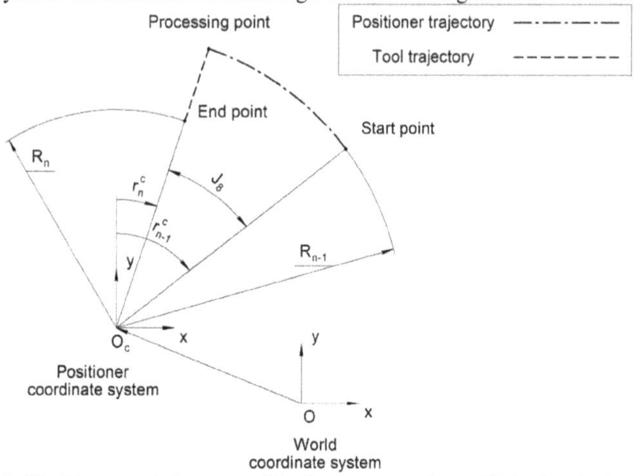

Fig. 1. *Working travel due to combined movement of an additive head of a robot with rotation of a positioner table.*

The initial data for the calculation are the coordinates of the start X_0 and end X_1 points of a trajectory, the coordinates of robot base O relative to the world coordinate system, the coordinates and direction of the axis of rotation of the table of the positioner O_c relative to the world, which are represented by vectors in the following form:

$$X_0 = \begin{bmatrix} x_0 \\ y_0 \\ z_0 \\ i_0 \\ j_0 \\ k_0 \end{bmatrix}; X_1 = \begin{bmatrix} x_1 \\ y_1 \\ z_1 \\ i_1 \\ j_1 \\ k_1 \end{bmatrix}; O = \begin{bmatrix} x_o \\ y_o \\ z_o \\ 0 \\ 0 \\ 0 \end{bmatrix}; O_c = \begin{bmatrix} x_c \\ y_c \\ z_c \\ 0 \\ B \\ C \end{bmatrix},$$

where B, C are the angles of rotation of the axis of the positioner table.

First, the coordinates of the start and end points are converted to the positioner coordinate system:

$$X_{0C} = [M_Y]\cdot(X_0 - (O_c - O)) = \begin{bmatrix} x_{0C} \\ y_{0C} \\ z_{0C} \end{bmatrix},$$

$$X_{1C} = [M_Y]\cdot(X_1 - (O_c - O)) = \begin{bmatrix} x_{1C} \\ y_{1C} \\ z_{1C} \end{bmatrix},$$

where $[M_Y] = \begin{bmatrix} \cos(p) & 0 & -\sin(p) \\ 0 & 1 & 0 \\ \sin(p) & 0 & \cos(p) \end{bmatrix}$ is the transformation matrix when rotated by an angle p relative to the Y axis of the positioner coordinate system, to which the positioner is to be rotated if the tool axis does not coincide with the Z axis.

Then the possible radii of rotation of the positioner relative to the Z axis are determined:

$$R_0 = \sqrt{x_{0C}^2 + y_{1C}^2};$$

$$R_1 = \sqrt{x_{1C}^2 + y_{1C}^2}.$$

The relative error is calculated and compared with the specified accuracy:

$$\delta = \left|1 - \frac{R_0}{R_1}\right| \le \delta_{\text{req}},$$

if the condition is not met, then the table rotation is not required, otherwise the calculation continues.

The guiding orts i_0, j_0, i_1, j_1 are located in the positioner's coordinate system.

Guide orts of the start point:

$$L_0 = \sqrt{x_{0C}^2 + y_{0C}^2 + z_{0C}^2};$$

$$i_0 = \frac{x_{0C}}{L_0};\ j_0 = \frac{y_{0C}}{L_0}.$$

Guide orts of the end point:

$$L_1 = \sqrt{x_{1C}^2 + y_{1C}^2 + z_{1C}^2};$$

$$i_1 = \frac{x_{1C}}{L_1};\ j_1 = \frac{y_{1C}}{L_1}.$$

The rotation angles are determined.

For the start point:

if $i_0 \ge 0$ and $j_0 \ge 0$, then 1 and 4 quadrants,

$$r_{C0} = \arctan\left(\frac{j_0}{i_0}\right) + \pi;$$

if $i_0 < 0$ and $j_0 \geq 0$, then quadrant 2,

$$r_{C0} = \arctan\left(\frac{j_0}{i_0}\right) + 2\pi;$$

if $i_0 < 0$ and $j_0 < 0$, then quadrant 3,

$$r_{C0} = \arctan\left(\frac{j_0}{i_0}\right).$$

For the end point:

if $i_1 \geq 0$ and $j_1 \geq 0$, then 1 and 4 quadrants,

$$r_{C1} = \arctan\left(\frac{j_1}{i_1}\right) + \pi;$$

if $i_1 < 0$ and $j_1 \geq 0$, then quadrant 2,

$$r_{C1} = \arctan\left(\frac{j_1}{i_1}\right) + 2\pi;$$

if $i_1 < 0$ and $j_1 < 0$, then quadrant 3,

$$r_{C1} = \arctan\left(\frac{j_1}{i_1}\right).$$

Next, the coordinates of the end point are calculated when the positioner table is rotated by an angle r_C relative to the Z axis:

$$X_{0C1} = M_z \cdot X_{1C},$$

where M_z – is the transformation matrix when rotated by an angle $(r_{C1} - r_{C0})$ relative to the Z axis,

$$M_z = \begin{bmatrix} \cos(r_{C1} - r_{C0}) & \sin(r_{C1} - r_{C0}) & 0 \\ -\sin(r_{C1} - r_{C0}) & \cos(r_{C1} - r_{C0}) & 0 \\ 0 & 0 & 1 \end{bmatrix}.$$

Then the coordinates of the robot tool movements are calculated to compensate for the displacements when the positioner table is rotated.

$$X_{1C1} = [M_y]^{-1} \cdot X_{0C1} + \left(\begin{bmatrix} x_c \\ y_c \\ z_c \end{bmatrix} - \begin{bmatrix} x_o \\ y_o \\ z_o \end{bmatrix}\right) = \begin{bmatrix} x_{1C1} \\ y_{1C1} \\ z_{1C1} \end{bmatrix},$$

where $[M_y]^{-1}$ – inverse transformation matrix for the tilt of the positioner table axis,

$$[M_y]^{-1} = \frac{[M_y]^T}{\det[M_y]}.$$

The angle of rotation of the positioner around the Z axis:

$$C_1 = C - \frac{r_{C1} - r_{C0}}{\pi} \cdot 180.$$

The angle J_1 can be found from the equation:

$$J_1 = \arctan\left(\frac{y_{1C1}}{x_{1C1}}\right),$$

where x_{1C1}, y_{1C1} – tool position coordinates.

The projections of the robot's arms on the X and Z axes give a system of equations:

$$\begin{cases} L_{35}\cos(J_3 - J_2) + L_{56}\cos(J_5 - J_3 + J_2) + \\ + L_{12}\cos\alpha + L_{23}\sin(J_2) - \dfrac{x_{1C1}}{\cos(J_1)} = 0; \\ L_{35}\sin(J_3 - J_2) - L_{56}\sin(J_5 - J_3 + J_2) + \\ + L_{12}\sin\alpha + L_{23}\cos(J_2) - z_{1C1} = 0. \end{cases}$$

Since the tool must have a vertical orientation to ensure the uniformity of the powder spray, the condition is introduced into the system of equations:

$$J_2 - J_3 - J_5 = 0.$$

To determine the angles J_2 и J_3 , methods for solving systems of nonlinear equations are used. In particular, the fsolve function was used in the application software package for solving technical problems of MATLAB R2012b. The angles J_4 и J_6 are assumed to be 0.

Thus, we get the coordinates of the recalculated points and the vector of the values of the rotation angles of the robot and positioner links:

$$J = [J_1, J_2, J_3, J_4, J_5, J_6, J_7, J_8].$$

Simulation Results

To confirm the correctness of the created mathematical model, a simulation of processing was performed in the academic version of the Fanuc ROBOGUIDE program. Next, a NC-program was generated with the positioner fixed.

In the MATLAB R2012b program, an executable module was created that reads the input text file, recalculates the coordinates, and outputs the output text file. Input and output file record structure:

X_0 – row 6 of the coordinates of the start point;
X_1 – row 6 of the coordinates of the end point;
O_t – coordinates of the tool's coordinate system relative to the world coordinate system;
O_p – coordinates of the positioner's coordinate system relative to the world coordinate system;

Next, the new NC-program is loaded back into Fanuc ROBOGUIDE, where the processing process is re-simulated in order to verify the correctness of the coordinate recalculation.

Testing of Deposition Modes

A research of the modes is based on the analysis of statistical data, which consider the macro-and microstructure of samples obtained via DMD. The DMD modes during the experiment are shown in Table 1.

Table 1. *Modes of metal deposition of powder material*

Mode parameter	Sample no.							
	1	2	3	4	5	6	7	8
	Value							
Laser radiation power P, W	1000	1200	1400	1600	1000	1200	1400	1600
Side step of deposition S, mm	1,6				1,4			
Powder consumption, g/s	31							
Deposition speed V, mm/min	1200							
Vertical deposition pitch t, mm	0,4							

The maximum size of single defects found in the sample material is shown in Table 2.

Table 2. *Microhardness, type and maximum size of defects in samples*

Sample no.	Maximum size of defect				Average microhardness value, $HV_{0,1/5}$
	Area of looseness, mm^2	Average pore diameter, mm	Maximum pore diameter, mm	Maximum junction length, mm	
1	0,0042	0,0475	0,055	0,13	317,5
2	0	0,03425	0,047	0,4	313
3	0,000174	0,0275	0,045	0,1	301
4	0,0064	0,02375	0,05	0,16	301,5
5	0	0,03125	0,035	0,08	293,5
6	0,0485	0,02225	0,04	0,086	305,5
7	0	0,019	0,03	0,17	282,5
8	0	0,03	0,03	0	272,5

Discussion

On this complex, the part can be processed in two ways: by simply moving the robot, which coordinates the position of the additive head, or by involving a positioner, which allows larger part of movements to be used, which allows the robot to reduce the number of movements and, thereby, reduce positioning errors.

Since the NC-programs for growing complex parts have a much larger volume compared to the NC-programs for palletizing, the robot controller is often unable to implement such programs. Therefore, further development in the field of optimization will go to reduce the volume of the NC-program by using loops and subprograms.

Conclusion

This article presents the results of the development of a mathematical model that considers the collaborative work of a six-axis robot manipulator and a two-axis positioner. In addition, the model takes into account the accuracy set by the user, which makes it possible to rationalize the recalculation of coordinates and adjust the degree of involvement of the positioner in the execution of the NC-program.

In the Fanuc ROBOGUIDE a robotic technological complex was modeled, on which the simulation of working out NC-programs took place. As a result of the recalculation of the coordinates, a significant proportion of the movements began to work out the positioner, while the manipulator made movements of the tool, correcting deviations of the specified trajectory from the trajectory of movement along the arc of the positioner.

Analysis of technological modes of the DMD according to the generalized quality factor shows that the most preferable deposition modes are 5 and 8 modes.

References

[1] Nof S. Y. Handbook of Industrial Robotics, 2nd ed. New York: John Wiley & Sons, 1999, p. 1378. https://doi.org/10.1002/9780470172506

[2] V. F. Filaretov, A.V. Zuev, A.S. Gubankov, Manipulators control when performing various technological operations, (in Russian). Moscow: Nauka, 2018, p.229.

[3] A.A. Ivanov, "Industrial robots in welding production" in Innovative mechanical engineering, Moscow, 2013, no. 4, pp. 22-25.

[4] R.R. Gabitov, I.I. Khabibullin, "Assessment of the possibility of robotic welding of the low-pressure turbine support body" in Molodezhnyi vestnik UGATU, Ufa, 2020, no. 2, pp. 19-22.

[5] E.A. Drobina, "Intelligent control system of a robotic-controlled painting complex" in Innovative development of science and education, collection of articles of the IX International scientific-practical conference: in 2 parts, Penza, 2020. pp. 93-95.

[6] Y.G. Kozyrev, Application of industrial robots, Moscow: KNORUS, 2010, 488 c.

[7] Robotic-Assisted Surgery [Online]. Available: https://www.intuitive.com/en-us, Accessed: 2020, Sep. 15.

[8] T. Cvitanic, V. Nguyen, S.N. Melkote, "Pose optimization in robotic machining using static and dynamic stiffness models" in Robotics and Computer-Integrated Manufacturing Volume 66, 2020. https://doi.org/10.1016/j.rcim.2020.101992

[9] A.V. Balyakin, V.A. Demenok, V.P. Kurbatov, "Influence of processing by milling on the surface layer of titanium alloy VT20 obtained by the method of direct laser growth" in Modern high technologies, 2018, no. 6, pp. 14-19.

[10] P. Xu, et al. Novel hybrid robot and its processes for precision polishing of freeform surfaces in Precision Engineering Volume 64, 2020, p. 53-62. https://doi.org/10.1016/j.precisioneng.2020.03.013

[11] Institute of Laser and Welding Technologies [Online]. Available: http://ilwt-stu.ru, Accessed: 2020, Sep. 16.

[12] Khaimovich A.I., Stepanenko I.S., Smelov V.G., Optimization of selective laser melting by evaluation method of multiple quality characteristics // IOP Conference Series: Materials Science and Engineering, C. 012067. p. 7, 2018. https://doi.org/10.1088/1757-899X/302/1/012067

[13] S.P. Murzin at al., "Creation of technology for high-speed manufacturing of parts and components of aircraft engines by methods of heterophase powder metallurgy: Report IiI on the component of applied research and experimental development", Samara, 2016, p. 350.

[14] Glukhov V.V., Turichin G.A., "Economic efficiency of direct laser sintering technology in industry", in Theoretical foundations of industrial policy formation. St. Petersburg: Peter the Great St. Petersburg Polytechnic University, 2015, pp. 176-190.

[15] Bykovsky D.P., et al., "Application of direct laser sintering technology for the manufacture of stainless steel samples and investigation of their mechanical properties", in Additive

technologies: present and future, collection of reports of the II International conference, Moscow, 2016.

[16] CAD/CAM/CAE Observer #6 (50) / 2009 [Online]. Available: http://www.cadcamcae.lv, Accessed: 2020, Sep. 16.

[17] S.L. Zenkevich, A.S. Yushchenko, "Fundamentals of manipulation robots control", Moscow: MGTU, 2000. p. 400.

[18] Peng J., et al., Smoothness-oriented path optimization for robotic milling processes // Science China Technological Sciences, 2020, p. 15. https://doi.org/10.1007/s11431-019-1529-x

[19] Huynh H., et al. Modelling the dynamics of industrial robots for milling operations // Robotics and Computer-Integrated Manufacturing Volume 61, 2020, p. 16. https://doi.org/10.1016/j.rcim.2019.101852

Modern Technologies for Producing Foamed Phosphate Glass for Oil Sorbents

K.G. Karapetyan[1,a*] and D.O. Sobyanina[1]

[1]Saint-Petersburg Mining University, St. Petersburg, Russia

[a] karapetyan_kgv@pers.spmi.ru

Keywords: Composite Materials, Foamed Glassy Phosphate Materials, Biosorbents, Surface Modification, Method of Molecular Layering, Purification of Water and Soil from Hydrocarbons

Abstract. The paper considers applications of foamed glassy phosphate materials as carriers of biologically active substances. The atomic layer deposition method allowed chemically synthesizing surface-modified composite materials based on graphite and carbon fibers, which maximally preserved the activity of enzymes and biologically active substances. The synthesis process of a mono-layer titanium-containing coating on the surface of graphite and carbon fibers by processing them with $TiCl_4$ vapors is considered. The main characteristics of the graphite surface before and after modification and characteristics of the porous structure and surface of carbon fibers are obtained. Biologically active substances by adsorption from a solution to the surface of both the original and modified carriers were applied. biological activity and temperature stability of the obtained composite materials were studied. It is shown that the activity of a biologically active substance depends on the chemical composition and state of the surface of carbon-based carriers.

Introduction

Environmental problems have now acquired a global character and are expressed in irreversible changes in the conditions of human existence. In this regard, there are tasks of eliminating oil spills, products of condensation of exhaust gas aerosols in megacities, cleaning and restoration of technogenically destroyed territories, which include auto enterprises, technology parks, garages and parking lots, territories adjacent to highways and tunnels.

The solution of the above problems is possible by using the technology of complex biological cleaning of territories from oil pollution [1].

Currently used technologies for cleaning territories and eliminating oil spills have a number of disadvantages: firstly, the toxic effect of sorbents on the soil, and therefore the need to collect the sorbent from the surface to be cleaned and its disposal, and secondly, the mandatory implementation of a number of measures for additional soil restoration [2].

Taking into account these shortcomings and on the basis of previously performed studies [3], the biosorbent "BIAVA" was developed and patented, which is a porous sorbent carrier on the surface of which there are microorganisms-destructors of hydrocarbons and catalysts of biooxidation processes [4].

Methods

A foamed glassy phosphate material (foam glass) (Fig.1, a) with micro-additives of boron, cobalt, iron, copper, manganese, and molybdenum, which provide optimal conditions for the vital activity of microorganisms for a long time, was chosen as the carrier sorbent. The internal porous structure of the sorbent is shown in the photo (Fig.1, b) obtained using an electron microscope.Glasses for

Materials Research Proceedings **21** (2022) 358-363 https://doi.org/10.21741/9781644901755-62

biosorbents were synthesized in the system at 1200-1300°C in the system (K2O-CaO-MdO-P2O5) from the charge in quartz crucibles for one hour with stirring. Good results of cooking in gas-heated pot furnaces were achieved by using a liquid charge obtained from orthophosphoric acid.

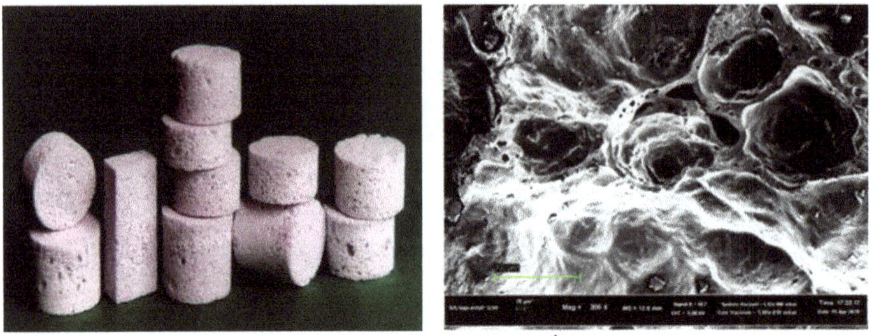

a b

Fig. 1. *foamed glassy phosphate material (a), internal porous structure of the sorbent (b)*

Also, the main raw materials used were magnesite concentrate, potassium carbon dioxide with a K2CO3 content of at least 98 wt.%, and apatite concentrate. The charge was loaded at a temperature of 1100-1150°C in portions. Each new portion was loaded onto the charge that had been boiled and was already in the crucible. Heating was carried out at a temperature of (1200-1250 °C). The production of glass was carried out on metal plates and it`s cooled in the air.

For the foaming of glass, a laboratory device was made, including a horizontally arranged tubular furnace, the temperature along the axis of which varies in the range of 150 – 700 – 150°C. The foaming of the pellets inside the furnace was carried out in a "fluidized bed" with the tilt of the furnace, ensuring the continuous movement of the pellets at the desired speed. To maintain the pellets in a suspended state, a pipe was introduced into the working area of the furnace, through which air was blown from the compressor. During the research, the foaming temperature in the furnace varied within 520-620 °C with an interval of 20 °C.

When obtaining foamed samples based on phosphate glasses under laboratory conditions, the initial mixture was a mixture containing 66.7 wt.% of the phosphate glass powder fraction ≤ 1.25 microns and 33.3 % of the foaming agent. The mixture obtained in this way was moistened, and granules of 5-8 mm in size were pressed from it. Foaming was carried out in a gradient furnace, which was heated to 650 °C (at the inlet-220°C, at the outlet-200 °C) (Fig. 2) .

The production of foam glass granules includes the following stages:
- preparation of the initial mixture for foaming;
- forming of pellets;
- foaming of pellets.

The initial components of the mixture (glass powder and foaming agent) were ground in a ball mill to obtain a specific surface area of 1000-4000 cm2 / g. The specific surface area was calculated based on the determination of the air permeability of the porous material. The analytical composition of the glasses was determined by the atomic emission method with inductively coupled plasma. It is shown that the difference in the compositions of glasses synthesized in both laboratory and industrial conditions does not exceed 1.0 – 1.5 %.

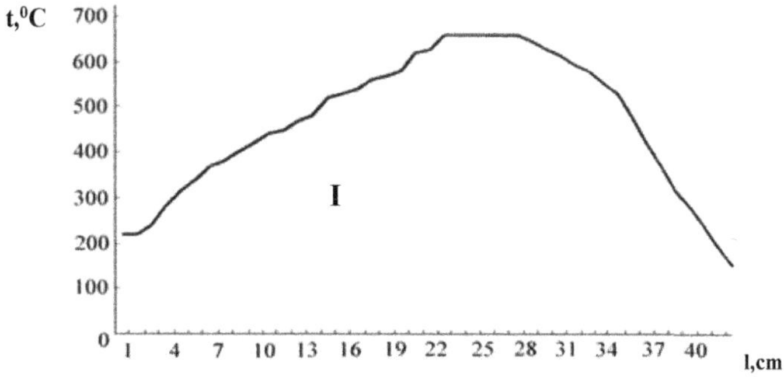

Fig. 2. Calibration curve: 1 - drying zone; 2 - hot zone; 3 - annealing zone

In order to change the chemical nature of the graphite surface, we synthesized monolayer coatings based on elemental layers of titanium and zirconium. One of the effective methods for their synthesis is the atomic layer deposition (ALD) method [5 - 10]. The main advantages of the ALD technology are: control the thickness of the resulting film (a coating of several nanometers), reagents don't interact with each other until they reach the substrate, obtaining a uniform film thickness on the substrate.

The route allows forming covalently-bound monolayer coatings on the substrate surface. Importantly, the synthesis process allows depositing a self-assembled element oxychloride monolayer, since the treatment time required to achieve a maximum coverage is determined by exhaustion of the active surface sites. As a result, a highly reproducible and conformal coating can be obtained [11].

The proteolytic enzyme trypsin was chosen as a model system of biologically active substances. Trypsin is an enzyme preparation with a proteolytic effect, has pronounced anti-inflammatory and decongestant properties, is able to break down dead tissue, fibrinous formations, viscous secrets and exudates. In relation to healthy tissues, the enzyme is inactive and safe. This drug can be used as a part of coatings of artificial organs, vessels during implantations. Immobilization of trypsin on a carrier will allow maintaining the enzyme activity at a sufficiently high level, prologging its action and increasing resistance to destabilizing factors.

Trypsin immobilization on the surface of carbon materials was carried out by adsorption. The catalytic activity of trypsin was determined spectrophotometrically from the decomposition products of casein.

Results and discussion

Evidently (Table 1), the concentration of oxygen-containing functional groups increases significantly after oxidation and slightly decreases as a result of successive cycles of formation of titanium oxide layers.

The studied carbon fibers are characterized by a small volume of macropores and a specific surface area of not more than 50 m^2/g (table 2). These results indicate a low sorption activity of the studied carbon fibers.

The changes in the surface characteristics of the initial and modified graphite correspond to the dependence of the immobilized trypsin amount on the number and chemical composition of the titanium and zirconium oxide layers (Fig. 3, curves 1′ and 2′). With an increase in the number of cycles of graphite surface treatment by the ALD method, the number of active functional groups of the surface decreases, which leads to a decrease in the amount of adsorbed enzyme.

Table 1. *Characteristics of the surface of original and modified graphite*

Sample	Static exchange capacity for NaOH [μmol/m^2]	Specific surface [m^2/g]
Original graphite	70	16.0
Oxidized graphite	3.00	17. 0
Graphite modified with titanium oxide groups (after 1. Treatment cycle)	2.85	15.5
Graphite modified with titanium oxide groups (after 2. Treatment cycle)	2.80	15.3
Graphite modified with titanium oxide groups (after 3. Treatment cycle)	2.65	15.2
Graphite modified with titanium oxide groups (after 4. Treatment cycle)	2.50	14.9

Table 2. *Characteristics of the porous structure and surface of carbon fibers*

Sample of carbon fibers	Heat treatment temperature [K]	pore volume, m^3/g	Static exchange capacity for NaOH [μmol/m^2]	Specific surface [m^2/g]
gralen	2073	02	05	15
uglen	1173	16	12	50
licron	773	03	03	18

At the same time, samples modified with titanium and zirconium oxide layers show an increase in the activity of the immobilized enzyme (Fig. 3, curves 1 and 2). Obviously, samples based on graphite modified after 4 cycles of sequential deposition of zirconium oxide layers with an enzyme activity of about 80% were obtained (Fig. 3, curve 1).

As a result of surface modification of carbon fiber uglen by zirconium containing coatings, composite materials with the following surface characteristics were obtained (Table 3).

Table 3 shows that it was impossible to significantly increase the biological activity of carbon fiber-based composite materials. Apparently, the interaction of the enzyme molecule with the surface of the carbon fiber changes the tertiary structure of the molecules, which leads to a decrease in catalytic activity. The main reason for the decrease in catalytic activity is usually the formation of associates of enzyme molecules on the surface and blocking of the active center of the enzyme.

Table 3. *Characteristics of the surface of original and modified carbon fibers uglen*

Sample of carbon fibers	Zirconium content in the sample [μmol/m^2]	Specific surface [m^2/g]	Biological activity A [%]
original uglen	0	50	32
uglen modified with zirconium oxide groups (after 1st treatment cycle)	4	47	39
uglen modified with zirconium oxide groups (after 2nd treatment cycle)	9	38	45
uglen modified with zirconium oxide groups (after 3ed treatment cycle)	1.3	30	47
uglen modified with zirconium oxide groups (after 4th treatment cycle)	1.7	25	49

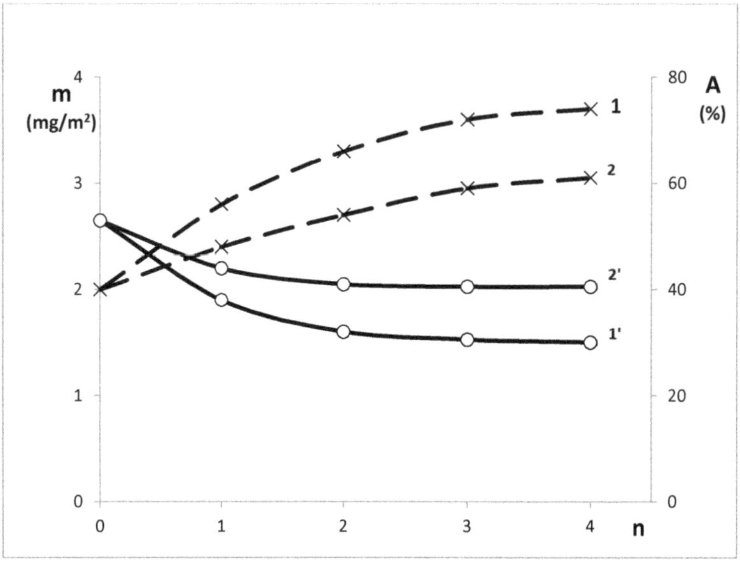

Fig. 3. *Dependences of the amount of immobilized trypsin* ***m*** *(1′, 2′) and biological activity* ***A*** *(1, 2) on the number and chemical composition of element-oxide layers* ***n***

Summary

Thus, the decrease in trypsin activity during adsorption on carbon carriers can be due to two factors: the influence of the chemical nature of the active surface centers, as well as the formation of inactive surface associates of the enzyme at high degrees of filling of the carrier surface.

When trypsin was immobilized on a graphite surface with a zirconium-containing coating, samples were obtained with biological activity of about 80 % of the activity of the free enzyme in solution, possessing resistance of the biologically active substance to elevated temperature and pH, the capability of repeated use, which enables further use of these composite materials for the production of electrochemical biosensors.

Thus, changing the chemical nature of the graphite surface by modifying the surface with monolayer titanium and zirconium-containing coatings allows for directional control of the adsorption activity of the surface and production of materials with predictable properties and high catalytic activity for biomedical sensors and other industrial purposes.

References

[1] V.E. Kogan, K.G. Karapetyan, Phosphate glasses as a basis of ecologically safe fertilizers of prolonged action and biosorbents-hydrocarbon destructors, Int. multidisciplinary scientific geoconference surveying geology and mining ecology management, SGEM, 18(5.1) (2018) 637–644. https://doi.org/10.5593/sgem2018/5.1/S20.082

[2] P'Yanova, L.G.,, Sedanova, A.V.,. Drozdetskaya, M.S. Synthesis and Research of Modified Carbon Sorbents with Hydroxy Acids (2016) International Conference on Oil and Gas Engineering, OGE 2016; Omsk State Technical UniversityOmsk; Russian Federation vol 152 pp. 639-646. https://doi.org/10.1016/j.proeng.2016.07.668

[3] P'Yanova, Kornienko, N.V., Suprunyuk, V.A., Study of the adsorption properties of molded carbon sorbents in relation to organic dyes in model conditions. AIP Conference Proceedings (2019) Volume 2143 N 020006. https://doi.org/10.1063/1.5122905

[4] I.V Pleskunov, A.G. Syrkov, Development of research of low-dimension metal-containing systems from P.P. Weymarn to our days, J. of mining instit. 231 (2018) 287–291.

[5] Basso A. Industrial applications of immobilized enzymes: A review/ Basso A., Serban, S.// Molecular Catalysis –2019 – V.479–110607. https://doi.org/10.1016/j.mcat.2019.110607

[6] T. Kääriäinen, A. Cameron, M.-L. Aäriäinen, K.A. Sherman, Atomic Layer Deposition: Princples, Characteristics, and Nanotechnology Applications, Book, John Wiley and Sons, 2013. https://doi.org/10.1002/9781118747407

[7] V. Miikkulainen, M. Leskelä, M. Ritala, R. Puurunen, Crystallinity Of Inorganic Films Grown By Atomic Layer Deposition: Overview And General Trends, J. Of Applied Physics 113(2) (2013) 021301. https://doi.org/10.1063/1.4757907

[8] Johnson, R.W., Hultqvist, A., Bent, S.F. A brief review of atomic layer deposition: From fundamentals to applications. (2014) Materials Today, 17 (5), pp. 236-246. https://doi.org/10.1016/j.mattod.2014.04.026

[9] Yatsenko, E.A., Gol'tsman, B.M., Kosarev, A.S., Karandashova, N.S., Smolii, V.A., Yatsenko, L.A. Synthesis of Foamed Glass Based on Slag and a Glycerol Pore-Forming Mixture // Glass Physics and Chemistry, 2018, Vol. 44(2), P. 152-155. https://doi.org/10.1134/S1087659618020177

[10] Yatsenko, E.A., Zubekhin, A.P., Smolii, V.A., Grushko, A.S., Gol'tsman, B.M. Resource-Conserving Technology of Heat-Insulation-Decorative Glass-Composite Material Based on Ash-Slag Wastes // Glass and Ceramics, 2015, Vol. 72(5-6), P. 216-219. https://doi.org/10.1007/s10717-015-9759-6

[11] Denisova, O.V., Rastvorova I.I. (2020) Carbon materials for immobilization of biologically active substances. Key Engineering Materials, 2020, 836 KEM, 52-57. https://doi.org/10.4028/www.scientific.net/KEM.836.52

Influence of Grain Shape in Grinding Wheels on the Machining Quality of Bearing Rings

Alexander Korotkov[1,a*], Vitaly Korotkov[1] and Nikolay Prokaev[1]

[1]T.F. Gorbachev Kuzbass State Technical University, Chair of Metal-Cutting Machines and Tools, 28 Vesennyaya str., Kemerovo, Russian Federation, 650000

[a]korotkov.a.n@mail.ru

Keywords: Shape of Abrasive Grains, Shape Factor, Grinding of Bearing Rings, Evaluation of Microstructure, Microhardness, Roughness of Machined Surfaces, Increasing the Resistance of Grinding Wheels

Abstract. The quality of grinding of bearing races is related to their performance and durability. The aim of this work is to establish how the machining quality of bearing rings depends on the characteristics of grinding wheels, in particular, on such a parameter as the shape of abrasive grains. Several batches of experimental grinding wheels were made, containing grains of different shapes (from isometric to lamellar varieties). Quantitatively, the shape of the grains was estimated by the shape factor parameter (S_F), which is equal to the ratio of the diameters of inscribed and circumscribed circles in the contours of the used grains. The shape factor was determined using a scanner or a digital microscope and a special computer program. The tests were carried out on circular, plunge, and finishing grinding, using coolant, on a SIW 4E machine in a bearing factory. The machining quality of the bearing rings was assessed by studying the microstructure, microhardness and roughness of their bearing races. The durability of grinding wheels was determined by the number of machined rings before maximum tool wear. It has been established that by a differentiated approach to the choice of grain shapes in the wheel, it is possible to significantly increase its operational capabilities: increase the microhardness of the ground surfaces, reduce the roughness of processing, and increase the durability of grinding wheels.

Introduction

High requirements are imposed on rolling bearings, since performance of the units and mechanisms where they are used depends on their quality. Requirements are made for static and dynamic strength, quality of the material from which they are made, quality of heat treatment, as well as condition of their working surfaces - roughness and processing accuracy, presence of defects in the surface layers of the metal. In the case of insufficiently high quality of the tool during grinding process, burns, increased roughness of the machined surfaces and a small number of machined parts (low tool life) are possible. Reliability and durability of operation largely depends on the condition of surfaces of rolling bearings, therefore it is important to know what factors and to what extent affect formation of the quality of these surfaces.

This work is aimed at studying the quality of the machined surfaces of bearing rings made of heat-hardened 52100 steel (ANSI) after fine grinding with experimental wheels with a controlled grain shape [1]. The wheels were made according to the standard technology for ceramic-bonded tools. They included mixtures of clay, feldspar, pegmatine, boron glass, talc and other materials [2]. A distinctive feature of the wheels was the presence in their composition of abrasive grains of white electrocorundum (25A), which have a different shape [3]. To obtain such grains, the initial mass of free abrasive was subjected to separation on a vibrating separator into a number of fractions with different shape factors (S_F) [4]. The separation of

Materials Research Proceedings **21** (2022) 364-369 https://doi.org/10.21741/9781644901755-63

grains was carried out on a vibrating inclined plane (vibrodeck), where they were poured in a laminar flow into the upper part of the deck and scattered along it in a diverging fan, depending on their shape (coefficient of friction of grains with the deck) [5]. The abrasive collected in the lower part of the deck was used further for manufacture of experimental grinding wheels according to the standard technology. In this way, six batches of grinding wheels were made with grains of different shapes in their composition. As an objective parameter for the quantitative assessment of the grain shape, we used the shape factor (S_F) equal to the ratio of the diameters described and inscribed in the contours of individual grains and circles. At least one hundred measurements were carried out for each grain fraction; the obtained coefficients were averaged and the average shape factor was formed, which characterized the shape of the grains of this fraction [6, 7]. The manufactured batches of experimental grinding wheels had the following shape factors: S_{F1}=1.19, S_{F2}=1.51, S_{F3}=1.7 (standard wheels from unsorted grains), S_{F4}=2.4, S_{F5}=2.39, S_{F6}=2.88.

The closer the S_F value is to 1.0, the more grains in such wheels approach the isometric shape, and the more significantly S_F differs from 1.0, the more grains correspond to the lamellar varieties.

The experimental wheels were tested on circular, plunge, and fine grinding of the bearing races of the outer rings of bearings (307 and 205) in a bearing plant on a SIW 4E machine [8, 9].

Methodology for preparing and conducting tests.

In the manufacture of experimental wheels, all necessary operations were sequentially performed: preparation of a mixture of grains with a binder in accordance with the required structure, molding of wheels and their firing at a temperature of 1280 °C, followed by sulfurization. Along with the listed actions, an operation was added to sort the initial mass of abrasive into a number of fractions with the same grain shape. During wheel manufacture, their visual and dimensional control was carried out (Russian standard GOST 2424-83), as well as control of their hardness after firing (GOST 25961-83).

As a result, the experimental wheels produced had the following characteristics:

Type1 63x12x16 25A F120 [S_F] K 5 V;

Type1 40x7x6 25A F120 [S_F] K 5 V,

where [S_F] is the grain shape factor, which varied discretely in different batches of wheels in the range from 1.19 to 2.88;

The bearing rings to be machined were made of heat-hardened 52100 steel (ANSI). They were hardened, followed by low tempering to a maximum hardness of HRC 64-66. Cutting modes for grinding rings were selected in accordance with the recommendations for this type of processing and had the following parameters: t = 0.173 mm, S_{fine} = 0.12 mm/min, V_{cut} = 50 m/s [10]. Grinding was carried out using coolant. When using the plunge-in method (without longitudinal feed), after each machined ring, the experimental abrasive wheel was straightened with a diamond pencil in order to restore the required tool profile, corresponding to the cross-sectional profile of the bearing ring race.

To automate the process of measuring the shape factor on each individual grain, a specially developed computer program was used, according to which the inscribing and describing the circles in the contours of the grains was carried out by a computer using a scanner, and for fine grain sizes - by a digital microscope [11].

Test results

The results of studying the microstructure and roughness of the surfaces of the bearing races after grinding with wheels having different grain shapes are shown in Fig. 1, Fig. 2, Fig. 3 and Fig. 4 [12, 13].

Materials Research Proceedings **21** (2022) 364-369
https://doi.org/10.21741/9781644901755-63

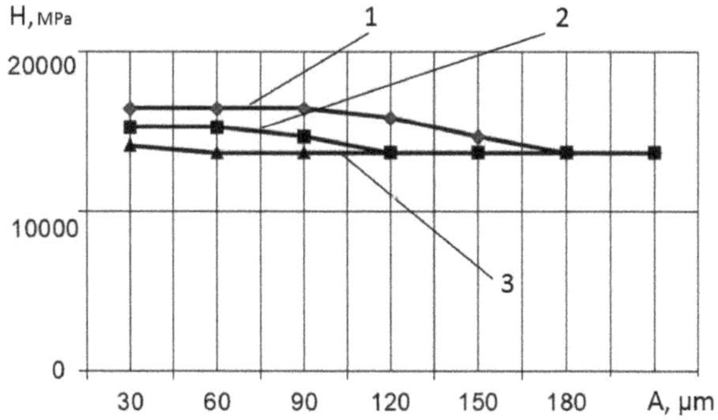

***Fig. 1.** Distribution of microhardness (H, MPa) on the ground surface of the bearing race inside the part to the depth (A): 1-wheel $S_F = 1.19$; 2-wheel $S_F = 1.51$, 3-standard wheel $S_F = 1.70$.*

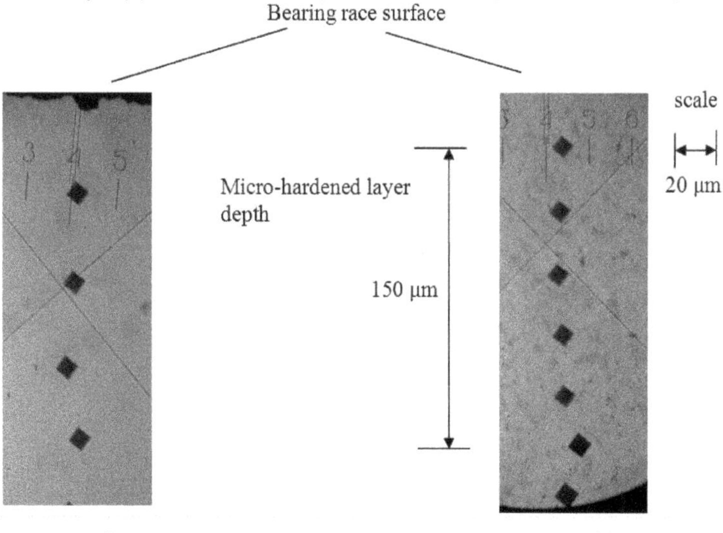

***Fig. 2.** Distribution of microhardness in the surface layer of 307 bearing race after wheel processing: a) Type1 63x12x16 25A F120 [S_F] K 5 V [standard wheel, S_F =1.70]; b) Type1 63x12x16 25A F120 [S_F] K 5 V [S_F=1.19].*

The graph in Fig. 1 shows that the shape of the grains in the wheel significantly affects the microhardness of the treated surface. Thus, changing the value of the shape factor S_F from 1.70 (standard wheel) to S_F = 1.19, it is possible to increase the microhardness of the ground surfaces of bearing rings 305 and 205 from 13500 MPa to 16500 MPa [14].

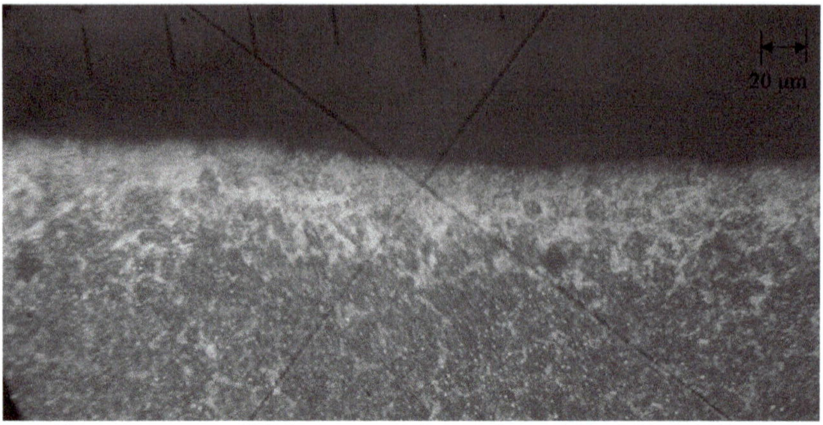

Fig. 3. *The microstructure of 307 bearing race after grinding with Type1 63x12x16 25A F120 [S_F] K 5 V (magnification x500)*

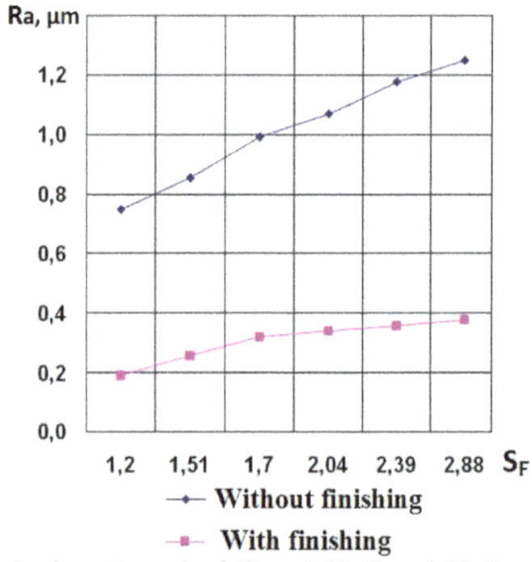

Fig. 4. *Influence of the grain shape in a wheel (S_F = 1.19; S_F = 1.51; S_F = 1.7; S_F = 2.04; S_F= 2.39; S_F = 2.88) on the roughness (Ra, µm) of the bearing races of the outer ring of bearing 307 (grinding wheels Type1 63x12x16 25A F120 [S_F] K 5 V)*

The durability of experimental wheels was assessed by a quantitative indicator - the number of ground rings in one experimental wheel up to the maximum wear determined by the machine control system.

The results of resistance tests are shown in Fig. 5, from which it can be seen that about 20% more rings were processed by wheels with isometric grains (S_F = 1.19) compared to standard wheels (S_F = 1.7). Grinding wheels with lamellar grains (S_F = 2.88), on the contrary, grinded about 15% fewer bearing rings compared to standard wheels.

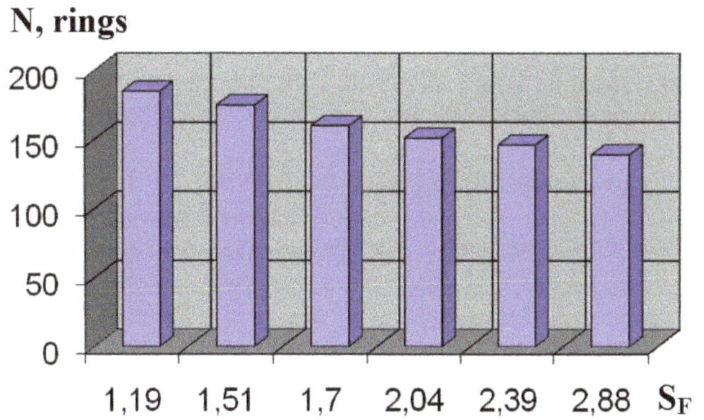

Fig. 5. *Number of ground rings (up to the specified dimensional wear of the wheel) depending on the shape factor of the grains in the wheel*

In addition to resistance tests, the quality of the treated surface was assessed. It was controlled by the following parameters - waviness, facetedness, roughness, microhardness of the polished surface, microstructure of the polished metal, and the presence of burns [15].

Studies have shown the possibility of improving the machined surface quality of bearing races. Thus, deliberately changing the shape of abrasive grains in the wheel, it was possible to reduce the roughness of the ground surfaces by about 43% (without preliminary aging) when processing with isometric grain wheels (S_F=1.19).In this case, the phenomenon of cold hardening was observed in the surface layers of parts at a depth of 150 μm, accompanied by an increase in microhardness from 14000 MPa to 16500 MPa (by 20%). Due to low temperature in the surface layers of parts, burns and other changes in the structure associated with decomposition of the structure of hardened steel - martensite were not revealed. As for macrofactors - waviness and facetedness, the negative influence of the shape factor in this case was not revealed.

Conclusions

In general, studies have shown that by purposefully selecting the shape of abrasive grains, it is possible to improve the quality of processing products by significantly reducing the roughness of the processed surface (up to 43%), increasing the microhardness of the ground surface (up to 22%).At the same time, the service life of the grinding tool can be increased (by 20%). In the future, it is also possible to increase cutting conditions to reduce technological processing time, or it is possible to correct errors obtained as a result of previous operations.

Thus, the use of ceramic-bonded grinding wheels with a differentiated approach to the choice of the grain shape makes it possible to increase the tool life when bearing races of the outer rings of bearings made of heat-hardened 52100 (ANSI) steel. Selection of the grain shape, as shown by tests, in its significance can be comparable to changing the grade of abrasive to a more expensive one or reducing the grain size to a finer one.

References

[1] Korotkov, A. Improving operational capability of grinding tools: Monograph / KuzSTU Kemerovo, (2005).

[2] Abrasive materials and tools. Catalog (M .: VNIITEMR, 1986)

Materials Research Proceedings 21 (2022) 364-369 https://doi.org/10.21741/9781644901755-63

[3] Korotkov, A. Operational properties of abrasives: Monograph Publishing House of Krasnoyarsk. University Press, (1992).

[4] Patent. 224885136303 RF MPK7. V07V1/40, 13/007/08. Vibratory separator / Korotkov A., Dubov G. Bashtanov V. № 2003129945; statements. 08. 10. 03; published. 27. 03. 05; Bulletin. № 9.

[5] Taschenbuch Maschinenbau (Verlag Technik Berlin, 1978)

[6] Certificate of official registration of the computer number 2006610153. Shape sanding grains / Korotkov A., Rilov G. № 2005612738; statements. 25. 10. 05, published. 10. 01. 06.

[7] Certificate of official registration of the software for computer №2003611938. Determination of geometrical characteristics of grinding grains / Korotkov A., Zakharov D. № 2003611454; statements. 2. 07. 03; published. 22. 08. 03.

[8] A.Hirsch, Werkzeugmaschinen. Grundlagen, Viewegs Fachbuecher der Technik (2000). https://doi.org/10.1007/978-3-322-93837-4

[9] Verein Deutscher Werkzeugmaschinenfabriken e. V. (Hrsg.): Werkzeugmaschinen Bezugsquellenverzeichnis, (Ausgabe, Frankfurt am Main, 1997)

[10] Handbuch der Metallbearbeitung (Verlag Europa-Lehrmittel, 2000)

[11] Certificate of official registration of the software for computer №2006613051. Form sanding grains in volume / Korotkov V., G. Rilov, № 2006612327; statements 04/07/06, published. 1. 09. 06

[12] E. Berghof-Hasselbächer, Atlas of Microstructures (DGRT, Berlin, 2013)

[13] G. Zlateva, Microstructure of Metals and Alloys: An Atlas of Transmission Electron Microscopy Images (Taylor&Francis Group, London, 2008). https://doi.org/10.1201/9781420075571

[14] B. Bramfitt, Metallographer's Guide: Practices and Procedures for Irons and Steels (ASM International, Manchester, 2002). https://doi.org/10.31399/asm.tb.mgppis.9781627082587

[15] Degner W., Lutze H., Smejkal E. Spanende Formung: Theorie, Berechnung, Richtwerte, (Carl Hanser Verlag Muenchen Wien,1993)

Towards Electron-Beam-Driven Soft / Polymer Fiber Microrobotics for Vacuum Conditions

Oleg V. Gradov[1,a*], Margaret A. Gradova[1,b], Irina A. Maklakova[1,2,c] and Svetlana N. Kholuiskaya[1,d]

[1]Semenov Federal Research Center for Chemical Physics of the Russian Academy of Sciences, 119991, Kosygin Street 4, Moscow, Russia

[2]Institute of Biochemical Physics named after N.M. Emanuel of the Russian Academy of Sciences, 119991, Kosygin Street 4, Moscow, Russia

[a]o.v.gradov@gmail.com, [b]m.a.gradova@gmail.com, [c]maklakosha@yandex.ru, [d]s_n_khol@mail.ru

Keywords: Polymer Fibers, Electron Beam, Microrobotics, Vacuum Microrobotics, Elastic Cantilever, Polymer Fiber Spring, Electron Beam Driven MEMS, Electron Beam Control, 2D Fourier Spectra, Integral Spatial Characteristics, Time Resolved Correlation, Real Time Correlation-Spectral Analysis

Abstract. The possibility of creating vacuum robotics based on the polymer structures irradiated by an electron beam, in particular, polymer fibers, which provide high functional flexibility and a variety of states, is discussed. The possibility of using polymer fibers as different types of MEMS-like electromechanical elements is demonstrated - from elastic cantilevers to springs that change their state under the electron beam. Experimentally proved the presence of different functional types of fibers, correlating with their thickness, as well as the phenomenon of the fiber break. A number of exotic forms of dynamics have been demonstrated and a method for their detection has been developed using 2D Fourier spectra, integral spatial characteristics, time resolved correlograms and wavelet transforms (visualized as the scaleograms / scalograms). Access barcodes for the full video records of the corresponding experiments are provided.

Introduction

Vacuum, ultra-high vacuum and cryovacuum microrobotics are at the frontiers of the modern mechatronics development. Design of the robot machine for ITER (International Thermonuclear Experimental Reactor) vacuum vessel assembly and maintenance assumes that the robot can operate in vacuum vessel conditions [1-6]. Recently, similar work is underway in China aimed at the design of a fusion reactor CFETR (Chinese Fusion Engineering Test Reactor [7]). The authors of the last cited work [7] emphasize that "the assembly of vacuum vessel (VV) is a vital task for the fusion reactor" and any violation of the technological conditions during the VV assembly or operation is an emergency situation. A similar situation is observed in microelectronics, where the robots for use in ultrahigh vacuum are also useful for wafer-transfer manipulations [8-12], and the disruption of the processing modes leads to multimillion-dollar losses from any defective batch of the mass chips. Accordingly, the controllability of robots in vacuum conditions is of high importance. In the case of vacuum robots for microelectronics, the most widely used are magnetic transmission technology or coaxial twin-shaft magnetic fluid seals applied in vacuum wafer-handling robots [13]. Operation stability at the macroscales is achieved by controlling the magnetic field since magnetic switching for microrobotics is a much more difficult task. The problem of position-sensitive control in vacuum microrobotics is still poorly developed.

Materials Research Proceedings 21 (2022) 370-383 https://doi.org/10.21741/9781644901755-64

At the same time, for microrobotic tasks that do not allow working with cables (in contrast to macroscales as in a cable-driven robotized arm for tracking in TOKAMAK vacuum vessels [14]), the size of which may exceed the cross-sections of the actuated structures, it is advisable to create conditions for the contactless signal transduction at the scales from the order of hundreds of nanometers up to several microns.

For a “cable” approach application, another method of solving the above problem is possible – combining a vacuum “microcable” and an extended micromanipulator in a single structure.

However, the question of supplying a signal to the input unit (sensor and actuator) remains open. In this case, it is possible to propose the use of the antenna properties of the cable to receive the input signal, especially for transmission of the signal to the cable using a focused microbeam corresponding in size at the projection point to the size of the antenna and the actuated structure. The use of radio frequency signals for actuation at these scales is not possible, since the “antenna" dimensions are much smaller than the wavelength (excluding the terahertz range, where the physics of metamaterials operates).

Therefore, it is proposed to use the excitation of "antennas" integrated with the "microcables" for signal transduction and fibrillar or flexible actuators (powered by the energy of the signal supplied to the antenna), by the focused electron microbeam of the medium energy that does not disrupt the structure, but provides kinematics of its movement. Application of metal wire structures for these purposes is not suitable both due to the metallographic reasons and their fractographic consequences, and due to the high densities of the beam energy flux, which must be supported to the material in order to ensure its deformation and movement of the metal wires (this task is more close to the electron beam melting than to the problem of the material deformation under the electron beam).

Therefore, perhaps the only rational approach is the use of polymer / fibrillar-composite materials, the extended components of which can serve both as an antenna and an actuator, and also possess simple cable or membrane-mimetic properties, which ensure the signal transduction along the surface / "membrane" of the actuators [15]. It is noteworthy that despite the vacuum robotics history starting from the last quarter of the XX century [16-18], the above approach integrating all functions in a single structure in accordance with TRIZ requirements, has not been found in the literature. This indicates the novelty and promising nature of the approach proposed and opens the prospects of its wide practical application.

Materials and Methods

Polyhydroxybutyrate (PHB) 16F was purchased from BIOMER Company and used as received. PHB molding solutions with a PHB concentration of 7 wt% in chloroform were used to obtain ultra-thin fibers by electrospinning (ESP) method.

Position-sensitive registration of the dielectric charging process and fiber dynamics was carried out using scanning electron microscope JEOL JSM-T330A at the accelerating voltage of 10 kV. In order to observe the dielectric charging phenomena no metal sputtering has been applied on the sample surface before the SEM studies.

Photo- and video registration of the experiments was performed using a system developed by P. Alexandrov (Shemyakin-Ovchinnikov Institute of Bioorganic Chemistry of the Russian Academy of Sciences) and O. Gradov. Access codes for the full video records of all the experiments in Code-128 standard are given at the bottom of the corresponding tables and figures.

2D Fourier spectra and integral spatial characteristics (ISC) were obtained in real time using QAVIS software based on the FFTW library, developed in V.I. Il’ichev Pacific Okeanological Institute (Far Eastern Branch of the Russian Academy of Sciences).

Materials Research Proceedings **21** (2022) 370-383 https://doi.org/10.21741/9781644901755-64

Charge distribution mapping for "integrated antenna and cable" structures was performed using a discrete differentiation operator, computing an approximation of the gradient of the image intensity function - isotropic 3x3 image gradient operator, or Sobel – Feldman operator (from SAIL - Stanford Artificial Intelligence Laboratory).

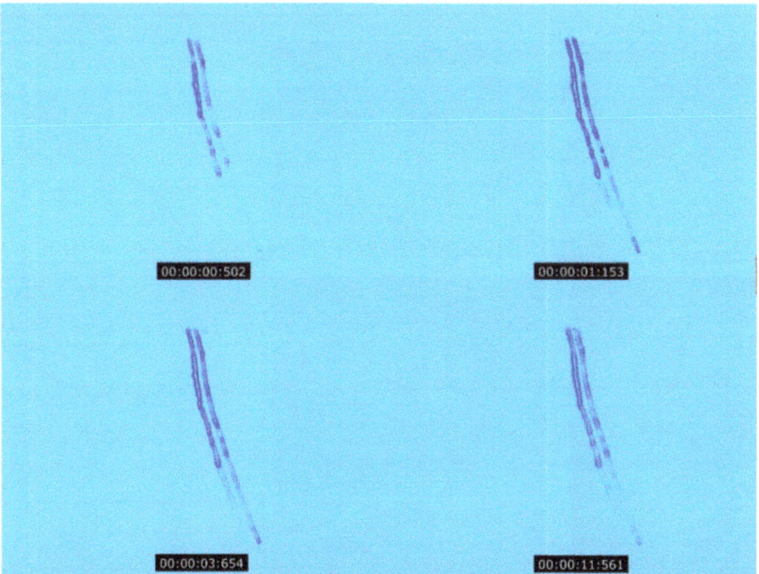

***Fig. 1**. Signal "receiving" and transduction along the polymer fiber.*
(Full video recording of this experiment is available at https://youtu.be/epWhGKAzCtE).

Results

Initially we studied the ability of a non-sputtered PHB fiber to "receive" an electron beam signal, which was estimated by the fact of dielectric charging of the single fiber surface (see Fig. 1). The possibility of the discrete signal propagation (i.e., charge distribution) over the dielectric surface and the presence of attenuation phenomena were also observed. The contribution of the surface inhomogeneities to maintaining a discrete structure of the charge distribution was also demonstrated.

After evaluating the ability of a non-sputtered microfiber to "receive" an electron beam signal, we studied the micromechanics of its dynamics and response to the changes in the operation modes of the electron microscope (including the electron beam focusing, magnification, accelerating voltage, etc.).

As you can see from the table 1, after increasing the power density, polymer fibers begin to move rapidly under the electron beam and "relax to a metastable state" in a little more than half a minute. Up to this point, a complete reconfiguration of the fiber directions occurs, depending on their position under the beam, as can be seen from the comparison of 2D Fourier spectra and Integral Spatial Characteristics (ISC) at different times. After that, reversible fiber oscillations also occur, but the whole network topology remains stable, which is consistent with the stability of the ISC symmetry axes, the angular ratios of which vary by only a few degrees.

Table 1.

Time-lapse microphoto	2D Fourier spectrum	ISC
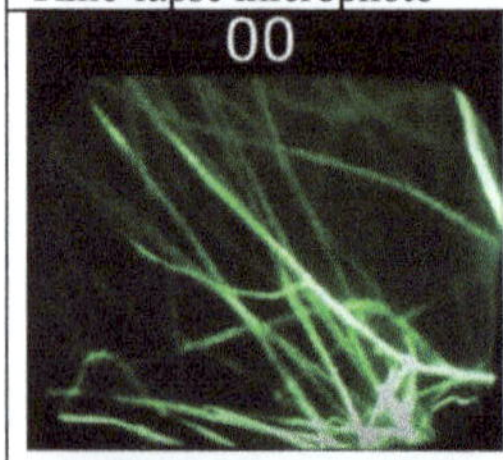	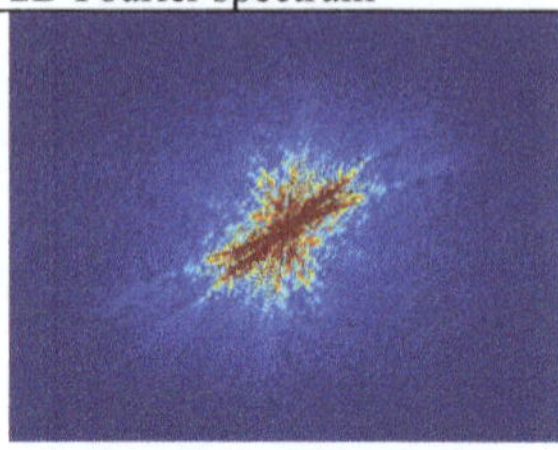	
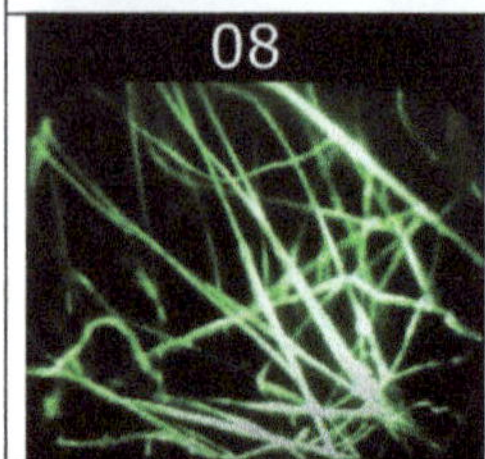	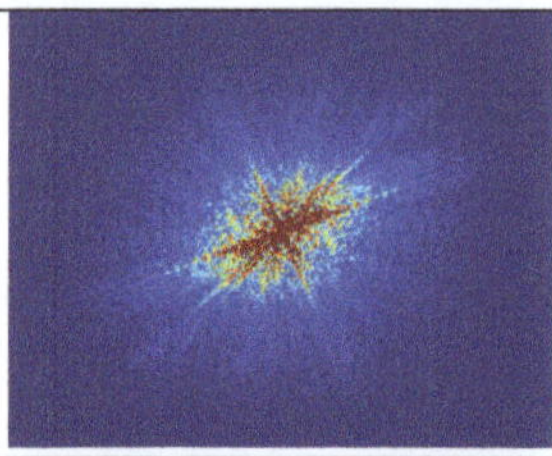	
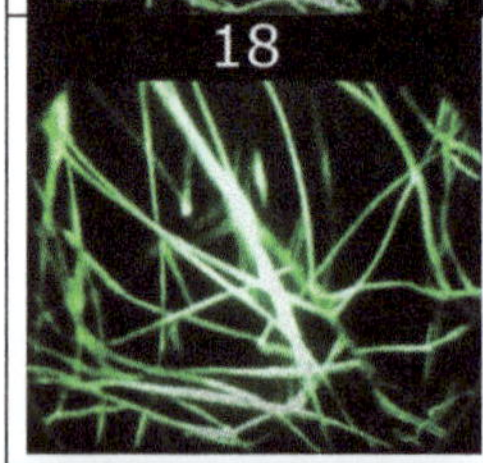	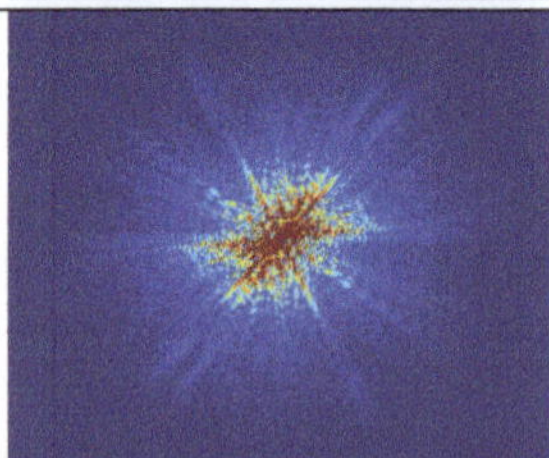	
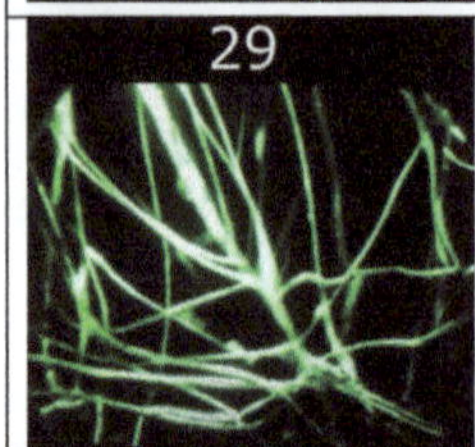	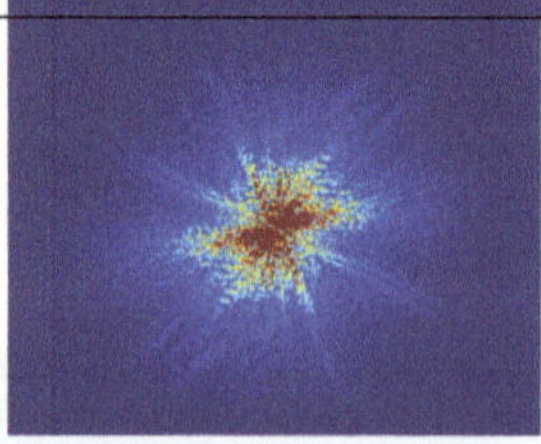	

Video recording of the experiment is available at https://youtu.be/uHBLeWysbAA.

Materials Research Proceedings **21** (2022) 370-383 https://doi.org/10.21741/9781644901755-64

In some cases, not smooth movements are observed, leading to a change in the network configuration under the electron beam, but spontaneous instantaneous changes accompanied by a sharp increase or disappearance of ISC components, as shown in Table. 2.

Table 2.

Time-lapse microphoto	2D Fourier spectrum	ISC
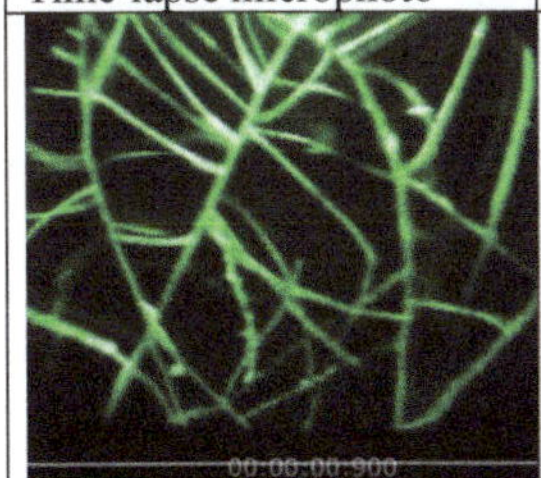	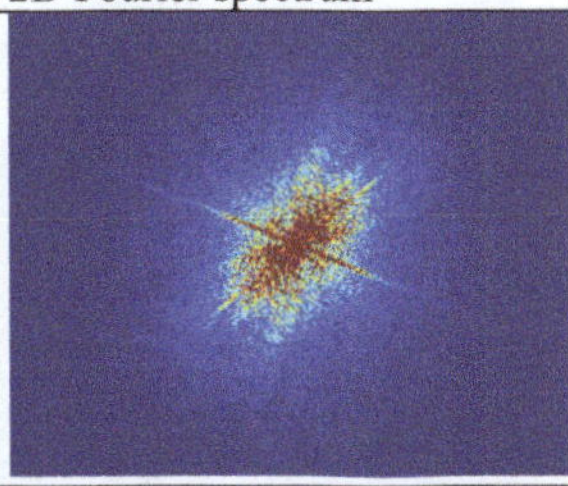	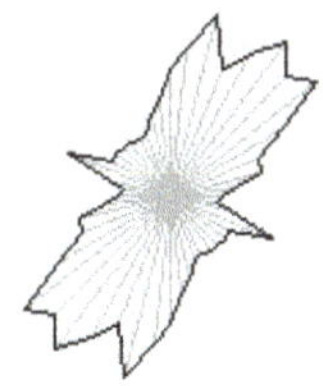
	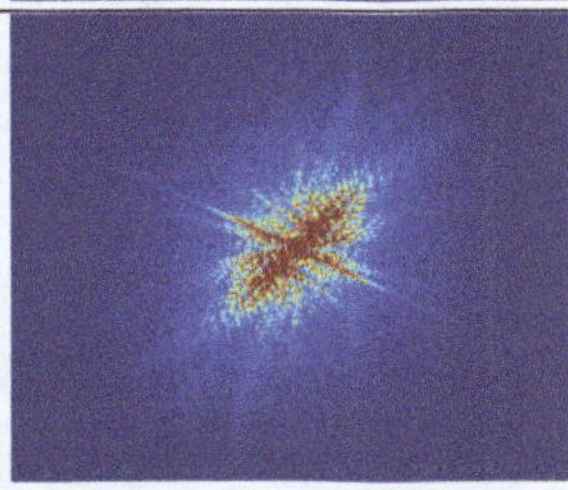	
	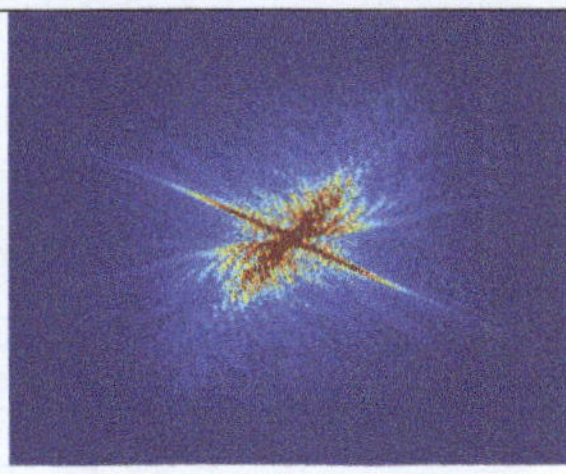	
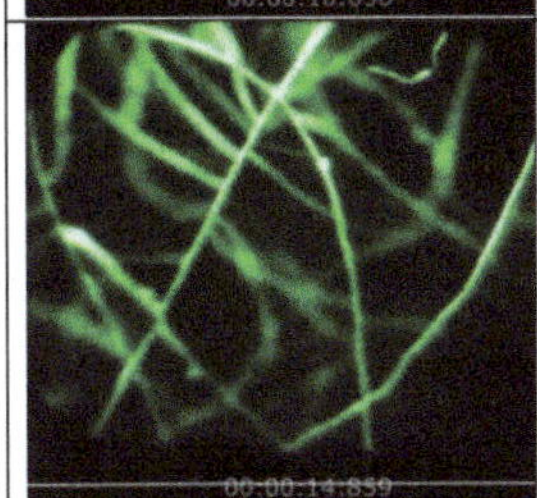	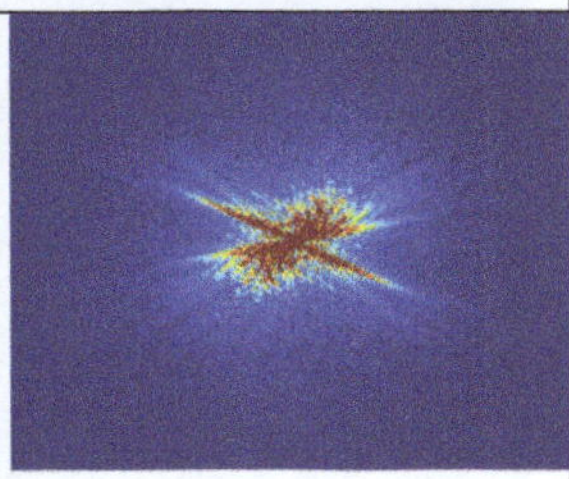	

Video recording of the experiment is available at https://youtu.be/cl5NegKEng0 .

Materials Research Proceedings **21** (2022) 370-383 https://doi.org/10.21741/9781644901755-64

Table 3.

SEM microphotography	Correlogram (0.4 s.)	Time code
	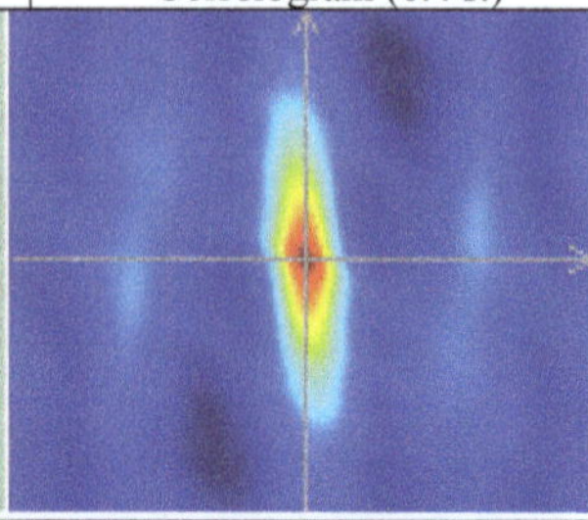	Start
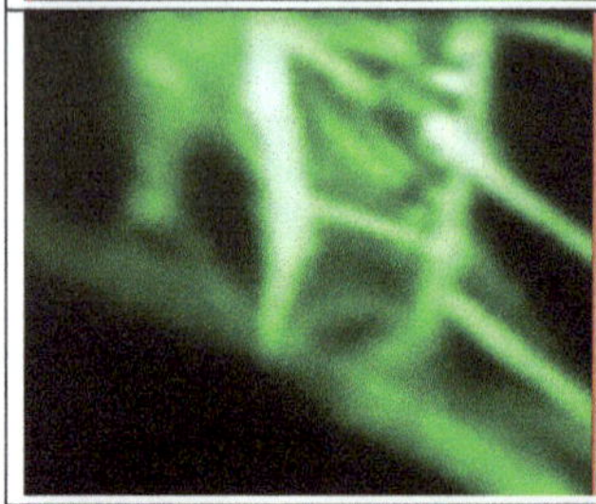	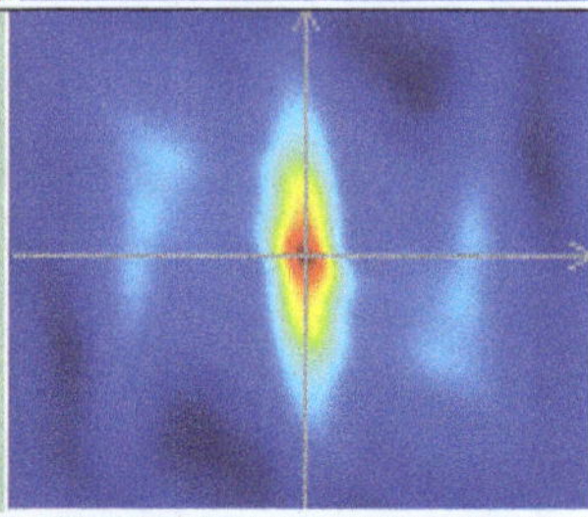	03:590
	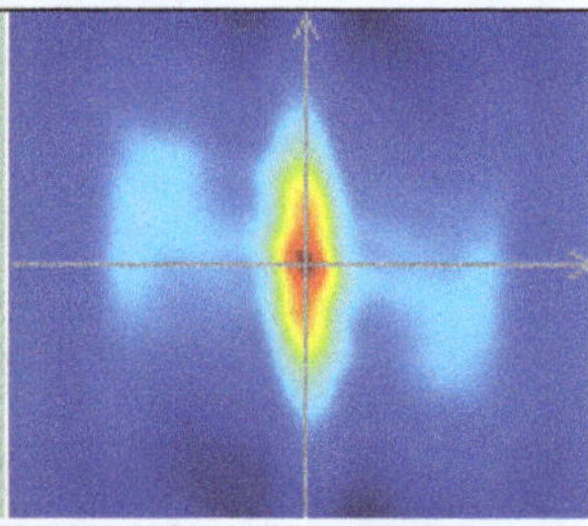	04:444
	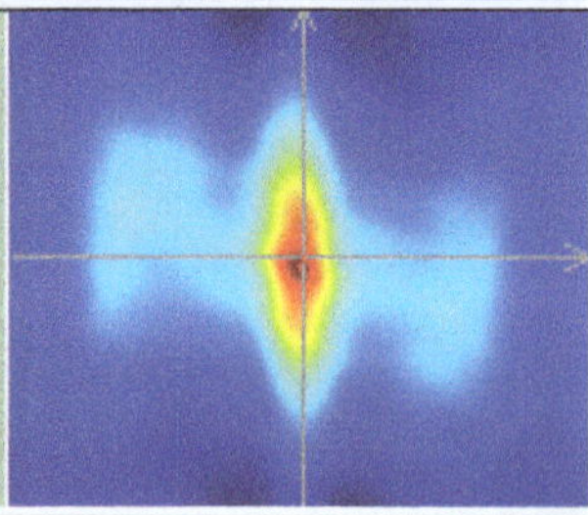	05:893

Video recording of the experiment is available at https://youtu.be/A9tVG6Zu2sI .

In some cases, a reorganization of the entire structures is observed, which can be applied for the electron-beam-controlled microrobotics. Thus, table 3 includes micrographs and correlograms taken with a temporal resolution of 0.4 seconds for such a reorganized area. It is obvious that exposure to an electron beam induces mechanical changes in the cross-linked micromechanical structure of the polymer. In this case, different fibers work as "jacks", "drive belts", "springs", "elastic beams", "pushers", etc. The mechanical effect observed depends on the irradiated area, that is, the effective cross-sectional area of the scanning electron beam.

It is possible to confirm the assumption that some (thick) fibers perform the functions of bearing elements of mechanical strength, while the others (thin) operate as the flexible elements controlled by the electron beam. Let us consider the experiment with two different structures. In table 4 one can see the movement of a thin flexible fiber (the thread in the upper part of the micrograph) against the stability of the large thick one (in the center). The fiber movement features can be tracked using the data of the wavelet transform (Morlet wavelet) in the Re and Im regions (one can observe the drift of the extremum on the "blue" representation).

The measurements were carried out using the scalographic / scaleographic tool of the QAVIS program (see http://oias.poi.dvo.ru/qavis/rus/scalogram.html), which makes it possible to use the appropriate mathematical formalization for their interpretation, excluding the ineffective Gaussian approach.

If some fibers oscillate, move, contract like springs (thin fibers), while others perform a frame function (thick fibers), then the micromechanical properties of the polymer fibers should be assessed in dynamics. It is obvious that a fiber break is a possible cause of damage for the polymer-based controlled microelectromechanical structures, including those for microrobotic or MEMS applications. Therefore, mechanical strength tests should be performed for each chemical or structural type of polymer fibers under the electron beam with different parameters. An example of a PHB fiber break clearly observed on the SEM image is shown in Fig. 2.

Discussion

Where can the above described concept be applied? Obviously, its implementation is possible in different systems operating under vacuum conditions with the priority application area being space technologies. Polymer materials in vacuum are used in dozens of space applications, including vacuum-based 3D printing for in-space satellite assemblies [19]. Possible damages of polymer materials in outer space, including the effect of cosmic rays and elementary particles, are well known and predictable [20-24]. Therefore, the development of polymer robotics for space technologies is not a problem from modern materials science. From the point of view of mechanics, it is necessary to simulate vibrations in a vacuum without taking into account the earth's gravitational force [25], but this is rather a simplification than a complication of the problem.

Table 4.

Materials Research Proceedings **21** (2022) 370-383
https://doi.org/10.21741/9781644901755-64

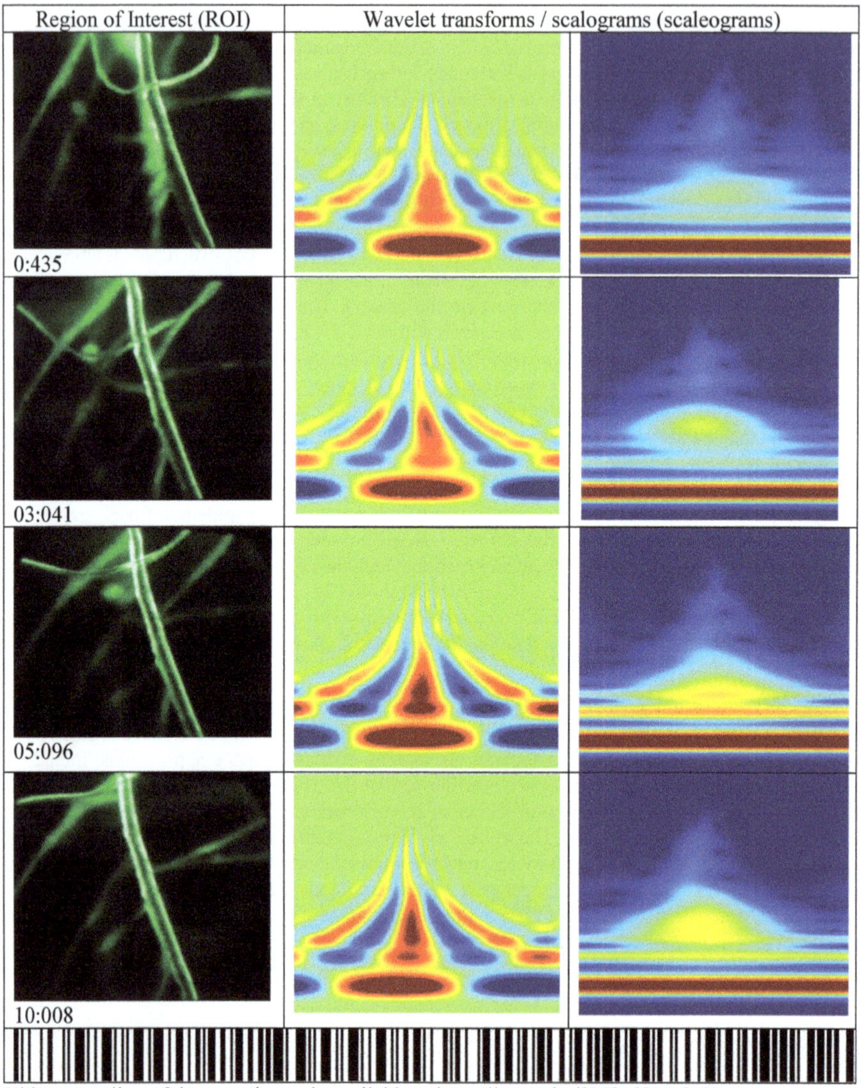

Video recording of the experiment is available at https://youtu.be/9M3y3VAt9zY

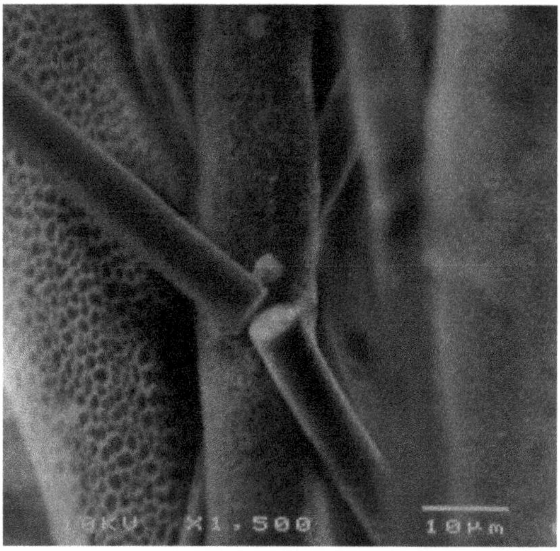

Fig. 2*. Example of the broken polymer fiber.*

Speaking about the application of robotics in space technologies, it is necessary to take into account its long history - from the late 1960s (*sensu lato*, taking into account the automated Babakin spacecrafts in the USSR and a number of similar developments in the USA) or, at least, since the mid-1970s (*sensu stricto*, considering robotic manipulators with automatic decision making function) for interstellar research [26,27]. The main fundamental modern requirements for interstellar robotics are formulated in departmental international meetings of the late 20th - early 21st centuries [28,29]. The difference between the extended interpretation of space robotics and its narrow interpretation is that both mechanically active systems for sensing the external environment and passive mapping systems [30,31] are included in the first, while the narrow interpretation includes only active mechatronic elements such as robotic arms \ manipulators [32].

In our concept, a narrow interpretation prevails, since the dynamics of microfibers, cantilevers under the particle beam makes sense only if there are moving elements - actuators in the spacecraft construction. A similar interpretation is also typical for environment – robot interactions for planetary or terrestrial subsurface explorers / robots based on earthworm locomotion [33,34] and hopping robots for planetary or asteroid surface exploration [35-37]. It should be noted that asteroids do not have an atmosphere and sufficient gravitational force, while planets usually have it. Therefore, in the first case, we can speak about the vacuum systems, which are essentially equivalent to the experimental vacuum conditions in SEM and TEM, and in the second case, only about the similarity with ESEM (Environmental Scanning Electron Microscopy) and ASEM (Atmospheric Scanning Electron Microscopy).

We are not considering here aerial robot explorers for planetary exploration [38], which by definition are not vacuum devices. For the same reason robots with manipulators for exploring planets are beyond the scope of this paper [39-42]. Therefore, in this case, we can only talk about the applications in "open space" and non-atmospheric bodies: meteorites and others [43]. As practice shows, this indirect and not always accessible way of solving problems (the transition from macroscale to microscale robotics and the introduction of non-standard interpretations despite the inconsistency with the high-budget mainstream) can help in obtaining non-standard results using spacecraft, as was the case decades ago [44-46]. A fibrillar manipulator cannot replace a macromanipulator, but for nanosatellites and picosatellites (such as "CanSat"), sets of cooperative fibrillar manipulators with the electron beam control may be the only possible means of achieving efficiency in many mechanical applications including pulsed one-time manipulations (such as opening a mirror or a solar battery) [47-57].

Launching CanSat into orbit is a separate technical problem, since the suborbital trajectories do not meet the requirements for sufficient evacuation degree and weightlessness. That is, we can speak about the applicability in open space beyond the Karman line (>100 km), and not beyond the Armstrong physiological line of 19 km. However, vacuum studies of potentially applicable polymers have not yet been carried at all the necessary gradations of heights and characteristics of the medium (from atmospheric to vacuum ones). Thus, one can only postulate the presence of a kinematic effect on polymer fibers under the electron beam and speculate about its potential applicability for the purposes of space technology. It should be noted that not all the polymers used in our studies are resistant to the conditions of outer space during long-term exposure. Therefore, this material can be considered only as the first step in this direction, illustrating the physical effect, but not specifying its chemical implementation.

Acknowledgements
The authors are grateful to V.K. Fischenko and A.A. Goncharova from the V.I. Il'ichev Pacific Okeanological Institute of Far Eastern Branch of the Russian Academy of Sciences for providing the original software. We also gratefully acknowledge M. K. Filippov and P. L. Aleksandrov for automation of the electron microscope, and A.V. Ratnovskaya for assistance in SEM image and video registration.

References

[1] H. Wu, H. Handroos, J. Kovanen, A. Rouvinen, P. Hannukainen, T. Saira, L. Jones, Design of parallel intersector weld/cut robot for machining processes in ITER vacuum vessel, Fusion Engineering and Design 69(1-4) (2003) 327-331. https://doi.org/10.1016/S0920-3796(03)00066-8

[2] H. Wu, H. Handroos, P. Pessi, Mobile parallel robot for assembly and repair of ITER vacuum vessel, Industrial Robot 35(2) (2008) 160-168. https://doi.org/10.1108/01439910810854656

[3] J.C. Hatchressian, V. Bruno, L. Gargiulo, D. Keller, Y. Perrot, P. Bayetti, J.J. Cordier, J.P. Friconneau, J.D. Palmer, F. Samaille. Development of an inspection robot under ITER relevant vacuum and temperature conditions, Journal of Physics: Conference Series 100(6) (2008) 062031. https://doi.org/10.1088/1742-6596/100/6/062031

[4] M. Li, H. Wu, H. Handroos, G. Yang, Software design of the hybrid robot machine for ITER vacuum vessel assembly and maintenance, Fusion Engineering and Design 88(9-10) (2013) 1872-1876. https://doi.org/10.1016/j.fusengdes.2013.04.023

[5] H. Wu, Y. Wang, M. Li, M. Al-Saedi, H. Handroos, Chatter suppression methods of a robot machine for ITER vacuum vessel assembly and maintenance, Fusion Engineering and Design 89(9-10) (2014) 2357-2362. https://doi.org/10.1016/j.fusengdes.2014.02.007

[6] M. Li, H. Wu, H. Handroos, G. Yang, Y. Wang, Software protocol design: Communication and control in a multi-task robot machine for ITER vacuum vessel assembly and

maintenance, Fusion Engineering and Design 98 (2015) 1532-1537. https://doi.org/10.1016/j.fusengdes.2015.05.058

[7] S. Moradkhani, Y.S. Hagh, H. Wu, H. Handroos, Dynamic analysis and control of a fusion reactor vacuum vessel assembly robot, Fusion Engineering and Design 154 (2020) 111532. https://doi.org/10.1016/j.fusengdes.2020.111532

[8] M. Kanetomo, H. Kashima, T. Suzuki, Wafer-transfer robot for use in ultrahigh vacuum, Journal of Vacuum Science and Technology A: Vacuum, Surfaces, and Films 15(3) (1997) 1385-1388. https://doi.org/10.1116/1.580546

[9] M. Kanetomo, H. Kashima, T. Suzuki, Robot for use in ultrahigh vacuum, Solid State Technology 40(8) (1997) 63-68.

[10] H.C. Chen, S.M. Huang, Design and analysis of the power drive module for ultrahigh vacuum wafer-transfer robot, Journal of Energy and Power Engineering 4(6) (2010) 55-59.

[11] M. Cong, T. Li, Design and application of magnetic coupling used for ultra-high vacuum robot, International Journal of Intelligent Systems Technologies and Applications 8(1-4) (2010) 231-246. https://doi.org/10.1504/IJISTA.2010.030202

[12] M.J. Chung, S.J. Lee, Development of automatic wafer centering system for vacuum transfer robot using for semiconductor manufacturing, Applied Mechanics and Materials 607 (2014) 782-785. https://doi.org/10.4028/www.scientific.net/AMM.607.782

[13] M. Cong, H. Wen, Y. Du, P. Dai, Coaxial twin-shaft magnetic fluid seals applied in vacuum wafer-handling robot, Chinese Journal of Mechanical Engineering 25(4) (2012) 706-714. https://doi.org/10.3901/CJME.2012.04.706

[14] L. Wang, S. Xu, W. Zhang, Y. Liu, Y. Xia, A cable-driven robot arm for visual tracking in tokamak vacuum vessel, IEEE International Conference on Robotics and Biomimetics (ROBIO-2018) (2018) 1125-1131. https://doi.org/10.1109/ROBIO.2018.8665129

[15] O.V. Gradov, M.A. Gradova, A.A. Olkhov, A.L. Iordanskii, Charge propagation along the polymer fiber of polyhydroxybutyrate: Is it possible to apply the cable model? Key Engineering Materials 869 (2020) 246-258. https://doi.org/10.4028/www.scientific.net/KEM.869.246

[16] N. Zaporozhets, Application of air microejector in vacuum gripping device of industrial robot., Mekhanizatsiya i Avtomatizatsiya Proizvodstva (Moscow, USSR) 12 (1986) 25.

[17] S. Belinski, W. Trento, R. Imani-Shikhabadi, S. Hackwood, Robot design for a vacuum environment, Proceedings of the WSTL, 1 (1987) 95-103.

[18] M. Shirazi, Development and testing of a vacuum-compatible robot, Manuf. Rev. 1(4) (1988) 259-264.

[19] J. McCrea, J.T. Cerri, C.R. Hartsfield, Design of a zero-gravity, vacuum-based 3D printer robot for use of in-space satellite assembly, 2018 AIAA Aerospace Sciences Meeting (2018) 2201. https://doi.org/10.2514/6.2018-2201

[20] E. Grossman, I. Gouzman, Space environment effects on polymers in low earth orbit, Nuclear Instruments and Methods in Physics Research Section B: Beam Interactions with Materials and Atoms 208 (2003) 48-57. https://doi.org/10.1016/S0168-583X(03)00640-2

[21] B. Basu, S. Biswas, S. Dey, A. Maulik, A. Mazumdar, S. Raha, S. Saha, S. K. Saha, and D. Syam, Polyethylene terephthalate polymers at mountain altitude as cosmic ray heavy particle

detector, Radiation Measurements 43 (2008) S262-S265. https://doi.org/10.1016/j.radmeas.2008.04.021

[22] U.H. Hossain, W. Ensinger, Experimental simulation of radiation damage of polymers in space applications by cosmic-ray-type high energy heavy ions and the resulting changes in optical properties, Nuclear Instruments and Methods in Physics Research Section B: Beam Interactions with Materials and Atoms 365 (2015) 230-234. https://doi.org/10.1016/j.nimb.2015.06.007

[23] D. Bója, A. Major, The influence of cosmic radiation on the thermal properties of different polymers, Gradus 4(2) (2017) 351-356.

[24] A.A. Major, D. Boja, What can cosmic radiation cause in polymers? IOP Conference Series: Materials Science and Engineering 448(1) (2018) 012057. https://doi.org/10.1088/1757-899X/448/1/012057

[25] D.I. Park, C.H. Park, Y. Yoo, Vibration simulation of hybrid type substrate handling robot in the vacuum environment, 12th International Conference on Control, Automation and Systems (2012) 2131-2134).

[26] A.C. Clarke, An optimum strategy for interstellar robot probes, Correspondence in Journal of the British Interplanetary Society 31 (1978) 438.

[27] G.L. Matloff, Robosloth: A slow interstellar thin-film robot, Journal of the British Interplanetary Society 49(1) (1996) 33-36.

[28] R. Noble, M.V. Sykes, Small body exploration technologies as precursors for interstellar robotics, Journal of the British Interplanetary Society 66 (2013) 15-24.

[29] E. Rhodes, A.J. Mauceri, M.M. Clarke, T.S. Lindsay, Paper Session IC-Robotics for Interstellar Missions, The Space Congress Proceedings 7 (1993) 35-41.

[30] D. Clery, Robot detector to map cosmos for clues to dark energy's force, Science 365(6458) (2019) 1066. https://doi.org/10.1126/science.365.6458.1066

[31] S. Vernov, B. Tverskoi, V. Yakovlev, E.V. Gorchakov, P. Ignatev, G. Lyubimov, N.V. Peresleg, O.A. Marchenko, T.E. Shvidkov, N. Kontor T. Morozova, Cosmic-ray measurements by Mars-2 robot space station, Izvestiya Akademii Nauk SSSR Seriya Fizicheskaya 38(9) (1974) 1859-1862.

[32] V.I. Gouliaev, T.V. Zavrazhina, Dynamics of a flexible multi-link cosmic robot-manipulator, Journal of Sound and Vibration 243(4) (2001) 641-657. https://doi.org/10.1006/jsvi.2000.3409

[33] A. Ellery, Environment-robot interaction-the basis for mobility in planetary micro-rovers, Robotics and Autonomous Systems 51(1) (2005) 29-39. https://doi.org/10.1016/j.robot.2004.08.007

[34] H. Omori, H. Kitamoto, A. Mizushina, T. Nakamura, T. Kubota, Satellite, planetary or terrestrial subsurface explorer robot based on earthworm locomotion, Journal of Robotics and Mechatronics 26(5) (2014) 660-661. https://doi.org/10.20965/jrm.2014.p0660

[35] L. Bai, W.J. Ge, X.H. Chen, M. Zhang, Research on hopping robot for planetary exploration, Robot, 31(4) (2009) 311-319.

[36] L. Bai, W. Ge, X. Chen, R. Chen, Design and dynamics analysis of a bio-inspired intermittent hopping robot for planetary surface exploration, International Journal of Advanced Robotic Systems 9(4) (2012) 109. https://doi.org/10.5772/51930

[37] L. Bai, W. Ge, X. Chen, X. Kou, Design and implementation of a bio-inspired intermittent hopping robot for planetary surface exploration, Robot 34(1) (2012) 32-37. https://doi.org/10.3724/SPJ.1218.2012.00032

[38] A.A. Pankine, K.M. Aaron, M.K. Heun, K.T. Nock, R.S. Schlaifer, C.J. Wyszkowski, A.P. Ingersoll, R.D. Lorenz, Directed aerial robot explorers for planetary exploration, Advances in Space Research 33(10) (2004) 1825-1830. https://doi.org/10.1016/j.asr.2003.07.049

[39] R. Bertrand, B. Schaefer, M. Van Winnendael, R. Rieder, European tracked micro-robot for planetary surface exploration, IFAC Proceedings Volumes 31(33) (1998) 37-44. https://doi.org/10.1016/S1474-6670(17)38384-2

[40] H. Das, X. Bao, Y. Bar-Cohen, R. Bonitz, R.A. Lindemann, M. Maimone, I.A. Nesnas, C.J. Voorhees, Robot manipulator technologies for planetary exploration, Smart Structures and Materials 1999: Smart Structures and Integrated Systems 3668 (1999) 175-182. https://doi.org/10.1117/12.350698

[41] A. Seeni, B. Schäfer, G. Hirzinger, Robot mobility systems for planetary surface exploration–state-of-the-art and future outlook: a literature survey, Aerospace Technologies Advancements 492 (2010) 189-208. https://doi.org/10.5772/6930

[42] I. Nanyageev, I. Shardyko, I. Dalyaev, Motion specification algorithms for both platform and arms of a mobile robot for planetary research, IOP Conference Series: Materials Science and Engineering 747(1) (2020) 012090. https://doi.org/10.1088/1757-899X/747/1/012090

[43] M. Maurette, M. Mellini, J. Silen, I. Tabacco, A. Morbidelli, R. Chatila, Meteorite at dome C? A project for the automated search for meteorites with a planetary exploration robot equipped with two radars, Meteoritics and Planetary Science Supplement 33 (1998) A100.

[44] R. Bertrand, B. Schaefer, R. Reider, European tracked micro-robot for planetary surface exploration, Space Technology 20(2) (2000) 55-64.

[45] G. Musser, Planetary science-NASA's robot rover scouts unknown terrain on the Angry Red Planet, Scientific American 290(3) (2004) 52-57. https://doi.org/10.1038/scientificamerican0304-52

[46] P.J. Westwick, Planetary exploration in extremis: JPL's robot explorers are the pride of NASA, but the lab nearly got shut down in the budget-cutting early'80s, Engineering and Science 69(4) (2006) 32.

[47] K. Ui, K. Nakaya, K. Konoue, H. Sawada, S. Tsurumi, M. Mori, R. Hodoshima, N. Maeda, H. Okada, N. Miyashita, M. Iai, O. Mori, S. Matunaga, Titech CanSat Project 2000: Report of Sub-orbital Flight and Balloon Experiment, in: M. Rycroft, N. Crosby (Eds.), Smaller Satellites: Bigger Business? Space Studies, 6, Springer, Dordrecht, 2000, pp. 417-418. https://doi.org/10.1007/978-94-017-3008-2_62

[48] N. Sako, Y. Tsuda, S. Ota, T. Eishima, T. Yamamoto, I. Ikeda, H. Ii, H. Yamamoto, H. Tanaka, A. Tanaka, S. Nakasuka, CanSaT suborbital launch experiment-university educational space program using can sized pico-satellite, Acta Astronautica 48(5-12) (2001) 767-776. https://doi.org/10.1016/S0094-5765(01)00039-X

[49] T. Eishima, Y. Nakamura, S. Nakasuka, Space outreach program using CanSat-kit, Transactions of the Japan Society for Aeronautical and Space Sciences, Space Technology 7(26) (2009) 19-23. https://doi.org/10.2322/tstj.7.Tu_19

[50] S.H. Won, H.Y. Jun, S.H. Kim, S.R. Lee, Very small satellite program for expending the space technology base: CanSat competition, Journal of the Korean Society for Aeronautical and Space Sciences 40(7) (2012) 636-645. https://doi.org/10.5139/JKSAS.2012.40.7.636

[51] R. Kawashima, CanSat leader training program: Past, present and future, Ciencia UANL 19(81) (2016) 76-82.

[52] A. Colin, B. Bermudez Reyes, G.E. Morrobel, G.A. Lira Ibarra, D.M. Zúñiga Rosales, L.A. Avalos de la Cruz, M. Villarreal Méndez, J. Mendoza Martínez, B. Alvarez Arce, Construcción de un picosatélite CanSat, Ciencia UANL 19(81) (2016) 34-38.

[53] A. Colin, A pico-satellite assembled and tested during the 6th CanSat Leader Training Program, Journal of Applied Research and Technology 15(1) (2017) 83-91. https://doi.org/10.1016/j.jart.2016.10.003

[54] L.A., Anchino, A.F., Torti, E.M., Dovis, E. Bernardi, R. Podadera, Implementacion de una Plataforma de Desarrollo CanSat Multiproposito, Elektron 3(2) (2019) 120-127. https://doi.org/10.37537/rev.elektron.3.2.93.2019

[55] H.-U. Oh, H.-I. Kim, J.-K. Kim, J.-S. Choi, S.-H. Kim, Smartphone CanSat for actualization of real-time streaming video calls using remote screen touch system with shape memory alloy actuator, Transactions of the Japan Society for Aeronautical and Space Sciences 62(5) (2019) 256-264. https://doi.org/10.2322/tjsass.62.256

[56] J. Nakaya, T. Takada, Y. Kajimura, H. Tsuchiya, N. Uezono, Y. Sasaoka, S. Ueta, M. Wakabayashi, K. Kitamura, Development of CubeSat Ground Model Extended from CanSat: Application to Space Education at KOSEN, Transactions of the Japan Society for Aeronautical and Space Sciences, Aerospace Technology Japan 18(5) (2020) 281-287. https://doi.org/10.2322/tastj.18.281

[57] J.S. Rodríguez, A.Y. Botero, D.V. Lopera, J.G. Serna, F. Botero, Experimental approach for the evaluation of the performance of a satellite module in the CanSat form factor for *in situ* monitoring and remote sensing applications, International Journal of Aerospace Engineering 2021(81) (2021) 1-28. https://doi.org/10.1155/2021/8868797

Nonlinear Mathematical Models of Metamaterials Defined as "Mass-in-Mass" and "Damper-in-Mass" Chains

Daniil A. Kolesov[1,a*], Vladimir I. Erofeev[1,b], Alexey O. Malkhanov[1,c] and Ashot V. Shekoyan[2,d]

[1]Mechanical Engineering Research Institute of the Russian Academy of Sciences – Branch of Federal Research Center "Institute of Applied Physics of the RAS" ,85, Belinsky Street, Nizhny Novgorod, 603024, Russia

[2]Institute of Mechanics of the National Academy of Sciences of the Republic of Armenia, 24B, Baghramyan Ave., Yerevan, 0019, Republic of Armenia

[a]kolesov.d.a@yandex.ru, [b]erof.vi@yandex.ru, [c]alexey.malkhanov@gmail.com, [d]ashotshek@mechins.sci.am

Keywords: Metamaterial, Chains of Elastic and Viscoelastic Elements, Nonlinearity, Deformation

Abstract. To describe dynamic processes in an acoustic (mechanical) metamaterial, there are proposed models that are a one-dimensional chain containing the same masses connected by linearly elastic (or nonlinearly elastic) elements (springs) with the same stiffness. In this case, it is assumed that each mass contains inside itself a series connection of another mass and an elastic element or viscous element (damper).

Introduction

An adequate description of the physical and mechanical properties of metamaterials within the framework of the classical theory of elasticity is impossible. Recently, generalized micropolar theories of the Cosserat continuum type have become widespread for modeling structurally inhomogeneous materials [1]. However, these theories include a large number of material constants that require experimental determination and the connection of which with the structure of the material is not clear. An alternative direction, structural modeling, is devoid of such a drawback [2].

Nonlinear elastic "mass-in-mass" chain

In [3], to describe the dynamic properties of a metamaterial, a one-dimensional chain was considered containing the same masses m_1 connected by elastic elements (springs) with the same stiffness k_1, while each mass inside itself contained another mass m_2 and one more elastic element - a spring with stiffness k_2 (Fig. 1). This model is called a "mass-in-mass chain".

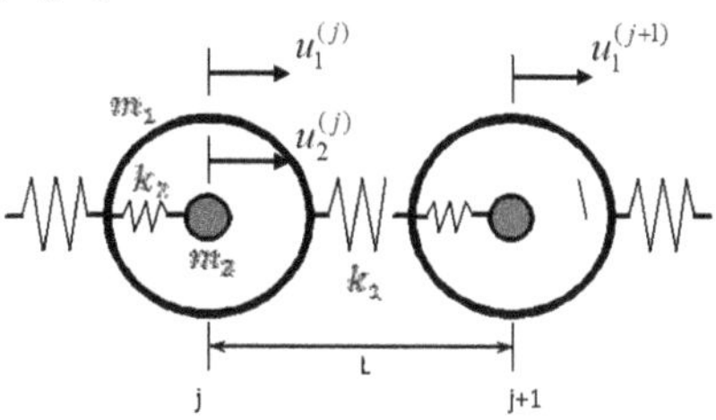

Fig. 1. The mechanical model of elastic metamaterial

Let us generalize the model [3] by taking into account the quadratic nonlinearity of the external and internal elastic elements.

The potential energy of the unit cell of the "mass-in-mass" chain will be written as:

$$W^{(j)} = \frac{1}{2}\left[k_1\left(u_1^{(j+1)} - u_1^{(j)}\right)^2 + k_2\left(u_2^{(j)} - u_1^{(j)}\right)^2 + h_1\left(u_1^{(j+1)} - u_1^{(j)}\right)^3 + h_2\left(u_2^{(j)} - u_1^{(j)}\right)^3\right], \tag{1}$$

and its kinetic energy in the form:

$$T^{(j)} = \frac{1}{2}\left[m_1\left(\dot{u}_1^{(j)}\right)^2 + m_2\left(\dot{u}_2^{(j)}\right)^2\right]. \tag{2}$$

Let us suppose that $u_1(x)$ and $u_2(x)$ are continuous functions that give offsets of all m_1 and m_2, respectively. Using the Taylor series expansion of displacements, and restricting ourselves to two terms, we obtain

$$u_1^{(j+1)} = u_1(x+L) = u_1(x) + \frac{\partial u_1}{\partial x}L = u_1^{(j)} + \frac{\partial u_1}{\partial x}L. \tag{3}$$

The displacement decomposition technique in (3) was effectively used by I.A. Kunin [4] in the transformation of multi-mass discrete systems into a quasi-continuum.

The potential and kinetic energy densities for the equivalent continuum obtained from (5) and (6) are written as:

$$W = \frac{1}{2L}\left[k_1\left(\frac{\partial u_1}{\partial x}L\right)^2 + k_2(u_2 - u_1)^2 + h_1\left(\frac{\partial u_1}{\partial x}L\right)^3 + h_2(u_2 - u_1)^3\right], \tag{4}$$

$$T = \frac{1}{2L}[m_1(\dot{u}_1)^2 + m_2(\dot{u}_2)^2]. \tag{5}$$

Let us compose from (4) and (5) the Lagrangian $\mathcal{L} = T - W = \mathcal{L}(\dot{u}_1, \dot{u}_2, u_{1x}, u_1, u_2)$ and use the equations of analytical mechanics

$$\frac{\partial}{\partial t}\left(\frac{d\mathcal{L}}{d\dot{u}_1}\right) + \frac{\partial}{\partial x}\left(\frac{\partial \mathcal{L}}{\partial u_{1x}}\right) - \frac{\partial \mathcal{L}}{\partial u_1} = 0,$$
$$\frac{\partial}{\partial t}\left(\frac{d\mathcal{L}}{d\dot{u}_2}\right) - \frac{\partial \mathcal{L}}{\partial u_2} = 0.$$

to obtain a system of equations in the displacements of equations (4), (5), we get:

$$\frac{m_1}{L}\ddot{u}_1 - k_1 L\frac{\partial^2 u_1}{\partial x^2} - 3h_1 L^3\frac{\partial u_1}{\partial x}\frac{\partial^2 u_1}{\partial x^2} - \frac{k_2}{L}(u_2 - u_1) - \frac{3h_2}{2L}(u_2 - u_1)^2 = 0,$$
$$\frac{m_1}{L}\ddot{u}_2 + \frac{k_2}{L}(u_2 - u_1) + \frac{3h_2}{2L}(u_2 - u_1)^2 = 0. \tag{6}$$

Further we will consider a particular case of system (6), where $h_1 \neq 0$, $h_2 = 0$, i.e.:

$$\frac{m_1}{L}\ddot{u}_1 - k_1 L\frac{\partial^2 u_1}{\partial x^2} - 3h_1 L^3 \frac{\partial u_1}{\partial x}\frac{\partial^2 u_1}{\partial x^2} - \frac{k_2}{L}(u_2 - u_1) = 0,$$
$$\frac{m_2}{L}\ddot{u}_2 + \frac{k_2}{L}(u_2 - u_1) = 0. \tag{7}$$

System (7) can be rewritten as a single equation:

$$\frac{\partial^2 u_2}{\partial t^2} - \frac{k_1 L^2}{m_1 + m_2}\frac{\partial^2 u_2}{\partial x^2} + \frac{m_1 m_2}{k_2(m_1+m_2)}\frac{\partial^4 u_2}{\partial t^4} - \frac{k_1 L^2 m_2}{k_2(m_1+m_2)}\frac{\partial^4 u_2}{\partial x^2 \partial t^2}$$
$$- \frac{3h_1 L^4}{m_1+m_2}\left(\frac{\partial u_2}{\partial x}\frac{\partial^2 u_2}{\partial x^2} + \frac{m_2}{k_2}\frac{\partial^2 u_2}{\partial x^2}\frac{\partial^3 u_2}{\partial t^2 \partial x} + \frac{m_2}{k_2}\frac{\partial u_2}{\partial x}\frac{\partial^4 u_2}{\partial t^2 \partial x^2}\right.$$
$$\left. + \left(\frac{m_2}{k_2}\right)^2 \frac{\partial^3 u_2}{\partial t^2 \partial x}\frac{\partial^4 u_2}{\partial t^2 \partial x^2}\right) = 0. \tag{8}$$

Let's move on to dimensionless time, coordinate and displacement:

$$\tau = \frac{t}{T}, y = \frac{x}{X}, u_2 = u_0 u\,. \tag{9}$$

The transformed equation (8) with the substitutions (9) takes the following form:

$$\frac{\partial^2 u}{\partial \tau^2} - \frac{k_1 L^2}{m_1+m_2}\frac{T^2}{X^2}\frac{\partial^2 u}{\partial y^2} + \frac{m_1 m_2}{k_2(m_1+m_2)}\frac{1}{T^2}\frac{\partial^4 u}{\partial \tau^4} - \frac{k_1 L^2 m_2}{k_2(m_1+m_2)}\frac{1}{X^2}\frac{\partial^4 u}{\partial y^2 \partial \tau^2}$$
$$- \frac{3h_1 L^4}{m_1+m_2}\frac{T^2 u_0}{X^3}\left(\frac{\partial u}{\partial y}\frac{\partial^2 u}{\partial y^2} + \frac{m_2}{k_2}\frac{1}{T^2}\frac{\partial^2 u}{\partial y^2}\frac{\partial^3 u}{\partial \tau^2 \partial y} + \frac{m_2}{k_2}\frac{1}{T^2}\frac{\partial u}{\partial y}\frac{\partial^4 u}{\partial \tau^2 \partial y^2}\right.$$
$$\left. + \left(\frac{m_2}{k_2}\frac{1}{T^2}\right)^2 \frac{\partial^3 u}{\partial \tau^2 \partial y}\frac{\partial^4 u}{\partial \tau^2 \partial y^2}\right) = 0. \tag{10}$$

We require that all coefficients (10) be finite or small values. Let us choose them so that among the nonlinear terms it is possible to single out only one, the main term.

All further discussions are valid under two conditions:

$$\frac{k_1 L^2}{m_1+m_2}\frac{T^2}{X^2} = 1, \frac{m_1 m_2}{k_2(m_1+m_2)}\frac{1}{T^2} = \varepsilon$$

where $\varepsilon << 1$.

When these conditions are met in equation (10), some of the terms can be neglected, since they have a higher order of smallness and do not have a significant effect on dynamic processes. Thus, equation (10) takes the form:

$$\frac{\partial^2 u}{\partial \tau^2} - \frac{\partial^2 u}{\partial y^2} + \varepsilon\frac{\partial^2}{\partial \tau^2}\left[\frac{\partial^2 u}{\partial \tau^2} - \alpha\frac{\partial^2 u}{\partial y^2}\right] = \delta\frac{\partial u}{\partial y}\frac{\partial^2 u}{\partial y^2}, \tag{11}$$

where $\frac{m_1 m_2}{k_2(m_1+m_2)}\frac{1}{T^2} = \varepsilon << 1, \frac{3h_1 L u_0 \sqrt{\varepsilon\alpha}}{k_1\sqrt{\frac{k_1 m_2}{k_2 m_1}}} = \delta << 1, \alpha = \frac{m_1+m_2}{m_1} > 1.$

Returning to the original dimensional variables in equation (11), we obtain a simplified equation (8) in the form:

$$\frac{\partial^2 u_2}{\partial t^2}-\frac{k_1 L^2}{m_1+m_2}\frac{\partial^2 u_2}{\partial x^2}+\frac{m_1 m_2}{k_2(m_1+m_2)}\frac{\partial^4 u_2}{\partial t^4}-\frac{k_1 L^2 m_2}{k_2(m_1+m_2)}\frac{\partial^4 u_2}{\partial x^2 \partial t^2} -\frac{3h_1 L^4}{m_1+m_2}\frac{\partial u_2}{\partial x}\frac{\partial^2 u_2}{\partial x^2}=0. \quad (12)$$

Visco-elastic "damper-in-mass" chain

It is not possible to study the dissipative properties of a metamaterial within the framework of a purely elastic formulation of problem (12). To solve this problem, we replace the elastic element with stiffness k_2 by a viscous element (Fig. 2).

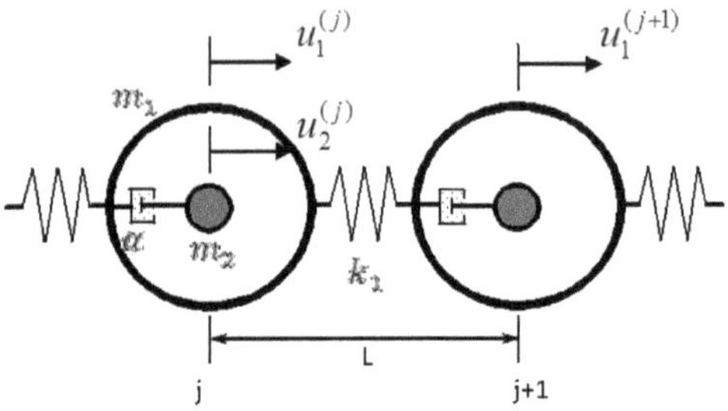

***Fig. 2.** The mechanical model of a viscoelastic metamaterial*

The dynamics equations of the modified "mass-in-mass" chain in the long-wavelength range will have the form:

$$\frac{m_1}{L}\frac{\partial^2 u_1}{\partial t^2}-k_1 L\frac{\partial^2 u_1}{\partial x^2}-\frac{\alpha}{L}\frac{\partial}{\partial t}(u_2-u_1)=0, \quad (13)$$

$$\frac{m_2}{L}\frac{\partial^2 u_2}{\partial t^2}+\frac{\alpha}{L}\frac{\partial}{\partial t}(u_2-u_1)=0. \quad (14)$$

Note that system (13), (14) can be reduced to one equation for displacement:

$$\frac{\partial^2 u_1}{\partial t^2}-C_0^2\frac{\partial^2 u_1}{\partial x^2}+\frac{m_1 m_2}{L\alpha(m_1+m_2)}\frac{\partial^3 u_1}{\partial t^3}-\frac{C_0^2 m_2}{L\alpha}\frac{\partial^3 u_1}{\partial x^2 \partial t}=0. \quad (15)$$

If we introduce in (15) the dimensionless displacement $U=\frac{u_1}{u_0}$, coordinate $X=\frac{x}{a}$ and time $T=\frac{t}{b}$, where $b=\frac{m_2}{L\alpha}$, $a=C_0 b$, then this equation can be rewritten as:

$$\frac{\partial^2 U}{\partial T^2}-\frac{\partial^2 U}{\partial X^2}+\delta\frac{\partial^3 U}{\partial T^3}-\frac{\partial^3 U}{\partial X^2 \partial T}=0. \quad (16)$$

Here $\delta=\frac{m_1}{m_1+m_2}$. This parameter belongs to the interval $\delta=[0,1]$, which includes two limiting cases: $\delta\to 1$, if $m_1 \gg m_2$ and $\delta\to 0$, if $m_2\gg m_1$.

We consider equation (16) with the following initial conditions:

$$U(X,0)=A sech(\gamma X)=\frac{2A}{e^{\gamma X}+e^{-\gamma X}}, \quad (17)$$

$$\frac{\partial U(X,0)}{\partial t} = 0, \tag{18}$$

where A – is an amplitude, γ – is a spatial parameter.

The development of the initial ($T_0 = 0$) disturbance (17), (18) can be traced over the next three time instants (Fig. 3).

The solution to the problem is symmetric with respect to $X = 0$, because the initial value (17) is an even function. Solutions obtained at $\delta = 0.05$, are plotted to the left of the axis of symmetry (dashed line), and the solutions obtained at $\delta = 0.5$, are plotted to the right of the axis.

Comparison of these cases shows the difference in dispersion. The character of the attenuation of disturbances can vary and depends on the value of δ. In the case of a small value of parameter δ the attenuation is much faster than when δ is greater. The initial sections (at $T = T_0$) in both cases are qualitatively similar. This is explained by the fact that in both cases anomalous dispersion takes place at large values of the wavenumber k. The main difference between the presented cases arises when considering the tail of the curves. At small values of parameter δ the solution behaves more like a solution of the diffusion equation, but for large values of parameter δ the solution behaves similarly to the solution of the wave equation. The presence of a more bulky terms at low values of parameter δ due to the superposition of the effects of normal dispersion and negative group velocity.

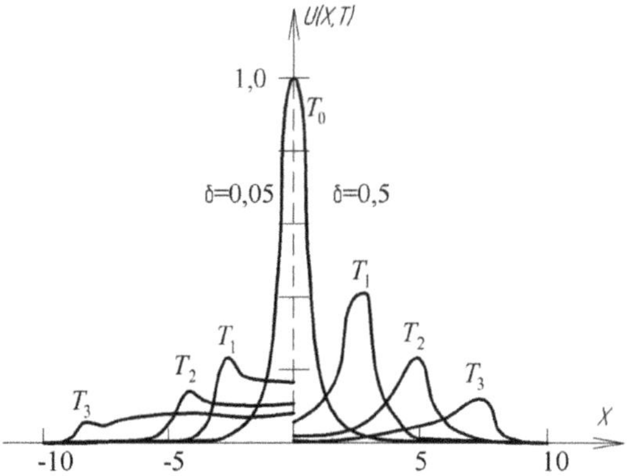

Fig. 3. *Instant wave profiles at* $A = 1$ *and* $\gamma = 3$, *at the moments* $T_0 = 0, T_1 = 7/3, T_2 = 14/3, T_3 = 21/3$, *which calculated for two values of parameter* δ.

In the Fig. 4 it is shown that the wave profiles for asymmetric development for four consecutive time instants. The initial value consists of the sum of two perturbations that have different fundamental frequencies. The initial perturbation has the form:

$$U(X,0) = A sech(\gamma X) + B sech[g(X+1)], \tag{19}$$

$$\frac{\partial U(X,0)}{\partial t} = 0. \tag{20}$$

Here, the main perturbation, marked in Fig. 4 as a, has an amplitude A = 1 and the spatial parameter $\gamma = 1.6$ (corresponds to a disturbance with a low fundamental frequency). The secondary disturbance, which is marked in Fig. 4 as b, shifts to the left with a certain step with respect to the main disturbance. The secondary disturbance has an amplitude of $B = 0.55$ and a spatial parameter of $g = 10$, which corresponds to a high fundamental frequency.

In the Fig. 4 it is shown that the high-frequency perturbation b propagates faster than the main perturbation, which has a lower frequency. For example, in position b_2 and at the corresponding moment in time T_2, the maximum of perturbation b_0 reaches the maximum of the main perturbation, and in position b_3 the maximum of perturbation b_0 is ahead of the maximum of the main perturbation. This phenomenon is explained by anomalous dispersion, which is expressed in the fact that the group velocity exceeds the phase velocity.

From the results of dispersion analysis, it follows that the high-frequency wave components must also decay faster than the low-frequency components. Indeed, this statement is confirmed in Fig. 4. The peak value of the main perturbation decreased from the initial amplitude $A = 1$ at time T_0 to $A \approx 0.2$ at time T_3. On the other hand, the perturbation amplitude b_0 decreases more significantly, from $B = 0.55$ at time T_0 to $B \approx 0$ at time T_3.

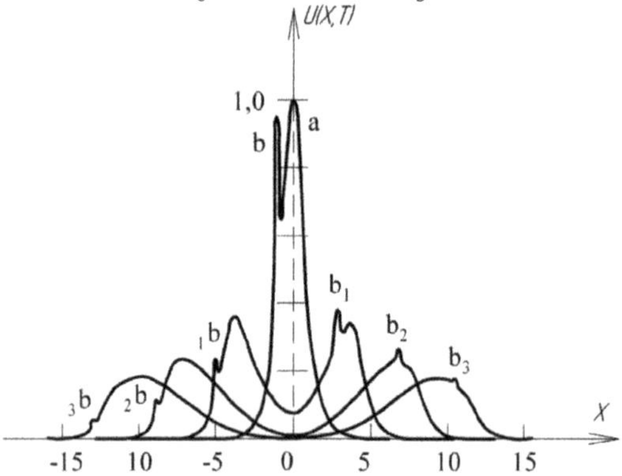

Fig. 4. *Instant wave profiles at* $\delta = 0.5, \gamma = 1.6, g = 10, A = 1, B = 0.55,$ *at the moments* $T_0 = 0, T_1 = 15/4, T_2 = 30/4, T_3 = 45/4$. *Markers Маркеры* b_i *depict the place of the peak* b_0 *when the disturbance propagates the to right,* $_ib$ *– when the disturbance propagates the to left.*

As a result of the studies, it was shown that the longitudinal wave in the viscoelastic metamaterial, defined as the mass-in-mass chain, has dispersion and frequency-dependent attenuation. The evolution of the wave profile is analyzed, both in the low-frequency and in the high-frequency ranges.

Summary

The models "mass-in-mass" and "damper-in-mass" constructed in this work are free from the drawbacks inherent in a number of other mechanical models of metamaterials: they free from the need to endow deformable bodies with negative mass, density and (or) negative modulus of elasticity.

Acknowledgments
The research was carried out under the financial support of the Russian Scientific Foundation (Project No 21-19-00813).

References

[1] I. Vardoulakis, Cosserat Continuum Mechanics., Springer, Cham, Switzerland, 2019. https://doi.org/10.1007/978-3-319-95156-0

[2] V.I. Erofeev, I.S. Pavlov, Structural Modeling of Metamaterials, Springer Nature Switzerland, 2021. https://doi.org/10.1007/978-3-030-60330-4

[3] I.A. Kunin, Elastic Media with Microstructure. 1: One-Dimensional Models, Springer, Berlin, 1982. https://doi.org/10.1007/978-3-642-81748-9_1

[4] H.H. Huang, C.T. Sun, G.L. Huang, On the negative effective mass density in acoustic metamaterials. Int. J. Eng. Sci. 2009. Vol. 47. P.610-617. https://doi.org/10.1016/j.ijengsci.2008.12.007

Modem Trends in Manufacturing Technologies and Equipment Materials Research Forum LLC
Materials Research Proceedings **21** (2022) 391-395 https://doi.org/10.21741/9781644901755-66

Influence of Internal Defects of Basic Metal of the Equipment on Its Operational Properties

S.I. Valeev[1,a*]

[1]Kazan National Research Technological University, Kazan, Republic of Tatarstan, Russian Federation

[a] vsi73@mail.ru

Keywords: Defect, Accident, Stress, Operation, Delamination Of The Base Metal, Technical Diagnostics

Abstract. The paper deals with the issue of safe operation of technological equipment with a defect in the base metal of the delamination type. Mathematical modeling of the stress-strain state in the defect zone by the finite element method in ANSYS software package is carried out. The article presents and analyzes the obtained graphs of the distribution of stress intensity factors. Based on the analysis of the obtained dependences, it is shown that stratification of the base metal does not have a significant effect on the performance of the process equipment.

Currently, equipment for potentially hazardous chemical, petrochemical, oil and gas processing industries is worn out by 80-90%. This leads to a decrease in the reliability and durability of equipment and, in turn, to an increase in industrial accidents [1,2,6,7,11].

Equipment for chemical, petrochemical, oil and gas processing industries is made mainly of various types of metals. The causes of emergencies are mainly associated with formation and development of defects that may appear not only during operation, but may also be initially present in the metal of the equipment that was not noticed during the incoming inspection [1,4,7,8]. All this can lead to depressurization of technological equipment, release and spill of explosive and fire hazardous substances.

Defects and destruction of materials can be caused by metallurgical reasons, when there are defects in the workpieces.

One of the most dangerous defects of metallurgical origin, leading to accidents, is a crack-like defect such as delamination of the base metal or welded joint [4,9,10].

In literary sources, there are several definitions of delamination, among which the most accurate and complete is the following: delamination is an internal discontinuity of the base metal of the equipment wall, oriented in the direction of the fiber, which occurs during processing of an ingot with pressure, which had shrinkage cavities or looseness, as well as when rolling of non-metallic inclusions and gas bubbles.

According to the normative technical documentation, the main measures to ensure safe operation of technological equipment are their timely technical diagnostics during manufacture, during operation and at the end of their service life [2].

When carrying out technical diagnostics, an important step is to carry out non-destructive testing of the base metal and welded joints of technological equipment. Non-destructive testing during technical diagnostics allows detecting defects without violating their suitability for intended use at all stages of the equipment life cycle.

As noted earlier, the most acceptable method for detecting delamination is the ultrasonic method of non-destructive testing [9, 10].

In accordance with federal norms and rules in the field of industrial safety, if a crack-like defect of delamination type is detected during technical diagnostics of technological equipment, then operation of such equipment must be prohibited, since this defect is unacceptable.

Usually, the operation of such equipment is possible after repair by removing the defective area and welding in its place a so-called patch from a similar material or replacing this equipment. Both options are associated with withdrawal of the equipment for a long time from technological process and shutdown of production, which is not always the correct and reasonable decision, especially in modern economic conditions.

The accumulated experience in the field of fracture mechanics and automated analysis enables modeling and assessing the possibility of conditions for further operation of technological equipment with a delamination type defect [3,5,9,10].

Mathematical modeling of the stress-strain state in the defect zone was carried out by the finite element method implemented in ANSYS software package.

ANSYS is a unique finite element system that covers many phenomena of a different physical nature, such as fluid dynamics, strength, electromagnetism, and thermal physics. It has the ability to solve related problems that combine all the listed types of phenomena. The standard calculation procedure is divided into three stages: building a model; application of loads and obtaining a solution; viewing and analysis of results. It has been established that the main criterion parameter for analyzing brittle and quasi-brittle fractures, where the operating temperature does not exceed the first critical one, is the stress intensity factor K (SIF).

For example, with the help of ANSYS software package [9], the possibility of safe operation of equipment with delamination of the base metal, depending on the ratio of delamination length to the length of the shell of the equipment body, was considered. The calculation results showed that the maximum values relative to other stress intensity factors reach the stress intensity factors of the second type - KII.

In work [10], the possibility of safe operation of equipment with stratification of the base metal, depending on the depth of its occurrence in the base metal, was considered, modeling was also carried out in ANSYS software package. The data obtained show that the maximum values relative to other stress intensity factors are reached by the stress intensity factors of the second type - KII (longitudinal shear crack) located closer to the inner wall of the shell.

According to the results of the above calculations, it was established [3,5,9,10] that the values of the stress intensity factors were small compared to the critical ones for a given material, further growth of a delamination-type defect was excluded, which meant that delamination did not significantly affect the load-carrying capacity of the equipment.

It should be noted that the above works were carried out under the same conditions, simulation parameters and were interrelated [9, 10]. In particular, a model of a half shell was considered with the corresponding boundary conditions recorded on the plane of symmetry. To simulate the reaction of bottoms, the corresponding tensile stresses were applied to the end planes. The modeling did not take into account the effect of stress concentrators (joints, fittings, supports, etc.). To create the correct mesh in the delamination front area, 2-D elements with offset central nodes (in order to implement the shape function with a root feature) PLANE82 and 3-D elements SOLID186 with the addition of contact elements CONTA174 and TARGE170 were used. A nonlinear contact static problem was solved without taking into account plastic deformations in the delamination front zone.

The purpose of this work is to continue the development of a method for determining parameters of fracture mechanics KI, II, III and to create a system for their automated calculation to predict the resource of technological equipment with crack-like defects such as delamination, other things being equal, a different ratio of delamination axes relative to the shell axis is considered.

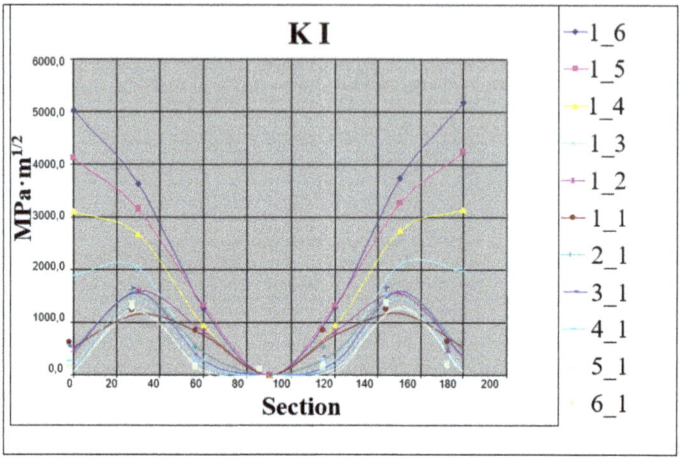

***Fig. 1.** Calculated values of stress intensity factor KI for different ratios of delamination axes relative to the shell axis.*

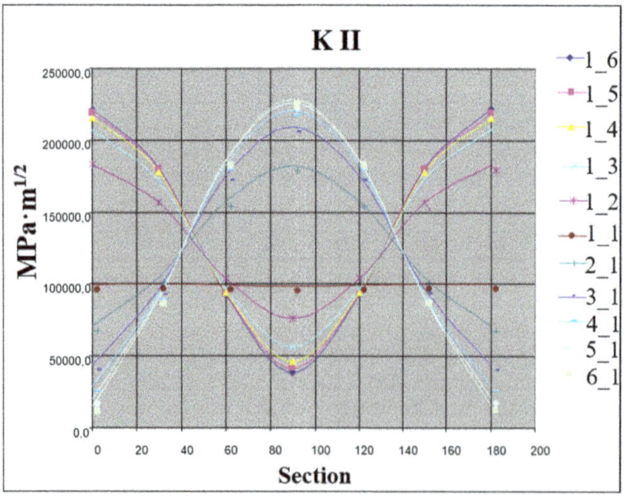

***Fig. 2.** Calculated values of stress intensity factor KII for different ratios of delamination axes relative to the shell axis*

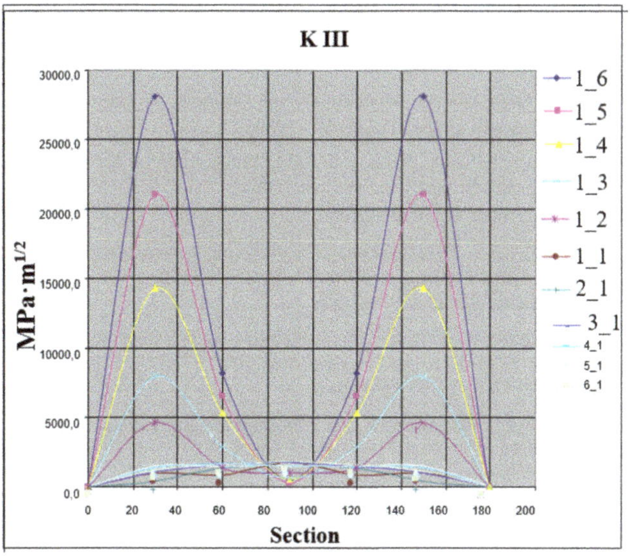

Fig. 3. Calculated values of stress intensity factor KIII at various ratios of delamination axes relative to the shell axis

Figures 1, 2, 3 show the calculated values of stress intensity factors (SIF) KI, KII and KIII for 7 points along the delamination front located every 30 deg from 0 to 90 deg relative to the delamination axis passing along the normal to the plane through its center. The values of stress intensity factors were obtained at various ratios of delamination axes (relative to the shell axis) from 1/6 to 6/1. The obtained calculation results showed that the maximum values of the stress intensity factors relative to others were reached by the KIII-transverse shear crack oriented in the annular direction.

Analysis of the obtained dependences and the previously obtained results [9, 10] enables one to conclude that, all other things being equal, the stress intensity factors SIF II and III prevail, which means that a delamination type defect does not have a significant effect on the bearing capacity of the equipment while it is local and does not spread.

References

[1] Ainsworth, R.A., The Treatment of Thermal and Residual Stresses in Fracture Assessments / Engineering Fracture Mechanics, Vol. 24, No. 1, pp. 65-76, 1986. https://doi.org/10.1016/0013-7944(86)90008-1

[2] Federal Law dated 21.07.1997, No. 116-FL "On Industrial Safety of Hazardous Production Facilities", as amended;

[3] Kaplun A.B. ANSYS in the hands of an engineer. Practical Guide / A.B. Kaplun, E.M. Morozov, M.A. Olfer'eva // M.: Editorial Publ., URSS. - 2003. - 272 p.

[4] Kharlamov I.E., Valeev S.I. Calculation of strength of shell suffering exfoliation / Chemical and Petroleum Engineering, Vol. 54, Nos. 3-4, pp. 278-282, July 2018. https://doi.org/10.1007/s10556-018-0475-2

[5] PNAE G-7-002-86 Standards for calculation of the strength of equipment and pipelines of nuclear power plants. Moscow, ENERGOIZDAT, 1989.

[6] Ponikarova A S, Zotov M A and Salin A A 2020 Management model of sustainable development of petrochemical industry enterprises J. Phys.: Conf. Ser. 1679 052031. https://doi.org/10.1088/1742-6596/1679/5/052031

[7] Shipilov S.A. and May I.L. Structural integrity of againg buried pipelines having cathodic protection//Engineering Failure Analysis.-2006.-13.-pp.1159-1176. https://doi.org/10.1016/j.engfailanal.2005.07.008

[8] Valeev, S.I. Investigation of cylindrical hydrocyclone for emulsion separation / IOP Conference Series: Materials Science and Engineering, 2020,971(5),052007. https://doi.org/10.1088/1757-899X/971/5/052007

[9] Valeev, S.I. Risks during the operation of technological equipment at a hazardous production facility / Journal of Physics: Conference Series, 2020, 1679(4),042028. https://doi.org/10.1088/1742-6596/1679/4/042028

[10] Valeev, S.I., Kharlamov, I.E. Operation of process equipment with crack-like defects / IOP Conference Series: Materials Science and Engineering, 2020, 971(4),042037. https://doi.org/10.1088/1757-899X/971/4/042037

[11] Zotov M A, Ponikarova A S, Kadeeva E N 2020 Balanced management of innovative sustainable development of the petroleum and gas chemical complex IOP Conference Series: Materials Science and Engineering 971(5) 052023. https://doi.org/10.1088/1757-899X/971/5/052023

Investigation of Structure and Physical-Mechanical Properties of Metal in Different Zones of Welded Trunk Pipelines Made of X80 Strength Grade Steel

Evgeniia Putilova[1,a,*], Sergey Zadvorkin[1] and Kristina Kryucheva[1]

[1]Institute of Engineering Science, Ural Branch of the Russian Academy of Sciences (IES UB RAS), 620049, 34 Komsomolskaya st., Ekaterinburg, Russia

[a]tuevaevgenya@mail.ru

Keywords: High-Strength Pipe Steel, Steel Grade X80, Structure, Properties, Non-Destructive Magnetic Testing

Abstract. The development of trunk pipelines network, taking into accont their operation in difficult climatic conditions, leads to the creation and the development of non-destructive testing methods to control the stress-strain state of the pipe itself, as well as welded joints. This work presents the results of studies of the microstructure, as well as the mechanical and magnetic characteristics of the metal cut from different zones of the welded joint of X80 strength class pipe steel.

Introduction

The use of high-strength pipe steels with strength grade X80 and higher, can reduce the metal consumption of main gas pipelines and reduce costs during their construction [1]. Currently, high-strength pipe steels of strength grade X80 are increasingly used in the construction of gas pipelines in areas with harsh climatic conditions and unstable soils, including seismic zones [2-5].

It is well known that most often the centers of destruction of metal structures, including main pipelines, are precisely welded joints, in particular the heat-affected zone (HAZ). Violation of the strength of welded joints of pipelines occurs because of the two main reasons. The first one is the presence of defects. And the second - due to the action of stresses [6-9].

Several zones could be distinguished in welded structures: the base metal, the weld and the HAZ. Differences in their structure, physical and mechanical properties, and the level of residual stresses determine the fact that in the process of manufacturing and operation, the material of different sections of welded structures will react differently to the action of applied loads. In this regard, the development of methods for assessing changes in the stress-strain state in various zones of welded steel structures at the stages of their manufacture and operation is an urgent problem of non-destructive testing.

In this work, we investigated the structure and mechanical and magnetic properties of the metal of various zones of the welded joint of steel of strength grade X80 under the conditions of applied loads in order to determine the possibility of using magnetic methods to assess the changes occurring in the material of welded pipes of large diameter during manufacture and operation.

Materials and methods

The studies were carried out on samples cut along the rolling direction, i.e. along the weld joint, from base metal, HAZ material and weld metal. This pipe steel was produced by controlled rolling method and with the strength grade of X80 (1420 mm diameter and 21.6 mm wall thickness) (according to API standard). Samples cut from the HAZ material were cut at a distance of about 10 mm from the weld. The chemical composition of the investigated materials is given in Table 1.

Materials Research Proceedings **21** (2022) 396-402
https://doi.org/10.21741/9781644901755-67

The appearance of the welded joint is shown in Fig. 1. Dotted lines in fig. 1 these are the lines along which the microhardness distribution was measured. The microhardness HV0.05 of the samples in the cross section was determined on a Shimadzu microhardness tester with a load on the indenter of 0.49 N. The results of microhardness measurements will be shown below on Fig. 4.

To carry out microstructural studies, thin sections were prepared on the cross section of the samples. The microstructure of the obtained samples was investigated using optical and electron microscopy techniques, as well as by electron backscattered diffraction (EBSD analysis). To perform these tasks, a NEOPHOT-21 optical microscope and a Tescan scanning electron microscope equipped with Advanced AZtecHKL diffraction (EBSD) analysis system were used.

Table 1. *Chemical composition of the investigated materials (mas. %)*

Zone	C	Si	Mn	P	S	Cr	Ni	Mo	Nb	V	Ti
Base metal and HAZ metal	0,083	0,21	1,77	0,009	0,001	0,101	0,243	0,144	0,040	0,002	0.009
Weld material	0,077	0,35	1,66	0,011	0,002	0,079	0,286	0,338	0,02	0,003	0.011

Mechanical properties (yield stress σ_Y, ultimate strength σ_U, relative elongation δ and relative contraction ψ) of the materials under the study were determined in accordance with GOST 1497-84 by carrying out uniaxial tensile tests of longitudinal specimens on a Tinius Olsen SL-60 testing machine at room temperature. The tensile elongation of the samples was determined contactlessly using a La Vision video extensometer.

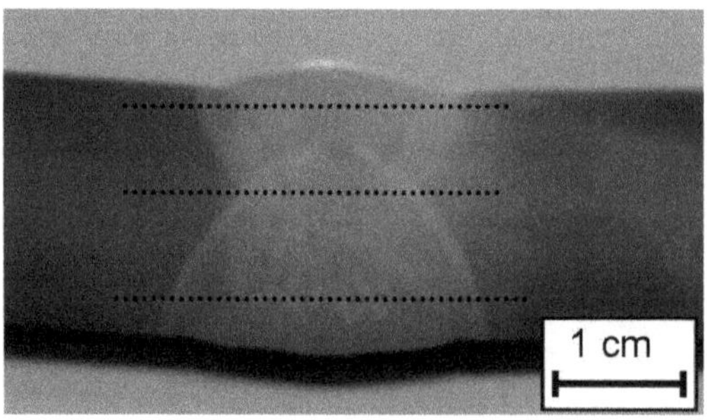

Fig. 1. *The appearance of the investigated welded joint of steel of strength class X80*

The influence of elastic deformation according to the uniaxial tension scheme on the behavior of the magnetic characteristics of the metal of separate zones of the welded joint of steel of strength class X80 was investigated.

The tests were carried out at room temperature on a universal testing machine with a maximum tensile force of 50 kN. Upon reaching a certain degree of deformation, the loading process was suspended without unloading the sample, and magnetic hysteresis loops were recorded using a Remagraph C-500 magnetic measuring complex. Magnetic measurements were carried out in a closed magnetic circuit according to the permeameter scheme. A magnetic field with an intensity of up to 500 A / cm was applied along the axis of the sample. The values of the coercive force H_c, residual induction B_r, and maximum magnetic permeability μ_{max} were determined from the magnetic hysteresis loops. The error in measuring the field and induction did not exceed 3%. Before and after each measurement of the magnetic properties, the sample was demagnetized. By differentiating the descending branches of the magnetic hysteresis loops, the field dependences of the differential magnetic permeability μ_{diff} were obtained.

Results and its discussion

The results of metallographic studies are shown in Fig. 2.

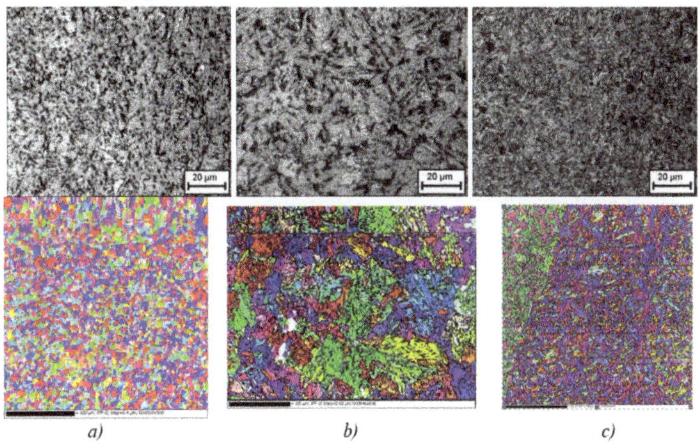

Fig. 2. *Microstructure of steel with the strength grade X80: a –Base metal; b – HAZ metal; c- weld material.*

The structure of the base metal of a pipe made of steel of strength grade X80 consists of a mixture of fine quasi-polygonal ferrite, and globular bainite (see Fig. 2a). As it is shown in [10], to create steel with such a morphology, a fine austenite grain structure should be formed, which is achieved by microalloying of niobium, vanadium, and titanium. Nb, V and Ti leads to the formation of carbides that contribute to precipitation hardening of steel [1]. For steels of strength grades X80-X100, and higher a fine-dispersed structure mostly consists of bainite, sometimes ferrite and bainite [1] is characteristically. And such structure helps to achieve high strength and good cold resistance even at large thicknesses of rolled products. But to get these some conditions have to be met [2, 10]: reduction of carbon content (0.04-0.08%); the introduction of such alloying elements Mn, Mo, Ni, Cr, Cu; complex microalloying with Nb, Ti, V; reduction of the content of harmful impurities (≤0.002% S; ≤0.010% P); the use of accelerated cooling after rolling in controlled modes.

The HAZ material of the studied steel is characterized by the presence of a zone of coarser grain (Fig.2b). The reason for the appearance of a coarse-grained structure immediately adjacent to the fusion line is heating to temperatures at which the majority of carbide particles dissolve, which are effective barriers to inhibiting grain growth [11]. In addition to lath bainite, the structure contains acicular and globular bainite. The weld material (Fig. 2c) of steel of strength class X80 is characterized by greater dispersion in comparison with the base metal and HAZ material, and has such components as acicular bainite, ferrite and martensite-austenitic component. Due to the greater dispersion and differences in the chemical composition (see Table 1) of the weld material, the hardness of the weld material exceeds the values of the hardness of the base metal and the HAZ material (see Table 2). Fig. 2 also demonstrates the results of EBSD analysis of the investigated materials. And it is obvious that all the conclusions made from the optical microscopy is proved by these pictures.

Table 2 shows the mechanical (strength, microhardness and plasticity) and magnetic (coercitive force, remanente, maximal magnetic permeability, and saturation magnetization), characteristics of the metal of various zones of the welded pipe. For a pipe made of steel of strength class X80, the strength characteristics (σ_Y and σ_U) of the weld metal are almost 10% higher than the similar characteristics of the base metal. Different zones of the welded joint of pipes have practically the same value of σ_Y / σ_U, which indicates the equal strength of the welded joint in this material.

In fig. 3 shows the distribution of hardness over the section of the weld. It is clearly seen from the distribution that in the region of the transition from the weld material to the HAZ zone, a decrease in hardness values is observed even below the level of the hardness values of the base metal. The level of strength characteristics for these metals is practically the same.

The level of magnetic characteristics measured in the initial state on all the samples under study is also shown in Table 2. According to classical concepts, the level of magnetic properties correlates well with magnetic parameters [12]. The largest value of the coercive force (and the smallest for the residual induction and maximum magnetic permeability) is observed just for the weld material with the highest values of strength properties.

Table 2. *Mechanical and magnetic properties of investigated materials*

Zone	σ_Y, [MPa]	σ_U, [MPa]	δ, %	σ_Y / σ_U	$HV_{0,05}$	H_c, [A/cm]	B_r, [T]	μ_{max}
Base metal	595	650	14,7	0,91	260	7,1	1,125	744
Weld material	640	710	14,8	0,90	320	10,6	1,063	544
HAZ metal	600	680	14,2	0,88	235	7,1	1,127	760

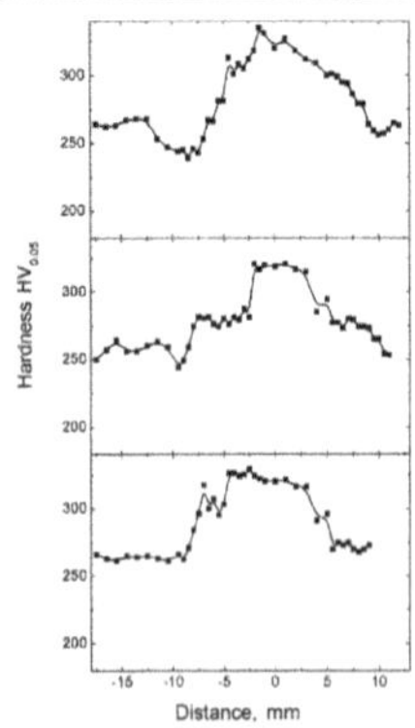

Fig. 3. *The hardness distribution across the weld joint. The measurement lines are shown on Fig.1.*

Fig. 4 shows the field dependences of the differential magnetic permeability of the material of individual zones of the welded joint. It is clearly seen that the peaks from the HAZ material and the base metal are located in the same fields, since they have practically equal strength properties, and the values of their coercive forces coincide. The peak in the field dependence of the differential magnetic permeability, corresponding to the material of the weld of steel X80, lies in stronger fields, possibly due to the greater alloying and dispersion of the structure of this material. There by, the application of field dependences and it's parameters makes it possible to obtain the necessarily information about the state of individual components of the welded joint. If to compare with the information from main characteristics determined from the magnetic hysteresis loop (H_c, B_r and μ_{max}) – they reflect an integral state from the entire analyzed volume.

In fig. 5 shows the dependences of the magnetic characteristics on the level of applied normal stresses during elastic deformation.

As can be seen from Fig. 5, the values of the magnetic characteristics of the materials of the heat-affected zone and the base metal of the steel are approximately at the same level. The values of the coercive force of the weld metal, both in the initial state and in the entire range of applied stresses, exceed the values of the coercive forces for the base metal and HAZ material. Perhaps this is due to the greater dispersion of the structural components of the weld.

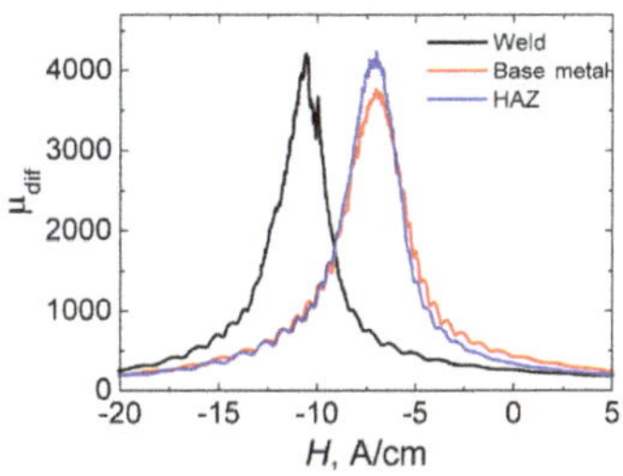

Fig. 4. *Field dependences of differential magnetic permeability of different zone of welded pipe with strength grade X80.*

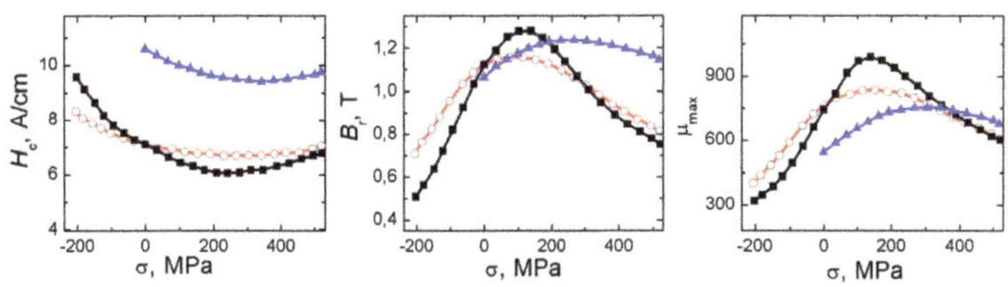

Fig. 5. *The dependences of magnetic characteristics of base metal (■), weld material (▲), and HAZ metal (○) on the normal applied stresses.*

The dependences H_c (σ), B_r (σ) and μ_{max} (σ) of all the samples under study exhibit extrema in the region σ ≈ 150–250 MPa. The experimentally obtained non-monotonic dependence of H_c (σ), B_r (σ) and μ_{max} (σ) in the elastic region can be represented as a result of the action of a number of factors. In particular, stretching of samples in the elastic region leads to the formation of a magnetic stress texture, also called induced magnetic anisotropy [13]. Thus, the studied magnetic characteristics of all investigated materials unambiguously change at normal stress values in the range from -200 MPa to +200 MPa, and these parameters can be used for magnetic control of normal stresses and elastic deformations in trunk pipelines. It should be noted that the established range of applied stresses is several times higher than the working pressure for modern pipelines, which, according to [14], is 10-15 MPa.

Conclusion

It is shown that the weld material of the investigated steel has a greater dispersion in relation to the base metal, and even more so to the HAZ material, where a coarse-grained structure is formed.

It was found that the values of the field of the maximum differential permeability on the field dependences correlate with different structural states and the level of physical and mechanical properties of individual zones of welded joints. This can be the basis for further development of a methodology for diagnosing the deviation of the structure and level of physical and mechanical properties of the material from the required state and identifying the most dangerous zones of the welded joint from the point of view of the possibility of brittle fracture.

At the level of normal stresses from -200 to 200 MPa, not exceeding 0.4 of the yield point, the coercive force, residual induction and maximum magnetic permeability of the metal from all zones of the welded joint behave unambiguously, which makes it possible to use them as parameters for assessing the stress the deformed state of individual zones of welded joints of large-diameter pipes.

Ackowledgement

The research was partially supported by RFBR grant No. 20-08-00392_a. The equipment of the Plastometriya collective use center was used in the research.

References

[1] V.V. Rybin, V.V. Malyshevsky, E.I. Khlusova, V.V. Orlov, Y. Kh. Shakhpazov, Iu. D. Morozov, S. Yu. Nastich, M. Yu. Matrosov, High-strength steels for the main pipelines, Voprosy materialovadenia. 3(59) (2009) 127-137.

[2] Z.P. Zhao, G.Y. Qiao, G.P. Li, et al., Fatigue properties of ferrite/bainite dual-phase X80 pipeline steel welded joints, Sci Technol Weld Join. 22 (3) (2017) 217-226. https://doi.org/10.1080/13621718.2016.1219120

[3] V.V. Rusakova., T.L. Lobanova, Prospects for the use of high-strength pipes of strength category K65 (X80) for long-distance gas transportation projects, Science and technique in gas industry. 1 (2009) 4-7.

[4] W. Qian, T. Han, Y. Wang, et al., Microstructure and mechanical properties of X80 girth welds with different welding processes, Mater Sci Forum. 4502 (1796) (2017) 1079-1087. https://doi.org/10.4028/www.scientific.net/MSF.898.1079

[5] Jie Ning, Zi-ShuYu, Kun Sun, Mei-Juan Hu, Li-Xu Zhang, Yan-Bin Zhang, Lin-Jie Zhang, Comparison of microstructures and properties of X80 pipeline steel additively manufactured based on laser welding with filler wire and cold metal transfer, Journal of Materials Research and Technology. 1-2 (2021), 752-768. https://doi.org/10.1016/j.jmrt.2020.12.021

[6] O.I. Sleptsov, A.V. Lyglaev, A.M. Bol'shakov, S.A. Sintsov, Diagnostics and safety of aging large mechanical systems operating under conditions of the far north: the problem and ways to

solve it, Russian Journal of Nondestructive Testing. 6(44) (2008) 395-402. https://doi.org/10.1134/S1061830908060053

[7] V.A. Kanaikin, V.E. Loskutov, A.F. Matvienko, B.V. Patramanskii, A pig introscope for inspection of main gas pipelines, Russian Journal of Nondestructive Testing. 7(43) (2007) 423-426. https://doi.org/10.1134/S1061830907070017

[8] E.V. Poyarkova, On the issue of solving the problem of increasing the reliability of equipment in the oil and gas industry on the basis of quality control of welded joints, Neftegasovoe delo. 1(6) (2008) 278-283.

[9] D.I. Vichuzhanin, V.A. Khotinov, S.V. Smirnov, The effect of the stress state on the ultimate plasticity of steel X80, Diagnostics, Resource and Mechanics of Materials and Structures. 1 (2015) 73-89. https://doi.org/10.17804/2410-9908.2015.1.073-089

[10] L. Lan, C. Qiu, D. Zhao, et al., Analysis of microstructural variation and mechanical behaviors in submerged arc welded joint of high strength low carbon bainitic steel, Mater. Sci. Eng. A. 558 (2012) 592-601. https://doi.org/10.1016/j.msea.2012.08.057

[11] Y. Ci, Z.Z. Zhang, Simulation study on heat-affected zone of high-strain X80 pipeline steel, J. Iron. Steel Res. Int. 24 (9) (2017) 966-972. https://doi.org/10.1016/S1006-706X(17)30140-1

[12] V.E. Scherbinin, E.S. Gorkunov, Magnetic control of the metal quality, Ekaterinuburg, UB RAS, 1996.

[13] S.V. Vonsovski, Ya.S. Schur, Ferromagnetizm, Leningrad, OGIZ, 1948.

[14] A.V. Kushnarev, E.V. Shekhovtsov, E.S. Kapustina, S.A. Remigo, I.V. Kostenko, Production of pipe steel of strength classes X80 and X70, Steel in Translation. 6(38) (2008) 490-491. https://doi.org/10.3103/S0967091208060168

Materials Research Proceedings 21 (2022) 403-409 https://doi.org/10.21741/9781644901755-68

Automatic Registration of the Nascence and Propagation of Cracks in Electrically Conductive Materials

Vyacheslav M. Kiiko[1,a*], Kirill A. Khvostunkov[2,b] and Konstantin A. Fedotov[2,c]

[1]Institute of Solid State Physics of the Russian Academy of Sciences, Chernogolovka Moscow district, 142432 Russia

[2]Moscow State University, Moscow, 119991 Russia

[a] kiiko@issp.ac.ru, [b]khvostunkov@gmail.com, [c]fedotishn@mail.com

Keywords: Electrically Conductive Specimen, Equipotential Lines, Crack, Conformal Mapping, Automatic Fracture Control

Abstract. A correlation is established between the length of the propagating crack in electrically conductive flat specimens and the change in the field of electric potentials over the surface of the specimen when an electric current is passed through it. The experimental data correspond to the analytical solution in a two-dimensional formulation obtained by means of a conformal mapping. Work can be attributed to the field of converting mechanical values into electrical ones. This transformation simplifies the process of recording mechanical processes and provides a convenient form of their control. The work is aimed at automatic control of the process of fracture of electrically conductive materials.

Introduction

At present, theoretical and experimental methods of fracture control are being actively developed, for example, based on the analysis of changes in the electromagnetic field [1] or lines of equal potential when passing an electric current through the material [2].In principle, such methods of converting mechanical quantities into electrical ones are capable of providing automatic control of the process of mechanical testing of specimens for fracture toughness and control over the occurrence and propagation of cracks in structural elements operating under load.

This work is aimed at calculating and automatic registration of the occurrence and position of a crack in electrically conductive materials or materials with such a coating. The combination of analytical solutions and automatic crack registration is the main difference from the existing techniques in this area [3–6]. The basis of the work is to establish a one-to-one correspondence between the position of the crack and the geometry of the distribution of equipotential lines in the specimen when an electric current is passed through it. The crack is assumed to be rectilinear, inclined at an arbitrary angle to the boundary of the strip (hereinafter referred to as "sample") made of conductive material. The geometry of the equipotential lines does not depend on the electrical resistance of the material and the magnitude of the current passed through it, since these quantities are not included in the Laplace equations describing the positions of the equipotential lines [2]. The technique does not require preliminary experimental calibration of the crack position; it is provided by calculations performed on the basis of analytical solutions by the method of conformal mappings. Experimental modeling was carried out on sheet thermally split graphite, and laboratory tests – on sheet material made of aluminum alloy.

Analytical solution

Consider a rectangular strip $L \times H$, of electrically conductive material, located with its extended side L along the axis x, containing a rectilinear edge crack directed at an angle α to the axis x and with length *l*, *l* <<*H*<<*L*. Assume that the crack size *l* is small compared with the characteristic width of the strip *H*. It is known that the crack increases in its length. It is necessary to determine the dependence of the change in the geometry of equipotential lines during the growth of the crack at a constant value of the potential difference at the ends of the strip (Fig. 1a, Fig. 1b).

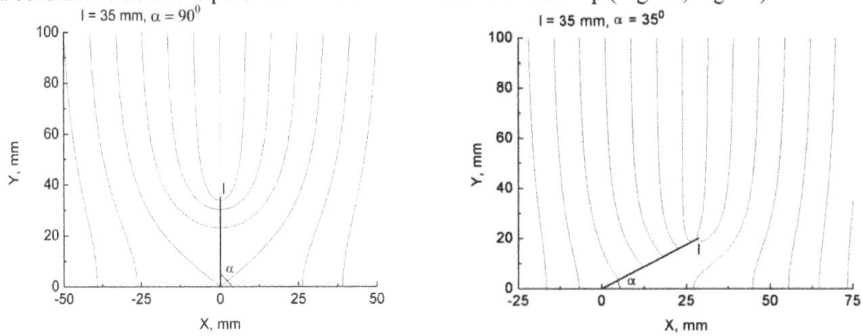

(a) (b)

Fig. 1. *(a) – the case of a crack perpendicular, α = 90º, to the sample boundary, (b) – the case of an arbitrary inclination α of the crack to the boundary.*

Due the *l* <<*H*, we will assume that the equipotential lines at the upper boundary of the band are parallel and we will further consider the half-plane. As in the case of the perpendicular crack in Fig. 1a [2], we will use the conformal mapping method.

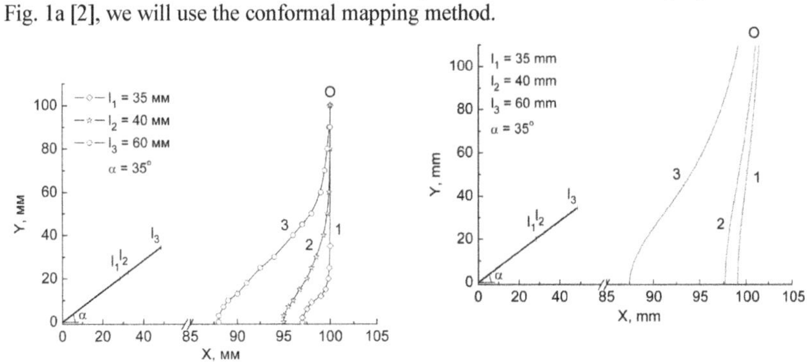

(a) (b)

Fig. 2. *(a) – experimentally obtained equipotential lines passing through point O (100, 100), at different lengths l of an oblique crack in a specimen of sheet thermally split graphite 100 mm vertically wide, 320 mm horizontally long and 1.5 mm thick; (b) – calculated lines, O (100, ∞).*

The conformal mapping, which translates the half-plane into our desired half-plane with the crack, has the form [7, 8]: $w = l(z-1)^{\frac{\alpha}{\pi}}\left(1-\frac{\alpha z}{\pi-\alpha}\right)^{1-\frac{\alpha}{\pi}}$, $w = x+iy,\quad z = \xi + i\eta$. It is known that the solution of the Laplace equation for equipotential lines is invariant under conformal mapping. Fig. 2a and Fig. 2b present examples of the behavior of the fixed line as the crack length increases.

The theoretical part of the work consists in determining the velocity of the point N along the axis x at the level of the upper edge of the sample (Fig. 3b, Fig. 4a) belonging to the equipotential line passing through a predetermined fixed point M, (Fig. 4a). For this purpose, let us define the real and imaginary parts of mapping (Eq. 1):

$$x(\xi,\eta) = l\rho_1^{\frac{\alpha}{\pi}}\rho_2^{1-\frac{\alpha}{\pi}}\cos\left(\frac{\alpha}{\pi}\theta_1+\left(1-\frac{\alpha}{\pi}\right)\theta_2\right),\quad y(\xi,\eta) = l\rho_1^{\frac{\alpha}{\pi}}\rho_2^{1-\frac{\alpha}{\pi}}\sin\left(\frac{\alpha}{\pi}\theta_1+\left(1-\frac{\alpha}{\pi}\right)\theta_2\right)$$
$$\rho_1 = \sqrt{(\xi-1)^2+\eta^2},\ \theta_1 = arctg\left(\frac{\eta}{\xi-1}\right),\ \rho_2 = \frac{\alpha}{\pi-\alpha}\sqrt{\left(\frac{\pi}{\alpha}-1-\xi\right)^2+\eta^2},\ \theta_2 = arctg\left(\frac{\eta}{\frac{\pi}{\alpha}-1-\xi}\right). \tag{1}$$

Each equipotential line is parameterized by η as a one-parameter curve with an individual number $\xi = Const$ .Then it is necessary to determine which curve passes through point M for a given crack length l, and which coordinate x it will have at the level of $y = H$. From the condition that the equipotential line passes through both points M and N, we find the position along x at the level of $y = H$.

$$tg\left(\frac{\alpha}{\pi}arctg\left(\frac{\eta}{\xi-1}\right)+\left(1-\frac{\alpha}{\pi}\right)arctg\left(\frac{\eta}{\frac{\pi}{\alpha}-1-\xi}\right)\right) = A,\quad A = \frac{y_M}{x_M}$$
$$\left(\frac{l}{x_M}\right)^2\left(\sqrt{(\xi-1)^2+\eta^2}\right)^{2\frac{\alpha}{\pi}}\left(\frac{\alpha}{\pi-\alpha}\sqrt{\left(\frac{\pi}{\alpha}-1-\xi\right)^2+\eta^2}\right)^{2-2\frac{\alpha}{\pi}} = 1+A^2 \tag{2}$$

In order to solve the system (Eq. 2), let us assume specific values for the positions of control points at the outer edge and at a distance H from it, $\eta_M = 0,\quad \eta_H = 1,\quad A = 0,\quad a = \frac{\pi}{\alpha}-1$

$$\left(\frac{l}{x_m}\right)^{a+1} = \frac{1}{(\xi-1)\left(1-\frac{\xi}{a}\right)^a},\quad \left(\frac{x}{x_m}\right)^2 = \left(\frac{l}{x_m}\right)^2\left((\xi-1)^2+1\right)^{\frac{1}{a+1}}\left(\left(1-\frac{\xi}{a}\right)^2+\left(\frac{1}{a}\right)^2\right)^{\frac{a}{a+1}} - \left(\frac{H}{x_m}\right)^2.$$

On the interval of strict monotonicity at $\xi \in (1, \frac{2a}{a+1})$ the dependence $(l/x_m)^{a+1} = (\xi-1)^{-1}\left(1-\frac{\xi}{a}\right)^{-a}$ is reversible and we can express the number of the equipotential line passing through the x-axis as a function of the dimensionless crack length: $\xi = f\left(l/x_m\right)$. In each special case, the function f can be obtained numerically, and in the case of $a = 2$, (it corresponds to a crack slope angle of 60 degrees), it can be calculated analytically by the Ferrary-Cardano method.

As a result, we obtain the dependence of the coordinate of the second end of this equipotential line at distance H from the bottom edge:

$$\frac{x}{x_m}=\frac{l}{x_m}\left\{\left[\left(f\left(\frac{l}{x_m}\right)-1\right)^2+1\right]^{\frac{1}{a+1}}\left[\left(1-\frac{1}{a}f\left(\frac{l}{x_m}\right)\right)^2+\left(\frac{1}{a}\right)^2\right]^{\frac{a}{a+1}}-\left(\frac{H}{l}\right)^2\right\}^{\frac{1}{2}}.$$

Experiments

As already mentioned, the distribution of electric potentials in the specimen is established by an analytical solution based on the use of a conformal mapping in a two-dimensional formulation, and the calculation results are compared with the experimental data (Fig. 2, Fig. 3). This opens up the possibility of establishing a one-to-one correspondence of the geometry of the developing crack with the geometry of the distribution of electric potentials and automatic control of experiments related to the propagation of a crack in the material of the specimen [2], and, in the future, in real structure elements.

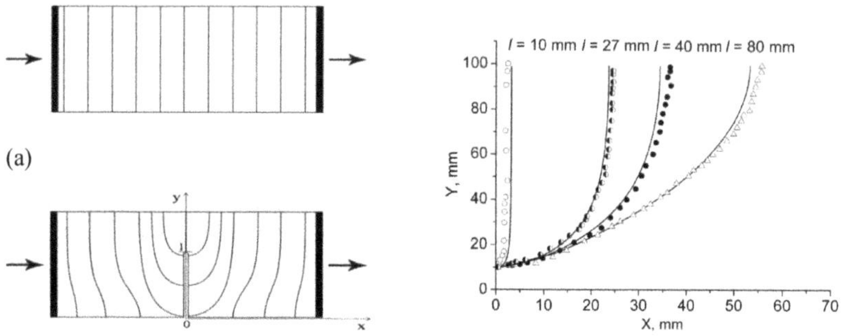

(b) (c)

Fig. 3. Scheme of the arrangement of equipotential lines over the surface of a flat specimen when passing through it an electric current, shown by arrows: (a) – a specimen without a crack, (b) – with a lateral crack of length l, perpendicular to the lower edge of the specimen; (c) – the position of the equipotential lines passing through the point (0.10) – in millimeters, – depending on the lengths of the lateral crack (indicated in the graph field) with the specimen width (along the height of the figure) 100 millimeters: solid lines – the results of the analytical solution, " points " – determination of the position of equipotential lines experimentally.

The scheme of tracking the change in the coordinate x of the "exit" of the equipotential line passing through the fixed point of contact M located at the beginning of the crack to the upper edge of the sample is shown in Fig. 4a. A fixed measuring contact is placed at point M. By moving the movable measuring contact along the upper edge of the specimen, its position is found at which the voltage between the points M and N (x_1) is zero, which is controlled by the readings of the voltmeter V. In this case, points M and N will be on the same equipotential line. An increase in the crack length from l_1 to l_2 by Δl leads to a change in the geometry of the potential field and, therefore, the equipotential line passing through the point M will take the position MN′. On the edge there will be a new point N′, different in position from point N, such that there will be no voltage between M and N′. The new position N′ (x_2) of the movable contact is found by moving the contact from point N to Δx along the edge of the specimen. Thus, a one-to-one correspondence is established between the increment in the crack length $\Delta l = l_2 - l_1$ and the displacement of the contact $\Delta x = x_2 - x_1$. Moving a contact can be done either manually or automatically. In Fig. 4b shows the results obtained manually, automatically, and by calculation.

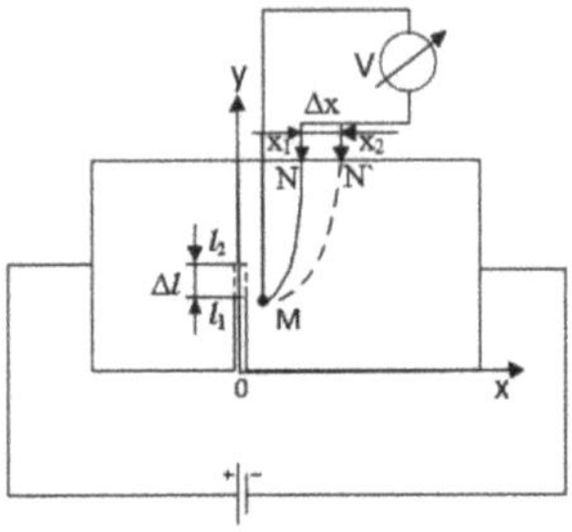

(a)

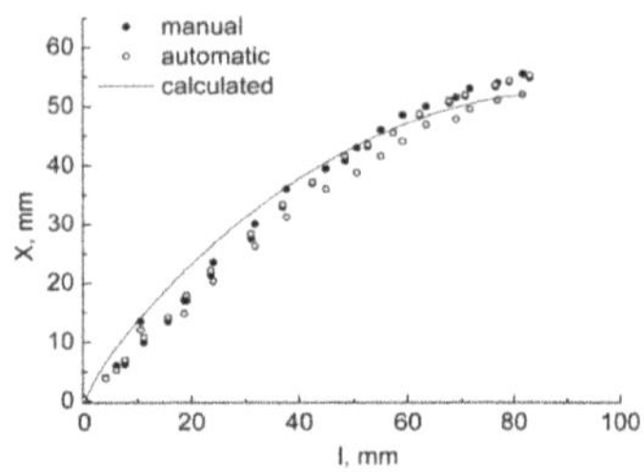

(b)

Fig. 4. *Scheme for converting the value of the growth of the crack length Δl into the movement of the movable contact Δx (a); (b) – dependences of the displacement x of the movable measuring contact on the length of the lateral crack l, obtained manually, automatically and by calculation.*

The dependencies shown in Fig. 4b are, in essence, calibration.

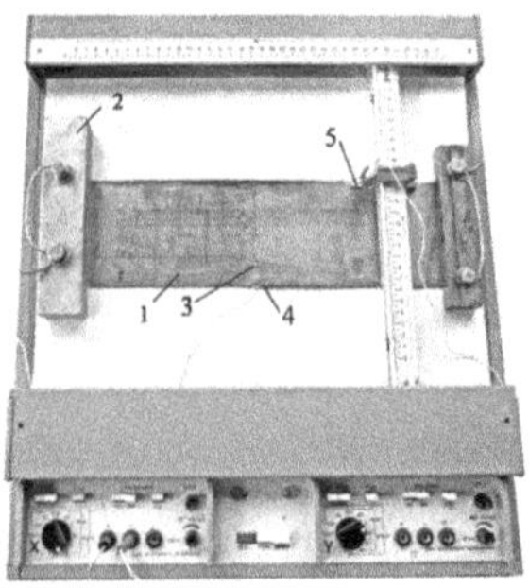

(a)

(b)

Fig. 5. *A device for carrying out model experiments related to determining the distribution of electrical potentials over the surface of the specimen (a); (b) – electro-mechanical part of a laboratory device for automatic recording of the crack length. 1 –specimen, 2 – current leads, 3 – crack, 4 – fixed contact, 5 – movable contact, 6 – reversible electric motor.*

The experimental obtaining of "points" on the graph (Fig. 4b) was carried out in a model version on a strip of sheet thermally split graphite. A direct current was passed through the sample using current leads 2 (Fig. 5a) providing a uniform current distribution over the ends of the specimen. The lateral crack was simulated by cut 3."Manually" the position of the movable contact was determined with a caliper, and the corresponding points are plotted on the graph in Fig. 4b. The automatic device for experiments is based on an industrial two-coordinate recorder H307, adapted for the tasks under consideration. Fixed contact 4 was fixed at the beginning of the crack, through which the initial line of equal potential passed. The other "end" of this line went out to the upper (according to Fig. 1 – 4a) edge of the specimen. Its position was also determined automatically. The voltage supplied from the fixed - movable contacts to the input "x" of the recorder controlled the movement of the recorder carriage with the movable contact fixed on it and brought it by the modified recorder circuit to the equipotential line passing through the fixed contact, every time the crack length changed. A specially added multi-start potentiometer, connected with the carriage movement mechanism, included in the bridge electrical circuit, made it possible to obtain at the output of the system electrical voltages corresponding to the position of the carriage. These voltages are a function of the crack length. Carriage positions in the form of "points" are shown in Fig. 4b ("automatic" measurements).

Similar to the model version, (Fig. 5a) a laboratory device was manufactured (Fig. 5b) for automatic measurement of the crack length in sheet specimens made of conductive materials or with such a coating. Tests with automatic registration of the fatigue crack length in a specimen of aluminum alloy D16 are shown in Fig. 5b.

Summary

In this work, solutions of the tasks of the distribution of equipotential lines in flat specimens by passing an electric current through them are obtained in a two-dimensional formulation. Analytical solutions are based on conformal mappings for specimens with different orientations of edge cracks.

The distributions of equipotential lines were also obtained experimentally on similar specimens of electrically conductive thermally split graphite on a specially made simulator.

The numerical results obtained on the basis of analytical solutions and the results obtained experimentally coincide with the accuracy acceptable for practice. These coincident results make it possible to abandon the procedure for calibrating specimens before testing.

Developed, manufactured and used one of the possible electronic-mechanical device for automatic registration of the crack length in conductive materials or with such a coating. Laboratory tests with a developing fatigue crack were carried out on sheet D16 aluminum alloy.

Acknowledgments

The work was carried out within the framework of State Assignment No. 0028-2019-0020. Also the authors are grateful to R K Odzhaev and D G Pizhenin for help in computational and experimental work.

References

[1] Ni Qing-Qing, J. Hong, P. Xu, Xu Zhenzhen, K.A. Khvostunkov, H. Xia, Damage detection of CFRP composites by electromagnetic wave nondestructive testing (EMW-NDT), J.Composites Science and Technology. 211 (2021) 108839. https://doi.org/10.1016/j.compscitech.2021.108839

[2] V.M. Kiiko, K.A. Khvostunkov, R.K. Odzhaev, D.G. Pizhenin, S.A. Abashkin, Automatic registration of the length of the propagating crack in conductive materials, in V.V. Rubanik (Ed.), Actual problems of strength, Publishing “Pobeda”, Molodechno, 2020, pp. 251-253. (in Russian).

[3] F.R. Sosnin, K.V.Podmasterev,Unbrakable control, Moscow, 2004. (in Russian)

[4] P.N. Shkatov, P.I. Chernenko, Measurement of the depth and angle of inclination of surface cracks using the electropotentialmethod, Vestnik MGUPI, 44 (2013). (in Russian)

[5] L.N. Litvinov, F.G.Tsipushtanov, V.A. Parkov, V.N. Bakumov,Electropotential Crack Depth Gauge, Patent RF, 1408205 (1988). (in Russian)

[6] D.A.Ozetskiy, V.V. Tsvelev, Method for measuring crack propagation in conductive specimens, Patent RF, 1834491 (1966). (in Russian)

[7] M.A. Lavrentiev, B.V. Shabat, Methods of the theory of functions of a complex variable, Moscow, 2002. (in Russian)

[8] Y.V. Sidorov, M.V. Fedoryuk, M.I. Shabunin, Lectures on the theory of functions of a complex variable, Moscow, 1989. (in Russian)

Investigation of the Operational and Technological Reliability of the Small-size Internal Threading Process

Aleksander Kharchenko[1,a*], Andrey Kharchenko[1,b] and Ekaterina Vladetskaya[1,c]

[1] Sevastopol State University, Universitetskaya St., 33, Sevastopol, Russian Federation

[a] khao@list.ru, [b] a.a.kharchenko@sevsu.ru, [c] vladetska@rambler.ru

Keywords: Operational and Technological Reliability, Internal Small-Sized Thread Processing, Technological Elements of the Module, Flexible Manufacturing Module (FMM) for Thread Processing, Simplest Flow, Markov Chain, Probability of Failure-Free Operation

Abstract. The article presents the results of research in the direction of improving technological equipment for internal threading of parts in a flexible automated production. Methods for assessing the operational and technological reliability of a flexible manufacturing module (FMM) of thread processing are considered, which can be used as the basis for the developed method of synthesis of its elements. It is proposed to consider the technological system of the flexible manufacturing module (FMM) of thread processing as a system in which transitions from state to state occur under the action of the simplest flows with parameters of the transition probabilities of a continuous Markov chain. The developed mathematical model describing the state of the FMM taking into account the failures of the functioning of its elements, parametric failures, as well as taking into account the recovery after these types of failures, makes it possible to reflect the influence on the operation of the module of the parameters of the flows of failures and restorations of the tool, machine tool, fixture, loading device. The solution of the obtained systems of equations of final probabilities allows for given (or experimentally obtained) intensities of failure streams (functioning and parametric) for FMM threading to obtain the values of the probability of failure-free operation, as well as the probabilities of finding the system in an inoperative state due to corresponding failures. The measures taken make it possible to solve the synthesis problem at the level of structural and layout optimization, so that at the stage of parametric synthesis to determine the elements that are vulnerable according to the developed criterion, the improvement of which will lead to the creation of the most efficient system.

Introduction

The creation of effective technological equipment for flexible manufacturing modules (FMM) for automated thread processing based on such modern technological elements as a machine tool, a tool, a fixture, a loading device should be carried out at the level of new requirements and is associated with obtaining an economic effect by increasing processing productivity and improving the quality of products. When creating elements of flexible manufacturing modules (FMM) for small-sized internal threading, one should take into account the effect of failure streams (parametric and functional) on the output of the system. It is necessary to describe the process of formation of indicators of operational and technological reliability as a result of their interaction in the real conditions of the operation of threading equipment. The impact of operational reliability on productivity and economic efficiency is assessed by introducing the availability factor or the technical utilization factor [1, 2]. Obtaining this estimate is of practical interest, but it does not take into account all aspects related to parametric failures of various elements.

In [2], a performance model is proposed based on taking into account sudden failures of elements of a technological system, assuming Poisson reliability laws, and an exponential distribution of the recovery time. At the same time, steady-state probabilities exist and the known results of the mathematical theory of queuing were used to determine them [3]. Such an approach to the construction of mathematical models is useful at the preliminary stage of system synthesis, but for assessing the efficiency of the FMM at the last stages of parametric synthesis, such an indicator is clearly insufficient due to the assumptions about the homogeneity of the system elements in terms of failures and maintenance.

Methods for analytical assessment of the technological reliability of elements are usually based on mathematical models of changes in processing accuracy over time [4, 5]. Models [1, 6] have been developed for the analysis and forecasting of reliability, which creates the possibility of finding ways to reduce the influence of errors on the change in reliability and the impact on it not only by organizational, but also by design methods. It is advisable to use this approach in the synthesis of FMM for thread processing at the stage of assessing the intensity of parametric failures of technological elements by choosing the main types of initial errors. For the processes of cutting internal threads, it is necessary to determine the main parameters of processing accuracy, to develop a methodology for their assessment.

Studies of the technological reliability of thread-cutting equipment are given in [7], where a hypothesis about the mechanism of size formation when working with a tap is proposed, according to which the profile on the product is formed as a result of the "transfer" of the tool size during cutting. The author theoretically and experimentally proved that at the achieved level of accuracy of the technological system, cutting of precise threads can be performed if it is possible to create optimal designs of taps, tap holders, bearing spindles and the machine tools themselves. The approach used by researchers [8, 9] allows predicting the accuracy of threaded holes when processing various materials and developing practical recommendations. The methods for determining the precision parameters of the thread can be taken as a basis for determining the characteristics of the technological reliability of the FMM elements for thread processing.

Studies of the operational reliability of threading equipment [10, 11, 12] have shown that as a result of breakage, 50...70% of cutting taps with sizes M2...M6 fail. Ranking by the specific weight of failures indicates that tools are in the first place, then, respectively, mechanical units, electrical and hydraulic equipment [13]. As measures to improve the reliability of the tool, it is necessary to increase its durability, strength, pay more attention to the choice of its design, operating modes. Currently, many designs of taps and cutting patterns are known [14]. A number of studies of the process of thread rolling with chipless, plastic deforming taps (PDT) have been carried out. The difference between cutting and rolling is that in the first case, the thread is formed by cutting out the metal, and in the second – by extrusion, that is, without the formation of chips, which is one of the advantages of the rolling method, especially when processing blind threads. Research and development [15], as well as standards for chipless taps, based on the results of [11], played a significant role in the popularization and implementation of the new method. Contradictory opinions can be found on the issue of PDT reliability. Some authors do not get an increase in the durability of a rolling tool in comparison with a cutting one, while others get an increase of 10 ... 20 times. Studies [12] found that thread rolling is the more efficient, the smaller the thread size. For example, when rolling M3...M4 threads in workpieces made of viscous materials, it is possible to obtain a tenfold increase in tool life, but with an increase in diameter, the difference in resistance sharply decreases and for M10...M12 threads, no advantages in durability are observed.

Thus, the recommendations of the researchers are mainly intended and effective for narrow areas, which makes it difficult to use them in the new conditions that are characteristic of flexible automated production. These features should be taken into account when creating PDT elements for internal threading.

Main part

FMM threading is a technical dynamic system, as it has the following features: the module consists of separate interacting subsystems; the state of the subsystems and the module as a whole changes over time; each subsystem and module as a whole is characterized by its current state, history of development [16, 17].

On the basis of the assumptions made, to assess the options for the selected structures, it is necessary to consider the features of the occurrence of operational failures and parametric failures, that is, the operational and technological reliability of the module. Each element of the technological system can have different recovery rates μ, as well as different failure rates λ, therefore, different reliability. In this case, a more complex element with the same λ can have a lower value of μ (lower reliability), or with the same μ a larger value of λ than a less complex element. This must be taken into account when choosing the criteria for evaluating the effectiveness of the module.

Considering the threading module as a discrete system, let us assume that the elements fail independently of each other, and each element is restored after failure. We will assume that the initial properties of an element do not in any way affect the reliability of other elements. The moments of failure of each element form the recovery process, and by virtue of our assumptions, these processes are independent. Let us denote by $F_k(t)$ the distribution law of the lifetime of the k-th technological element. We assume that these laws have a continuous density $f_k(t)$ and that there is an average lifetime of a technological element T_k and its variance σ_k^2.

Since the elements in the module system are connected in series (in terms of reliability), the failure of any element causes the system to fail. The appearance of failures in one section of time does not change the probability of occurrence of any number of failures in another section that does not intersect with the first. From these physical considerations it follows that under the above assumptions, there should be no aftereffect in the flow of failures of the technological system of the module. In addition, we assumed above that the distribution laws $F_k(t)$ have continuous densities. Hence it follows that $H(t)$ – the average number of failures – is continuous and the flow of failures of the module system is ordinary, that is, the probability of two failures occurring simultaneously is zero. In queuing theory, it is proved that an ordinary flow with a continuous function $H(t)$, in which there is no sequence, is the simplest non-stationary flow. This means that the probability of occurrence of n failures in any section (t_1, t_2) is expressed by the formula

$$P_n(t_1,t_2) = \frac{[H(t_2)-H(t_1)]^n}{n!} e^{-[H(t_2)-H(t_1)]} \quad (1)$$

and for any system of non-overlapping intervals, the events consisting in the appearance of a given number of failures at each interval are independent. So, it has been shown that under fairly general conditions the flow of failures of the technological system of the module will be the simplest non-stationary flow. It can be seen from the previous formula that the number of failures at each interval is distributed according to Poisson's law (Poisson flow with a variable parameter). This variable is the failure rate $\lambda(t)$. Then the formula Eq. 1 can be written as:

$$P_n(t_1,t_2)=\frac{[\int_{t_1}^{t_2}\lambda(t)dt]^n}{n!}e^{-\int_{t_1}^{t_2}\lambda(t)dt}. \quad (2)$$

If on the segment (t, $t+\tau$) the function $\lambda(t)$ is approximately linear, then a good agreement for these probabilities gives the equality:

$$Pn(t_1,t_2)\approx\frac{[\lambda(t+\frac{\tau}{2})\tau]^n}{n!}e^{\lambda(t+\frac{\tau}{2})dt}. \quad (3)$$

If the function $\lambda(t)$ on this segment is approximately constant, $\lambda(t)\approx\lambda$, then

$$P_n(t_1,t+\tau)\approx\frac{(\lambda\tau)^n}{n!}e^{-\lambda\tau}. \quad (4)$$

Since $\lambda(t)$ almost does not change in the sections between adjacent failures, the average time until the first failure of a technological element is expressed by the approximate formula:

$$\overline{Tp}\approx 1/\lambda(t).$$

As a rule, the operation of the FMM technological system continues for a long time, as a result of which the elements (machine, tool, fixture, loading device) can fail several times. So, the tool in the process of weekly operation fails on average 120 ... 140 times. Under these conditions, the recovery processes become stationary and, therefore, the flow of system failures is also steady-state.

Since the recovery densities for each element tend to the limits $\lim_{t\to\infty}\mu_k(t)=1/\overline{Tk'}$, the failure rate of the system has a limit $\lim_{t\to\infty}\lambda(t)=\lambda_0=\sum_{k=1}^{m}1/\overline{Tk'}$, where m is the number of technological elements. In this case, the flow of the system becomes a Poisson flow with a constant parameter, that is, the simplest flow. For this flow, the probability of occurrence of n failures in a section of time of duration τ does not depend on the position of this section and is expressed by the formula:

$$P_n(\tau)=\frac{(\lambda_0\tau)^n}{n!}=e^{-\lambda_0\tau}. \quad (5)$$

In particular, the probability of failure-free operation of the module system during time τ is equal to $P_0(\tau)=e^{-\lambda_0\tau}$, therefore, the mean time of failure-free operation is $\bar{T}=1/\lambda_0$. Since the number of failures occurring in a given time interval has a Poisson distribution, the time between failures has an exponential distribution, that is, the reliability of each technological element in the module system obeys an exponential law:

$$F_k(t)=1-e^{-\lambda_k t}. \quad (6)$$

The recovery process for each element will form the simplest flow and the module system recovery flow as the sum of the simplest flows will also be the simplest flow.

Materials Research Proceedings **21** (2022) 410-420 https://doi.org/10.21741/9781644901755-69

The above arguments and reasoning are quite fair when considering the operational reliability (failures of functioning) of the technological system of FMM thread processing. To analyze the technological reliability (parametric failures) of the system, it is necessary to take into account the effect of such factors as cutting forces (fast processes), wear of the tool, fixture, loading device (medium speed processes), wear of the table guides, spindle bearings, fixing elements of the fixture (slow processes).

Taking into account the above arguments, we will consider the technological system of FMM threading as a physical discrete system that can be in the states S_0, S_1, S_2, S_3, S_4. State S_0 characterizes the normal functioning of the module in the absence of failures.

In the S_1 state, the replacement of a failed tool is required due to the flow of failures of its functioning with the intensity λ_1. In this case $\lambda_1=1/\overline{T\text{и}}'$, where $\overline{T\text{и}}'$ is the average time between two failures of functioning, *min*. The S_2 state is characterized by the failure of the machine (including the die head) due to the flow of failures with the intensity λ_2. In this case, $\lambda_2=1/\overline{Tc}'$, where $\overline{Tc}'$ is the average time between two failures in the operation of the machine, *min*.

In the S_3 state, it is necessary to repair the broken device due to the flow of failures with the intensity λ_3. In this case $\lambda_3=1/\overline{T\text{п}}'$, where $\overline{T\text{п}}'$ is the average time between two failures of the device functioning, *min*.

State S_4 is characterized by the failure of the loading device due to the flow of failures of functioning with intensity λ_4. In this case $\lambda_4=1/\overline{T\text{з}}'$, where $\overline{T\text{з}}'$ is the average time between two failures in the functioning of the loading device, *min*.

The quantities $\overline{T\text{и}}'$, $\overline{Tc}'$, $\overline{T\text{п}}'$, $\overline{T\text{з}}'$ are determined by the general formula:

$$\bar{T}' = \sum_{j=1}^{m} t_j/m\,, \qquad (7)$$

where m is the number of failures in the functioning of the corresponding technological element of the FMM threading; t_j – time interval between $(j-1)$ -th and j-th orders.

Each technological element of the module undergoes recovery immediately after failure. The recovery rate, respectively, for each technological element will be μ_1, μ_2, μ_3, μ_4, assuming that the recovery time, being a random variable, obeys Poisson's law.

In this case $\mu_1=1/\overline{T\text{ви}}'$, $\mu_2=1/\overline{T\text{вс}}'$, $\mu_3=1/\overline{T\text{вп}}'$, $\mu_4=1/\overline{T\text{вз}}'$, where $\overline{T\text{ви}}'$ – average recovery (replacement) time of the tool, *min*; $\overline{T\text{вс}}'$ – average time of restoration (repair) of machine elements, including the threading head, *min*; $\overline{T\text{вп}}'$ – average recovery (repair) time of the device, *min*; $\overline{T\text{вз}}'$ – average recovery (repair) time of the boot device, *min*.

To describe the labeled graph of states of the technological system of FMM threading, taking into account the failures of functioning (Fig. 1), we denote:

P_0 – is the probability of the correct operation of the module, $P_0=P\ (S_0)$;

P_1, P_2, P_3, P_4 – respectively, the probability of breakage (failure) of the tool, machine elements and threading head, fixture, loading device.

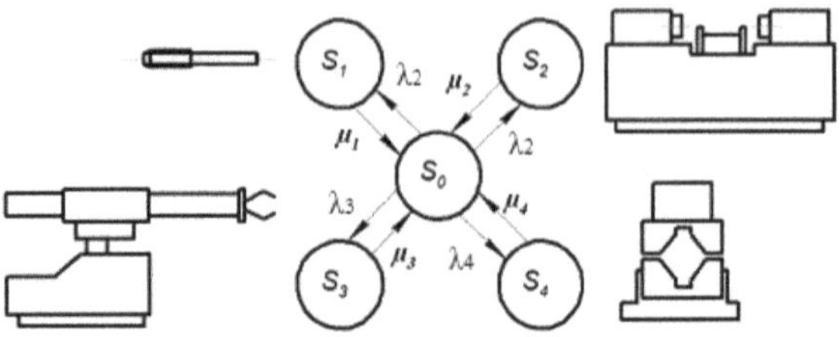

Fig. 1. *The marked graph of states of the technological system of the FMM thread processing, taking into account the failures of functioning*

A. Kolmogorov's equations for the final probabilities P_0, P_1, P_2, P_3, P_4 of states S_0, S_1, S_2, S_3, S_4 are represented as a system of equations:

$$\left.\begin{aligned} &P_0(\lambda_1+\lambda_2+\lambda_3+\lambda_4)=P_1\mu_1+P_2\mu_2+P_3\mu_3+P_4\mu_4\\ &P_1\mu_1=P_0\lambda_1\\ &P_2\mu_2=P_0\lambda_2\\ &P_3\mu_3=P_0\lambda_3\\ &P_4\mu_4=P_0\lambda_4^{'} \end{aligned}\right\} \qquad (8)$$

Solving the resulting system of equations using the normalization condition $P_0+P_1+P_2+P_3+P_4=1$ allows one to obtain expressions for the probability of failure-free operation (P_0), as well as the probabilities of the system being in an inoperative state due to corresponding breakdowns (failure) of the tool (P_1), machine elements (P_2), fixture (P_3), loading device (P_4):

$$\left.\begin{aligned} &P_0=(1+\lambda_1/\mu_1+\lambda_2/\mu_2+\lambda_3/\mu_3+\lambda_4/\mu_4)^{-1}\\ &P_1=P_0\lambda_1/\mu_1\\ &P_2=P_0\lambda_2/\mu_2\\ &P_3=P_0\lambda_3/\mu_3\\ &P_4=P_0\lambda_4^{'}/\mu_4 \end{aligned}\right\} \qquad (9)$$

The technological system of FMM threading as a discrete system, due to parametric failures resulting from tool deformations, positioning errors of the loading device, fixture, spindle runout, changes in the temperature of the machine tool and control device elements, loses serviceability, not ensuring machining accuracy.

FMM as a result, it can be in the states S_{00}', S_{10}', S_{20}', S_{30}', S_{01}', S_{02}', S_{03}', S_{11}', S_{21}', S_{31}', S_{12}', S_{22}', S_{32}', S_{13}', S_{23}', S_{33}'.

State S_{00}' characterizes the correct operation of the module in the absence of parametric failures.

In state S_{10}', parametric tool failures with intensity λ_{01} are observed. In this case $\lambda_{01}=1/\overline{Tи}''$, where $\overline{Tи}''$ is the average time between two parametric tool failures, *min*.

In state S_{20}', inspection and readjustment of the machine, the die head is required due to the flow of failures with the intensity λ_{02}. In this case, $\lambda_{02}=1/\overline{Tc''}$, $\overline{Tc''}$ is the average time between two parametric machine failures, *min*.

State S_{30}' is characterized by the simultaneous presence of parametric failure streams for tool λ_{01} and machine tool λ_{02}.

In state S_{01}', inspection and readjustment of the device is required with the intensity of incoming parametric failures λ'_{01}. In this case $\lambda'_{01}=1/\overline{T\text{п}''}$, where $\overline{T\text{п}''}$ is the average time between two parametric failures of the device, *min*.

The state S_{02}' is characterized by the presence of parametric failures of the loading device due to the flow with the intensity of requests λ'_{02}. In this case $\lambda'_{02}=1/\overline{T\text{з}''}$, where $\overline{T\text{з}''}$ is the average time between two parametric failures of the loading device, *min*.

In the S_{03}' state, inspection and readjustment of the device and the loading device is required due to the flows of incoming parametric failures with intensities λ'_{01} and λ'_{02}.

State S_{11}' is characterized by the simultaneous presence of streams of parametric failures of the tool with the intensity λ_{01} and the fixture with the intensity λ'_{01}.

In the state S_{21}', inspection and readjustment of the machine and the fixture is required, caused, respectively, by the flows of failures with the rates λ'_{02} and λ'_{01}.

State S_{31}' is characterized by the simultaneous presence of streams of parametric failures of tool λ_{01}, machine tool λ_{02} and fixture λ'_{01}.

In the S_{12}' state, inspection and readjustment of the tool and the loading device is required due to the flows of parametric failures with intensities λ_{01} and λ'_{02}.

State S_{22}' is characterized by the simultaneous presence of parametric machine failures with intensity λ_{02} and a loading device with intensity λ'_{02}.

In the S_{32} state, inspection and readjustment of the tool, machine tool and loading device is required, caused, respectively, by the failure streams with rates λ_{01}, λ_{02} and λ'_{02}.

State S_{13}' is characterized by the presence of streams of parametric failures of the tool with the intensity λ_{01}, the fixture with the intensity λ'_{01} and the loading device with the intensity λ'_{02}.

In the S_{23}' state, inspection and readjustment of the machine, the fixture and the loading device is required due to flows of parametric failures with intensities λ_{02}, λ'_{01} and λ'_{02}, respectively.

State S_{33}' is characterized by the presence of streams of parametric failures of a tool with an intensity λ_{01}, a machine tool with an intensity λ_{02}, a fixture with an intensity λ'_{01} and a loading device with an intensity λ'_{02}.

The values $\overline{T\text{и}''}$, $\overline{T\text{с}''}$, $\overline{T\text{п}''}$, $\overline{T\text{з}''}$ are determined by the general formula:

$$\overline{T''} = \sum_{j=1}^{k} t_j/k\,, \tag{10}$$

where k is the number of parametric failures of the corresponding technological element during the operation of the FMM for thread processing; t_j – time interval between (j –1)-th and j-th failures.

Each technological element of the FMM thread processing undergoes restoration immediately after parametric failure. Recovery rates, respectively, for each technological element μ_{01}, μ_{02}, μ'_{01}, μ'_{02}. In this case, $\mu_1=1/\overline{T\text{ви}''}$, $\mu_{02}=1/\overline{T\text{вс}''}$, $\mu'_{01}=1/\overline{T\text{вп}''}$, $\mu'_{02}=1/\overline{T\text{вз}''}$, where $\overline{T\text{ви}''}$ is the average recovery time (inspection and adjustment) of the tool, *min*; $\overline{T\text{вс}''}$ – average time of inspection and adjustment of the machine, *min*; $\overline{T\text{вп}''}$ – average time of inspection and adjustment of the device, *min*; $\overline{T\text{вз}''}$ – average time of inspection and adjustment of the loading device, *min*.

To describe the marked-up graph of states of the FMM technological system of thread

Materials Research Proceedings **21** (2022) 410-420 https://doi.org/10.21741/9781644901755-69

processing, taking into account parametric failures (Fig. 2), the following designations are adopted:

P_{00}' – the probability of the module operation in the absence of parametric failures, $P_{00}' = P(S_{00}')$;

P_{10}', P_{20}', P_{30}' – respectively, the probability of parametric failures of the tool, machine, tool and machine at the same time;

P_{01}', P_{02}', P_{03}' – the probabilities of parametric failures, respectively, of the device, the loading device, the device and the loading device at the same time;

P_{11}', P_{21}', P_{31}' – respectively, the probability of parametric failures of the tool and fixture; machine tool and fixtures; tool, machine and fixture at the same time;

P_{12}', P_{22}', P_{32}' – the probabilities of parametric failures of the tool and the loading device; machine and loading device; tool, machine and loading device;

P_{13}', P_{23}', P_{33}' – the probabilities of parametric failures of the tool of the fixture and the loading device; machine tool, fixture and loading device; simultaneously all technological elements - tool, machine tool, fixture and loading device.

Equations for final probabilities P_{00}', P_{10}', P_{20}', P_{30}', P_{01}', P_{02}', P_{03}', P_{11}', P_{21}', P_{31}', P_{12}', P_{22}', P_{32}', P_{13}', P_{23}', P_{33}' states are represented as a system of equations:

$$\left.\begin{aligned}
&P_{00}(\lambda_{01}+\lambda_{02}+\lambda^{I}{}_{01}+\lambda^{I}{}_{02}) = P_{10}{}^{I}\mu_{01}+P_{20}{}^{I}\mu_{02}+P_{01}{}^{I}\mu^{I}{}_{01}+P_{02}{}^{I}\mu^{I}{}_{02}\\
&P_{10}{}^{I}(\lambda_{02}+\lambda^{I}{}_{01}+\lambda^{I}{}_{02}+\mu_{01}) = P_{00}{}^{I}\lambda_{01}+P_{30}{}^{I}\mu_{02}+P_{11}{}^{I}\mu^{I}{}_{01}+P_{12}{}^{I}\mu^{I}{}_{02}\\
&P_{20}{}^{I}(\lambda_{01}+\lambda^{I}{}_{01}+\lambda^{I}{}_{02}+\mu_{02}) = P_{00}{}^{I}\lambda_{02}+P_{30}{}^{I}\mu_{01}+P_{21}{}^{I}\mu^{I}{}_{01}+P_{22}{}^{I}\mu^{I}{}_{02}\\
&P_{30}{}^{I}(\lambda^{I}{}_{01}+\lambda^{I}{}_{02}+\mu_{01}+\mu_{02}) = P_{20}{}^{I}\lambda_{01}+P_{10}{}^{I}\lambda_{02}+P_{31}{}^{I}\mu^{I}{}_{01}+P_{32}{}^{I}\mu_{02}\\
&P_{01}{}^{I}(\lambda_{01}+\lambda_{02}+\lambda^{I}{}_{02}+\mu^{I}{}_{01}) = P_{11}{}^{I}\mu_{01}+P_{21}{}^{I}\mu_{02}+P_{00}{}^{I}\lambda^{I}{}_{01}+P_{03}{}^{I}\mu^{I}{}_{02}\\
&P_{11}{}^{I}(\lambda_{02}+\lambda^{I}{}_{02}+\mu^{I}{}_{01}+\mu_{01}) = P_{01}{}^{I}\lambda_{01}+P_{10}{}^{I}\lambda^{I}{}_{01}+P_{31}{}^{I}\mu_{02}+P_{13}{}^{I}\mu^{I}{}_{02}\\
&P_{21}{}^{I}(\lambda_{01}+\lambda^{I}{}_{02}+\mu_{02}+\mu^{I}{}_{01}) = P_{01}{}^{I}\lambda_{02}+P_{20}{}^{I}\lambda^{I}{}_{01}+P_{31}{}^{I}\mu_{01}+P_{23}{}^{I}\mu^{I}{}_{02}\\
&P_{31}{}^{I}(\lambda^{I}{}_{02}+\mu_{01}+\mu_{02}+\mu^{I}{}_{01}) = P_{30}{}^{I}\lambda^{I}{}_{01}+P_{21}{}^{I}\lambda_{01}+P_{11}{}^{I}\lambda_{02}+P_{33}{}^{I}\mu^{I}{}_{02}\\
&P_{02}{}^{I}(\lambda_{01}+\lambda^{I}{}_{01}+\lambda_{02}+\mu^{I}{}_{02}) = P_{00}{}^{I}\lambda^{I}{}_{02}+P_{03}{}^{I}\mu^{I}{}_{01}+P_{22}{}^{I}\mu_{02}+P_{12}{}^{I}\mu_{01}\\
&P_{12}{}^{I}(\lambda^{I}{}_{01}+\lambda_{02}+\mu_{01}+\mu^{I}{}_{02}) = P_{02}{}^{I}\lambda_{01}+P_{10}{}^{I}\lambda^{I}{}_{02}+P_{32}{}^{I}\mu_{02}+P_{13}{}^{I}\mu^{I}{}_{01}\\
&P_{22}{}^{I}(\lambda_{01}+\lambda^{I}{}_{01}+\mu_{02}+\mu^{I}{}_{02}) = P_{02}{}^{I}\lambda_{02}+P_{20}{}^{I}\lambda^{I}{}_{02}+P_{32}{}^{I}\mu_{01}+P_{23}{}^{I}\mu^{I}{}_{01}\\
&P_{32}{}^{I}(\lambda^{I}{}_{01}+\mu_{01}+\mu_{02}+\mu^{I}{}_{02}) = P_{22}{}^{I}\lambda_{01}+P_{12}{}^{I}\lambda_{02}+P_{30}{}^{I}\lambda^{I}{}_{02}+P_{33}{}^{I}\mu^{I}{}_{01}\\
&P_{03}{}^{I}(\lambda_{01}+\lambda_{02}+\mu^{I}{}_{01}+\mu^{I}{}_{02}) = P_{02}{}^{I}\lambda^{I}{}_{01}+P_{01}{}^{I}\lambda^{I}{}_{02}+P_{13}{}^{I}\mu_{01}+P_{23}{}^{I}\mu_{02}\\
&P_{13}{}^{I}(\lambda_{02}+\mu_{01}+\mu^{I}{}_{01}+\mu^{I}{}_{02}) = P_{03}{}^{I}\lambda_{01}+P_{11}{}^{I}\lambda^{I}{}_{02}+P_{12}{}^{I}\lambda^{I}{}_{01}+P_{33}{}^{I}\mu_{02}\\
&P_{23}{}^{I}(\lambda_{01}+\mu_{02}+\mu^{I}{}_{01}+\mu^{I}{}_{02}) = P_{03}{}^{I}\lambda_{02}+P_{22}{}^{I}\lambda^{I}{}_{01}+P_{21}{}^{I}\lambda^{I}{}_{02}+P_{33}{}^{I}\mu_{01}\\
&P_{33}{}^{I}(\mu_{01}+\mu_{02}+\mu^{I}{}_{01}+\mu^{I}{}_{02}) = P_{23}{}^{I}\lambda_{01}+P_{13}{}^{I}\lambda_{02}+P_{32}{}^{I}\lambda^{I}{}_{01}+P_{31}{}^{I}\lambda^{I}{}_{02}
\end{aligned}\right\} \quad (11)$$

Materials Research Proceedings **21** (2022) 410-420 https://doi.org/10.21741/9781644901755-69

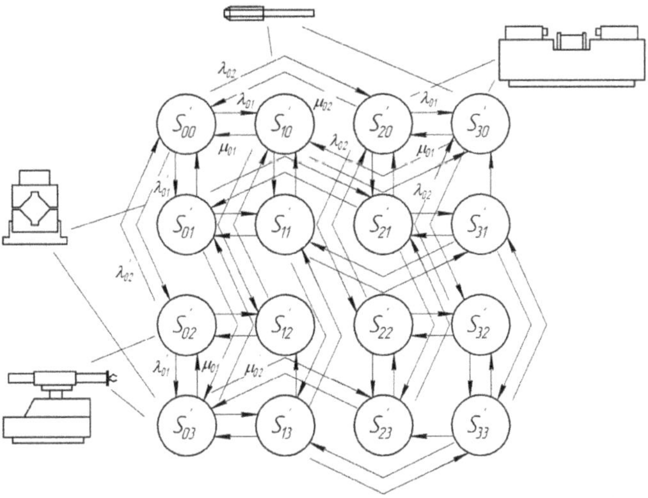

Fig. 2. *The marked graph of states of the FMM technological system of thread processing, taking into account parametric failures*

Solving this system of equations using the normalization condition $P_{00}' + P_{10}' + P_{20}' + P_{30}' + P_{01}' + P_{02}' + P_{03}' + P_{11}' + P_{21}' + P_{31}' + P_{12}' + P_{22}' + P_{32}' + P_{13}' + P_{23}' + P_{33}'=1$ allows you to obtain expressions for the probability of failure-free parametric operation (P_{00}'), as well as the probabilities of finding the system in a faulty state due to the corresponding parametric failures of one or more (in different combinations) technological elements of the FMM threading (tool – P_{10}', machine elements – P_{20}', fixtures – P_{01}', loading device – P_{02}').

Summary

The carried out theoretical studies made it possible to develop a mathematical model for describing the states of the FMM thread processing, taking into account the failures of functioning and parametric failures of its technological elements, as well as recoveries after these failures. The model makes it possible to take into account the operational and technological reliability of the tool, the power threading head, the basic elements of the machine, the fixture, the loading device when choosing the optimal structure of the module.

The structure of the model and the labeled graph of the states of the system can be improved as the number of parameters and characteristics is refined. For complete information and an objective assessment of the preferred option for use in FMM conditions, it is necessary to take into account the stochastic processes occurring in the system under real operating conditions, as well as to develop an optimality criterion and an integral evaluation function, taking into account, in addition to reliability (operational and technological), flexibility (adaptability) and economic efficiency of the module [18-22].

The measures taken make it possible to solve the synthesis problem at the level of structural and layout optimization, and then, at the stage of parametric synthesis, to determine technological elements that are vulnerable according to the developed criterion, the improvement of which will lead to the creation of the most efficient system.

References

[1] A.S. Pronikov, Metal-cutting machines and automatic machines, Mashinostroyeniye, Moscow, 1981, 479 p.

[2] V.E. Push, R. Pigert, V.L. Sosonkin, Automatic machine tools, Mashinostroyeniye, Moscow, 1982, 319 p.

[3] E.S. Wentzel, L.A. Ovcharov, Probability Theory and Its Engineering Applications, Vysshaya shkola, Moscow, 2000, 480 p.

[4] G. D. Grigoryan, Elements of reliability of technological processes, Vishcha school, Vishcha shkola, Kiev-Odessa, 1984, 214 p.

[5] A.A. Matalin, Precision Machining and Process Design, Mashinostroyeniye, Moscow, 1970, 380 p.

[6] A.S. Pronikov, Precision and reliability of numerical control machines, Mashinostroyeniye, Moscow, 1982, 279 p.

[7] V.V. Matveyev, Precision thread cutting, Mashinostroyeniye, Moscow, 1978, 88 p.

[8] Yu.L. Frumin, High Performance Threading Tool, Mashinostroyeniye, Moscow, 1977, 183 p.

[9] A.S. Yamnikov, Determination of dynamic errors in threading with a measuring tool, Research in the field of technology for the formation of external and internal threads, thread-forming tools, machine tools and methods for controlling threads, Tula: Publisher TPI, 1974. – pp. 68-72.

[10] A.A. Grudov, Ways to increase the accuracy of the cut thread and the durability of taps, Publisher VNII, Moscow, 1971. – 128 p.

[11] V.M. Menshakov, G.P. Urlapov, V.S. Sereda, Chipless taps, Mashinostroyeniye, Moscow, 1976. – 167 p.

[12] V.G. Yakukhin, Optimal thread manufacturing technology, Mashinostroyeniye, Moscow, 1976, 167 p.

[13] G.I. Melamed, B.M. Tursunov, Flexible automatic production. CNC machines and robots, Publisher Belarus, Minsk, 1986, 159 c.

[14] S.M. Bratan, F.N. Kanareyev, P.A. Novikov, A.O. Kharchenko, Improving the accuracy of forming small-sized threads with taps in aluminum alloys, Vuzovskiy uchebnik: INFRA-M, Moscow, 2017, 164 p.

[15] E.V. Ryzhov, O.S. Andreychikov, A.E. Steshkov, Thread rolling, Mashinostroyeniye, Moscow, 1974, 122 p.

[16] Yu.K. Novoselov, A.O. Kharchenko, Choice of the Optimal Structure of a Flexible Production Cell, Soviet engineering research. 7(2) (1987), pp. 48-52.

[17] A.O. Kharchenko, Development of flexible production modules taking into account the reliability of technological systems, Publisher Society " Znaniye", Kiev, 1989, 20 p.

[18] E.G. Nahapetyan, Diagnostics of flexible automated manufacturing equipment, Publisher Nauka, Moscow, 1985, 224 p.

[19] A.O. Kharchenko, S.M. Bratan, E.A. Vladetskaya, S.I. Roshchupkin, Analysis and Synthesis of Structures of Modern Multi-Operational Machine Tools, Publisher Tsentrkatalog, Moscow, 2018, 144 p.

[20] V.P. Dolgin, A.O. Kharchenko, Reliability of technical systems, Publisher Vuzovskiy uchebnik, INFRA-M, Moscow, 2015, 157 p.

[21] A.O. Kharchenko, A.A. Kharchenko, Structural and layout synthesis of subsystems of modules for internal thread processing, Fundamental and applied problems of engineering and technology, Orel: Publisher "OSU named after I.S. Turgenev". 4-2 (342) (2020), pp. 85-94. https://doi.org/10.33979/2073-7408-2020-342-4-2-85-94

[22] A. Kharchenko, Progressive threading of small holes based on mechanisms of parallel structures (MPS), Materials Today: Proceedings, Volume 38, Part 4. (2021), pp. 2038-2042. https://doi.org/10.1016/j.matpr.2020.10.040

Influence of the Technology of Obtaining the Material of the Cathode of the Cu – Fe System at the Depth of Penetration of Ions into the Titanium Target

Viktor V. Ovchinnikov[1,a], Elena V. Luk'yanenko[1,b], Irina A. Kurbatova[1,c], Svetlana V. Yakutina[1,d] and Nadezda V. Uchevatkina[1,e*]

[1]Moscow Polytechnic University, B. Semenovskaya st., 38 Moscow, 107023, Russian Federation

[a]vikov1956@mail.ru, [b]evlukyanenko20@yandex.ru, [c]kia0904@mail.ru, [d]svyakut@yandex.ru [e]uchevatkina@yandex.ru

Keywords: Titanium Alloys, Ion Implantation, Implanter Cathode, Penetration Depth, Cluster Ions

Abstract. The article presents the results of the influence of the technology of obtaining the material of the cathode of the implanter of the Cu – Fe system on the penetration depth of the titanium alloy VT20. It is shown that the use of 50% Cu – 50% Fe material as the material of the cathode of the implanter, obtained by alloying copper and iron, leads to a better increase in the thickness of the ion-doped layer than the use of the cathode obtained by powder metallurgy.

Introduction

Since its early development, ion implantation has shown, in addition to tremendous possibilities, such as structuring or selective surface doping of materials, limitations such as the maximum thickness that can be changed, or simple flat geometry that allows uniform modification of the target. This prompted active development and research, which led to new tools for material modification. As is known, the nature of the dislocation structure and its quantitative characteristics in implanted metals and alloys depend on the initial state of the target, the type and energy of ions and the remaining dose, and can vary greatly, therefore, the study of various mechanisms of defect formation under conditions of ion implantation [1].

Materials and research methods

To study the effect of the technology of obtaining the material of the cathode of the implanter of the Cu – Fe system, the method of melting and the method of powder metallurgy were used.

In the first case, a sample of a given composition was melted in graphite crucibles. By the method of directional crystallization using a seed in the form of a copper rod, a rod with a diameter of 15–18 mm was obtained from an alloy 50% Cu – 50% Fe (Fig. 1, a).

To prevent oxidation of the melt during melting and holding at a given temperature, the crucible was closed with a tightly ground-in lid made of the same material. Heating was carried out to a liquid state of 1100 ° C, the holding time of the melt at a given temperature was 45 minutes.

Then the rod was cut into blanks, which were pressed in a heated state to obtain washers with a diameter of 50 mm and a thickness of 10 mm. After pressing, the washers were subjected to turning, resulting in disc cathodes of an implanter 40 mm in diameter and 5 mm thick (Fig. 1, b).

In fig. 2 shows the microstructure of the implanter cathode material. You can clearly distinguish between the immiscible phases of copper (in the photo, golden tone) and iron (greenish areas).

The structure of an iron-copper alloy with a 50/50 component content in the cast state is a set of dendrites of a supersaturated solid solution of copper in iron, evenly distributed in a matrix of a supersaturated solid solution of iron in copper.

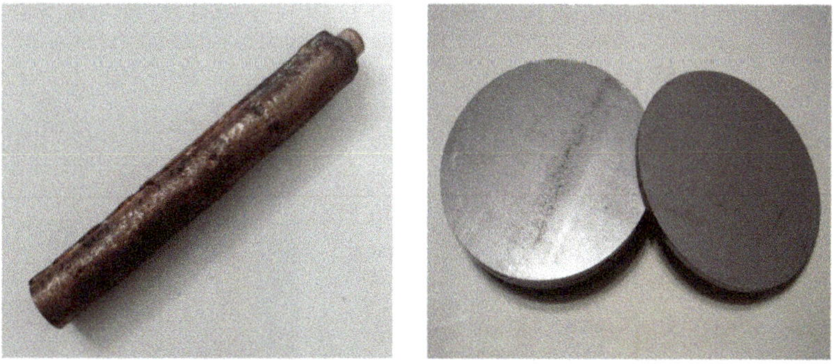

a b

Fig. 1. *Rod made of 50% Cu-50% Fe alloy (a) and disk cathode of the implanter (b)*

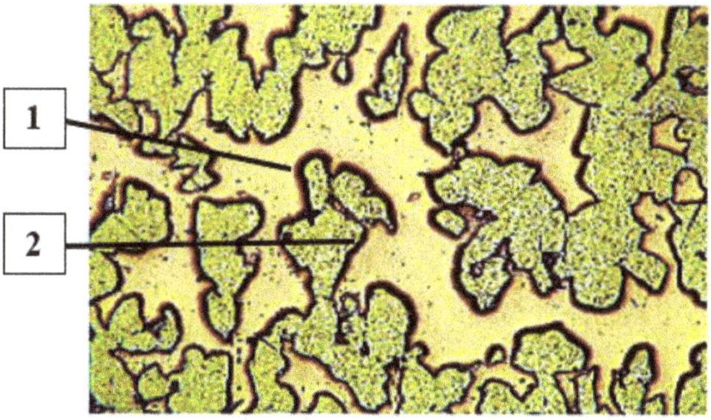

Fig. 2. *Microstructure of the material of the cathode of the implanter made of alloy 50% Cu – 50% Fe (x230): 1 - copper; 2 – iron*

In the second case, the prepared mixture of copper and iron powders was pressed to obtain implanter cathode blanks with a size of 60x10 mm, which were then subjected to sintering at a temperature of 820–850 ° C. After sintering, the resulting billets were turned by turning to obtain the required dimensions of the implanter cathode.

To obtain comparative data on the effect of the technology of obtaining the material of the cathode of the implanter with the use of the obtained cathodes, the target was implanted from the titanium alloy VT20 at fluence values of 10^{17}, $5 \cdot 10^{17}$, and 10^{18} cm $^{-2}$. After implantation, the

Materials Research Proceedings **21** (2022) 421-426
https://doi.org/10.21741/9781644901755-70

thickness of the ion-doped layer was determined by Auger spectroscopy. The research results are shown in Fig. 3.

An increase in the fluence value in the investigated range when using 50% Cu – 50% Fe material, obtained by powder metallurgy, as the implanter cathode material, leads to an increase in the thickness of the ion-doped layer from 205 to 320 nm (by 56%). At the same time, when using the implantation cathode, obtained by alloying copper and iron, this increase was 212%.

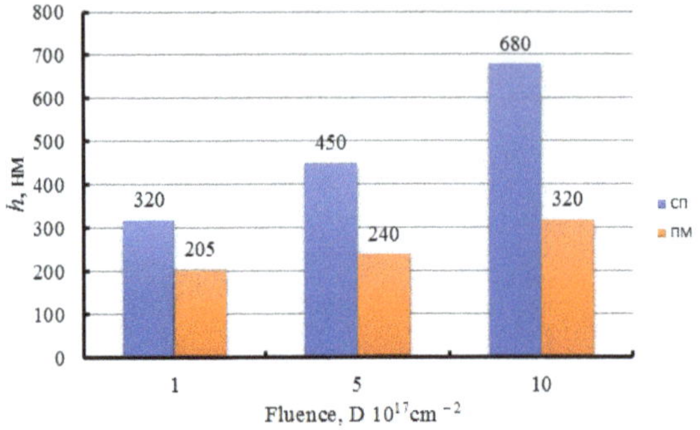

Fig.3. *Influence of the implantation cathode manufacturing technology from 50% Cu – 50% Fe material on the thickness of the ion-doped layer: SP - material obtained by fusion of components; PM - material obtained by pressing and sintering copper and iron powders.*

Considering the same conditions for the parameters of the implantation process, the difference in the thickness of the ion-doped layer can be associated with the presence of more massive ion blocks, the so-called clusters, when using a cathode made of material obtained by fusing the components.

The formation of cluster ions can be assumed based on the conditions for the formation of the ion flow based on the study of the behavior of the cathode spot of the arc on the surface of the composite cathode (Figure 4).

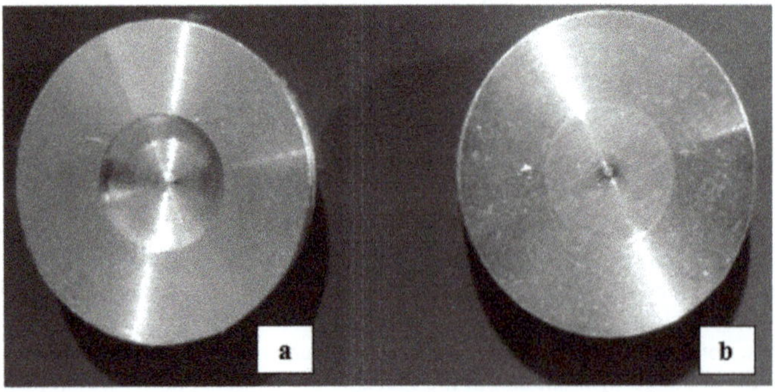

Fig. 4. *Composite experimental iron-based cathode with a central fused copper insert: a - workpiece; b - cathode after treatment*

The composite model cathode is an Armco iron disk 40 mm in diameter in the central part of which a circular hole is made, which is fused with copper. The diameter of the indicated hole is 10–12 mm. After crystallization and cooling of copper, the end working surface of the cathode was carefully machined on a lathe.

The design of the composite cathode makes it possible to estimate the speed of movement of the active arc spot over the cathode surface. It was 18-23 m / s.

The charge distribution of ions was measured using a magnetic mass separator (GSI) and a time-of-flight spectrometer (LBNL). In both cases, the ions were accelerated from the plasma at a voltage of 40 kV using a three-electrode multi-aperture extraction system. To exclude the effect of the arc pulse duration, all measurements of the charge distribution of ions were carried out every 100 ms after ignition of the discharge. During the measurements, the pressure of the residual gas in the working chamber of the setup was maintained at a level of 10–5 mm Hg.

The studies were carried out on an ion implantation unit with an IGMI-50 source operating in a pulsed mode and capable of generating polyenergetic beams of metal ions. Irradiation was carried out with pulsed beams of copper and lead ions. The average beam current was 0.1 A, the pulse duration was 300 μs, and the pulse repetition rate was 17 Hz.

Results and discussion

Studies have shown that short beams have a duration of about 5 μs, then the beam current pulses were cut out with a step of 5 μs. For the first 15 μs of arc burning, the beam spectrum contains only iron ions. After 20 μs of arc burning, both iron ions and copper ions are already present in the beam. Starting from 25 μs from the onset of arc burning, only copper ions are present in the spectrum.

The result obtained can be explained by the directed movement of the cathode spot from the peripheral sections of the composite cathode to its center. Therefore, it can be argued that the cathode spot passes from a more refractory metal to a less refractory metal. In this case, the reverse transition is not fixed.

This is explained by the fact that it is energetically more favorable for the arc to burn on a less refractory surface [2]. Indeed, regardless of the surface on which the arc is currently burning, the arc current remains constant, since a modulator with arc current stabilization is used to power the arc discharge. In this case, the voltage drop across the arc for a less refractory material will be less than for a more refractory metal. The presence of this voltage jump on the arc during the transition of the cathode spot from iron to copper is confirmed by the presence of a beam of Cu2 + ions in the charge spectrum at this moment, which then disappear from the spectrum. That is, at this moment, the voltage drop across the arc is greater than during further work from the copper surface of the composite model cathode.

It should be noted that when working with a purely copper cathode, Cu2 + ions were observed only during the duration of the ignition pulse, while in the results obtained, they are observed after the end of this pulse.

Using the results obtained, an estimate was made of the smallest velocity of the cathode spot movement [3]. Based on the assumption that the cathode spot is based on the cathode edge at the point closest to the copper impregnation, and taking into account that there are no copper ions on the oscillogram corresponding to a 25 μs delay, the minimum speed of the cathode spot is 18 m / s. It should be noted that this value was obtained on the assumption of a rectilinear motion of the cathode spot from the place of formation to the copper insert.

Observation of the movement of the cathode spot of the arc showed that when using the composite cathode of the implanter, the cathode spot mainly moves along the radial part of the cathode disk from its peripheral part to the central part. The image of the movement of the cathode spot of the arc over the cathode surface of the ion implantation device is shown in accordance with Fig. 5.

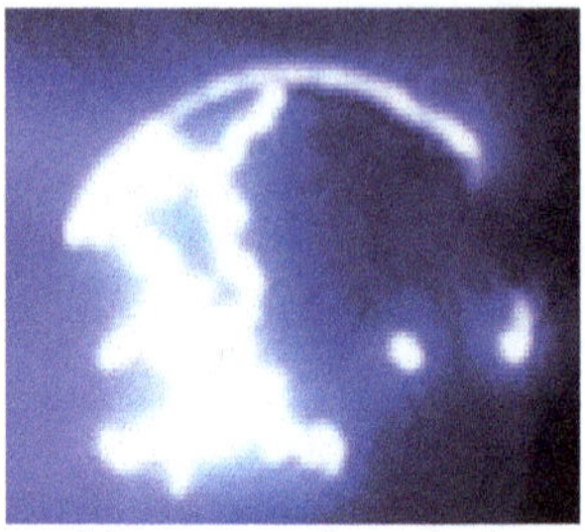

Fig. 5. *Moving the cathode spot of the arc over the cathode surface of the ion implantation device when using a composite cathode with a central hole.*

This is explained by the fact that it is energetically more favorable for the arc to burn on a less refractory surface. Indeed, regardless of the surface on which the arc is currently burning, the arc current remains constant, since a modulator with arc current stabilization is used to power the arc discharge. In this case, the voltage drop across the arc for a less refractory material will be less than for a more refractory metal. The presence of this voltage jump on the arc during the transition of the cathode spot from iron to copper is confirmed by the presence at this moment in the charge spectrum of a beam of Cu^{3+} ions, which then disappear from the spectrum. That is, at this moment, the voltage drop across the arc is greater than during further work on a copper surface.

The appearance of large-mass cluster ions significantly changes the pattern of the effect of the ion flux on the irradiated target surface (Fig.6); destruction of the cluster is observed when it interacts with the target surface.

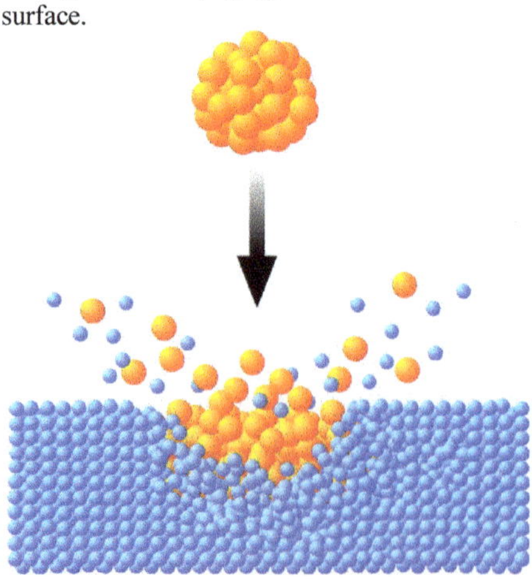

Fig. 6. *Scheme of irradiation of a titanium target with cluster ions*

When a cluster is destroyed, some of the atoms that make up the cluster are reflected from the target surface together with the surface sputtered atoms. At the same time, due to the greater mass of the cluster, it acquires a large energy impulse, which ultimately provides a greater penetration of the ions that make up the cluster into the target material. In this case, a sublayer with a strongly deformed atomic structure is formed directly under the ion-doped target layer.

Summary

Thus, on the basis of the research results, it can be concluded that an increase in the fluence value in the investigated range of 10^{17}–10^{18} cm $^{-2}$ when using 50% Cu – 50% Fe material obtained by powder metallurgy as the cathode material of the implanter leads to an increase in the thickness of the ionic doped layer from 205 to 320 nm (56%). At the same time, when using an implanter cathode obtained by alloying copper and iron, this increase was 212% and reached 680 nm. When working with a purely copper cathode, Cu2 + ions were observed only during the duration of the ignition pulse, while in the results obtained they are observed after the end of this pulse.

Acknowledgement

This study is conducted with financial support from the Ministry of Education and Science of the Russian Federation (projekt No. FZRR-2020-0023/code 0699-2020-0023).

References

[1] Pogrebnyak A D Erdybaeva N K Bratushka S N et al.2008 Effect of high-dose implantation of metal and gas ions on the physicomechanical properties of titanium alloys Questions of atomic science and technology No 1 P 81-92

[2] S.N. Brother and L.V. Malikov, Problems of Atomic Science and Technology: (2011) 126-140

[3] V.V. Ovchinnikov, N V Uchevatkina I A Kurbatova E V Lukyanenko and S V Yakutina Periodico Tche Quimica: 16 (32) (2019) 945-966. https://doi.org/10.52571/PTQ.v16.n32.2019.963_Periodico32_pgs_945_966.pdf

[4] Sharkeev Yu P Ryabchikov A I Kozlov E V Kurzina I A Stepanova I B Bozhko I A Kalashnikov M P Fortuna S V Sivin D O 2004 High-intensity ion implantation - a method for the formation of fine intermetallic compounds in the surface layers of metals. Izvestiya vysshikh educational institutions. Physics: 9 (2004) 44-52. https://doi.org/10.1007/s11182-005-0033-1

[5] Rissel H. Ruge I. Ion implantation. Moscow: Nauka (1983) 360 p.

[6] Nikitenkov N N Ion-immersion implantation of aluminum in VT1-0. Bulletin of the Russian Academy of Sciences. Physical series 80(2) (2016) 131 p. https://doi.org/10.3103/S1062873816020222

Modern Trends in Manufacturing Technologies and Equipment | Materials Research Forum LLC
Materials Research Proceedings **21** (2022) 427-435 | https://doi.org/10.21741/9781644901755-71

Influence of Heat Treatment on Structure and Durometric Properties of Coatings Obtained by Surfacing with *CSR-04CR27NI7MO3CU2T* Cast Rods

Aleksandr Nazarko[1,a*] and Roman Plomodyalo[1,b]

[1]Kuban State Technological University, 2, Moskovskaya Str., Krasnodar, 350072, Russia

[a]nazarkoaleksandr@rambler.ru, [b]sutk_kubstu@mail.ru

Keywords: Argon Arc Surfacing, Cast Rod, Heat Treatment, Hardness, Structure

Abstract. The effect of heat treatment modes on the structure and durometric properties of coatings obtained by surfacing with *CSR-04CR27NI7MO3CU2T* cast rods, is considered. It is found that the temperature of 800°C and soaking time of 5 hours are optimal to increase the deposited metal hardness. It is shown that such a phenomenon results from the formation of the austenitic structure hardened by the precipitates of the σ-phase (FeCr), chromium carbides (Cr_3C_2) and titanium carbides (TiC). The heat treatment modes proposed can be applied in the wear-resistant surfacing technology of chemical equipment parts.

Introduction

Parts of chemical equipment operating under conditions of carbamide fusion (H_2NCONH_2) with pressure indicators of up to 20 MPa and temperature indicators of 200°C, gaseous and liquid ammonia, ammonium nitrate are subject to intense wear, as well as intergranular and general corrosion.

To surface the parts of carbamate pumps, throttle and shut-off elements (saddles, valves), the following procedures are most often used – manual arc surfacing with *OK 61.25* electrodes, manual and automatic submerged-arc surfacing with the following fluxing agents – *OK Autrod 316 L Si* wire [1]. Such surfacing materials ensure the production of coatings of the austenitic and ferritic class with increased wear resistance in a corrosive environment.

At the same time, the hardness of the coatings deposited with such materials is insufficient to ensure the required wear resistance. To increase the hardness of such a deposited metal, it is necessary to heat it. It should be taken into account that the modes of such heat treatment have been studied insufficiently.

Based on the above mentioned, the aim of this paper is to study the effect of heat treatment modes on the structure and durometric properties of coatings obtained by surfacing with cast rods.

Research Methods

The surfacing was carried out on the *AISI 316L* steel plates with the dimensions of 40x15x10 mm with experimental *CSR-04CR27NI7MO3CU2T* cast rods [2] with their diameter of 5 mm and length of 270 mm, in the argon atmosphere, in four layers, with the strength of current equal to 180 A.

The research of the deposited metal microstructure was carried out with the help of the *ZEISS EVO HD 15 MA* scanning electron microscope. The microstructure was revealed by the pickling with the aqua regia reagent – a mixture of concentrated nitric acid HNO_3 (65-68 wt. %) and hydrochloric acid HCl (32-35 wt. %) taken in a ratio of 1:3 by volume (with the mass ratio about 1:2 in terms of pure substances).

The hardness and microhardness were measured on metal samples after surfacing and heat treatment using the *TK-2M* hardness tester in accordance with the method of Rockwell and the *DuraScan-80* microhardness tester in accordance with the method of Vickers. Microhardness was measured along the cross section of the deposited coating, starting from the base metal with a step of 0.2 mm [3].

The deposited samples heat treatment was carried out in the *ПМ-12М2П-1200* muffle furnace in the modes established experimentally.

Research Results and Discussion

After surfacing, the coating hardness was researched in accordance with the method of Rockwell, and it made 29-32 HRC. However, in order to ensure high wear resistance, it is necessary to increase the hardness of the deposited metal. For this purpose, after surfacing, the samples under study were subjected to heat treatment.

Three modes of heat treatment were researched – at heating temperatures of 800°C, 900°C and 1,000°C, with the soaking time from 1 hour to 6 hours. The results of measuring the hardness over the cross-section of the deposited metal are given in Table 1.

Table 1. *Hardness Distribution in Coating Obtained by Surfacing with Cast Rods after Heat Treatment*

Heat treatment temperature, [°C]	**Soaking time, [hours]**	**Hardness distribution in accordance with deposited metal layers, [HRC]**				
		1	**2**	**3**	**4**	**Average value**
800	1	32	40	36	35	35.75
	2	36	41	42	43	40.5
	3	38	42	43	43	41.5
	4	41	46	48	47	45.5
	5	42	47	48	48	46.25
	6	40	45	46	46	44.25
900	1	33	39	37	36	36.25
	2	37	42	41	43	40.75
	3	39	40	40	41	40
	4	38	45	47	47	44.25
	5	40	46	47	46	44.75
	6	38	43	44	45	42.5
1000	1	34	40	40	37	37.75
	2	35	41	42	41	39.75
	3	37	39	41	40	39.25
	4	41	47	46	46	45
	5	41	47	48	47	45.75
	6	39	44	45	43	42.75

The studies showed that the soaking time of 5 hours was optimal to obtain the required hardness of the deposited coating (Fig. 1), and in the temperature range from 800°C to 1,000°C in the deposited metal there were practically no changes in hardness (Fig. 2). Thus, to increase the hardness of the metal deposited with the *CSR-04CR27NI7MO3CU2T* cast rods of the coatings, it is possible to recommend the following heat treatment modes –

1. Heating up to 800 °C with a furnace.
2. 5 hours of soaking time.
3. Cooling by ambient air.

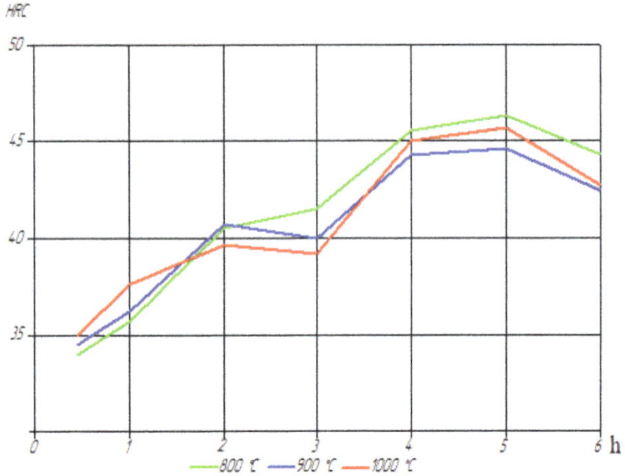

Fig. 1. *Influence of Soaking Time on Hardness of Coating (Average Values) Obtained by Surfacing with CSR-04CR27NI7MO3CU2T Cast Bars in Temperature Range of 800-1,000°C*

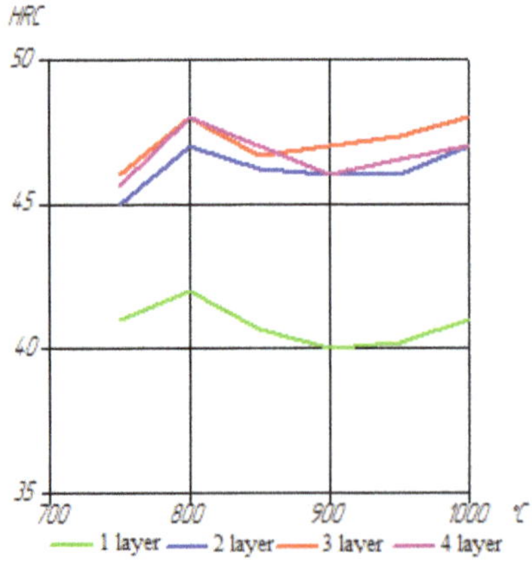

Fig. 2. *Influence of Heat Treatment Temperature (with Soaking Time of 5 Hours) on Hardness of Coating Obtained by Surfacing with CSR-04CR27NI7MO3CU2T Cast Rods*

Materials Research Proceedings **21** (2022) 427-435 https://doi.org/10.21741/9781644901755-71

For a complete representation of the distribution of hardness over the cross-section of coatings after surfacing and heat treatment (800°C – 5 hours), the total hardness of the metal was researched in accordance with the method of Rockwell. The results are shown in Fig. 3.

It can be seen that after heat treatment, the hardness is uniformly distributed over the deposited metal cross-section in the range of 45-48 HRC. This hardness of the coating ensures high wear resistance of the deposited coatings.

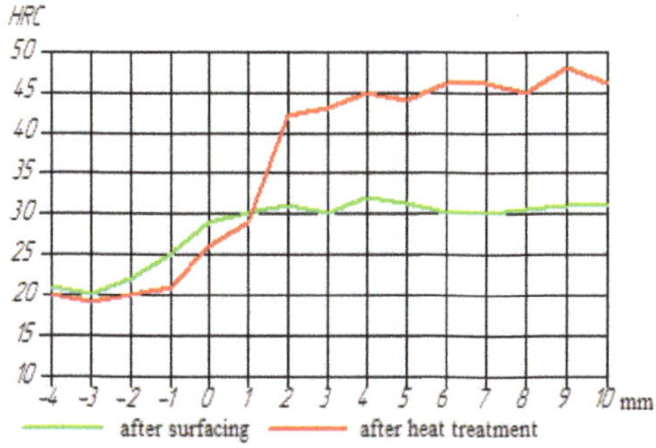

Fig. 3. *Distribution of Metal Hardness over Cross-Section of Coating Obtained by Surfacing with CSR-04CR27NI7MO3CU2T Cast Rods*

The results of measuring the microhardness over the cross section of the deposited metal after heat treatment are shown in Fig. 4. There, the distribution of microhardness in the metal of coatings after the procedure of surfacing is also presented.

Analyzing the results obtained, it is possible to note that the microhardness in the metal after surfacing varies within 315-404 HV. After heat treatment (800°C – 5 hours), the microhardness significantly increases to the values from 415 HV to 570 HV.

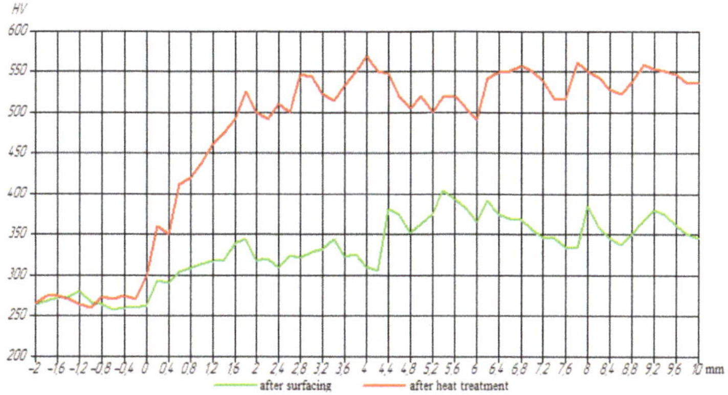

Fig. 4. *Distribution of Microhardness over Cross-Section of Coating Obtained by Surfacing with CSR-04CR27NI7MO3CU2T Cast Rods*

Materials Research Proceedings **21** (2022) 427-435
https://doi.org/10.21741/9781644901755-71

The microstructure of the metal deposited with *CSR-04CR27NI7MO3CU2T* cast rods, is represented by the mixture of ferrite (α-Fe) and austenite (γ-Fe) (Fig. 5). The accumulation of strengthening phases is observed along the grain boundaries. This mixed structure has a hardness of 29-32 HRC.

The places of measurement and the values of the microhardness of the structural components of the metal after surfacing are shown in Fig. 6 and in Table 2.

As we can see, the microhardness of the matrix varies from 315 to 355 HV, and that of the hardening phases – from 364 to 451 HV.

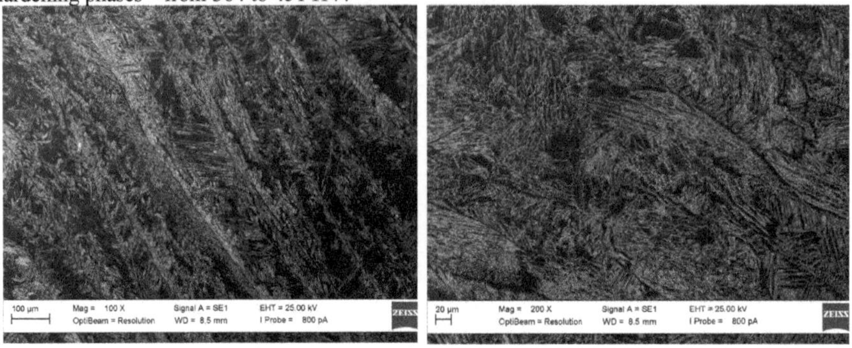

Fig. 5. *Microstructure of Metal Deposited with CSR-04CR27NI7MO3CU2T Cast Rods*

Table 2. *Microhardness $HV_{0,01}$ of Structural Components of Metal Deposited with CSR-04CR27NI7MO3CU2T Cast Rods*

Puncture No.	**1**	**2**	**3**	**4**	**5**	**6**	**7**	**8**	**9**	**10**	**11**	**12**
Microhardness value, HV	315	404	451	439	421	338	364	445	355	346	322	383

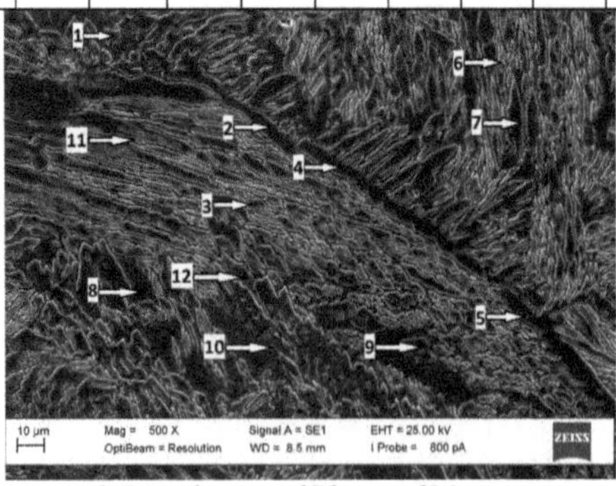

Fig. 6. *Microstructure with Distribution of Places of Measurement of Microhardness of Structural Components of Metal of Coating Deposited with CSR-04CR27NI7MO3CU2T Cast Rods*

Heat treatment of the studied coating at the temperature of 800°C and soaking for 5 hours led to changes in the microstructure of the deposited metal. As a result of heat treatment in the deposited metal, the structural components decomposition occurred (Fig. 7), i. e., all ferrite (α-Fe) passed into the σ-phase (FeCr), which caused the hardness increase from 31 HRC to 48 HRC.

The microhardness of the structural components of the deposited metal after heat treatment of the coating under research increased significantly (Fig. 8, Table 3).

As we can see, the microhardness of the deposited metal matrix after heat treatment increased significantly – from 415 HV to 566 HV, and the microhardness of the strengthening phases changed from 539 HV to 786 HV. Such structure changes caused a significant increase in hardness.

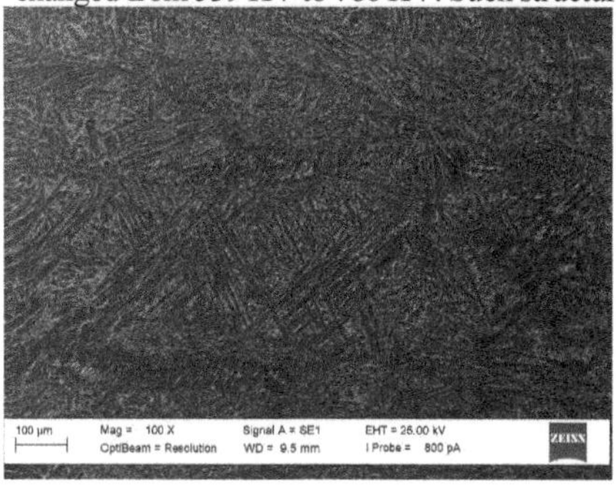

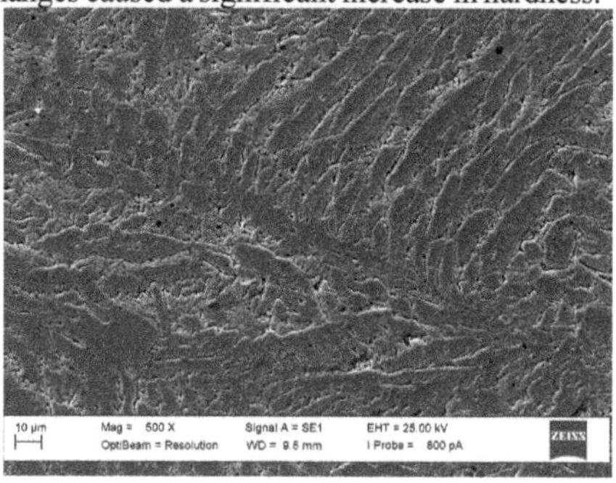

Fig. 7. *Microstructure of Metal of Coating Deposited with CSR-04CR27NI7MO3CU2T Cast Rods after Heat Treatment at 800°C during 5 Hours.*

Table 3. *Microhardness $HV_{0.01}$ of Structural Components of Metal Deposited with CSR-04CR27NI7MO3CU2T Cast Rods after Heat Treatment*

Puncture No.	**1**	**2**	**3**	**4**	**5**	**6**	**7**	**8**	**9**	**10**	**11**	**12**
Microhardness value, HV	461	415	426	504	539	473	634	469	560	539	786	515

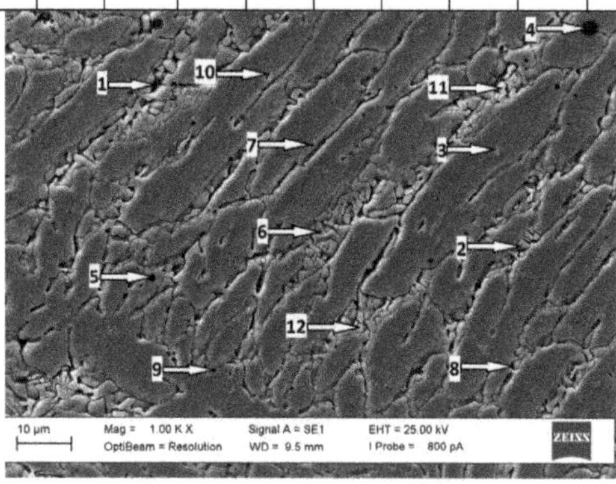

Fig. 8. *Microstructure with Distribution of Places of Measurement of Microhardness of Structural Components of Metal of Coating Deposited with CSR-04CR27NI7MO3CU2T Cast Rods after Heat Treatment*

To redefine the chemical composition of structural components in the deposited metal after heat treatment, studies were carried out with the help of a scanning electron microscope [3].

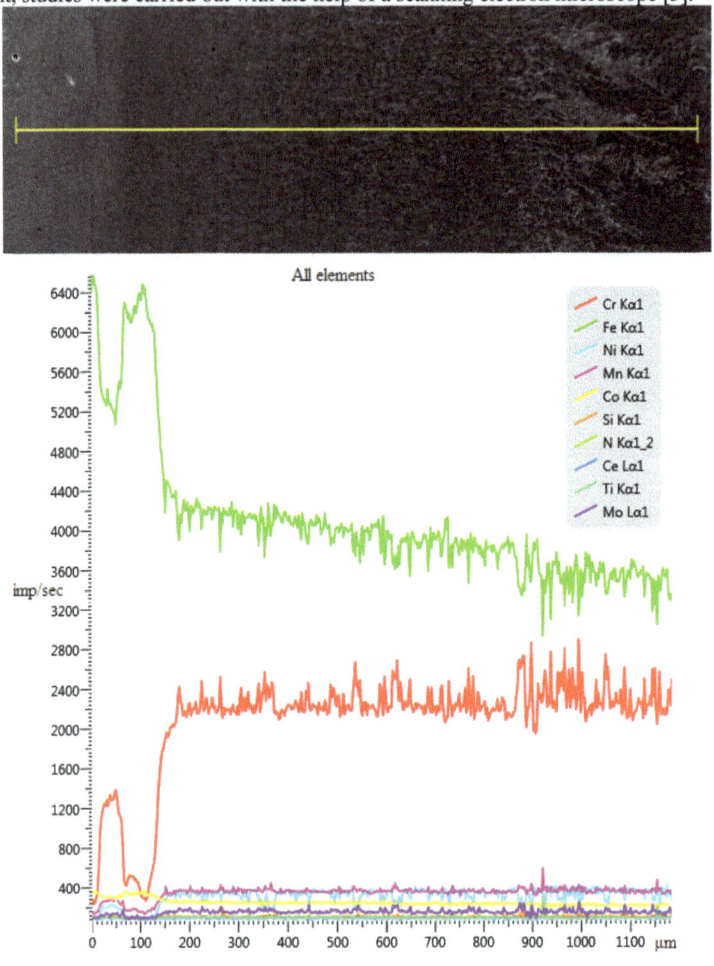

Fig. 9. *Results of Scanning Metal Structure over Cross-Section of Coating Deposited with CSR-04CR27NI7MO3CU2T Cast Rods after Heat Treatment*

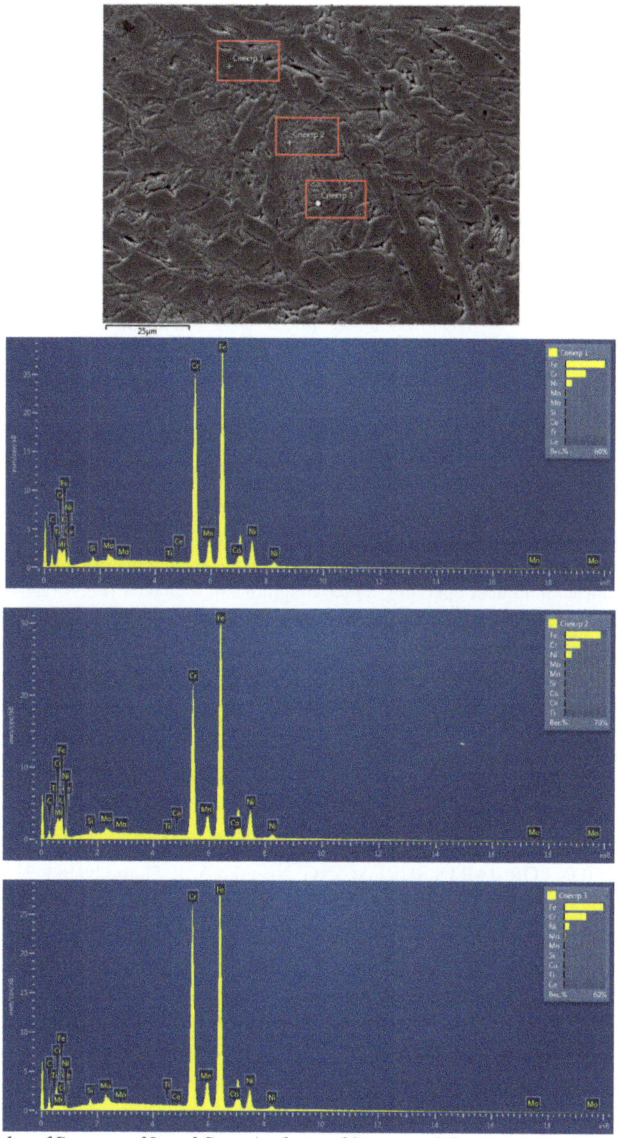

Fig. 10. *Results of Spectra of Local Spot Analysis of Structural Components of Metal of Coating Deposited with CSR-04CR27NI7MO3CU2T Cast Rods after Heat Treatment at 800°C during 5 Hours.*

Materials Research Proceedings **21** (2022) 427-435 https://doi.org/10.21741/9781644901755-71

Table 4. *Chemical Composition of Areas of Coating Deposited with CSR-04CR27NI7MO3CU2T Cast Rods, wt. %*

Spectrum	C	Si	Mn	Cr	Ni	Mo	N	Co	Ti	Ce	Cu	Fe
1	0.03	0.53	0.9	27.29	8.72	2.5	0.05	0.18	0.31	0.03	2.15	57.31
2	0.04	0.52	0.86	25.71	7.24	3.16	0.11	0.19	0.48	0.08	2.1	59.51
3	0.05	0.69	0.82	29.18	7.33	3.44	0.09	0.2	0.46	0.05	1.73	55.96

Based on the data obtained, reflecting the quantitative distribution of elements along the segment in the form of concentration spectrograms over the cross-section of the deposited coating, starting with the base metal (Fig. 9) and spectra of local spot analysis (Fig. 10, Table 4), it follows that the strengthening of the metal deposited with *CSR-04CR27NI7MO3CU2T* cast rods is mainly conditioned by the presence of the σ-phase precipitated along the grain boundaries of the intermetallic compound (FeCr), as well as chromium carbides (Cr_3C_2) [4, 5] and titanium carbides (TiC).

Conclusions

1. To increase the hardness of the deposited metal, it is advisable to carry out heat treatment at a temperature of 800°C and soaking for 5 hours. The hardness of this metal reaches 45-48 HRC, which ensures high wear resistance of the deposited coatings. This is explained by an increase in the microhardness of structural components due to the formation of an austenite structure strengthened by precipitates of the σ-phase (FeCr), chromium carbides (Cr_3C_2), and titanium carbides (TiC).

2. The proposed mode of heat treatment can be used in the technology of wear-resistant surfacing of parts of chemical equipment operating under conditions of urea melt (H_2NCONH_2), at pressures up to 20 MPa and temperature of 200°C, gaseous and liquid ammonia, ammonium nitrate.

Acknowledgments

The research was carried out with the financial support of the Kuban Science Foundation in the framework of the Scientific Project No. МФИ-20.1/26[2].

References

[1] ST CKBA 053-2008. *Armatura truboprovodnaya. Naplavka i kontrol kachestva naplavlennykh poverkhnostey. Tekhnicheskiye trebovaniya.* [Pipe Fittings. Surfacing and Quality Control of Welded Surfaces. Technical Requirements].

[2] COMPOSITION FOR PARTS SURFACING. Nazarko A.S., Plomodyalo R.L., Demonov M.S. Invention Patent RU 2711488 C1, 17.01.2020. Application No. 2018147605 dated 28.12.2018.

[3] Yeremin Ye.N., Losev A.S., Borodikhin S.A., Filippov Yu.O., Ponomarev I.A., Matalasova A.Ye. Effect of Heat Treatment on Structure and Properties of Coatings Based on Martensitic Chromium Steel Obtained by Surfacing with Flux-Cored Wire // Omsk Scientific Bulletin. Ser. Aircraft Missile and Power Engineering. 2017. Vol. 1, No. 2. Pp. 41–48.

[4] Bhambri S. K. Intergranular fracture in 13 wt% chromium martensitic stainless steel // Journal of Material Science. 1986. Vol. 21, Issue 5. Pp. 1741–1746.

[5] Schäfer L. Influence of delta ferrite and dendritic carbides on the impact and tensile properties of a martensitic chromium steel // Journal of Nuclear Materials. 1998. Vol. 258. Pp. 1336–1339.

Assessment of a Heterogeneous Environment

Alexander Zverovshchikov

Department of Technology and Equipment of Mechanical Engineering, Penza State University, 40 Krasnaya Street, Penza 440026, Russian Federation

azwer@mail.ru

Keywords: Heterogeneous Environments, Cutting, Coating, Metal, Technologies

Abstract. An approach to the assessment of a heterogeneous environment in accordance with the proposed concept of a heterogeneous environment is described. The classification of industrial heterogeneous media by technological feature is proposed. The processes described by the proposed approach include the process of metal cutting, surface plastic deformation, chemical-thermal treatment, formation of coatings, welding, most of the processes that change the level and gradient of hardness and internal stresses in the surface layer of the material. Application of the proposed principles makes it possible to find new directions in the study of such processes.

Introduction

The concept of heterogeneous environments is used in most scientific and technical branches of knowledge to describe a wide variety of systems. The basic concepts of a heterogeneous system or environment are described in works on physics, chemistry, sociology, economics, and biology [1,2]. The main features of a heterogeneous environment are the presence of areas with different characteristics of the state and boundaries between them [3].

There are a number of particular definitions of heterogeneous systems in materials science, physics, which are reduced to macroscopic thermodynamic inhomogeneity, differences in physical properties, chemical composition of parts (phases) [4], electrical properties [5]. In the definitions, in a number of special cases, there is a connection with dispersed systems, determined by the degree of development of the interfaces between the constituent phases [6]. A very large number of studies are devoted to the kinetics of heterogeneous media in chemistry, dynamics in mechanics [7,8,9,10], in particular in the study of the physics of combustion and explosion [11,12,13,14,15].

Attempts to create generalized universal models for describing and calculating heterogeneous media have been made repeatedly in various fields of knowledge [16].

The aim of this work is an attempt to generalize and localize the concept of a heterogeneous environment from the point of view of industrial technologies. The urgency of the problem is substantiated by the need for a systematic approach in the analysis of behavior and the creation of heterogeneous industrial environments.

Characteristics of Technological and Industrial Heterogeneous Environments

Let's highlight the main characteristics of technological and industrial heterogeneous environments as a basis for their classification. Considering a heterogeneous environment within the framework of the classical approach [17], we note that it is a system of elements separated by boundaries. The external boundaries of the system that define the area of analysis will be considered technical constraints depending on the production problem. Let's define the main characteristics that determine the difference between the elements of a heterogeneous technical environment:

- • Mechanical characteristics: hardness, elasticity, viscosity, optical permeability, etc .;
- • Geometric characteristics;
- • Electromagnetic characteristics.

Along with the above, it is advisable to introduce the concept of the spatial dimension of the medium. So, one-dimensional environments can be considered only in some contradiction with the term "environment" itself, which presupposes the environment, the scope of an object, in fact, presupposes multidimensionality. However, the consideration of such objects as, for example, an extended fiber, homogeneous in properties in the cross section, but variable in properties along the length, which is several orders of magnitude larger than the cross section size, allows one to consider a fiber, a thin conductor, and similar objects as a one-dimensional heterogeneous medium.

Two- and three-dimensional heterogeneous media are widespread in industrial engineering and technology. Examples include filter media, rough or abrasive surfaces, free abrasive media, construction and tool materials, and many other technical objects.

Considering the geometric characteristics of the elements of the medium, as the simplest, we can conclude that the characteristic of heterogeneity, even in the size of particles, when examined in detail, can be quite complex. For example, the medium of an abrasive powder with a single fraction differs significantly from a medium with a multifraction composition. Thus, we can talk about the presence of a second order or a level of heterogeneity. This, in turn, leads to the need to introduce a separation criterion according to the corresponding indicator for particles of the medium, and in some cases, its zones and regions. So, for abrasive media, the range of sizes of fractions is stipulated by the standards [18].

However, even in this case, the characterization of the particles will be incomplete, since the existing methods do not take into account the shape of the grain, which can radically differ within one fraction and, accordingly, affect the properties of the heterogeneous medium (for example, by changing the coefficients of internal friction).

Using this simplest example, one can hypothesize that the heterogeneity in each of the above groups of characteristics can be described by a multifactorial vector

$$Q_i = \begin{cases} X_1 \\ X_2 \\ X_i \ \text{x}, \\ \ldots.. \\ X_n \end{cases} \tag{1}$$

where X_i- a characteristic of an element of the environment. Moreover, the dimension of the vector will be determined both by the boundaries of the system under consideration and by the significance of the characteristic for the theoretical or applied problem being solved.

The level or order of heterogeneity may or may not be relevant when modeling the behavior of the environment at different boundaries and purposes of analyzing the system under study. Obviously, for the vector (1), an additional dimension is added, which complicates the study.

Let us further consider such an indispensable element of a heterogeneous environment as the boundary between elements with different properties. The boundary is characterized by a change in the properties of the environment, which must be assessed by quantitative criteria. In this case, even speaking about the spatial geometric boundaries of a heterogeneous medium, one can operate with such a characteristic or concept as the thickness of the boundary. For example, a heterogeneous medium in the form of an emulsion or suspension has clearly defined boundaries of elements at the atomic level. Solid solutions of metals, in which the elemental components can be clearly distinguished, have more diffuse boundaries due to the diffusion of materials.

Thus, if neighboring elements of the environment have differences in the X_i parameter of the heterogeneous environment, then the boundaries will be determined by the criterion difference

$$\Delta_i = X_i' - X_i'' > \Delta k, \tag{2}$$

where Δk– the difference between the element separation parameter, which is critical for the problem under consideration.

Thus, through Δk, it is possible to quantitatively determine the size (thickness) of the boundary, for example, zones of electrolyte solution of different concentrations, which form a heterogeneous medium according to the mentioned parameter. In addition, criterion (2) is inevitably associated with the spatial gradient $\frac{dX_i}{dL}$ of the considered parameter in the dimension of the space L, which allows us to introduce the concept of border contrast. Indirectly, criterion (2) allows one to characterize the level of interaction between elements of a heterogeneous environment.

In a number of cases, such an approach requires certain assumptions when assessing the boundaries of the components of a heterogeneous medium (let's call them monoelements). One of the striking examples of a high-contrast boundary, which, meanwhile, has a certain volume and gradient of the components of a heterogeneous medium is the boundary of a galvanic or other coating on the base material (Fig. 1).

On the one hand, the boundary between coating and base materials has a clear contrast in terms of chemical and structural composition. However, almost any surface has a complex of microroughnesses. If we take into account such a roughness indicator as the length of the supporting surface tp, the mechanical properties of the coating layer at a thickness corresponding to the height of the microroughness of the surface of the base material R_{max} will differ significantly. Thus, from the point of view of physical and mechanical properties, the boundary will be blurred and have a sufficiently large thickness, and this thickness, as well as the gradient of properties determined by the parameters R_{max} t_p.

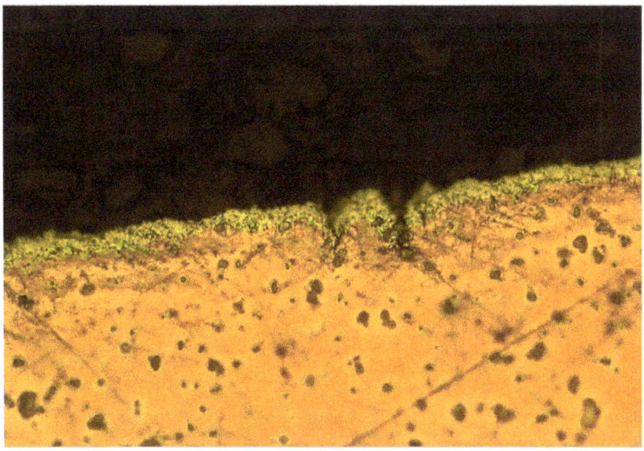

Fig. 1. *Boundary of a metal coating on a metal base*

When comparing the matrix of vectors Q_1 and Q_2 for different phases, it is possible to obtain a situation when the boundary characterized by the parameter X_c does not coincide with the boundary by the parameter $X_{n/}$

Or the boundary may be characterized by a critical change in only one parameter. For example, for electroplating, the presence of a certain concentration of microcracks and its change within (2) over the surface of the part indicates the heterogeneity of the coating on the surface, and all other indicators can be unchanged.

Thus, the localization of the border, depending on the tasks of research or modeling, can be formulated by one or more parameters of the vector (1), have a different gradient depending on the objectively present or subjectively specified criterion (2).

When considering the boundaries between the elements, one more key characteristic of the heterogeneity of the considered environment can be distinguished. This characteristic can be the length of the boundaries per unit of the dimensional characteristics of the medium. Taking into account the dimension of the medium, we can talk about point, linear, volumetric boundaries, as well as boundaries with a higher dimension. For example, the boundary between metals, in a simplified form, is considered as a certain line between grains in microscopic examination of a thin section. In reality, firstly, such a boundary is characterized by a certain criterion difference Δk, determined, as mentioned above, by the degree of diffusion of materials, and secondly, it is a complex three-dimensional surface. Then, the dimensional characteristic of the boundary will be the volume V_r, determined by the expression

$$V_r = S_r * \Delta k. \tag{3}$$

Accordingly, a characteristic of the heterogeneity of such an environment can be considered the indicator

$$Q = \frac{V_r}{V_a} \leq Q_k, \tag{4}$$

where V_a is the unit of the considered dimensional characteristic of a heterogeneous medium, for example, volume; Q_k is the critical ratio of the total dimensional characteristics of the boundaries and the total investigated volume for the problem under consideration.

The volume V_a for the case under consideration includes the components that can be considered homogeneous media for a particular study.

The considered case is a particular consequence of the proposed hypothesis for geometric three-dimensional boundaries.

Characteristics (3) and (4) are especially relevant precisely for geometric boundaries, since they make it possible to take into account the state of the surface and the development of the surface profile with the same dimensions of the elements of the components of the heterogeneous medium. In particular, for powder metallurgy, it is this indicator that will affect the degree of particle activation. For abrasive media with the same particle size - their surface area will determine the aggressiveness and abrasive ability during cutting,

Using the above approach, almost the majority of technical objects can be classified as heterogeneous media: structured materials, coatings, processing media, raw materials, materials obtained by using additive technologies. Even materials that are technically considered monolithic, for example, hypoeutectoid alloys of the iron-carbon diagram, can be classified as heterogeneous.

However, simplifying vector (1) i.e. removing from the vector (1) the corresponding indicators in the analysis of such materials and considering it conditionally monolithic, taking out of the analysis of the system the features of the structure and structure, the process of dispersing the material, for example, by the processes of abrasive or blade cutting, can be considered the synthesis of a heterogeneous structure from a homogeneous medium.

The above hypothesis makes it possible to form an almost universal approach to the study and modeling of a variety of processes associated with the formation of heterogeneous environments. For the technology of mechanical engineering, one of the basic ones is the processes of blade or abrasive cutting of metals and alloys. One way or another, but the essence of these processes is reduced to the dispersion of a monolithic solid material.

The process of dispersing a monolithic (conditionally monolithic) solid material can be considered as an increase in the degree of heterogeneity of the initial medium. The formation of cracks during destruction or dispersion is accompanied by the formation of interfaces and spaces between them. The presence of an initial crack - fulfills condition (2), at least the geometric characteristics of the formed elements appear in the vector (1). With an increase in the heterogeneity of the medium under consideration, its properties change dramatically.

If we consider various cutting methods and compare them according to the results of dispersion, then the resulting chips, as a heterogeneous medium, can be quantitatively characterized by the indicator (3). In this case, the increase in the surface of the boundaries of elements can be described by the function

$$Q = f(A, G, P, T, N), \quad (5)$$

where T is the cutting time, N is the power consumed by the process, G is the geometry parameters of the cutting wedge, A is the cutting modes, P is the complex of physical and mechanical properties of the material being processed.

In accordance with the theory of brittle fracture [19], the considered increase in heterogeneity in the form of processing a solid material into shavings requires energy consumption, which is determined by the properties of its own material and the characteristics of the elements. Thus, reducing the size of the elements requires higher energy consumption for dispersion, which does not contradict the known provisions of the theory of cutting.

If we generalize the influence of the dimensional characteristics of the boundaries and, accordingly, the parameter (5) on energy costs, then it will obviously be nonlinear, and the nonlinearity, depending on the parameter (4), will be changing (Fig. 2).

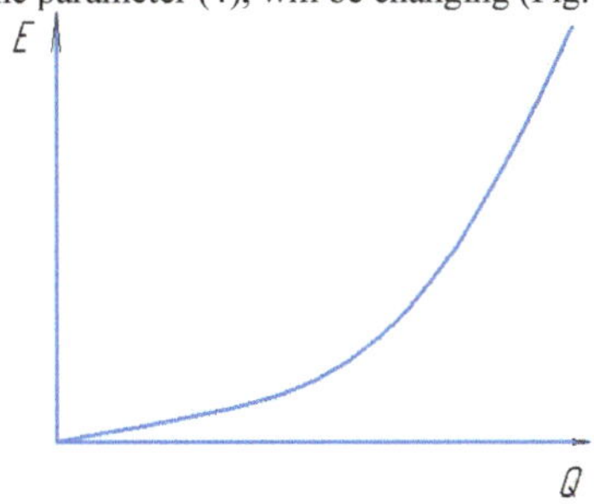

***Fig. 2.** Influence of the degree of development of the boundaries between the components of a heterogeneous environment on the energy consumption of its formation*

If we consider the processes of electroplating as the transformation of a two-component medium base-electrolyte into a three-component base-coating-electrolyte, then an increase in heterogeneity is also associated with the formation of another interface and with significant energy consumption.

There are known processes associated with the joining of metal surfaces by explosion welding [20]. In this case, the disappearance of two metal-gas-metal boundaries and the formation of a metal-metal boundary. Although the number of boundaries decreases, the new boundary is characterized by a sharp increase in volume due to the diffusion of metals. Thus, the heterogeneity of the environment in terms of indicator (3) increases, a spatial gradient $\frac{dX_i}{dL}$ of the boundary appears, respectively, this requires significant energy consumption, which is technologically confirmed.

A similar situation arises in the manufacture of suspensions, emulsions and other heterogeneous media consisting of substances in different phases.

Conclusion

Summarizing, we can put forward a hypothesis that increasing the heterogeneity of the environment requires energy costs and reduces the level of entropy.

The proposed hypothesis will obviously have many boundary conditions, for example, dispersion at the molecular level, processes related to nuclear interactions, in many cases may contradict the proposed hypothesis.

However, the processes described by the proposed approach include the process of metal cutting, surface plastic deformation, chemical-thermal treatment, formation of coatings, welding, most of the processes that change the level and gradient of hardness and internal stresses in the surface layer of the material. Application of the proposed principles makes it possible to find new directions in the study of such processes.

Acknowledgments

The work was carried out within the framework of the project No. 0748-2020-0013 "Scientific principles of the formation of heterogeneous structures by methods of physical and chemical dispersion" (state assignment to the university in the field of scientific activity. Customer: Ministry of Science and Higher Education of the Russian Federation).

References

[1] K Shannon, Mathematical theory of communication. «Works on Information Theory and Cybernetics, Moscow, Publishing of Foreign Literature Publishing House (1963) pp.243–332.

[2] B.A. Erznkyan, Technological and institutional development of the socio-economic system in a heterogeneous environment, JOURNAL OF INSTITUTIONAL STUDIES, Volume 4, No. 3. 2012.

[3] V.S. Igropulo, Heterogeneous environment with active components: physical system, mathematical model, real prototypes. Proceedings of Bulletin of Stavropol State University 38/2004 pp.113-116.

[4] V.I. Gorodnichenko and others, Materials Science: Workshop, Moscow – Logos Proceedings, 2003.

[5] G.I. Dulnev and V.V. Novikov, Transfer processes in heterogeneous media, Leningrad, Energoatomizdat Publishing, Leningrad Branch, 1991.

[6] Materials Science. Technology of structural materials: textbook. manual ed. B.C. Cherednichenko. - 2nd ed. - Moscow; Omega-L Publishing House, 2006.

[7] L.Barre, Kinetics of Heterogeneous Processes. - Moscow, Mir Publishing House, 1976.

[8] 8 D.A. Frank – Kamenetky, Diffusion and heat transfer in chemical kinetics, Moscow Publishing House Science, 1987.

[9] V.F. Kuropatenko, New models of continuum mechanics Engineering Physics Journal., 2011, Volume 84, No. 1. p. 74–92. https://doi.org/10.1007/s10891-011-0457-0

[10] V.F. Kuropatenko, Model of a multicomponent environment, Reports of the Academy of Sciences, 2005, Volume 403, No. 6. - p. 761–763

[11] V.Fortov, Powerful shock waves and extreme states of matter. Moscow, Bukos Publishing House, 2005. https://doi.org/10.1238/Physica.Topical.116a00007

[12] A.P. Aldushin, T.M. Martemyanova, A.G. Merzhanov, B.I. Khaikin and Shkadinsky K.G. Self-Oscillatory Propagation of a Combustion Front in Heterogeneous Condensed Matter, Journal of Combustion and Explosion Physics. 1973. Vol. 9, No. 5. p. 613-626. https://doi.org/10.1007/BF00742879

[13] V.G .Prokofiev and V.K. Smolyakov, Nonstationary regimes of combustion of a binary gasless mixture when ignited by a heated wall, Journal of Combustion and Explosion Physics. 2005.Vol. 41, No. 2. pp. 45-50. https://doi.org/10.1007/s10573-005-0018-y

[14] K.A. Aligozhin and A.G. Knyazeva, Modeling of transformation modes that are realized when joining dissimilar materials using solid-phase synthesis, Journal of Mathematical Modeling in Natural Sciences. 2013. No. 1. pp. 8-10

[15] A.Yu. Krainov and DS Shultz, Mathematical modeling of an ultrahigh-speed process in heterogeneous reacting powder mixtures, Journal of Computer Research and Modeling, 2011, Volume 3, No. 2, pp. 147-153. https://doi.org/10.20537/2076-7633-2011-3-2-147-153

[16] V.E. Fortov, Equations of state of matter: from ideal gas to quark-gluon plasma, Moscow, Fizmatlit Publishing House, 2012.

[17] R. I. Nigmatulin, Fundamentals of mechanics of heterogeneous media, Moscow, Science Publishing House. 1973.

[18] GOST R 52381-2005 Abrasive materials grain size and grain size composition of grinding powders. Grain composition control.

[19] V. M. Markochev and M. I. Alymov, On the theory of brittle fracture by J. Frenkel and A. Griffiths, CHEBYSHEVSKY Sbornik, 2017 Volume 18 Issue 3. https://doi.org/10.22405/2226-8383-2017-18-3-381-393

[20] Vladimir A. Grachev, Andrey E. Rozen, Yury P. Perelygin, Sergey Yu. Kireev, Irina S. Los Multilayer corrosion-resistant material based on iron–carbon alloys// Heliyon 6 (2020) e04039. https://doi.org/10.1016/j.heliyon.2020.e04039

Features of the Grinding Technology with the Use of Nanomodified Coolant

Igor I. Artemov[1,a], Vladimir D. Krevchik[1,b], Sergey A. Nesterov[1,c*], Dmitry A. Akimov[1,d] and Natalia E. Artemova[1,e]

[1]Research Institute for Fundamental and Applied Studies, Penza State University, 40 Krasnaya Street, Penza 440026, Russian Federation

[a]artemov@pnzgu.ru, [b]physics@pnzgu.ru, [c]nesterovs@list.ru, [d]info@zao-pztp.ru, [e]art.natalia2017@mail.ru

Keywords: Engineering and Technology, Cavitation, Nanomodification, Nanoparticles

Abstract. The results of the analysis of the possibility of constructing a system for ensuring the quality of mechanical engineering parts by nanomodification of the surface layer are presented. The implementation is considered on the example of grinding technologies with nanomodified lubricating coolant. Mathematical models describing the process of creating conditions for nanomodification of the surface layer and algorithms for practical implementation are presented.

Introduction

Formation of quality characteristics in the surface layers has always been the key problem of the production process during grinding operations. Being the finishing stage of processing, any deviations from the specified characteristics lead to failure of the entire process. Stable quality assurance in grinding operations, especially those of profile and deep types [1, 2], requires special attention to the design of operational technology. The paper discusses peculiarities of using coolant modified with various nanoparticles for grinding operations in order to improve quality characteristics of the surface layer by controlling surface defect structure of the material being processed.

Characteristics of Grinding Technology Using Nanomodified Lubricating Coolant

To control metal surface defect structure in designing machining to ensure the quality of the surface layer is an urgent and non-trivial problem. Nowadays, cavitation technologies for grinding intensification [3] and nanotechnologies to prepare lubricating coolants are widely used [4].

When developing grinding technology using nanomodified coolants, the following essential processing conditions must be satisfied: composition of nanomodified coolant (particle material, particle size, concentration); coolant supply conditions (pressure, flow rate, supply method); workpiece allowances for the grinding operation; processing modes (cutting speed, feed rate, cutting depth); material and type of grinding wheel (there is no available information on this condition yet).

To fulfill the above conditions, it is necessary to solve a number of scientific, engineering, and technological problems listed below.

Selection of Coolant Composition with Nanoadditives. The formation of cavitation bubbles within liquids allows solving a number of technological tasks for grinding, and change the quality of physical and chemical conditions of the coolant, namely: to reduce the cutting forces during processing, which provides for longer life of the wheels and increases the grinding ratio [4]; to improve cooling efficiency of the coolant [5]; to modify physical parameters of workpiece surface taking into account the activation of cavitation processes in the liquid, provided the presence of surface sediments in microelements [3, 6, 7]; to fill the arising microcracks in the surface layers of workpieces, thus increasing bearing capacity thereof [6].

Recommendations for using various nanoadditives [7, 8] in coolant composition to solve functional technological tasks are summarized in Table 1.

Table 1. *Applicability of nanoadditives to the coolant*

No.	*Functional technological task*	*Recommended additive*
1	Reducing the cutting force	Cu
2	Improving cooling efficiency	CuO, NiO, CNTs and SiC
3	Improving the efficiency of lubrication	TiO_2, SiO_2, Al_2O_3, MoS_2, ZnO and nanodiamond
4	Modifying surface physical parameters taking into account the activation of cavitation processes	Ni, Cu
5	Filling arising microcracks in the surface layers	Cu

Selection of Filler Composition in the Form of Nanodispersed Powder Additives is based on the adhesive properties of materials. There are practical recommendations on using nanopowder additives in coolant during processing with selection of material type in accordance with the grade being processed [5, 6].

For example, when grinding ferritic materials based on zinc manganese or copper compositions, there are recommendations [6] on the introduction of copper nanoparticles into coolant composition. Copper is chosen as a nanopowder material for grinding ferrite alloys due to the absence of influence of particles integrated into surface microcracks on the magnetic properties of materials. Besides, being filler in the coolant, copper is a versatile material, since it has high adhesion and penetrating properties to most metal materials processed by grinding. When grinding nickel-based superalloys, it is strongly recommended [6] to use nickel nanopowders as fillers.

Selection of Filler Particle Size in the Form of Nanodispersed Powder Additives is based on the key task to obtain penetration of particles into the deformed surface layer resulting from grinding. To accomplish this, one has to estimate the sizes of pores and microcracks on the surface, and select a particle size by three orders of magnitude smaller. For example, in [8], sizes of microcracks in ferrites are studied after being processed with diamond grinding wheels, and the crack width being 0.1-0.3 mm. In accordance with these data, it is recommended to use copper nanopowders with 50-100 nm particle size as additives. Damages to the surface layers in the form of microcracks with a width of 0.01-0.02 mm [9] were found during deep grinding of turbine blades made of superalloy. In accordance with [6], it is advisable to use nickel powders with 7-80 nm particle size.

Selection of Grinding Modes is based on the calculations recommended by the reference books, and theoretical calculation of the operating conditions of cavitation processes occurring in the nanomodified coolant.

During finishing processing by grinding with the use of nanomodified coolant, the surface of the processed material is actively saturated with nanoparticles from the liquid medium by means of the cavitation mechanism of the acoustocapillary effect with the use of metal nanoparticles contained as additives in the process liquid.

Cavitation can be realized in a coolant environment due to shock interaction of abrasive grains with the processed surface. The waves propagate through the abrasive grains, and then are transferred to the liquid. The collapse of cavitation bubbles near the surface layer leads to the formation of cumulative jets, which stimulate the penetration of process liquid with nanoparticles into the channels of microcracks. As a result, microcracks of the metal surface layer are filled with nanoparticles, and the effective surface Young's modulus is modified [10, 11, 12]. The use of nano-modified coolant in combination with cavitation processes could form the basis for a strengthening technology to form wear-resistant surface layers. To calculate the collapse time t_c of cavitation bubble, the Nolting-Nepayres equation was used. An analytical formula for t_c was obtained for incompressible liquid approximation when neglecting viscous losses:

$$t_c = \frac{1}{\omega}\arcsin\left[\frac{\omega_0^2 R_0^2 \rho_0}{p_m}\sqrt{\left(1-\Omega^2\right)^2+\left(\delta\cdot\Omega\right)^2}+\sin\varphi\right]-\frac{\varphi}{\omega}, \tag{1}$$

where ρ_0 is the density of the liquid; p_m is the amplitude of pressure fluctuations in the liquid; R_0 is the bubble radius; ω is the vibration frequency; ω_0 is the resonant frequency, $\omega_0 = \frac{1}{2\pi R_0}\sqrt{\frac{3\gamma p_0}{\rho_0}}$; p_0 is the atmospheric pressure; $\Omega=\omega/\omega_0$; δ is the logarithmic decrement; $tg\varphi = -\delta\cdot\Omega / \left(1-\Omega^2\right)$.

The value of t_c (10^{-8} s) has been estimated for the following parameters from Eq. (1) ($R_0=10^{-6}$ m, $p_0=10^5$ Pa, $\rho_0=103$ kg/m^3, $\gamma=1.4$, and $\delta=0.10$), constrained by:

$$\left|\frac{\omega_0^2 R_0^2 \rho_0 \delta}{p_m}\Omega+\sin\varphi\right|\leq 1. \tag{2}$$

It can be seen that t_c depends on the vibration frequency, the bubble radius, coolant density, amplitude of pressure fluctuation in the liquid, and on the logarithmic decrement.

The t_c value is bounded from above: $\omega\, t_c<1$, where ω is the vibration frequency. The cavitation bubble should collapse in a time shorter than the vibration period.

Let us analyze the potential of cavitation bubbles during grinding with wheels, taking into account the distribution of grains [13] in the structure of the wheel. Cavitation penetration of microparticles from the collapsing bubble into the formed crack occurs on the surface of the processed material at the moment of interaction between the grain and the formed microcrack. The vibration frequency should be calculated based on the distance parameters between the abrasive grains and the peripheral speed of the abrasive wheel.

The average distance between the abrasive grains for the wheels of the main grain fraction and structure can be calculated as:

$$l = \sqrt[3]{\frac{1000}{g\cdot n}} = \sqrt[3]{\frac{1000000}{\left(62-2\right)\cdot N\cdot\gamma\cdot n}}, \tag{3}$$

where γ is specific gravity of the abrasive material determined experimentally [14], being γ=3.85 for aluminium oxide and γ=3.10...3.20 for silicon carbide; N is the structure number of the grinding wheel; g is the weight of abrasive grains in 1 cm^3 of the wheel determined as $g=(62 - 2\cdot N)\cdot\gamma/100$; n is the amount of grains in 1g of grain (the n value is not stable, being 3,000 for 60-grit aluminium oxide grinding wheel [13]).

According to Eq. 3, the calculated average distance between the abrasive grains is 0.252 mm for the wheels of 60-grit and No. 4 structure.

The vibration frequency can be estimated by the classical formula $\omega=V_K/l$ during the grain passage over the processed surface.

Taking into account the real speed of the grinding wheel V_K of 35 m/s, we obtain the vibration frequency ω from 71 kHz to 146 kHz. Based on the condition for bubble collapse time t_c ($\omega\, t_c$ <1), it can be estimated that the calculated vibration frequencies match the limit parameters.

The performed estimation of the amount of nanoparticles N_H found in the microcrack has shown its dependence on the time of cavitation bubble collapse and geometric parameters of the microcrack. An analytical formula [15] for the effective Young's modulus of the metal surface layer has been obtained:

$$\frac{E_{eff}}{E_1} = \frac{4r_H^3}{r_0^2 L_{MC}} \cdot N_H \cdot \left(1 - \frac{16u_0}{\pi N_{MC}}\right) + \frac{16u_0}{\pi N_{MC}}, \tag{4}$$

where r_0 is the radius of the microcrack base (the geometric shape of a microcrack is modeled by a cone with a height L_{MC}, and the base radius r_0); E_1 is the Young's modulus in the volume of the structural material; u_0 is the distance dispersion between microcracks; N_{MC} is the amount of microcracks on the metal surface.

It is observed that the effective Young's modulus is a function of the number of nanoparticles found in the microcrack channel, and the size of nanoparticles and microcracks. The performed assessment has shown that the effective Young's modulus can exceed its initial value by about an order depending on the time of cavitation bubble collapse.

Thus, by varying t_c, it is possible to control such metal surface characteristics as roughness (due to pore and microcavity filling with metal nanoparticles), and microhardness (due to the change in the effective Young's modulus).

To test the method for selection of coolant composition, we consider the processing technology using an installation for coolant aeration (Fig. 1). Aeration of liquid is achieved by high-speed radial jets of compressed air throughout the entire volume of the liquid flow. Then the liquid is fed into device 8 for mixing it with hard-facing metal (nickel) nanopowder with a size of 7-80 nm.

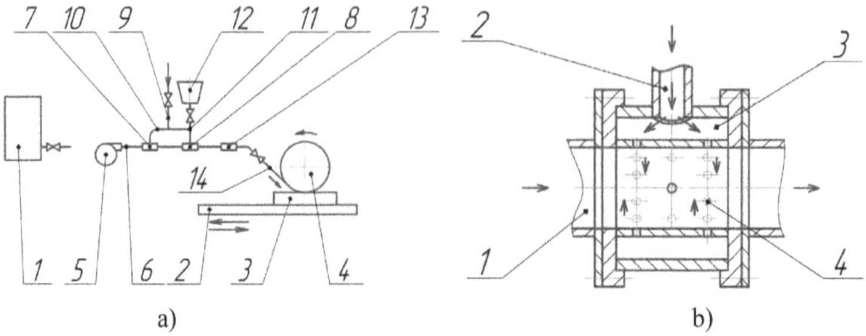

Fig. 1. *An installation for coolant aeration: a) overall scheme (1- lubricating coolant tank; 2 - grinding machine table; 3 - workpiece; 4 - abrasive tool, 5 - pump; 6 - lubricating coolant supply pipeline; 7- aerator; 8 - liquid and nanopowder mixing device; 9 - compressed air supply pipeline; 10,11 - air exhaust pipes; 12 - nanopowder container; 13 - hydrodynamic cavitation device for liquid flow treatment; 14 - coolant supply pipeline); b) aeration zone(1- liquid supply pipeline; 2 - compressed air supply pipe; 3 - annular chamber for compressed air spraying; 4 - radial air outlets).*

In device 8, the liquid is turbulized due to tangential air jet. Further, the aerated and turbulized fluid flow with nanopowder nanoparticles, uniformly distributed in the liquid medium, is fed to device 13 for hydrodynamic cavitation processing of liquid flow to form cavitation bubbles in the liquid and increase their total volume in the liquid to 45-60 vol. %. Nanoparticles of hard-facing metal (nickel) are concentrated along the boundaries of cavitation bubbles. The liquid saturated with cavitation bubbles and nanoparticles enters receiver 2, and its velocity and turbulence being decreased. Having been processed, the lubricating coolant is fed into the grinding area through pipeline 14 via the coolant supply system of the machine-tool. The implantation of hard-facing metal nanoparticles into microcracks and dislocations of the part surface is performed simultaneously with grinding [6].

Implementation of the Technology Design Concept

An algorithm for the RTP 2000 system procedure of computer-aided design of technological processes has been applied for grinding operation synthesis. The system has been developed at Penza State University (Penza, Russia) since the 2000s [16], and operates according to the principle described in [17]. An open architecture of this system allows users to independently implement their algorithms therein.

An example of grinding operational technology implementation in accordance with Russian standards for technology description is shown in Fig. 2. The example is given in Russian, since the software product is adapted to the Russian system of technological document management. Fig. 2 presents input box entries for the key parameters of a technological operation.

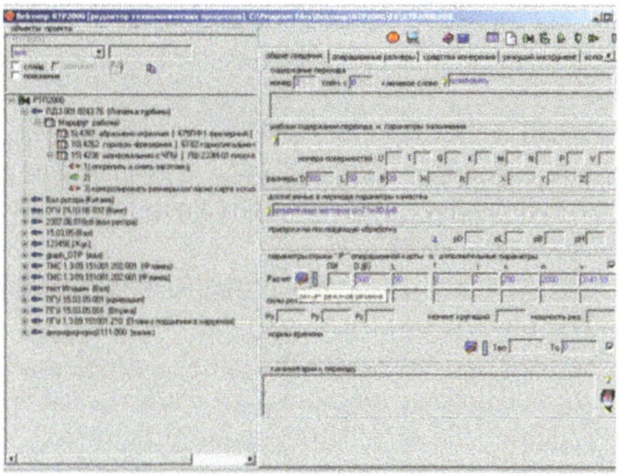

Fig. 2. *Input box entries for the key parameters of a technological operation.*

The software comprises the algorithms described above. There is an algorithm to select parameters and processing conditions in the system with built-in model. It allows to recommend a user the formation of an operational technology by creating the most adequate composition of the operation. Having analyzed the proposed option, the user is able to edit the content of operational transitions. The system gains new knowledge after user's edition, and considers the best algorithms to form the content and conditions for the considered data implementation in future.

Summary

Mathematical models for nanocluster formation mechanism have been created in the volume of metal surface layer microstructure under the conditions of its cavitation processing with a nanomodified lubricating coolant.

Methods and recommendations for regulating technological modes of finishing metal parts and assemblies with a nanomodified coolant under cavitation have been developed.

To create a technological system, a methodology and an algorithm for controlling a technological process to ensure quality of finishing metal products with a nanomodified lubricating coolant under cavitation have been developed and introduced into the RTP 2000 system of computer-aided design of technological processes.

Acknowledgements

The present study was supported by the Ministry of Education and Science of the Russian Federation (Project # 0748-2020-0012).

References

[1] G.A. Kardashev, Physical Methods for Intensification of Chemical Technology Processes, Chemistry Publishing, Moscow, 1990.

[2] S.S. Silin, B.N. Leonov, V.A. Khrulkov, V.A. Poletaev, A.V. Lobanov and N.S. Rykunov, Optimization of Creep-Feed Grinding Technology, Mechanical Engineering Publishing House, Moscow, 1989.

[3] M.A. Promtov, Prospects of cavitation technologies application for intensification of chemical technological processes, TSTU Transactions 14 (2008) 861-869.

[4] B. Shen, P. Kalita, A. Malshe and A. Shih, Performance of novel MoS2 nanoparticles based grinding fluids in minimum quantity lubrication grinding. Transactions of NAMRI/SME 36 (2008) 357-364.

[5] M. Yang, Ch. Li, Y. Zhang, D. Jia, R. Li and W. Ding, Thermodynamic mechanism of nanofluid minimum quantity lubrication cooling grinding and temperature field models, in: M.Sh. Kandelousi (Ed.), Microfluidics and Nanofluidics, IntechOpen Ltd., London, 2018. https://doi.org/10.5772/intechopen.74969

[6] I.I. Artemov, E.M. Kirin, V.D. Krevchik and N.P. Simonov, Method of hard facing of metal part, RU Patent 2560604 C1 (2015).

[7] F. Pashmforoush, R.D. Bagherinia, Influence of water-based copper nanofluid on wheel loading and surface roughness during grinding of Inconel 738 superalloy, J. Clean. Prod. 178 (2018) 363-372. https://doi.org/10.1016/j.jclepro.2018.01.003

[8] V.D. Krevchik, V.O. Sokolov and D.V. Vasin, Acoustically stimulated change in physical and mechanical properties of the surface layer of ferrite products, in: Proceedings of All-Russian Scientific and Technical Conference on Improving Efficiency of Machining Based on Modeling Physical Phenomena, RSATU, Rybinsk, 2009.

[9] A.P. Zhuk, V.A. Skryabin and A.G. Skhirtladze, Features of processing technologies for blade root of diesel turbochargers, Repair, Reconditioning, Modernization 11 (2012) 30-33.

[10] I.I. Artemov, V.D. Krevchik, Acoustic emission under “hidden” growth of microcracks, J. Mach. Manuf. Reliab. 4 (2005) 92-95.

[11] I.I. Artyomov, V.D. Krevchik and N.P. Simonov, The quantum nanocluster formation mechanism in the surface layer of a metal with the chain of microcracks in terms of acoustocapillary effect cavitation mode, University proceedings. Volga region. Technical sciences 4 (2011) 192-204.

[12] I.I. Artemov, V.D. Krevchik, A.V. Rudin, N.E. Artemova, A.A. Voinov, S.B. Menshova, A.V. Sokolov, N.P. Simonov and A.V. Lipov, Method to produce nanostructured layer on surface of metals under conditions of acoustocapillary effect, RU Patent 2547051 C2 (2015).

[13] P.I. Yashcheritsyn, A.G. Zaitsev, Improving the Quality of Ground Surfaces and Cutting Properties of Abrasive Diamond Tools, Science and Technology Publishing House, Minsk, 1972.

[14] E.N. Maslov, Basics of Metal Grinding Theory, Mashgiz Publishing House, Moscow, 1951.

[15] I.I. Artemov, V.D. Krevchik, N.P. Simonov and R.V. Zaytsev, Application of cavitation technologies and nano-modifiers of the process fluid in the surface engineering of machine parts, in: Proceedings of International Conference on Surface Engineering for Research and Industrial Applications (INTERFINISH-SERIA 2014), Novosibirsk, Russia, 30 June-04 July 2014, pp. 43-44.

[16] V.A. Belashov, A.V. Belashov and M.V. Belashov, Technological process editor, Certificate 940187 of Russian Agency for the Protection of Computer Programs, Databases and Topologies of Integrated Circuits (1994).

[17] V.A. Belashov, A.V. Belashov and M.V. Belashov, Operation technology synthesis automation based on the dimensional analysis, University proceedings. Volga region. Technical sciences, 2 (2007) 100-108.

The Structure and Properties of Ceramics, Obtained from Mixtures of the System ZrO_2–Al–C

S.I. Dzhelialov

Department of Electromechanics and Welding, Crimean Engineering-Pedagogical University named after F. Yakubov, Simferopol, Russia

dzhelyalov@mail.ru

Keywords: Ceramic Materials, Cutting Ceramic, Zirconium Oxide, Aluminum, Carbon, Hardness, Strength, Hot Pressing, Structural Analyses

Abstract. The paper presents the results of the analysis of the structures of materials obtained as a result of hot pressing of mixtures of the $ZrO_{2(m)}$ –Al – C system of various compositions. The nature of the change in the composition and physical and mechanical properties of the obtained materials was established depending on the concentration of aluminum and carbon in the initial mixture, as well as the parameters of hot pressing.

Introduction

The potential of the materials with which a metal-cutting tool is equipped is one of the key indicators that determine the quality and productivity of cutting processes. The constant expansion of the range of difficult-to-machine structural alloys and the need to increase cutting speeds entails the need to create new types of metal-cutting materials. In the traditional hierarchy of the latter, the class of mineral ceramics (ceramics) occupies its own niche, distinguished by its ability to provide high-speed finishing and semi-finishing of most structural alloys, including difficult-to-machine alloys. At the same time, ceramics are distinguished by their ability to most effectively resist chemical-thermal and adhesion processes occurring on the cutting edge of the tool during processing. A lot of research carried out by scientists from different countries [1–11] indicates the prospects for the development of these types of materials. It should be noted that in a number of works, researchers pay attention to eliminating one of the most important disadvantages of tool ceramics – reduced crack resistance.

The results of studying the regularities of the formation of the structure and properties of materials synthesized from powder mixtures of the $ZrO_{2(m)}$–Al–C [12, 13] system showed that it is promising for creating a material for instrumental purposes. The **purpose** of this work is to study the structure and phase composition of materials obtained by hot pressing (HP) from mixtures of this system and to establish the nature of their change depending on the concentration of aluminum and carbon in the initial mixture, as well as the parameters of hot pressing.

Experimental technique

When preparing a mixture to obtain a cutting ceramic material, we used powders of monoclinic zirconium dioxide produced by SRPE «Tsirkonii» (Dneprodzerzhinsk, Ukraine), manufactured according to TUU 14–10–098–99, and PAP-4 grade aluminum powder (GOST 5494–71) and lamp black grade S-1 (GOST 18307–78).

The levels and intervals of variation of factors accepted in the study, which correlated with the results of works [12, 13], are shown in Table 1.

Materials Research Proceedings **21** (2022) 450-459 https://doi.org/10.21741/9781644901755-74

Table 1. *Levels and intervals of variation of factors*

Factors	Variations Interval	Natural levels of factors corresponding to the coded				
		+2	+1	0	–1	–2
The concentration of the initial components of the mixture, wt %						
C	1	2,5	3,5	4,5	5,5	6,5
Al	2,5	16	18,5	21	23,5	26
Hot pressing process parameters						
P, MPa	3	18	21	24	27	30
T, °C	60	1400	1460	1520	1580	1640
Isothermal holding time, τ, min	5	5	12	19	26	33

The milling and mixing of the initial charge was conduct in a steel drum lined with hard alloy plates on a roller table. Carbide balls with a diameter of 10–12 mm were used as grinding solids. The portions of powder mixtures were pre-briquetted (cylinders with a diameter of 19 mm) and placed in a graphite mold with a cylindrical working chamber with a diameter of 20 mm, after which it was installed in the working zone of copper high-frequency current inductors. The hot pressing of prototypes was conduct on a unit equipped with a thyristor frequency converter of the ТПЧТ-120 type. To measure the heating temperature during hot pressing, a PPT-131 pyrometric transducer was used.

The hardness of the obtained samples was measured according to Vickers (HV_{15}) on a TP-2 hardness tester at a load of 15 kN and holding under load for 10 s (GOST 25172–82). Palmquist fracture toughness (K_{1c}) of the materials was determined on indentations obtained by indentation with a Vickers pyramid according to the method described in [14].

Metallographic studies of the structure of the obtained materials, their photographing were performed on a metallographic microscope "METAM-P1" manufactured by "LOMO", equipped with a digital video camera, in reflected light.

Fractographic studies in this work were performed using the method of scanning electron microscopy on an analytical field emission scanning electronic microscope "EVO 50XVP" (CarlZeiss). Detection and analysis of X-rays was performed using an X-ray energy dispersive spectrometer «INCA Penta FET-x3».

The discussion of the results

The results of work [12], which were performed in accordance with the intervals of variation presented in the Table 1, showed that the most pronounced effect on the structure of hot-pressed materials is exerted by a change in the concentration of carbon in the initial mixture.

In the Fig. 1 shows typical types of the structure of samples obtained by HP of mixtures of the composition ZrO_2–21Al–6.5C and ZrO_2–26Al–4.5C% (wt %). The first composition assumes the maximum possible carbon content in the initial charge from the specified variation intervals, the second assumes the maximum possible aluminum content (the corresponding carbon and aluminum concentrations are in the center of the variation interval). The Fig. 1, a shows that the phases included in the material structure obtained from a mixture of ZrO_2–21Al – 6.5C% (wt %) are unevenly distributed in the volume of the material.

According to the results of metallographic studies, materials obtained from mixtures of $ZrO_{2\,(m)}$ –18.5Al – 3.5C% (wt %) at a hot pressing pressure of 21 MPa have a structure containing local pore clusters and loose areas (Fig. 2, a).

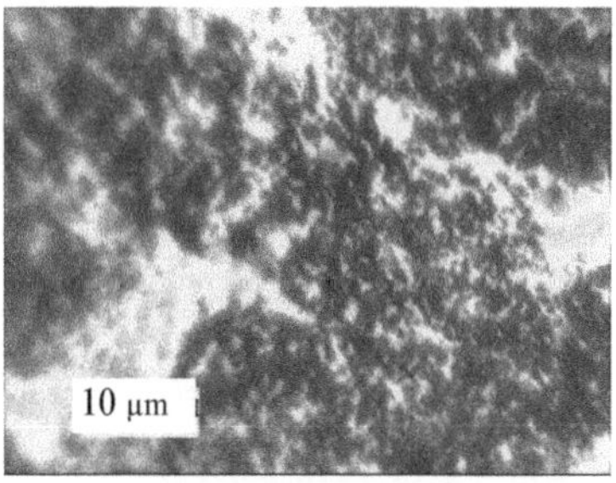

a

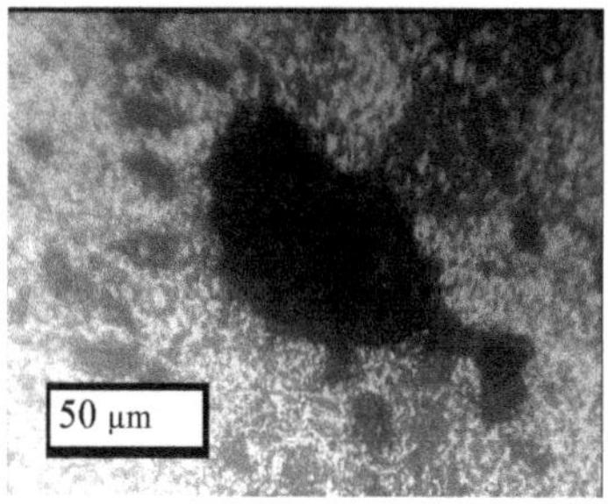

b

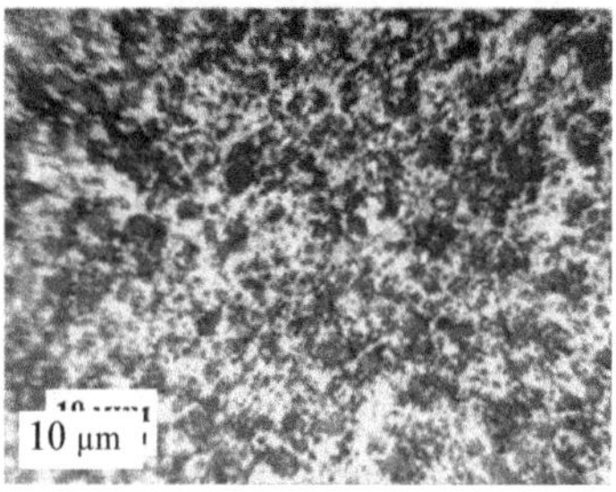

c

Fig. 1. *The structure of materials obtained by hot pressing mixtures: a, b – ZrO_2–21Al–6,5C; c – ZrO_2–26Al–4,5C % (wt %).*

The sizes of phase inclusions, considering the large grains, are in the range of 0.5–20 µm. The specified wide range of grains size creates a wide range of hardness values ($HV_{15}=18,7^{+0,5}_{-1,5}$ MPa) in the volume of the material. Also, in this structure, defects in the form of large pores are observed (see Fig. 1, b). The reason for the formation of pores can be the low compressibility of the sample due to the high carbon content in the initial powder mixture.

The structure of materials obtained from a mixture of the composition ZrO_2–26Al – 4.5C% (wt %) is distinguished by a significantly more uniform distribution of phase components and a defect-free structure, while the range of grain sizes is limited by narrower limits: 2–5 µm (Fig. 1, d). Taking into account the fact that the parameters of hot pressing for the considered mixtures were at the same level, the effect of an increase in the concentration of aluminum in the initial charge becomes obvious. At the same time, the relatively low hardness values of samples obtained from a mixture of the composition ZrO_2–26Al – 4.5C% (wt %) (on average HV_{15} = 17.7 MPa) are probably caused by the presence of intermetallic inclusions in its structure, the formation of which is characteristic of this system. (see the example in Fig. 3).

In Fig. 2 shows the structures of materials obtained from a mixture of $ZrO_{2\,(m)}$–Al–3.5C% (wt %), with different – 18.5 and 23.5% (wt %) aluminum content. It can be seen from the figure that, despite the differences in the Al content in the initial mixture, the grain size of the presented materials differs insignificantly (Fig. 2, c, f) and is in the range of 0.5–2 µm. In accordance with the data [12], an increase in the hot pressing pressure to 27 MPa leads to a decrease in the volume of the above-mentioned structural defects. However, the presence of randomly located single pores throughout the sample volume (Fig. 2, b, d) indicates that this is not enough for the formation of a continuous non-porous structure. This disadvantage is absent in materials obtained from mixtures

Materials Research Proceedings **21** (2022) 450-459 https://doi.org/10.21741/9781644901755-74

of $ZrO_{2(m)}$–23.5 Al–3.5C% (wt %). It is indicates that an increase in the aluminum content to 23.5% (wt %) improves the compressibility of the samples. and contributes to the formation of practically non-porous structures (Fig. 2, e).

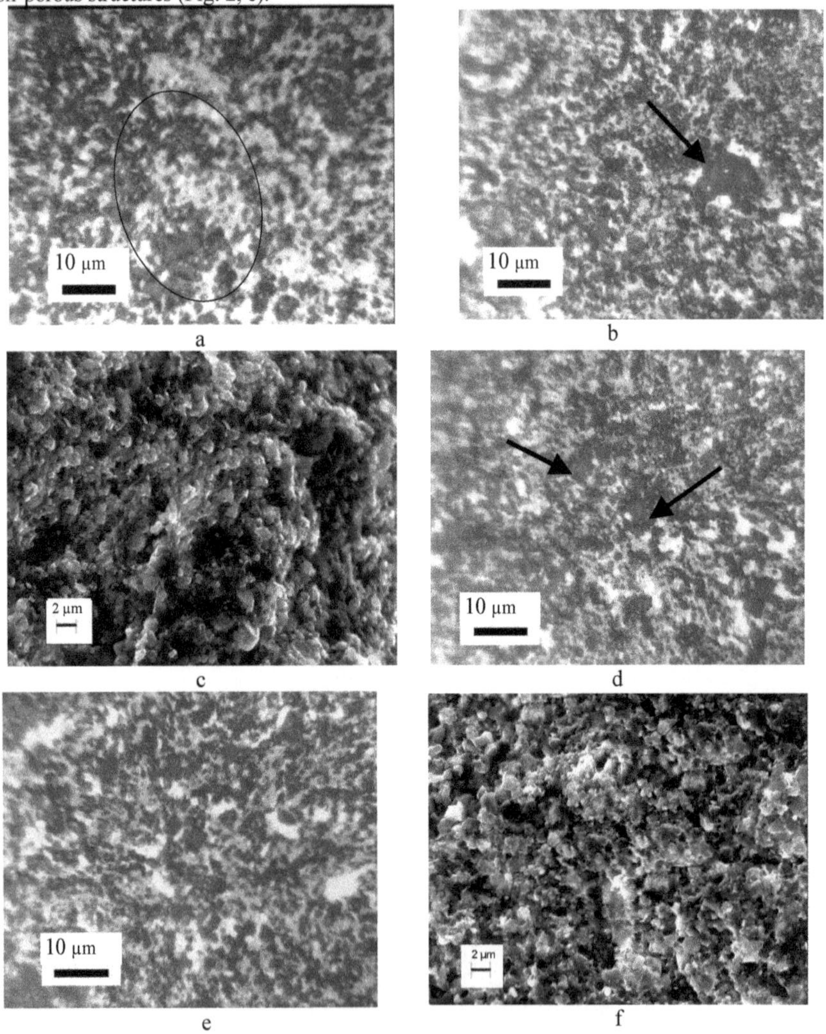

Fig. 2. *The structure of materials obtained by hot pressing a mixture of* $ZrO_{2(m)}$*–Al–3.5C% (wt %), with different content of aluminum in the initial mixture (% wt %): a, b, c, d – 18.5; e, f – 23.5; a – at P=21 MPa; b, d – at P=27 MPa.*

At the same time, it should be consider that an increase in the content of aluminum in the composition of the initial mixture also leads to an increase in the number of compounds based on it in the structure of hot-pressed samples. In Fig. 3 shows the distribution of elements in the structure of a hot-pressed material, from a mixture of $ZrO_{2\,(m)}$–23.5Al–3.5C% (wt %), obtained as a result of X-ray microanalysis.

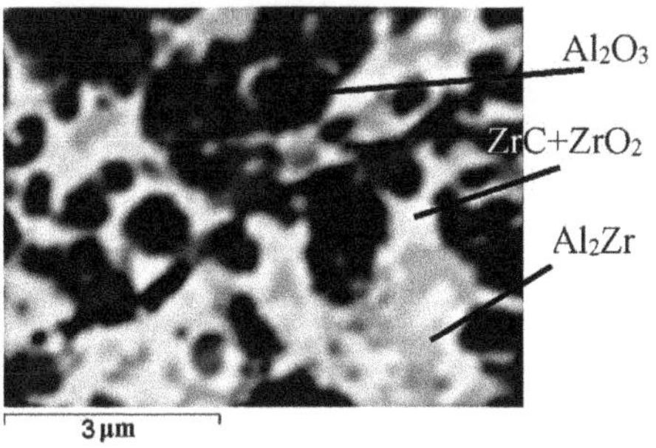

Fig. 3. *Microstructure of a thin section in phase contrast (a) of a hot-pressed sample obtained at T = 1500 °C, τ = 10 min, P = 24 MPa from a mixture of* $ZrO_{2\,(m)}$*–23.5Al–3.5C% (wt %).*

The presence of the intermetallic compound Al_2Zr (Fig. 3, see the gray phase), the amount of which in the composition of the materials under consideration is 2 times lower than the stoichiometric (ie, calculated according to the equilibrium equation), reduces the hardness indices.

Therefore, materials obtained even at high specified values of hot pressing parameters (temperature (1580 °C), pressure of 27 MPa, and isothermal holding time up to 26 min) have a relatively low level of hardness (up to HV_{15} = 16.3 GPa), which excludes consideration of its application as cutting ceramics. It should be noted that the structure of samples obtained from a mixture of $ZrO_{2\,(m)}$–18.5 Al–5.5C% (wt %) also indicates a weak compressibility of the material, which is expressed in the presence of areas with defects in the form of single pores and friability (Fig. 4).Studies of the microstructure of samples obtained at a temperature of 1580 °C (see Fig. 4, b) show that even an increase in the hot pressing temperature insignificantly reduces the percentage of structural defects and does not contribute to the production of high-density samples. This, in turn, may be caused by the low aluminum content, which reduces the tendency to compacting the compact both during the melting process and at subsequent stages, due to the viscous and plastic flow of lower-temperature (as opposed to compounds with zirconium and carbon) substances on its basis.

Materials obtained from a mixture of $ZrO_{2(m)}$–23.5Al–5.5C% (wt %) have a relatively defect-free structure, characterized by a fine-grained structure. However, in this case, the phase components are distributed unevenly in the bulk of the materials (see Fig. 5).

Their structure contains areas with accumulations of phases both based on aluminum (Fig. 5, a) and zirconium oxide and carbide (see Fig. 5, b). The latter factor creates a wide range of HV_{15} and K_{IC} indices (respectively $18{,}5^{+0{,}5}_{-1{,}4}-18{,}7^{+0{,}4}_{-1{,}3}$ GPa, 4.3–4.6 $MPa{\cdot}m^{-1/2}$). It should also be noted that the samples hot-pressed at a pressure of 27 MPa, according to statistics, have the highest average hardness values (about 18.7 GPa). However, at the same time, they have a relatively low level of crack resistance, which is probably caused by an increase in the content of the more brittle than the intermetallic phase ZrC, as well as an increase in the content of Al_2O_3.

Materials Research Proceedings **21** (2022) 450-459 https://doi.org/10.21741/9781644901755-74

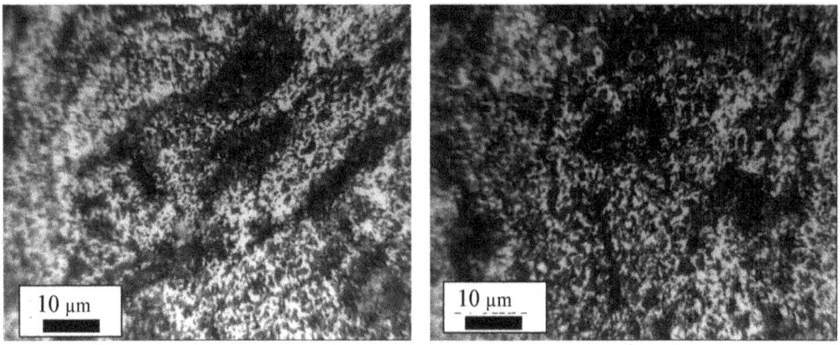

a b

Fig. 4. *The structure of some materials obtained by hot pressing a mixture of* $ZrO_{2(m)}$ *– 18.5Al–5.5C% (wt %), at P = 27 MPa and a – T = 1460 ° C, τ = 26min.; b – T = 1580 °C, τ = 12 min.*

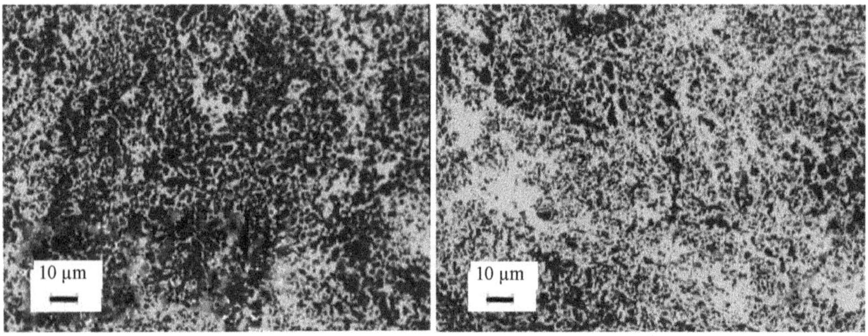

a b

Fig. 5. *Unevenness of the structure of materials obtained by hot pressing a mixture of* $ZrO_{2\,(m)}$*– 23.5Al–5.5C% (wt %): a – an accumulation of compounds based on aluminum (dark phase); b – accumulation of compounds based on zirconium (light phase).*

This is evidenced by a comparison of the structures of the samples of this material (Fig. 6) with those presented in Fig. 3 (where the carbon content in the initial mixture is 3.5% (wt %)) and the results of their X-ray microanalysis. A decrease in the amount of intermetallic compound occurs with an increase in the carbon content in the mixture, as a result of their interaction. In contrast to materials obtained from a mixture containing 5.5% C (wt %), the structure of materials with a carbon content of 4.5% (wt %) is primarily distinguished by a more uniform distribution of phases in the bulk of the material.

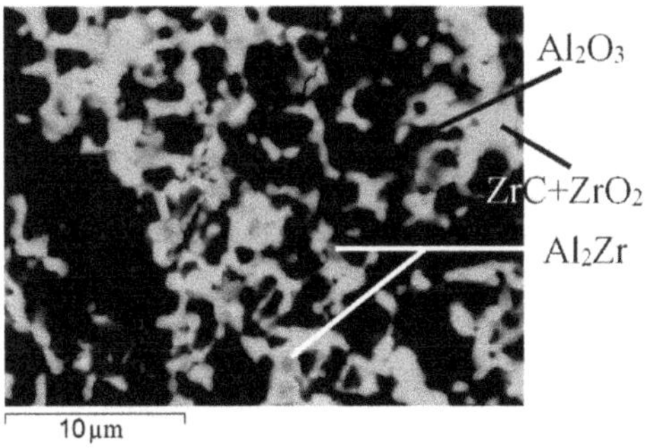

Fig. 6. *Microstructure of a thin section in phase contrast (a) of an hot-pressed sample obtained at T = 1580 °C, τ = 12 min, P = 27 MPa from a mixture of $ZrO_{2(m)}$–23.5Al–5.5C% (wt %)*

The low aluminum content in the composition of the initial mixture $ZrO_{2\,(m)}$ –16Al – 4.5C% (wt %) leads to poor compressibility of the mixture. The consequence of this is the presence of a small number of single pores in the material obtained from it (see Fig. 7, a). These materials are characterized by very low hardness values (average HV_{15}=15.5 GPa). According to studies [13], a decrease in the amount of aluminum leads to a decrease in the amount of intermetallic compound formed during the interaction of Al with $ZrO_{2(m)}$. This, in turn, reduces the amount of zirconium carbide, which is formed as a result of carbidization of the intermetallic compound, and prevents the growth of the aggregate hardness of the material. The structure of the latter is distinguished by a high content of the $ZrO_{2(m)}$ phase (see Fig. 7, b), which has not reacted, which also leads to a decrease in its physicomechanical characteristics.

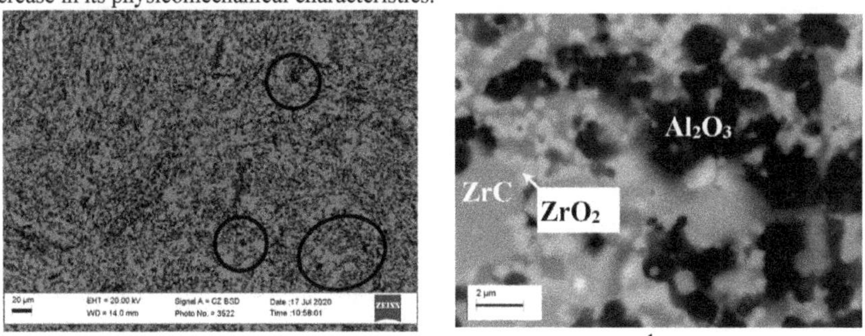

a b

Fig. 7. *The structure of the material obtained by hot pressing the mixture $ZrO_{2(m)}$–16Al–4.5C% (wt %), at T = 1520 °C, τ = 19 min, P = 24 MPa: a – general view; b – microstructure of a thin section in phase contrast.*

Of the entire group of materials containing 4.5% (wt %) carbon in the initial mixture, the samples obtained by hot pressing a mixture of $ZrO_{2(m)}$–21Al–4.5C% (wt %) at a temperature of 1520, 1640 °C under pressure of 24 MPa and above, with an isothermal holding time of at least 19 minutes, they have the highest hardness indicators. It should also be noted that, along with high hardness (HV15 = 18.6–18.7 GPa), they have a fairly high level of crack resistance (on average, in the range of 5.7–5.8 $MPa{\cdot}m^{-1/2}$). Their structure has no defects and an evenly distribution of phases, the range of values of which is within 0.5–2 μm (Fig. 8). The even nature of the distribution of phase components in the bulk of the material and the absence of defects in the samples shown in Fig. 8 indicates that the pressure and temperature of isothermal holding at the level of 24 MPa and 1520 °C are sufficient to obtain dense materials from this mixture.

The results of experiments on hot pressing of a mixture of $ZrO_{2\,(m)}$–21Al–4.5C (% wt %), at an operating pressure of 18 MPa, indicate that this level of compression is insufficient to obtain materials of maximum density. This is indicated by the presence in the structure of the obtained samples of pores (Fig. 9), distributed throughout the volume of the material. The similar structure has a sample obtained at a higher pressure (24 MPa), but a shorter time of isothermal holding (Fig. 9).

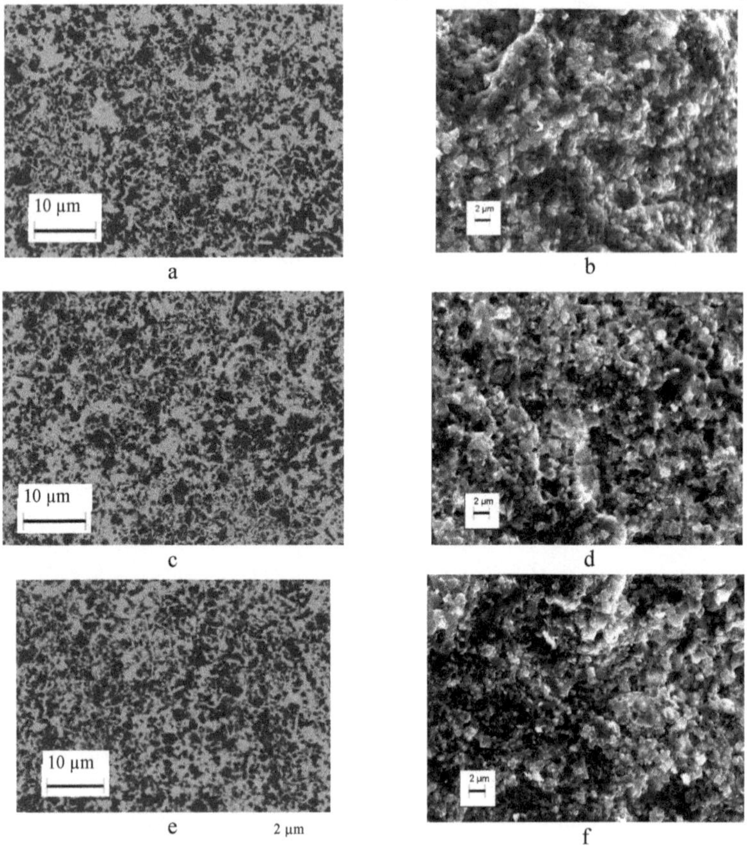

Fig. 8. *Microstructure in phase contrast (a, c) and fracture surface (b, d) of hot-pressed materials obtained from a mixture of* $ZrO_{2\,(m)}$ *–21Al – 4.5C% (wt %) at a –* $T = 1520$ *°C,* $\tau = 19$ *min,* $P = 30$ *MPa; b –* $T = 1640$ *°C,* $\tau = 19$ *min,* $P = 24$ *MPa; b –* $T = 1520°$ *C,* $\tau = 19$ *min,* $P = 24$ *MPa.*

The content of pores and looseness in such samples is somewhat lower. The presence of residual porosity is probably associated with a short time of isothermal holding, during which the sample does not have time to sufficiently compact due to plastic and viscous flow at the last stage of structure formation. The results of the analysis of the structures and physicomechanical characteristics of the last two samples indicate that a decrease in pressure to 18 MPa has a more significant effect on a decrease in the value of the relative density than a decrease in the isothermal holding time to 5 min. This is reflected in an increase in both the size and the number of pores.

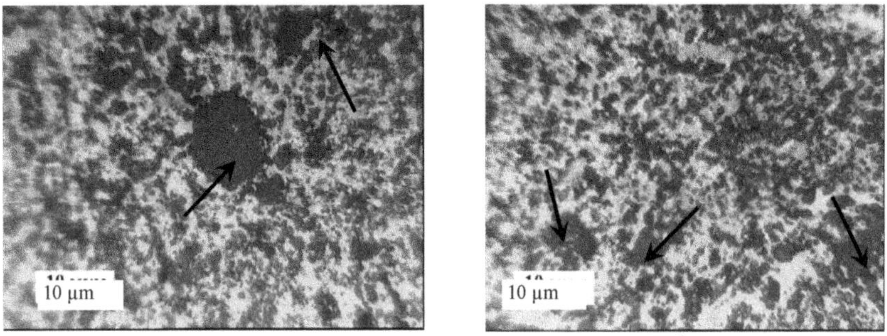

a b

Fig. 9. *Structure of hot-pressed materials obtained from a mixture of $ZrO_{2(m)}$–21Al–4.5C% (wt %) at T = 1520 °C: a – t = 19 min, P = 18 MPa; b – t = 5 min, P = 24 MPa.*

So, the analysis of the structures obtained as a result of those carried out according to the planning matrix based on the adopted levels and intervals of variation of factors allows us to draw the following **conclusions**.

The formation of the structure occurs through a combination of the phase constituent elements of the compounds Al_2O_3, Al_2Zr, ZrC, formed as a result of the initial charge components interaction, as well as residual $ZrO_{2(m)}$.

An increase in the aluminum concentration increases the compressibility of the samples and leads to an increase in the amount of aluminum oxide and intermetallic compounds formed during the interaction of Al with $ZrO_{2(m)}$.

An increase in the carbon content leads to a decrease in the Al_2Zr content and an increase in the amount of ZrC in the final material, which in continuous, defect-free zones of the material leads to an increase in the aggregate hardness of hot-pressed materials.

The formation of a defect-free structure of the final material in combination with physical and mechanical characteristics corresponding to cutting mineral ceramics is possible with an optimal combination of the levels of hot pressing parameters (P=24 MPa, T=1520 °C, τ=19 min) and the concentration of the initial components corresponding to the composition of $ZrO_{2(m)}$–21Al–4.5C. They are 18.5 ± 0.2 GPa and K_{IC} = 5.4 ± 0.2 MPa $m^{1/2}$. These samples have a continuous defect-free structure, which determines the suitability of using the material for the manufacture of cutting inserts and testing their cutting properties.

References

[1] N.Yu. Cherkasova, N.S. Stukacheva, R.I. Kuzmin, E.A. Puris, Development of compositions of aluminum-zirconium ceramics for the manufacture of cutting inserts. / Nanotechnology and Materials Science in Mechanical Engineering. 2019. pp. 347–351.

[2] N.A. Rejab, AZA Ahmad, M.M. Ratnam, Z.A. Ahmad, The effects of CeO2 addition on the physical, microstructural and mechanical properties of yttria stabilized zirconia toughened alumina (ZTA). Int. Journal of Refractory Metals and Hard Materials 2013; 36; pp. 162–166. https://doi.org/10.1016/j.ijrmhm.2012.08.010

[3] Liu B., C. Huang, X. Lu, M. Gu, H. Liu, In situ growth of TiC whiskers in Al2O3 matrix for ceramic machine tools. Ceramics International, 2007, Vol. 33, pp. 1475–1480. https://doi.org/10.1016/j.ceramint.2006.05.015

[4] L. Xikun, Q. Guanming, T. Qu. A12O3–Ti(CN)–Y2O3 composite prepared by HP and its cutting performance. Journal of Rare Earths. Jun. 2007, Vol. 25, pp. 37–41. https://doi.org/10.1016/S1002-0721(07)60519-5

[5] J. Vleugels, Fabrication, wear and performance of ceramic cutting tools. Advances in Science and Technology, 2006, vol. 45, pp. 1776–1785. https://doi.org/10.4028/3-908158-01-x.1776

[6] Q. Dong, Q. Tang, W. Li, Al2O3–TiC–ZrO2 nanocomposites fabricated by combustion synthesis followed by hot pressing. Materials Science and Engineering: A. 2008, No 475, pp. 68–75. https://doi.org/10.1016/j.msea.2007.01.158

[7] S.A. Suvorov, I.A. Turkin, M.A. Delovets, Strength of corundum-zirconium materials of microwave synthesis. Materialovedeniye (Materials Science). 2006, № 6, pp. 36–43.

[8] D.E. Clark, D.C. Floz, C.E. Folgar, M.M. Mahmoud, Microwave Solution for Ceramic Engineering. The American Ceramic Society. U.S.A., 2006, P. 494.

[9] K.Z. Gordashnik, A.A. Lebedev, G.K. Kozin, Trends in the creation of ceramic materials and their application in tool production. Instrumentalny svit (Instrumental world). 2007, No. 1 (33), pp. 23–26.

[10] Zhang Guo-Jun, Yang Jian-Feng, Ando Motohide, Ohji Tatsuki, Reactive hot pressing of alumina-silicon carbide nanocomposites. J. Amer. Ceram. Soc. 2004, № 2, pp. 299–301. https://doi.org/10.1111/j.1551-2916.2004.00299.x

[11] J. Zhao, The use of ceramic matrix composites for metal cutting applications. Advances in Ceramic Matrix Composites, 2014, 537–569. https://doi.org/10.1533/9780857098825.3.537

[12] S.I. Dzhelialov, Simulation of the Influence of the Hot-Pressing Parameters and the Mixture Compositions in the ZrO2(m)–Al–C System on the Hardness and the Crack Resistance of Ceramics. Inorganic Materials: Applied Research. 2016. Vol. 2 No. 2. 158–162. https://doi.org/10.1134/S2075113316020076

[13] S.I. Dzhelyalov, Interaction of components features of the reaction mixture ZrO2(m)-Al-C systems when heated. IOP Conference Series: Materials Science and Engineering, 2020, 889 (1), 012013. https://doi.org/10.1088/1757-899X/889/1/012013

[14] S.N. Dub, A.L. Maistrenko, A.A. Isakov, Peculiarities of Ceramic Crack Resistance Measurement by Indentation Method. Zavodskaya laboratoriya (Factory laboratory), 1993, No. 4, pp. 58–60.

Evaluation of the Manufacturability of "Shaft"–Type Parts with the Use of Complex Methods

Aleksey Sychugov[1,a], Yulia Frantsuzova[2,b*] and Vladimir Salnikov[2,c]

[1] Tula State University, Department of Information security, Tula, Russia

[2] Tula State University, Department of Computer technology, Tula, Russia

[a]xru2003@list.ru, [b]julianna_1204@mail.ru, [c]vladimirsalnikov95@yandex.ru

Keywords: Adaptability, Labor Intensity, Unification, Design, Shaft

Abstract. This work is devoted to improving the quality of engineering design solutions in mechanical engineering on the basis of the comprehensive qualimetric evaluation of 'shaft'-type part that takes into account cost, complexity, design and technological unification of the parts.

Introduction

In today's competitive environment, product quality is the ultimate goal of any manufacturer and determines its value in the eyes of the consumer. A machine-building product designed to meet certain needs has properties that form its quality.

The development of a new product is a complex design task associated not only with achieving the required functional, but also technological level.

The process of ensuring the manufacturability of the product is contradictory and difficult to formalize, and during its implementation, conflicts of professional interests are inevitable between the developers of the design solution and its external and internal consumers. For the objective resolution of such conflicts, it is necessary to introduce a qualimetric assessment of the quality of the design solution according to the criterion of manufacturability, taking into account the following parameters: labor intensity, cost and unification of the design.

An important quality indicator is the degree of unification of the designed part both in relation to its actual design, and in relation to the possibility of using a unified standard technology for its manufacture. The implementation of an operational assessment of the processability of a design solution in the context of computer-aided design is possible in the form of an appropriate intelligent CAD subsystem.

Thus, the task of revealing the regularity of the manifestation of information, economic and organizational relations between the design, technological and metrological aspects of machine-building production in the qualimetric assessment of the manufacturability of the design of parts of the "shaft" type based on the construction of mathematical and information models of their condition and quality dynamics is relevant.

The set of product properties that determine the adaptability of its design to achieve optimal resource costs during production and operation to achieve the specified quality indicators is the manufacturability of the product design.

Ensuring the manufacturability of the product design at the design stage is a significant and at the same time the most difficult task of the designer. If the conformity of the machine design to a given functional purpose is easy to assess with objective numerical indicators, then such a complex and multifactorial concept as manufacturability is very difficult to quantify. The absence of generally accepted methods of numerical evaluation of technological efficiency does not allow to objectively resolve the inevitable conflicts of professional interests between the designer as a supplier of design documentation (DD) and the technologist as its consumer.

It is obvious that each set of parameters of the part (shape and dimensions of surfaces, roughness, tolerances, design bases, the coefficient of workability of the material) corresponds to its own coefficient of manufacturability N_k. Criterion of adaptability of the part is exceeding the minimum permissible values $N_{доп}$, which is determined by technological possibilities of production, as a detail, quite technological for an advanced plant with modern equipment, can be unviable or unfeasible for a plant with outdated equipment.

Mathematical modeling

The number of processability indicators should be minimal, but sufficient to assess the processability.

The nomenclature of technological performance indicators and the method of their determination at each stage of CD development depends on the type of products and the type of production. Suggested list of indicators of technological items according to GOST Standard 14.201–83 "Integrating technology into product design" includes the complexity of manufacturing the product, the cost of technology and the coefficient of unification of the elements. The analysis of the research of Russian and foreign scientists in this subject area of science was carried out. The works of B. A. Yakimovich [10], V. I. Averchenkov [1], M. V. Novikova [5, 6], M. E. Popov [8] were studied. In these works, the complexity of manufacturing the product, the technological cost and the coefficient of unification of structural elements were considered.

The analysis of the above mentioned and works [2, 3, 7, 9] showed:

1. The authors did not consider the totality of all three indicators together;

2. To calculate the unification coefficient, only the applicability of structural elements in production and their standardization were taken into account, and the production capabilities of the enterprise and the unification of geometric dimensions were not taken into account. Its value is deterministic, either 1 or 0.

Since labor intensity has a direct monetary value, it is usually included in the cost price. Therefore, the proposed assessment of manufacturability includes two main components: technological cost C and unification, expressed by design k_{du} and technological k_{tu} coefficients. Then

$$N_k = \frac{2}{3}\|C\| + \frac{1}{3}(k_{du} + k_{tu}) \tag{1}$$

The weight coefficients 2/3 and 1/3 are introduced to reflect the fact that according to GOST Standard 14.201–83, cost, labor intensity and unification make an equal contribution to the manufacturability of the part.

Let's consider a mathematical model for determining the technological cost. The cost rate is calculated using the formula:

$$\|C\| = \frac{C_{nom,}}{C_T} \tag{2}$$

and is the ratio of the actual technological cost to the value $C_{nom,}$ that is either a technological cost of a detail-analog (previously manufactured item of similar size and design with a known cost), or nominal technology cost of a "simple" detail, for the calculation of which we take smooth shaft, dimensions of which coincide with the dimensions of considered details, and the surfaces have an average roughness Ra_{av}:

$$Ra_{av} = \sum_{i=1}^{n} \frac{Ra_i \cdot l_i}{n \cdot l} \tag{3}$$

where Ra_j – roughness of j-th treated surface; l_j – length of j-th surface; n – the number of treated surfaces; l – overall shaft length.

The technological cost is calculated using the formula [4]:

$$C_{\mathrm{T}} = C_{\mathrm{tot}}^{\mathrm{mat}} + T \cdot k_{\mathrm{mat}} \cdot (W_{\mathrm{w}} + C_{\mathrm{mh}}) \cdot (1 + K^{'}/100), \tag{4}$$

and to calculate the labor intensity T, an empirical dependence [6] was used:

$$T = M^y \cdot \sum_{i=1}^{n} S_i^a \cdot Ra_i^b \cdot t_i^c \cdot k_{\text{мат}} \tag{5}$$

where $C_{\mathrm{tot}}^{\mathrm{mat}}$ - the total cost of raw materials, RUB; W_{w} – the average hourly wage of the worker for the relevant production, rubles/hour; C_{mh} - cost of machine-hours RUB/hour; $K^{'}$ – average overhead costs, %; n – the number of design-technological elements (DTE); k_{mat} – coefficient that takes into account the machinability of the material; M – mass of the part, kg; S_i – the area of the shear layer of the i-th surface, mm2; Ra_i – the required roughness of the i-th surface, µm; t_i – the tolerance of the i-th surface, mm; n – the number of the processed surfaces; a, b, c, y – empirical power indicators determined by the regression analysis method.

Next, we will consider the definition of the coefficients of design and technological unification of a body of rotation part type. The coefficient of design unification k_{du} is defined as

$$k_{\mathrm{du}} = \frac{1}{n} \sum_{i=1}^{n} k_{\mathrm{du}_i} \cdot \delta_i \tag{6}$$

where n - the number of construction elements; $k_{\text{ку}_i}$ - the coefficient of design unification of the i-th structural element of the part, depending on the shape and the value of the landing size (the size responsible for the accuracy of the location of the part in the assembly):

for cylinder:

$$k_{\mathrm{du}_i} = \begin{cases} 1,\ if\ D_i = 1{,}6^N, \\ 0{,}75,\ if\ D_i = 1{,}25^N, \\ 0{,}5,\ if\ D_i = 1{,}12^N, \\ 0{,}25,\ if\ D_i = 1{,}06^N, \\ 0,\ if\ none\ of\ the\ conditions\ is\ met, \end{cases} \tag{7}$$

for cone:

$$k_{\mathrm{du}_i} = \begin{cases} 1,\ if\ \ 2\alpha_i \in A1, \\ 0{,}5,\ if\ \ 2\alpha_i \in A2, \\ 0,\ if\ none\ of\ the\ conditions\ is\ met, \end{cases} \tag{8}$$

where α_i is the angle at the top of the cone; $A1, A2$ - the series of preferred angles according to GOST Standard 8593-81. The given constants (1,6; 1,25...) express the series of preferred numbers for the cylinder diameter according to GOST Standard 8032-84.

For structural elements of other types (for example, shaped grooves, spherical surfaces), their presence in the catalog of analog parts is analyzed. If a similar element is found, then the coefficient $k_{\mathrm{du}_i} = 1$, otherwise - $k_{\mathrm{du}_i} = 0$.

In [4] it is indicated that from the point of view of manufacturability, it is necessary to observe a certain ratio between the dimensional tolerance on the surface and its roughness, due to the impossibility of achieving and measuring a rigid dimensional tolerance on an excessively rough surface. Therefore, for each surface to be treated, the following coefficient is introduced, taking into account the correctness of the roughness ratio and the tolerance field:

$$\delta_i = \frac{1}{n}\sum_{j=1}^{n}\delta_{ij}, \tag{9}$$

where $\delta_{ij} = \begin{cases} 0, roughness\ exceeds\ 5\%\ size\ tolerance, \\ 1, otherwise. \end{cases}$

According to the data collected from the enterprises of the Tula region, it was revealed that one of the main reasons for the return of DD (45% of the number of returns at the stage of technological preparation) is the inconsistency of the basing schemes. In some cases, due to incorrect design bases, the tolerances for the closing dimensions turned out to be impossible, which required the re-processing of the DD. Therefore, the coefficient of technological unification of the part should primarily assess the degree of unity of the design and technological bases as the most important indicator of manufacturability. This coefficient cannot be calculated only on the basis of the data presented in the DD – it does not specify the technological bases. Therefore, according to the drawing, a lot of design bases (structural elements from which dimensions with tolerances are set) of the part $M_{\text{КБ}}$ are identified . Further, on the basis of a suitable typical technological process, a set of hypothetical technological bases $M_{\text{ТБ}}$ is determined . The coefficient $k_{\text{ту}}$ is the ratio of the number of matched bases ($k_{\text{сов}}$) to the number of technological bases. ($k_{\text{ТБ}}$):

$$k_{\text{tu}} = \frac{k_{\text{match}}}{k_{\text{TB}}} \tag{10}$$

When generating a hypothetical technological process, we should select typical equipment for each operation, if there is no additional information about the features of the equipment of a particular production. The equipment depends on the content of the operation itself, the overall dimensions of the workpiece, the properties of the material, the accuracy of the resulting size and the serial production. Next, you need to select a base scheme for each installation. For bodies of rotation, it depends mainly on the overall dimensions of the workpiece. By expert estimates were identified based on schema-based (in the cartridge, in the cartridge with the rest, in the Chuck with a steady rest and center, in the cartridge with the center, floating center, in centers, in centers with steady rests) from the length and diameter of the workpiece. The coordinates of the distribution center ($D_{\text{ц}}, L_{\text{ц}}$) for the i-th base scheme will be calculated using the formulas:

$$D_{c_i} = \frac{1}{k_i}\sum_{j=1}^{k_i} d_j, \; L_{c_i} = \frac{1}{k_i}\sum_{j=1}^{k_i} L_j, \tag{11}$$

where i – number of a schema-based; k_i - the number of parts defined by experts, for which the i-th scheme-based is used; d_i, l_i – diameter and length of the i-th part.

The choice of the optimal layout $i_{\text{опт}}$ of the part with overall dimensions d, l d is based on the condition of the minimum difference in the distance of points in the phase space (D, L) for the selected material grade:

$$i_{\text{opt}} = \min_i \sqrt{(D_{ci} - d)^2 + (L_{ci} - l)^2}\,. \tag{12}$$

Summary

For each operation, we should choose a standard equipment. It depends on the content of the operation itself, the overall dimensions of the workpiece, the properties of the material, the accuracy of the resulting size and the serial production. Next, we need to solve the problem of building technological dimensional chains and checking them.

To check the adequacy of the model, a correlation coefficient was determined between expert estimates of the level of manufacturability of a number of parts of the type of bodies of rotation and estimates calculated according to the proposed method. The correlation coefficient was 0.84, which indicates the adequacy of the mathematical model.

To effectively manage the processability at the design stage, it is not enough to simply evaluate it – we need to give recommendations to the designer, specifying the most non-technological parameters of the part.

The paper proposes a mathematical model of the state and dynamics of the quality of machine parts at the design stage, characterized by a qualimetric assessment of the manufacturability of their manufacture, taking into account the cost, labor intensity and technological unification of the elements of the part.

The developed method of qualimetric assessment of the manufacturability of design solutions and the optimal assignment of bases for parts of the "shaft" type, implemented in the future in the form of an intelligent CAD subsystem, allows us to increase the level of manufacturability by analyzing the information obtained during the development of the CD.

References

[1] Averchenkov V. I. Formalization of the construction and selection of progressive technologies that provide the required quality of products: Abstract of the dis. ... doc. tekhn. science: 05.02.08. – Bryansk, 1990.

[2] Akimov, I. V. Self-learning system for the rapid assessment of the complexity of manufacturing of machine parts: author. dis. cand. tekhn. science: 05.02.08. – Tula, 1999.

[3] I. V. Akimov, A. N. Inozemtsev, N. I. Pas'ko, Substantiation of the adaptive method for fast regulation of the complexity of the goods machine tools, Izvestiya TSU. Ser. Mashinostroenie. - Tula, Issue 2 (1998) 216-221.

[4] Saati T. Decision-making: a method of analyzing hierarchies, Moscow, 1993.

[5] Amirov Yu. D Manufacturability of product design. Handbook, Moscow, 1990.

[6] Novikova M. V., Troitsky D. I. Integrated design and technological model of a part as a means of evaluating the labor intensity of its manufacture, Bulletin of Computer and Information Technologies, No. 1 (2006) 22-25.

[7] Volkhin K. A. Design documents and rules for their registration [Electronic resource]: textbook manual for students of technical sciences, Novosibirsk State Technical University. - Novosibirsk, 2004. - Access mode: http://graph.power.nstu.ru/wolchin/umm/eskd/index.htm.

[8] Popov M. E. Automated methods of improving the manufacturability of product design in an integrated CAD/ M. E. Popov, A. M. Popov, Journal of mechanical engineering, No. 10 (2003) 48-53.

[9] Fominykh R. L. Development of automated subsystems the definition of constructive-technological complexity, the complexity of the manufacture of parts and logistical level of multiproduct production: Author. dis. cand. tekhn. science: 05.13.06. – Izhevsk, 2003.

[10]. Yakimovich B. A., Kuznetsov A. P., Reshetnikov E. V. Model for calculating project costs for manufacturing machine-building products, Automation and Modern Technologies No. 5 (2003) 20-24.

Effect of Process Modes on Tangential Component of Cutting Force during Belt Rotary Grinding of Aluminium Alloy Blanks D 16

Tatiana Stadnik[1,a*], Denis Sidorov[1,b], Ilyas Temindarov[2,c] and Pavel Novikov[1,d]

[1]Polytechnical Institute, Sevastopol State University, 33, Universitetskaya, Sevastopol, 299053, Russia

[2]Crimean Engineering and Pedagogical University named after Fevzi Yakubov, per. Uchebnyy, 8, 295015, Russia

[a]tanya-stad@yandex.ru, [b]Sidorov-sntu@yandex.ru, [c]temindarov23@gmail.com, [d]PANovikov@sevsu.ru

Keywords: Grinding, Abrasive Belt, Rotary Belt Grinding, Cutting Force

Abstract. These days, in the manufacture of units and mechanisms of ships, aircraft and other technological machines, industrial robots, long-sized products from D 16 (Standart GOST-R) aluminum alloy are used, for the processing of which a complex for belt rotary grinding has been developed. The outcome measures of the rotary belt grinding process depend on the cutting forces generated during the processing process. According to cutting forces, process stability is diagnosed, values of surface roughness indices, temperatures and cutting modes are calculated according to displacement balance equation. The article is devoted to obtaining a mathematical model establishing the relationship between the tangential component of the cutting force and cutting modes during belt rotary grinding of D 16 aluminum alloy.

Introduction

In modern technology, in many industries such as aircraft engineering, instrument making, automotive, shipbuilding, etc., aluminum alloys are used to make parts. Aluminium alloys have high chemical activity and viscosity, which leads to the formation of growths on the tool [1]. D 16 (Standart GOST-R) aluminum alloy has gained widespread use. In view of its high lightness, strength and hardness, the alloy is used to make various metal rolled products.

Currently, in the manufacture of units and mechanisms of ships, aircraft and other technological machines, industrial robots use long-sized products from aluminum alloys, in particular from D 16 alloy. Long products include products made of rods, wire and tubular billets of rectilinear and curvilinear shape, which are used as guides of metal cutting machines, welding robots, conveyors.

Abrasive-diamond processing methods [2-10] are used for finishing a number of such articles. In the processing of such articles, there are a number of difficulties associated with the impossibility of rotating these parts during processing, especially in the case of curvilinear shaped articles. Therefore, for treatment of long axisymmetric non-rigid parts, where rotation of the machined part is impossible or difficult, such as rolled stock, pipes, straight and curvilinear wire with the most optimal machining scheme from the point of view of low surface roughness, a method of rotary belt grinding is proposed [11-14].

For the treatment of curvilinear (radius) blanks at the Department of Engineering Technology, Sevastopol State University, a complex for belt rotary grinding has been specially developed, which consists of a belt rotary grinding device 1 and a device for feeding the radius blanks 2 (Fig. 1).

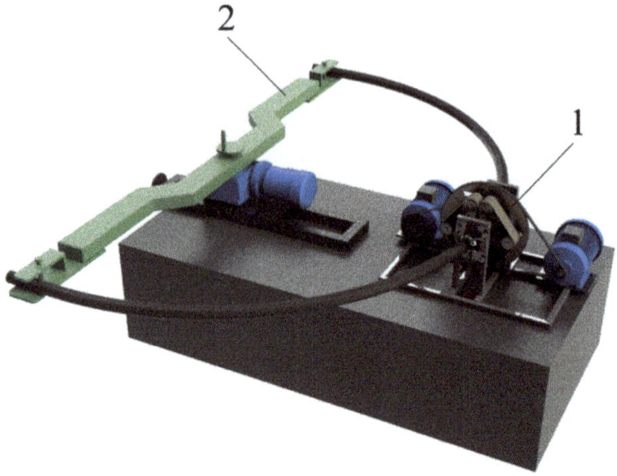

Fig. 1. *Model of the complex for belt rotary grinding of radial billets.*

The principle of operation of such a device is as follows: the radial blank is fixed in a rotary mechanism, the axis of rotation of which is installed from the axis of the belt grinding device at a distance equal to the radius of bending of the blank. The rotary mechanism with the blank in manual mode is set to the initial position - as close as possible to the edge of the blank. Rotation of the belt and the swivel mechanism is turned on. If it is necessary to move the workpiece back, reverse pendulum movement is activated.

It is known that the output of the belt rotary grinding process correlates with the cutting forces generated during the processing process. According to cutting forces, process stability is diagnosed, values of surface roughness indices and cutting modes are calculated according to displacement balance equation. According to the literature [15], the process of belt rotary grinding is significantly influenced by the grain and speed of the belt, as well as by the pressing force of the belt, which are suitably used as varying variable factors. On the basis of the above, the object of this work is to obtain a mathematical model of the tangential component of the cutting force depending on the speed of the belt, its grain size and the radial component of the cutting force in belt rotary grinding of D 16 aluminum alloy blanks.

Results and discussion

Experimental studies were conducted to build models. Studies were carried out on samples of round cross section, diameter 10-50 mm, normal manufacturing accuracy, length 3000 mm) from aluminum alloy D 16.

To measure cutting forces, dynamometric centers (front and rear) were used. They were equipped with wire-wound resistors (resistance sensors) between which the workpiece was fixed. Two dynamometer centers allow to record the cutting forces during the grinding process at any position of the grinding belt relative to the workpiece. Wire gauges, glued on the vertical planes of the front and rear centers, are connected in one bridge and are used to measure the force P_y (Fig. 2). Sensors glued to the horizontal planes of both centers are used to measure the force P_z.

The measurement scheme includes dynamometer centers, an amplifier, and an analog-to-digital

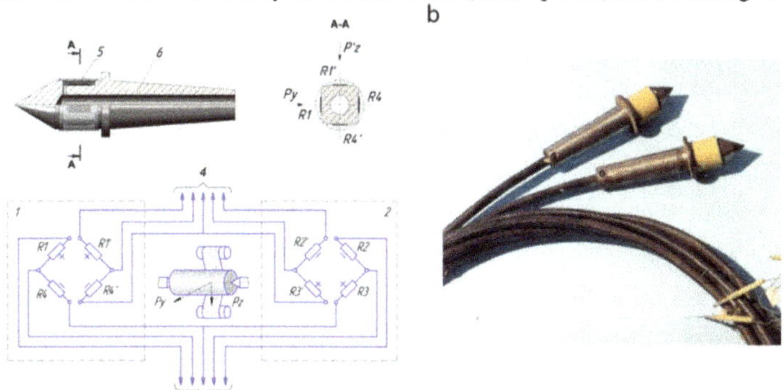

Fig. 2. *Tenzometric Centers (a) - a scheme to include wire resistance sensors to measure the forces of* P_y *and* P_z*. (1 - the front center; 2 - the rear center; 3 - wires to the device to measure the force of* P_y*, 4 - wires to the instrument to measure the force 5 - sensors; 6 - a hole for the output of wires); (b)-photography of the tensometric centers.*

Converter ЛА-50USB (Russian production) connected to a personal computer using a well-known technique. Calibration of strain centers was performed by static loading. The calibration scale was determined by the least squares approximation. The calibration graph was linear. The total error when using this method is 10%. The longitudinal movement of the dynamometric centers, with the workpiece fixed between them, during the measurement is carried out along the guides on the bed by means of a gear rack and a stepping motor (Fig. 3).

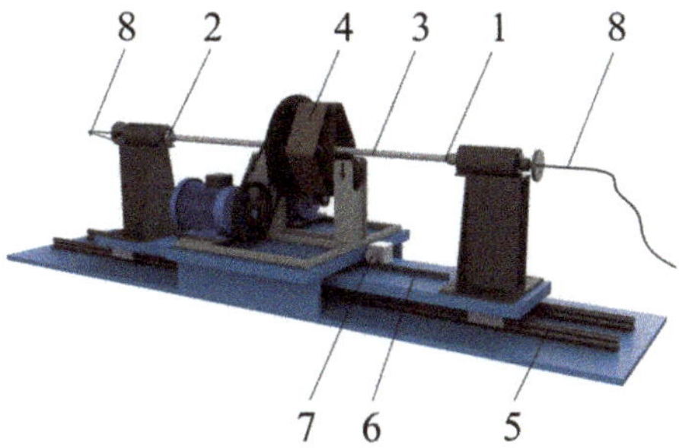

Fig. 3. *Installation scheme for measuring cutting forces: 1, 2 - front and rear tendometric centers, respectively; 3 - workpiece; 4 - installation for rotational belt grinding; 5 - guides to the mills; 6 - toothed rack; 7 - is a step-by-step electric motor; 8 - wires leading to the device to measure the forces of cutting.*

According to the analysis of the literature data [15], to simulate the values of the cutting force component P_z from variable factors: the radial component of the cutting force $-P_y$, the speed of the tape $-V_l$ and the grain of the abrasive tape $-X_H$ a regression dependence of the power form was adopted:

$$P_z = C_{Pz} \cdot P_y^{\alpha_1} \cdot V_l^{\alpha_2} \cdot X_H^{\alpha_3}, \tag{1}$$

where C_{Pz} – coefficient of proportionality; P_y – radial component of cutting force; V_l – belt speed; X_H – abrasive tape grain; α_1, α_2, α_3 – degree indices.

After taking the logarithm this equation 1 takes the form:

$$ln\, P_z = ln\, C + ln\, P_y + ln\, V_l + ln\, X_H. \tag{2}$$

Testing the adequacy hypothesis of the linear model when expressing the results of the experiment is carried out using a polynomial:

$$y = b_0 + b_1 x_1 + b_2 x_2 + b_3 x_3, \tag{3}$$

where y – is $ln\, P_z$; x_1, x_2, x_3 – coded values of factors; b_1, b_2, b_3 – coefficients of model parameters; b_0 – free term.

The encoded values of the factors were determined by the expression:

$$x_i = \frac{2\left(ln\, \breve{x}_i - ln\, \breve{x}_i^{в}\right)}{ln\, \breve{x}_i^{в} - ln\, \breve{x}_i^{н}} + 1, \tag{4}$$

where x_i – is the coded value of the i-th factor; $\breve{x}_i$ – is the natural value of the i-th factor; $\breve{x}_i^{в}$ – is the natural value of the upper level of the i-th factor; $\breve{x}_i^{н}$ – is the natural value of the lower level of the i-th factor.

The accepted factor levels are shown in Table 1.

As a result of statistical processing of experimental data, the dependence of the tangential component of the cutting force on the speed of movement of the belt, its grain and the radial component of the cutting force is obtained:

$$P_z = 14{,}48 \cdot P_y^{0{,}406} \cdot V_l^{0{,}04} \cdot X_H^{-0{,}612}. \tag{5}$$

Experimental studies and statistical processing are summarized in table 1.

Figures 4, 5 illustrate the variation of the force P_z from the above parameters.

Table 1. *Values of variability interval and results of processing of data on investigation of tangential force of cutting of rod from material D 16*

Controlled variables	P_y [N]	V_l [m/s]	X_H	Approximating constraint
Top Level	40	20	40	$ln P_z = 2{,}673 + 0{,}406 ln P_y + 0{,}04 ln V_l - 0{,}612 ln X_H$ $P_z = 14{,}48 \cdot P_y^{0{,}406} \cdot V_l^{0{,}04} \cdot X_H^{-0{,}612}$
Lower level	15	10	10	
Main level	27,5	15	25	
Variation interval	12,5	5	15	

Experiment planning matrix					Results of the experiment and variance of deviations							Check of homogeneity of dispersions		Adequacy check	
№ point	Controlled parameters														
	x_0	x_1	x_2	x_3	y_1	y_2	y_3	y_4	y_5	$\bar{y}$	S_v^2	$\sum_{v=1}^{N} S_v^2$	0,0104	$\sum_{v=1}^{N} (\bar{y}_v - y_v)^2$	$7{,}72 \cdot 10^{-6}$
1	+	+	+	+	1,98	2,01	1,96	1,94	1,97	1,97	$6{,}37 \cdot 10^{-4}$	$S^2_{v\,max}$	0,0029	S^2_{ad}	$9{,}65 \cdot 10^{-6}$
2	+	-	+	+	1,58	1,65	1,52	1,53	1,61	1,58	$2{,}87 \cdot 10^{-3}$	G	0,3185	F	$7{,}46 \cdot 10^{-3}$
3	+	+	-	+	1,96	1,93	1,95	1,97	1,91	1,94	$6{,}02 \cdot 10^{-4}$	q [%]	5	q_{ad} [%]	5
4	+	-	-	+	1,53	1,51	1,57	1,62	1,53	1,55	$1{,}83 \cdot 10^{-3}$	$V_{1в}$	4	V_{1ad}	4
5	+	+	+	-	2,86	2,83	2,87	2,81	2,88	2,85	$9{,}38 \cdot 10^{-4}$	$V_{2в}$	8	V_{2ad}	32
6	+	-	+	-	2,44	2,49	2,43	2,49	2,41	2,45	$1{,}48 \cdot 10^{-3}$	G_{cr}	0,3185	F_{cr}	2,69
7	+	+	-	-	2,82	2,81	2,77	2,86	2,85	2,82	$1{,}26 \cdot 10^{-3}$	Conclusion	$G < G_{cr}$	Conclusion	$F < F_{cr}$
8	+	-	-	-	2,45	2,39	2,39	2,43	2,45	2,42	$7{,}43 \cdot 10^{-4}$				

Coefficient b_i				Checking the significance of coefficients										Conclusion: All coefficients are significant
2,199	0,199	0,014	-0,437	$S^2\{b_i\}$	$S\{b_i\}$	q %	$V_{3н}$	t_{cr}	t_o	t_1	t_2	t_3		
				$3{,}24 \cdot 10^{-5}$	$5{,}69 \cdot 10^{-3}$	5	32	2,036	386,45	34,89	2,52	-76,82		

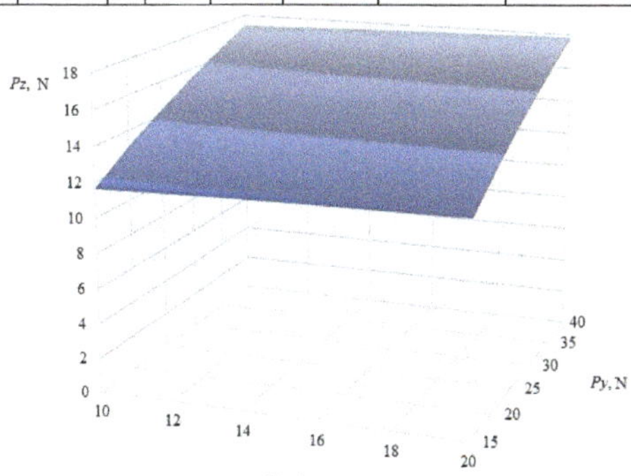

Fig. 4. *Effect of radial component of cutting force and cutting speed on tangential component of cutting force during treatment of D 16 material by abrasive tape with grain 10.*

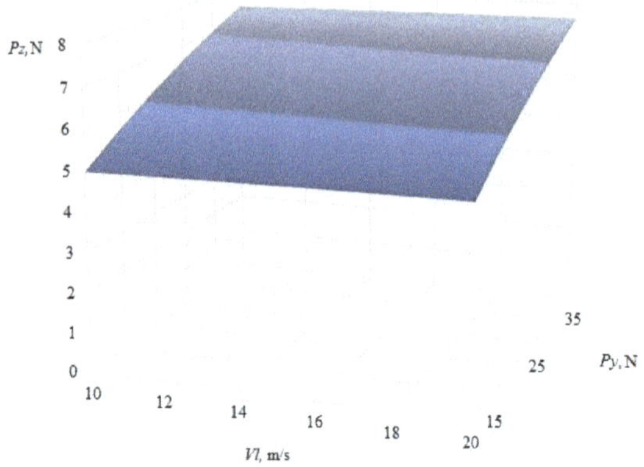

Fig. 5. *Effect of radial component of cutting force and cutting speed on tangential component of cutting force during treatment of D 16 material by abrasive tape with grain 40.*

Conclusions

The obtained dependence makes it possible to differentiate the effect of the belt speed, its grain size, and radial component of the cutting force on the tangential component of the cutting force and to judge the state of the tape rotary grinding process.

References

[1] S. Bratan, P. Novikov, S. Roshchupkin, Application of Combined Taps for Increasing the Shaping Accuracy of the Internal Threads in Aluminium Alloys, Procedia Engineering. 150 (2016) 802–808. https://doi.org/10.1016/j.proeng.2016.07.115

[2] S.M. Bratan, S.I. Roshchupkin, A.O. Kharchenko, A.S. Chasovitina, Probabilistic model of surface layer removal during grinding of brittle non-metallic materials, Metal processing (technology, equipment, tools). 23 No. 2 (2021) 6-16. (in Russ.). https://doi.org/10.17212/1994-6309-2021-23.2-6-16

[3] S.M. Bratan, S.I. Roshchupkin, A.O. Kharchenko, A.S., Chachavitina Simulation of allowance removal in the contact area during internal grinding of brittle non-metallic materials, Metal processing (technology, equipment, tools). 23 No. 2 (2021) 31-39. (in Russ.). https://doi.org/10.17212/1994-6309-2021-23.2-31-39

[4] S. Bratan, T. Stadnik, S. Roshchupkin, Investigation of the mechanism of interaction of cutting fluids with the material of the workpiece when grinding titanium parts in the presence of inorganic salt, Materials Today: Proceedings. 38 (2021) 2048–2051. https://doi.org/10.1016/j.matpr.2020.10.135

[5] A. Chasovitina, S. Bratan, S. Roshchupkin, The cutting emulsions for grinding ceramics and sitall parts, Materials Today: Proceedings. 38 (2021) 2083–2087. https://doi.org/10.1016/j.matpr.2020.10.153

[6] A. Kharchenko, A. Chasovitina, S. Bratan, Modeling of regularities of change in profile sizes and wear areas of abrasive wheel grains during grinding, Materials Today: Proceedings. 38 (2021) 2088–2091. https://doi.org/10.1016/j.matpr.2020.10.154

[7] S.M. Bratan, Influence of the tool working surface state on evaluation of the forming filter parameters, IOP Conference Series: Materials Science and Engineering. 709(2) (2020) 022009. https://doi.org/10.1088/1757-899X/709/2/022009

[8] S. Bratan, B. Bogutsky, S. Roshchupkin, Development of mathematical model of material removal calculation for combined grinding process, Lecture Notes in Mechanical Engineering. 9783319956299 (2019) 1759–1769. https://doi.org/10.1007/978-3-319-95630-5_189

[9] S. Bratan, E. Vladetskaya, Flat grinding model, considering internal dynamics of the process MATEC Web of Conferences. 224 (2018) 01002. https://doi.org/10.1051/matecconf/201822401002

[10] S. Bratan, S. Roshchupkin, Synthesis of lunberger stochastic observer for estimation of the grinding operation state, MATEC Web of Conferences. 224 (2018) 01133. https://doi.org/10.1051/matecconf/201822401133

[11] S.M. Bratan, T.V. Stadnik, A.G. Kolesov, Balance of movements in the technological system with belt rotary grinding, News of Tula State University, Technical Sciences, Issue 8, Part 1, Tula State University Publishing House, Tula, 2017, pp. 270-275. (in Russ.)

[12] S. Bratan, A. Kolesov, S. Roshchupkin, T. Stadnik, Theoretical-probabilistic model of the rotary belt grinding process, MATEC Web of Conferences. 129 (2017) 01078. https://doi.org/10.1051/matecconf/201712901078

[13] S. Bratan, T. Stadnik, V. Golovin, Evaluation of effectiveness of cutting fluids for belt grinding of long-length products of titanium alloys, Materials Today: Proceedings. 38 (2021) 2013–2018. https://doi.org/10.1016/j.matpr.2020.10.031

[14] S.A. Shilyaev, Development of a rotary belt grinding device to ensure accuracy of parts in assembly production, Assembly in mechanical engineering, instrument making. 3 (2008) 25-27. (in Russ.)

[15] S.A. Shilyayev Investigation of technological parameters of process of rotary belt grinding and obtaining mathematical relationships for quantitative analysis of cutting conditions, Izhevsk State Technical University Reporter. 2 (38) 2008 23-26. (in Russ.)

Keyword Index

About the Editors

Prof. Sergey Bratan is a head of Department of Mechanical Engineering, Sevastopol State University. His research interests include technological bases of quality assurance and improvement of stability of finish and fine grinding. He has made an important contribution into the development of a probabilistic approach for the modeling of abrasive operations of materials machining. Within 40 years of research work he has led many national research projects. In 2019, he was awarded the title of Professor of the Year at Sevastopol State University.

Dr. Stanislav Roshchupkin is an assistant professor of Department of Mechanical Engineering, Sevastopol State University. His research interests include additive manufacturing and abrasive methods of machining of non-metallic materials. He received a PhD in 2011 at Sevastopol State Technical University and MBA in 2019 at Eurasian Management and Administration School (EMAS). He also has extensive work experience at various manufacturing enterprises on in engineering and management positions.

www.ingramcontent.com/pod-product-compliance
Ingram Content Group UK Ltd.
Pitfield, Milton Keynes, MK11 3LW, UK
UKHW020659041026
15590UKWH00001B/366